ASTRONOMY AND ASTROPHYSICS

ASTRONOMY AND ASTROPHYSICS

Dr. Mohit Kumar Sharma
M.Sc., M.Phil., Ph.D. (Physics)
Young Scientist Awardee (ISCA, 2014)
National Postdoctoral Fellow, SERB, DST, New Delhi

Prof. (Dr.) Suresh Chandra
AvH Fellow (Germany), FRAS (London), FMAS (Pune)
Deputy Director & Professor
Amity Center for Astronomy & Astrophysics
Amity Institute of Applied Sciences
Amity University Uttar Pradesh
Noida, U.P., India

TECHSAR™
NEW DELHI

Published and Printed by
TechSar Pvt. Ltd.
4435-36/7, Ansari Road, Daryaganj
New Delhi-110 002 (India)
E-mail: tech_info@techsarworld.com
Website: www.techsarworld.com

ISBN: 978-93-90620-92-0

Reprint 2025

Dedicated to

Professor (Dr.) Wilhelm H. Kegel, Germany, who left for heavenly abode on 13 September 2019

Preface to the Second Edition

It is well accepted that the universe was created about 13.8 billion years ago through a BIG BANG and it is expanding continuously. Though the boundaries of the universe are not known, scientists believe that the universe consists of about one hundred billion (10^{11}) galaxies, and our galaxy, Milky Way, is one of them. Estimates show that our galaxy Milky Way has about one hundred billion (10^{11}) stars, and our Sun is one of them. It provides an idea, how small is our solar system in the universe. Our solar system has 8 planets, large number of minor planets, satellites and other bodies. Our Earth is a part of the solar system.

After having overwhelming response and support from all sections, including respected teachers and beloved students, from children to senior citizens, from all sexes, we felt encouraged to revised the book by adding various topics suggested to us. In particular, we have included role of India in Astronomy & Astrophysics. We have included new chapters on atomic spectra, molecules in interstellar medium, and waves in fluid plasma. This revised version of the book also can be used as a textbook in various universities in India and abroad.

We would greatly appreciate receiving again comments and suggestions from all the sections, which can be sent also to any of the e-mails:

mohitkumarsharma32@yahoo.in
schandra2@amity.edu; suresh492000@yahoo.co.in

Mohit Kumar Sharma
Suresh Chandra

Preface to the First Edition

We are aware of various subjects, like Physics, Chemistry, Mathematics, Botany, Zoology, etc. It is, however, obvious that all these subjects emerged out of a common topic, called the Science. Hence, in the early times, all the studies in Physics, Chemistry, Mathematics, Botany, Zoology, etc., were being carried out under one umbrella known as the Science. If we try to go still past, and think, how development of Science would have initiated. A reply may be that when the human being, after development of sense of recognizing things, looked upon some natural phenomena, such as the rising and setting of Sun each day; appearance of stars and the Moon in the night (in the absence of the Sun); solar as well as lunar eclipses; phases of Moon; etc., they would have intuitively started thinking, why and how these phenomena were taking place. Such events may probably be considered the beginning of the process of understanding of Science. Hence, the study (the process of thinking) of natural phenomena may be thought as the beginning of Science. The studies of natural phenomena in the present time are carried out under the subject, called the Astronomy & Astrophysics (abbreviated as A&A).

In a clear, dark night, we see sparkling points of light apparently situated on a vast sphere (called, the celestial sphere) in the sky. All these points are not necessarily the stars. They may be stars, star clusters, galaxies, planetary nebulae, etc. Because of large distances of these heavenly bodies (also known as the cosmic objects) from us, it is difficult to resolve them with the naked eyes and therefore, each of them appears like a star and we address them as the stars. The A&A is a branch of science which deals with the motions of heavenly bodies and with the science of their interior as well as the atmosphere.

This book deals with the Astronomy and Astrophysics, and can be used as a textbook for a course on Astronomy and Astrophysics. Some courses on A&A are being taught to M.Sc. (Physics or Maths.) students in various universities in India and abroad. In the latest syllabus of B.Sc., one course on the A&A is introduced by the University Grants Commission (UGC), New Delhi. Basically, the text presented here is the print form of the notes, which have been revised from time to time and are used for classroom teaching. One of the policies followed in this book is to discuss each topic in detail, in a systematic manner and easily understandable mode as well as language.

After briefing about the beginning of Astronomy, we have discussed about various coordinate system used to specify position of an object on the earth and in the space. Time is an important parameter and can help in having information about the relative occurrences of various events. The limitations of the observations with our eyes can be largely extended with the help of telescopes. Further, the telescopes help us for making observations of those radiations which cannot be seen by human eyes.

The fundamentals used for the study of various phenomena are summarized. Details about the Sun are discussed, as being the closest star, it has been studied in detail. Information about the solar family is described at length. Thermochemical reactions are responsible for creation of energy in the shining objects. Information about various types of thermochemical reactions is incorporated. Details of the galaxies and the universe are discussed. A brief description about the Cosmology – the science about the formation of universe is mentioned. The text cannot be considered complete in absence of a discussion about the possibility of the existence of life in the universe, other than that on the Earth.

While preparing the manuscript, we have consulted various books and got their help. we are grateful to the authors and publishers of those books. Some images are taken from the website of Google. We thankfully acknowledge the Google website. For this book, we have been helped, advised and encouraged by our friends working in various institutions/universities in India as well as abroad, and by our friends in the personal life. We are heartily thankful to all of them. Prof. Chandra would like to thank his students whom he taught the subject and got feedback for the improvement of the manuscript. We would greatly appreciate receiving comments and suggestions from the students, our friend teachers and others, which can be sent also to either of the e-mail addresses:

mksharma4@amity.edu; mohitkumarsharma32@yahoo.in

schandra2@amity.edu; suresh492000@yahoo.co.in

We are grateful to Hon'ble Dr. Ashok K. Chauhan, Founder President, Amity University, Hon'ble Dr. Atul Chauhan, Chancellor Amity University Uttar Pradesh, Prof. Balvinder Shukla, Vice-Chancellor Amity University Uttar Pradesh, Prof. Sunita Rattan, Director, Amity Institute for Applied Sciences, and Prof. S.N. Shukla, Dept of Applied Physics, Amity University, Noida for their moral support and encouragement.

Some part of the text was prepared during the visit of Suresh Chandra to the Max-Planck-Institut für Radioastronomie, Bonn, Germany, the Technical University, Berlin, Germany and the University of Kassel, Kassel, Germany. We are grateful to Prof. Dr. D. Breitschwerdt, Prof. Dr. W.H. Kegel, Prof. Dr. T.F. Giesen, and Prof. Dr. K.M. Menten and for their research groups for their warm and memorable hospitality. Thanks are due to the D.F.G., Germany and the Alexander von Humboldt Foundation, Germany for financial support for the visits to Germany. Suresh Chandra is specially thankful to his wife Mrs. Purnima Sharma for her valuable cooperation in his life and for sharing a major part of the responsibility of family affairs, so that he could spend his time for this book. Mohit Kumar Sharma is thankful to his wife Mrs. Shalini Sharma for her valuable cooperation. We are thankful to Dr. Monika Sharma and Mrs. Shalini Sharma for their help in preparation of the manuscript of this textbook. We are deeply grateful to our family members who always have been a source of inspiration and happiness for us. In the last, but not the least, we are highly thankful to the publisher for bringing out this book nicely.

Mohit Kumar Sharma
Suresh Chandra

Contents

1 Beginning of Astronomy

The eyes are one of the senses provided to human beings and the Earth's atmosphere is transparent for the visible part of electromagnetic spectrum, which is coming from the outer space. Combination of these two factors did not let the human being to remain ignorant about the beautiful phenomena occurring in nature. For example, from the inception, people would have noticed rising and setting of the Sun each day; change of the shape of the Moon and its position among the stars (phases of the Moon); solar and lunar eclipses; constellations of stars in the sky that looked like animals and people; occurrence of seasons, etc. As a natural habit of human being, after noticing such amazing phenomena, one would naturally ask questions from oneself or others: how and why these phenomena are occurring?

The phases of the Moon could be understood by considering spherical shape of the Moon and its revolution around the Earth. The appearance of day and night could be explained with the help of the spin motion of the Earth about its own axis. A big breakthrough in this field was to find the depth in the sky. That is, to find that the Moon and Sun are not at equal distances from the Earth. However, apparently, all the celestial objects, like the Sun, Moon, stars, etc., look as if they are situated on a common spherical surface (celestial sphere) around us. The difference in the distances of the Sun and Moon from the Earth and occurrence of total solar eclipse provided an idea about the different sizes of the Sun and Moon. Another breakthrough at par was to find spherical shape of the Earth.

One more milestone may be considered the idea of heliocentric system, replacing the geocentric system, which failed to resolve the retrograde motions of the Mars, Jupiter and Saturn when observed from the Earth. The aspects of the beginning of Astronomy have been discussed in this chapter.

1. Seven Ancient Civilizations in Astronomy

The human eye in conjunction with the transparency of the Earth's atmosphere attracted the human attention towards the beauty of nature available in the form of celestial bodies, like the Sun, Moon, stars, visible from the Earth and roaming from one place to another.

Appearance of such bodies naturally creates interest in learning about them. What are they? How they appear and disappear? How positions and shapes of the Moon change? Such stage may be treated as the beginning of Astronomy, which dates back to ancient times. People would have spent many centuries in planning and understanding the celestial bodies. As far as the development of Astronomy is concerned, seven ancient civilizations may be considered responsible for it.

1.1 Babylonian Astronomy

The Babylonians, dating back to 1800 B.C., may be considered as the first civilization who prepared document in detail about the movements of the Sun and Moon. These two celestial bodies are the largest in their appearance. Babylonians have also prepared detailed document of daily, monthly, and yearly positions of the celestial bodies. Such compilations had mystical value in terms of their use to suggest the king about some possible catastrophic events to take place.

People believe that the Babylonians are the first who documented the appearances of the famous Halley's comet, and divided the sky into various zones.

1.2 Greek Astronomy

Early Greek astronomers learned it from the Babylonians. However, it is beyond imagination to have any discussion about ancient Astronomy without including the Greeks first. They are, therefore, considered as the fathers of ancient Astronomy. They are believed to be the first who formulated theories and mathematical equations to understand the universe. Anaxagoras was the first explaining the phases of Moon. Eratosthenes is considered one of the most intelligent Greek scholars, who excelled not only in the field of Astronomy, but also in the field of Geography, Mathematics, poetry, and music. Several breakthroughs in Astronomy have been made by him.

One of his most important contributions may be his consideration of Earth as spherical in shape and calculation of the radius of the Earth. The diameter of the Earth was estimated to be twice that of the Moon. (Actual diameter of the Earth is 3.7 times that of the Moon.) Before that the Earth was considered a flat surface. His contributions are also the calculation of the tilt of the spin-axis of the Earth and the conceptualization of leap day.

Another Greek philosopher is Pythagoras, whose contribution is better known for Mathematics, but his contributions in Astronomy also are significant. He postulated that likewise other celestial bodies, the Earth is spherical in shape. He got this idea by seeing ships disappearing past the horizon as they sail. He is also considered the first to suggest that the movement of planets, the Sun, Moon, and stars could be expressed in numbers. He noticed that the so-called the evening-star and the morning-star were the same body, the planet Venus.

1.3 Indian Astronomy

Contribution of ancient India in the field of Astronomy is quite remarkable. The most notable one may be considered by Aryabhata I. It is because of him that Indian Astronomy went away from the mystical and religious considerations, towards the scientific approach. Although he worked following the geocentric system, many of his contributions are still important in modern Mathematics and Astronomy.

Contribution of Aryabhata is important as he was able to assume that the Earth rotates about its axis and that the Moon and other planets shine as they reflect the sunlight incident on them.

1.4 Mayan Astronomy

Mayan astronomers sought guidance from the sky. Their particular interest had been in the study of the motion of stars, the Sun, and other planets. The ancient Mayans could manage to observe and document these movements with the help of shadow-casting devices, which were invented by them. Using these observations, they developed the Mayan calendar to keep track of the passage of time.

1.5 Egyptian Astronomy

Ancient Egypt had one of the most advanced and affluent cultures. Their contribution to Astronomy of today is quite large. Similar to any ancient civilization, the movements and patterns of the sky developed the creation of myths to explain astronomical events. Similar concepts prevailed in Egypt as well. Based on the astronomical positions, they erected huge pyramids and temples. The Great Pyramid of Giza is one of the examples. It was built to align with the North Star, which was Thuban instead of Polaris at that time. In Egypt, The Nabta Playa is one of the most intriguing astronomical locations. One can find there a circular stone structure which is considered a giant calendar to identify the summer solstice.

The inclination of ancient Egyptians was not purely religious but practical too. The observations of the celestial bodies were used for predictions and they made predictions for the flooding of the Nile River. Egyptians had developed a solar calendar having 365 days in a year. The year was divided into 3 seasons, each of 120 days, plus an intercalary month of 5 epagomenal days, treated as outside of the year proper. Each season was divided into 4 months, each month having 3 weeks, and each week having 10 days.

Ancient Egyptians had great interest in the study of night sky. For example, they had drawn two bright stars which could be seen circling the North Pole always. Those stars are now known as Kochab, in the bowl of the Little Dipper (Ursa Minor), and Mizar, in the middle of the handle of the Big Dipper (Ursa Major).

1.6 Chinese Astronomy

Ancient Chinese civilization is well known for having one of the most detailed documentation of astronomical observations. In the ancient China, one of the most notable astronomers is Gan De, who is the first to take note of Ganymede. He described it as a small reddish star around the Jupiter. Shi Shen also created one of the most detailed and oldest catalogs of the stars, known as the *Star Catalogue of Shi*.

In several events, Chinese noticed a star which suddenly appeared among other fixed stars. It is now believed that what they observed were supernovae.

In a Buddhist cave in Dunhuang, China, a Star Atlas was discovered by an archaeologist. That is, Dunhuang Star Atlas was prepared by ancient Chinese, which is considered the earliest known preserved star map in the world which dates back before AD 700.

1.7 Persian Astronomy

Astronomy was highly popular during the post-Islam Persian civilization. One of the most brilliant astronomers of all time was Abd al-Rahman al-Sufi or commonly known as Azophi, who described first the Andromeda Galaxy in his book, entitled The Book of Fixed Stars. Some corrections and revisions were made on the original concept of constellations by Ptolemy.

Abu Mahmud Hamid ibn Khidr al-Khujandi was a brilliant astronomer. For calculating the earths axis, he built a giant sextant and due to large size of the sextant, his measurement was found very accurate, just off by two minutes.

Note on AD, BC, BCE and CE

AD stands for Anno Domini, Latin for in the year of the Lord, while BC stands for Before Christ. BC appears after the numerical year, while AD appears before it. For example, 995 BC, AD 967. CE stands for Common (or Current) Era, while BCE stands for Before the Common (or Current) Era. BCE and CE should both appear after the numerical year. For example, 995 BCE, 995 CE. Common Era (CE) is one of the year notations used for the Gregorian calendar. BCE and CE are alternatives to the BC and AD notations respectively. The two notation systems are numerically equivalent: "02022 CE" and "AD 2022" each describes the current year 2023; "400 BCE" and "400 BC" are each the same year.

2. Evolution of Astronomy

After noticing a number of phenomena occurring in nature, attempts would have been made to understand them. In the following sections, we have discussed how these phenomena could be understood and how further developments continued.

2.1 Phases of Moon

Apparent shape of the Moon was found to change from day to day, for example, 13 phases of the Moon are shown in Fig. 1. Sometimes, the Moon is full (full Moon night; central position). Then, it decreases in size and the shape changes, and disappears (new Moon night). The Moon again appears (first position on left), and its size increases and the shape changes in the reverse manner.

Fig. 1: Phases of Moon.

(*Courtesy:* Google website)

The explanation of phases of the Moon was given[1] that the Moon is spherical in shape and the sunlight is falling on it. Thus, always, half of the surface of the Moon is bright and the rest half is dark, as shown in Fig. 2, by the inner circle around the central body, the Earth. The Sun (not shown in figure) is on the left side. As the Moon revolves around the Earth, the area of its surface which reflects the sunlight, in the direction of the Earth, changes with time. Consequently, the phases of the Moon are seen, as shown in the outer circle around the Earth. This way, an acceptable explanation for the phases of the Moon was provided.

[1]By Anaxagoras (500 BC - 428 BC)

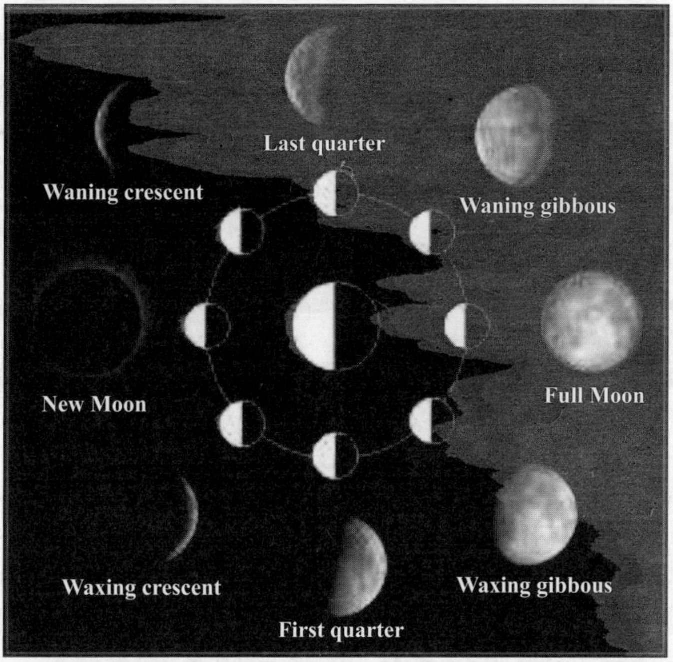

Fig. 2: Spherical body at the centre is the Earth. In the inner circle around the Earth, eight positions of the Moon are shown. The Sun (not shown in the figure) is on the left side. Half of the surface of the Moon is always bright. The outer circle around the Earth shows how the Moon looks like when seen from the Earth's surface. (*Courtesy:* Google website)

The shape of the Moon appears to change every day. On the full-moon night, the Moon appears as a full circular plate. After about two weeks, the Moon is not visible, and the night is known as the new-moon night. At the time of full Moon, the Earth is between the Sun and Moon, and at the time of new Moon, the Moon is between the Sun and the Earth.

2.2 Rising and Setting of Sun (Horizon)

For understanding the rising and setting of the Sun (or any celestial object), we need to define the horizon. For a person situated on the surface of the Earth, the horizon is a plane passing through the person and tangential to the surface of the Earth. The horizon obviously rotates along with the Earth from west towards east.

Now, an object (Sun) moving from the below side to the up side through the horizon is said the "rising of object" (Sun) (Fig. 3, left) and an object (Sun) moving from the up side to the below side through the horizon is said the "setting of object" (Sun) (Fig. 3, right).

2.3 Distances of the Sun and the Moon

For understanding the phases of the Moon, it was realized[2] that when the Moon appeared as half in size, the $\angle SME$ is 90°, as shown in Fig. 4. Here, S, M, E, respectively, denote the Sun, the

[2]Aristarchus (320 BC-250 BC) was probably the first person to realize.

Fig. 3: Rising of the Sun (left) and setting of the Sun (right).
(*Courtesy*: Google website)

Moon, and the Earth. After measuring the ∠SEM from his position (E) on the Earth, the ratio of the distances ES and EM could be worked out. It is not known how actually the ∠SEM to be 87° was measured; the distance ES was obtained to be 20 times more than the distance EM. (Accurate value of ∠SEM is 89° 51′ and the ratio ES/EM is 391.)

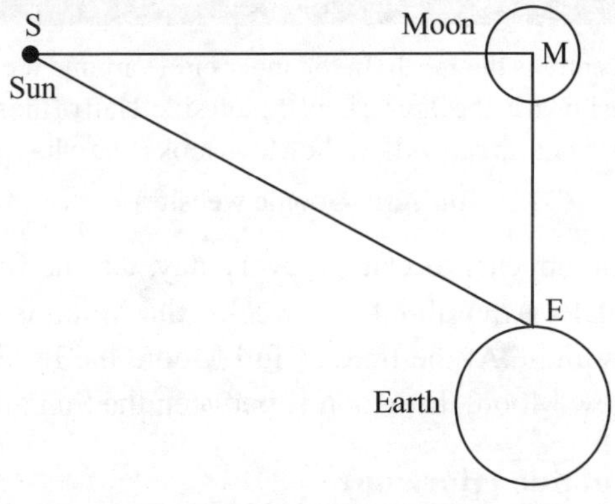

Fig. 4: Relative distances of the Sun (S) and the Moon (M) from the Earth (E); not to the scale.

This estimation was a remarkable discovery, as it was proved that the space has depth and the Sun and Moon do not move in a common surface around the Earth. (In those days all heavenly bodies, including the Sun and Moon, were assumed to revolve around the Earth and the Earth was considered the centre of whole universe. This arrangement is known as the geostationary system.)

Example 1.

By using the idea of phases of the Moon show that the distance of the Sun from the Earth is more than that of the Moon from the Earth.

Solution:

As shown in Fig. 3, when the Moon is half, ∠SME is nearly 90°. In a triangle, we can have only one right angle and other two angles are acute. Using the sine law, we have

$$\frac{\sin \angle SME}{SE} = \frac{\sin \angle SEM}{SM} = \frac{\sin \angle MSE}{ME}$$

Since ∠SME is larger than the ∠MSE, we have SE larger than ME. That is, the distance of the Sun from the Earth is more than that of the Moon from the Earth.

2.4 Sizes of the Sun and Moon

After having knowledge that from the Earth, the distance of the Sun is larger than that of the Moon, an explanation for occurrence of total solar eclipse was that at the time of eclipse, the Moon covers the Sun, as shown in Fig. 5. Since the duration of a total solar eclipse is not very large, it was thought that the angular sizes of both the Moon and Sun were nearly equal. Thus, we have

$$\text{Angle} = \frac{\text{Size of the Moon}}{\text{Distance of the Moon}} = \frac{\text{Size of the Sun}}{\text{Distance of the Sun}}$$

Therefore, we have

$$\text{Size of Sun} = \frac{\text{Distance of the Sun}}{\text{Distance of the Moon}} \times \text{Size of the Moon}$$

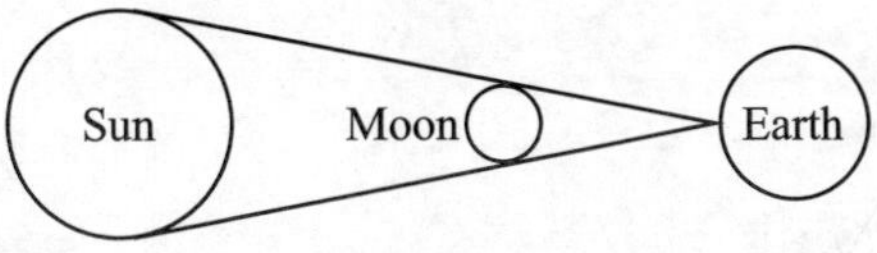

Fig. 5: During the total solar eclipse, the Moon covers the Sun.

Since the distance of the Sun is more than that of the Moon, it shows that the size of the Sun is larger than that of the Moon. From the estimation that ES = 20 EM, the size of the Sun was estimated to be 20 times larger than that of the Moon. (Accurate value is that the Sun is about 400 times larger than that of the Moon.) Aristarchus then posed a question about the size of the Earth as compared to that of the Moon.

Example 2.

Suppose, for a solar eclipse of very short duration, the ratio of the distances of the Sun and Moon from the Earth is 390. In the case of a solar eclipse of large duration, comment for the ratio of the distances of the Sun and the Moon from the Earth.

Solution:

The sizes of the Sun and the Moon are unchanged. For an eclipse of large duration, the angular size of the Moon is larger than that of the Sun. Hence, the Moon is closer to the Earth, as compared to its earlier position. Thus, the ratio of the distances of the Sun and the Moon from the Earth is more than 390.

Example 3.

Suppose, for a solar eclipse of very short duration, the ratio of the distances of the Sun and Moon from the Earth is 390. Comment for the ratio of the distances of the Sun and Moon from the Earth, so that the solar eclipse does not occur.

Solution:

The sizes of the Sun and Moon are unchanged. For no eclipse, the angular size of the Moon is smaller than that of the Sun. Hence, the Moon is farther away from the Earth, as compared to its earlier position. Thus, the ratio of the distances of the Sun and the Moon from the Earth is smaller than 390.

2.5 Shape of the Earth

It was later on discovered[3] that when the Sun was farthest north, on the day of solstice in mid-summer, it was vertically above the town of Syene and cast no shadow on the water in a well at that place when it was mid-day. At the same time, a pole planted vertically at Alexandria, 824 km to the north of Syene, cast a shadow which made an angle of 7.5° with the vertical, as shown in Fig. 6.

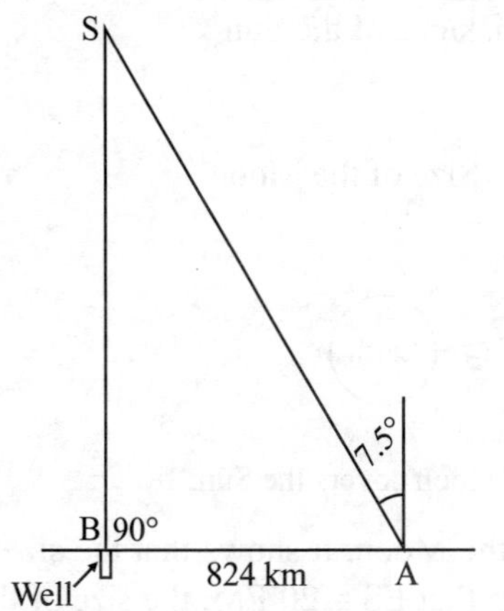

Fig. 6: Flat Earth between Syene B and Alexandria (A). S denotes the Sun.

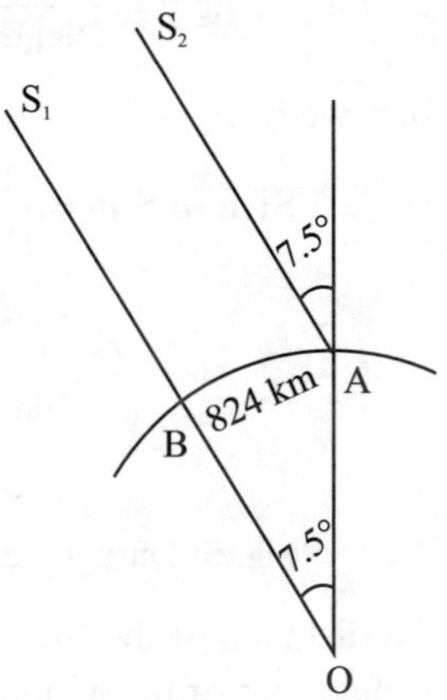

Fig. 7: Arc BA subtends at an angle 7.5° at the Earth's centre.

It was tried to understand the observations by assuming that the Earth is flat and the Sun (S) is vertically above the Syene (B), so that the $\angle SBA = 90°$, where A denotes Alexandria. At the same time, the Sun's rays subtend at an angle of 7.5° with the top of a vertical pole at Alexandria. Thus, $\angle SAB = 82.5°$. The calculated value of SB was found to be 6259 km. It was, however, already known that the Sun is very far away from the Earth.

It was argued that the Sun is so much away that its rays can be taken as parallel when they arrive the Earth[4]. Hence, although the Sun's rays are vertical at B, the angle of 7.5° between the rays and the vertical pole at A occurs because the Earth's surface is curved and not flat. (In the absence of curvature, the angle would be zero.)

---

[3]Some 60 or 70 years later, Eratosthenes (276 BC-194 BC) made another major step to know the shape of the Earth.

[4]Eratosthenes did not fall into the trap of assuming that the Earth is flat. He had sufficient proof to assume that the Earth is spherical in shape and not flat.

2.6 Size of the Earth

It was clear that a lunar eclipse occurs when the Moon moves into the shadow of the Earth, as shown in Fig. 8. During the lunar eclipse, the Earth is between the Moon and Sun. Accordingly, when the arc ABC of Earth's shadow is completed to describe the circle ABCDA, the diameter of this circle is equal to the Earth's diameter[5]. The diameter of the Earth was estimated to be twice that of the Moon. (Actual diameter of the Earth is 3.7 times that of the Moon.)

Thus, accordingly, the size of the Sun was 10 times larger than that of the Earth. (Actual diameter of the Sun is 100 times that of the Earth.) The importance of the work of Aristarchus lies in the fact that he proved for the first time that the Earth, although larger than the Moon, was relatively smaller than the Sun.

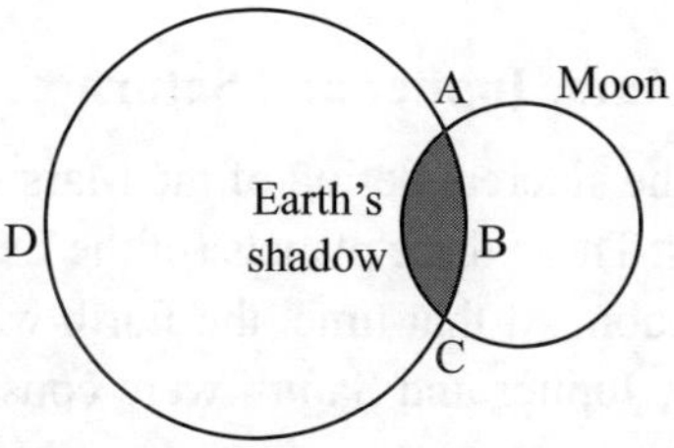

Fig. 8: Earth's shadow eclipses the Moon.

Note:

It was therefore reasonable to assume that the Moon being smaller than the Earth revolves around the Earth. But, it was not easy to accept that the Earth being smaller than the Sun should revolve around the Sun. Archimedes, who was younger to Aristarchus, mentioned in his writing that Aristarchus had a model that the Earth revolves around the Sun along with its five bodies (Moon, Venus, Mars, Jupiter and Saturn). The bodies Venus, Mars, Jupiter, and Saturn had been identified by that time. Aristarchus believed also correctly that the stars remain fixed as they are very far away from the Earth.

2.7 Seasons on the Earth

The angle between the Earth's equator and ecliptics[6] was also measured successfully[7]. The value was very close to the correct value 23.5°. It is this inclination of the Earth's equator to the ecliptic that is responsible for seasons on the surface of the Earth. In case, the angle of inclination had been zero, there had been no seasons (there had always been the same atmosphere) as the Earth's rotation axis would then be at right angle to the solar rays. There would had been hot strips on both sides of equator, with subsequent strips becoming progressively cooler as they were further away from the equator.

---

[5]Once again, Aristarchus had encountered a difficult problem.

[6]Information about the Earth's equator and ecliptic is discussed later on.

[7]by Eratosthenes

2.8 Day and Night on the Earth

The Earth spins about its own axis in a period of 24 hours. The day and night are due to the spin motion of the Earth. We have discussed about the rising and setting of the Sun. The duration from the rising to the setting of the Sun is called the day, and the duration from the setting to rising of the Sun is called the night.

2.9 Revolution of the Moon

The ancient Greeks knew that the Moon takes 29.5 days between one new Moon and the next. This period is known as the Moon's synodic period. To move once around the sky and to come back to the same position with regard to the stars took only 27.5 days. This period is known as sidereal period, discussed later on.

2.10 Sidereal Periods of the Mars, Jupiter and Saturn

The Greeks later on found that the sidereal period of the Mars was 687 days, of the Jupiter 12 years, and of the Saturn 30 years. These three planets (of the Earth) had to be much away from the Earth as compared to the Moon. At that time, the Earth was considered the centre of the universe, and Moon, Sun, Mars, Jupiter and Saturn were considered the planets of the Earth (Geocentric system).

2.11 Motions of the Mercury and Venus

Mercury and Venus did not show movement right across the sky, but were found to oscillate to and fro about the Sun. Venus never moved more than 47° away from the Sun and the Mercury never moved more than 28° away from the Sun. It was very difficult to explain these observations. A reasonable guess was that the Venus and Mercury were nearer to the Sun as compared to the Earth.

2.12 Positions of Stars

With the passage of time, even ages, positions of stars remained unchanged. It was understood in terms that the distances of stars were much larger than those of planets.

In early days, Greek astronomy reached at its height with Hipparchus in second century BC. He compiled a star atlas in which he registered 1000 stars. He very painstakingly measured the positions of stars to an accuracy of (1/15)°, and noticed that the positions of stars on his star charts differed from those on older charts by small amounts in a systematic manner. He deduced a brilliant idea that this was due to slow westward movement of a point (now called vernal equinox) where the equator intersects the ecliptic. The movement was estimated to be 40″ per year. At this rate, one circle is completed in 32,400 years. (We know that the precession of equinox is actually 50.3″ per year and the Earth's axis describes a circle in 25,765 years.)

2.13 Length of the Day

Hipparchus measured the length of a day to an accuracy of six minutes.

2.14 Magnitudes of Stars

Hipparchus classified stars according to their apparent brightness. The brightest stars were classified as having magnitude +1, and the faintest, that the eye could detect unaided, as having magnitude +6. The star having magnitude +6 was found about 100 times fainter than that having the magnitude +1. That is, all visible stars had magnitudes from +1 to +6.

3. Geocentric Model

The geocentric model for arrangement of celestial bodies is shown in Fig. 9. This model remained acceptable all over the world from the beginning of Astronomy until 1543, when the heliocentric system was proposed by Copernicus. According to the concept of geocentric model, our Earth is considered the supreme and all other celestial bodies, including the Sun, are considered to revolve around the Earth. Thus, the Earth is at the centre and the Moon, Sun, Mars, Jupiter and Saturn are revolving around the Earth, as planets. Further, the Mercury and Venus are revolving around the planet Sun, and are considered satellites of the Sun.

This model was supported by most of the scientists, including Ptolemy, Pathani Samanta and Tycho Brahe.

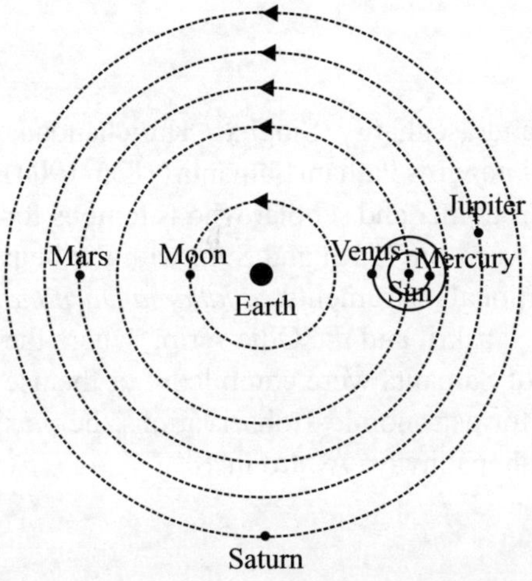

Fig. 9: Geocentric system of celestial bodies.

3.1 Ptolemy

Claudius Ptolemy (c.100-c.170 AD) (Fig. 10) was a mathematician, astronomer, astrologer, geographer, and music theorist. Out of about a dozen scientific treatises written by him, the three considered important are:

1. Astronomical treatise
2. Geography
3. Astrological treatise

Fig. 10: Ptolemy (c.**100**-c.**170** AD).
(*Courtesy*: Google website)

The Astronomical treatise is now known as the *Almagest*, although its original title was the *Mathematical Treatise*, and later known as *The Greatest Treatise*. The Geography is a thorough discussion on the maps and the geographic knowledge of the Greco-Roman world. The Astrological treatise is an attempt to adapt horoscopic astrology to the Aristotelian natural philosophy of his days.

3.2 Pathani Samanta

Mahamahopadhyaya Chandrasekhara Singha Harichandana Mahapatra Samanta, better known as Pathani Samanta (1835-1904) (Fig. 11), was an Indian astronomer and scholar who is famous for measuring the distance between the Earth and Sun, with the help of a bamboo pipe, and traditional instruments. *Siddhanta Darpana* is the book written in both Sanskrit and the Odia script, where the research and observations of Samanta were compiled. For the use of traditional instruments for astronomical observations, he was honoured with the Mahamahopadhyaya Award in 1893.

Fig. 11: Pathani Samanta (**1835-1904**).
(*Courtesy*: Google website)

3.3 Tycho Brahe

The first accurate determination of planetary positions was made by Tycho Brahe (Fig. 12). He did not accept the Copernicus' system of planets revolving around the Sun (heliocentric system). His argument was that if the Earth revolved around the Sun, the nearer stars would describe small circles against the background of more distant stars, as they will show the effects of parallax, as shown in Fig. 13. In Fig. 13, X is a nearby star and there are stars, situated at very large distances. Considering that the Earth is revolving around the Sun in an elliptical orbit ABCD. The star X is seen from different positions A, B, C, D on the Earth at different times. The projections of the star X on the celestial sphere are, respectively, denoted by *a*, *b*, *c*, *d*. Thus, the star X appears to move along a small ellipse denoted by abcd. The size of the ellipse abcd depends on the distance of star X from the Earth.

Fig. 12: Tycho Brahe (**1546-1601**).
(*Courtesy*: Google website)

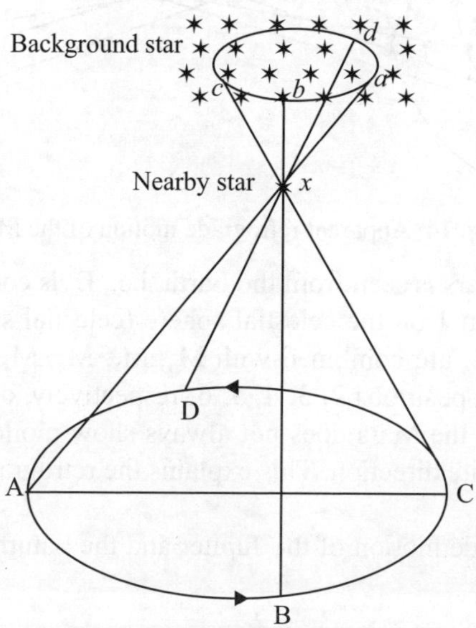

Fig. 13: Parallax of a nearby star. The ellipse ABCD denotes the elliptical orbit of the Earth revolving around the Sun.

He further argued that if the Mercury and Venus revolve in the orbits within that of the Earth, they should show phases, similar to those shown by the Moon, and the phases were not seen. (The reason for this was found later to be the limitation of instruments available in those days.)

Brahe always advocated the geocentric system shown in Fig. 14. Even if it is the case, the Mercury and Venus still should show the phases when they arrive between the Sun and Earth. But, since Brahe had no telescope, he could not see the phases of the Mercury and Venus.

However, Brahe rendered a great service to Astronomy by making accurate measurements of planetary positions, especially of the Mars. These data enabled Johannes Kepler (who believed in the heliocentric system) to formulate three famous laws of planetary motion.

4. Retrograde Motion of Mars

The retrograde motion of the Mars can be understood with the help of Fig. 14. Being at larger distance from the Sun as compared to the distance of the Earth from the Sun, the Mars moves with a slower speed as compared to that of the Earth. In Fig. 14, at six different increasing times, the positions of the Earth (E_1, E_2, E_3, E_4, E_5, E_6) and the corresponding positions of the Mars (M_1, M_2, M_3, M_4, M_5, M_6) are shown. Notice that when the Earth and the Mars are observed from the Sun, both are moving properly (along the continuous orbits, without any retrograde motion).

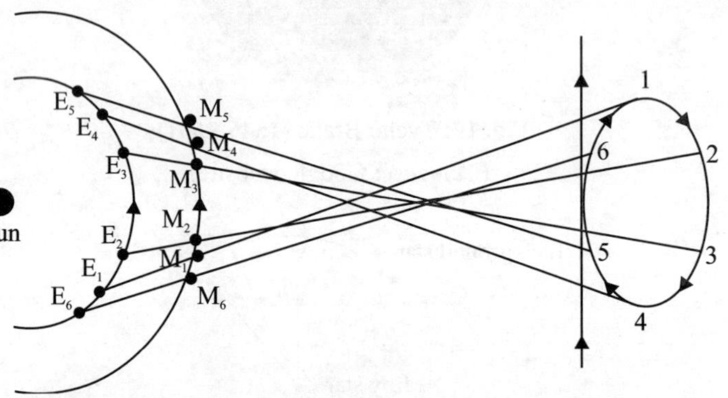

Fig. 14: Apparent retrograde motion of the Mars.

However, when the Mars is seen from the Earth, i.e., E_1 is combined with M_1, and further extension gives the position 1 on the celestial sphere (celestial sphere is discussed later on). Similarly, E_2, E_3, E_4, E_5, E_6 are combined with M_2, M_3, M_4, M_5, M_6, respectively, and their further extensions give the positions 2, 3, 4, 5, 6, respectively, on the celestial sphere. Thus, when seen from the Earth, the Mars does not always show motion in one direction, but also shows motion in the opposite direction. This explains the retrograde motion of the Mars when seen from the Earth.

Similarly, the retrograde motion of the Jupiter and the Saturn seen from the Earth can be understood.

4.1 Theory of Epicycles

Scientists attempted to understand the retrograde motion using the theory of epicycles. Figure 15 shows the deferent and epicycle. The deferent is a circle, whose centre is at the Earth. The epicycle is a circle moving on another circle. This model was first proposed by Apollonius of Perga at the end of the 3rd century BC. This is a geometric model, which was used to explain the variations in speed and direction of the apparent motion of the Moon, Sun, and planets. It was in particular used to explain the apparent retrograde motion of five planets known at the time. Further, it also explained the changes in the apparent distances of planets from the Earth.

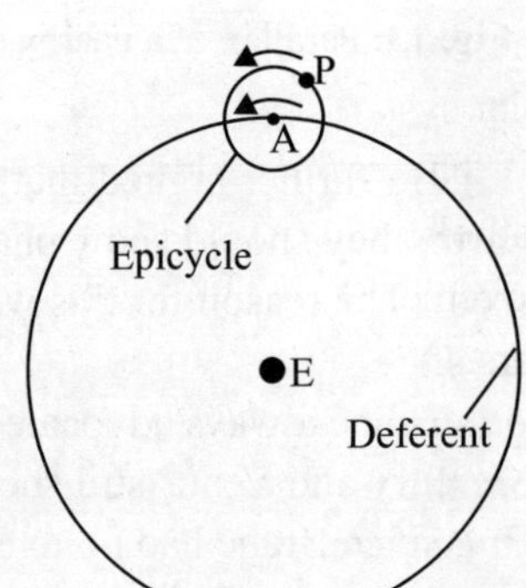

Fig. 15: Deferent and epicycle.

5. Heliocentric Model

The heliocentric system of celestial bodies, given by Copernicus is shown in Fig. 16. According to this system, the Sun is at the centre, and the Mercury, Venus, Earth, Mars, Jupiter and Saturn are revolving around the Sun, as planets. Further, the Moon is revolving around the Earth, as a satellite.

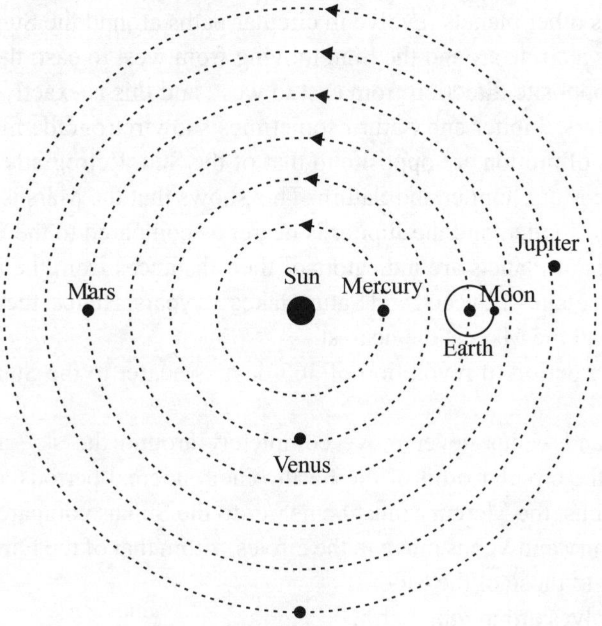

Fig. 16: Heliocentric system of revolving celestial bodies.

5.1 Copernicus Revolution

For 1600 years after Hipparchus, no remarkable progress was made in astronomy, and the geocentric system was applicable. The world had to wait until 1543 when Copernicus' book on solar system was published.

Nicolaus Copernicus (Fig. 17) was a priest in a Roman Catholic church. He spent many years on the study of the movement of the Sun, Moon and planets against the background stars. One of the main problems faced by him was the retrograde motion shown by the Mars, Jupiter and Saturn, when observed from the Earth.

Fig. 17: Nicolaus Copernicus **(1473-1543)**.
(*Courtesy*: Google website)

Based on his observations, Copernicus derived important conclusions. Some of them are as the following.

1. Earth is spherical in shape.

2. Earth spins from west towards east, and the atmosphere moves along with its surface. In case the atmosphere was not to move, there should have been a continuous wind blowing from east towards west. Because of Earth's spin motion, the Sun, Moon, and stars appear to rise in the east and to set in the west; in the opposite direction to that of the spin motion of the Earth.
3. Earth is not the centre of the universe, but the Sun is the centre of the universe and the Earth as well as other planets revolve in circular paths around the Sun.
4. Earth describes a circle around the Sun, moving from west to east; the stellar background must move in opposite direction from east to west, and this is exactly what is observed.
5. The planets, Mars, Jupiter and Saturn sometimes show retrograde motion. That is, when their directions of motion are opposite to that of the Sun. Retrograde motions decrease in sequence of the Mars, Jupiter and Saturn. This shows that the Mars is nearer to the Sun as compared to the Jupiter, and the Jupiter is nearer as compared to the Saturn.
6. Sidereal periods of planets are indicators of their distances from the Sun. The Mars takes 687 days, Jupiter takes 12 years, and Saturn takes 30 years. Hence, the Saturn is the farthest from the Sun and the Mars is the nearest.
7. The Earth has a period of revolution of 365 days is nearer to the Sun as compared to the Mars.
8. The Mercury and Venus never move completely around the sky, as they move in the circles within the circular orbit of the Earth. Their sidereal periods are 88 and 225 days, respectively. Thus, the Mercury must be nearer to the Sun as compared to the Venus.
9. Since the Mercury and Venus move in the circles within that of the Earth, they should show phases, similar to those of the Moon.
10. The Moon revolves around the Earth.

Copernicus system of revolving planets is shown in Fig. 16. This is known as heliocentric system. That is, the Sun is at the centre of orbits of revolving planets, including the Earth. Thus, it was for the first time when an idea of heliocentric system instead of geocentric system emerged. By 1530, Copernicus completed his writing, but he continued to revise and edit his work. In 1541, the first report about the work of Copernicus was published.

5.2 Reaction on the Concept of Copernicus

Soon after knowing about the Copernicus' system of revolving planets around the Sun, where the Earth also was revolving around the Sun, like other planets, attacks started against him by Roman catholic churches as well as by many others. In the meantime, the manuscript of the writing of Copernicus was in the hands of Osiander, who was a churchman. He took it to a publisher. Aware of attacks which had been made on the first report, Osiander added a preface to the effect that the ideas in the book were not to be taken as absolute truth, but merely a mathematical method for calculating planetary positions.

The book entitled *On the Revolution of the Celestial Spheres* was published in 1543, just two months before the death of Copernicus (24 May 1543). Copernicus acknowledged the contributions made by Aristarchus and stated that the stars do not show annual motion due to the parallax, because they are very far away from the Earth. Copernicus was the first to give the correct explanation of the fact that the Mercury and Venus do not move around the sky, but oscillate east and west of the Sun through 28° and 47°, respectively. It is because their orbits lie within that of the Earth.

5.3 Giordano Bruno

The opposition of church was something that had to be taken into account. Giordano Bruno (Fig. 18) was a keen supporter of the ideas of Copernicus, and he delivered lectures in France, England and Italy. He also said that the planets were the worlds like the Earth. In 1660, he was charged before the Inquisition and after a summary trial, was burnt alive at the stake.

Fig. 18: Giordano Bruno (**1548-1600**).
(*Courtesy*: Google website)

6. Johannes Kepler

For one or two years, Johannes Kepler (Fig. 19) worked with Tycho Brahe. They had acrimonious arguments about the Copernicus' system. Kepler was always in support of the Copernicus' system, whereas Brahe was not ready to accept it. After the death of Brahe in 1601, Kepler analyzed the Brahe's data for planetary positions and he was able to prove that the orbit of the Mars is not a circle but an ellipse. Finally, he formulated the following three laws of planetary motion.

Fig. 19: Johannes Kepler (**1571-1630**).
(*Courtesy*: Google website)

1. A planet revolves around the Sun in an orbit which is an ellipse having Sun as one of the foci.
2. The line joining the Sun to a planet sweeps equal area in equal interval of time.
3. The square of the time of revolution of a planet around the Sun is proportional to the cube of its average distance from the Sun.

These laws have been found applicable not only for the Mars, but for other planets as well. Moreover, these laws have been found applicable even for the motion of a satellite around the respective planet.

7. Galileo Galilee

While Kepler was busy in making these remarkable discoveries in planetary dynamics, Galileo Galilee (Fig. 20) got one of the first telescopes.

- When Galileo looked at the Jupiter, he saw four moons revolving around the Jupiter. It was like a solar system in miniature where the Kepler's laws could be applied also.
- Galileo also observed the phases of the Venus. Out of fear of being ridiculed, he stated this finding in the form of a mysterious riddle.

Fig. 20: Galileo Galilee (**1564-1642**).

(*Courtesy*: Google website)

- Galileo noticed that the Saturn appeared to have two ears, which a few years later seemed to have vanished. These were, of course, Saturn's rings which become invisible every 14.5 years when the Earth is in the plane of the rings.
- Galileo also noticed the craters on the surface of the Moon.

Galileo's discoveries changed a lot in support of the heliocentric system. In 1610, Galileo brought the issue before a wide audience. He continued his efforts, ever more boldly, and interpreted the problematical biblical passage in the book of Joshua to conform to a heliocentric cosmology. More importantly, he argued that the Bible is written in the language of the common person who is not an expert in Astronomy. The use of telescope, initiated by Galileo, started a revolution in Astronomy.

In 1633, Galileo was summoned before the Inquisition and was forced to recast his teachings, supporting the Copernicus system. Mindful of Bruno's fate in 1600, Galileo was very wise to recast. He was placed under house arrest, where he spent the last nine years of his life, working out *Principia* on which Isaac Newton, who was born in the year of the death of Galileo, was able to formulate his famous law of gravitation.

Fig. 21: Pisa Tower.
(*Courtesy*: Google website)

7.1 Pisa Tower and Equivalence Principle

The Leaning Tower of Pisa, or simply the Tower of Pisa (Fig. 21), is a tower of the cathedral of the Italian city of Pisa. It is famous worldwide for its nearly four-degree lean, which is the result of an unstable foundation. The height of the tower is 55.86 metres from the ground on the low side and 56.67 m on the high side, and is situated behind the Pisa Cathedral.

Galileo's experiment was performed by dropping simultaneously two cannon balls of different sizes and different masses freely from the top of the Tower of Pisa. Both the bodies arrived simultaneously on the surface of the Earth. It provided the equivalence principle. It may be interpreted that all bodies accelerate under a force towards the Earth at the same rate. The experiment was repeated several times for various bodies of different masses and different sizes. The result was common that a freely falling body was taking equal time for travelling equal distance.

That is, there is some force, and consequently, some acceleration, which does not depend on the mass of the body.

8. Newton

Isaac Newton (Fig. 22) was born within a year of the death of Galileo. Newton noticed that a body falling freely moves towards the surface of Earth, and the Moon is revolving around the Earth. He said that there is a relationship between the motion of the Moon and the motion of a freely falling body. Based on his dynamical and gravitational theories, he could understand and explain the Keplers laws of planetary motion, and established a modern quantitative science of gravitation. For that, he assumed the existence of an attractive force between all massive bodies, that can exist even when they are separated from one another, and the bodily contact between the bodies was not necessary. Making use of his law of inertia, Newton concluded that a force exerted by the Earth on the Moon is needed to keep it in a circular motion about the Earth. He realized that this force could be, at long range, the same as the force with which the Earth pulls objects on its surface downward. When Newton discovered that the acceleration of

Fig. 22: Isaac Newton (**1643-1727**).
(*Courtesy:* Google website)

the Moon is 1/3600 smaller than the acceleration at the surface of Earth, he related the number 3600 to the square of the radius of Earth. He calculated that the circular orbital motion of radius R and period T requires a constant inward acceleration A equal to the product of $4\pi 2$ and the ratio of the radius to the square of the time

$$A = \frac{4\pi^2 R}{T^2}$$

Newton noticed that the gravitational force between the bodies must depend on the masses of the bodies. Since a body of mass (M) experiencing a force (F) accelerates at a rate F/M, a force of gravity proportional to M would be consistent with the Galileo's observation that all bodies accelerate under gravity towards the Earth at the same rate, a fact that Newton also tested experimentally. In Newton's equation

$$F_{12} = G\frac{M_1 M_2}{r_{12}^2}$$

F_{12} is the magnitude of gravitational force acting between masses M_1 and M_2 separated by distance r_{12}. The constant G is the universal gravitational constant. This law of gravitation can be derived from the Kepler's laws of planetary motion, discussed later on.

8.1 Halley's Comet and Law of Gravity

Halley's Comet or Comet Halley is a short-period comet visible from the Earth every 7576 years. Halley is the only known short-period comet that is regularly visible to the naked eye from the Earth, and thus the only naked eye comet that can appear twice in a human lifetime. Halley last appeared in the inner parts of the solar system in 1986 and may next appear in mid-2061.

Halley's periodic returns to the inner solar system have been observed and recorded by astronomers around the world since at least 240 BC. But it was not until 1705 that the English astronomer Edmond Halley understood that these appearances were reappearances of the same comet.

It is here argued that Halley's comet had a pivotal role in triggering Newton's acceptance of the law of gravity.

8.2 Importance of Gravity as a Force in Astronomy

Because of gravity, the planets are held by the Sun and kept the planets revolving around the Sun. Similarly, the Moon is held by the Sun, and the Moon is kept revolving around the Earth. Gravity creates stars and planets by pulling together the material from which they are made.

9. Astronomy in Classical Literature

Mention of Astronomy may be found in classical literature. For example, often, Astronomy has played a role in deciding about the auspicious time (in terms of the positions of planets and stars) for doing important work or indicating about the disasters. In every era, people have noticed eclipses, comets etc., besides the rising and setting of the sun, and phases of the moon. Without any kind of preference, some events mentioning about astronomy are in the following literature.

1. *Mahabharata*
2. *Ramayana*
3. *Old Testament*
4. *New Testament*
5. *Puranas*
6. Kalidas's work
7. Khalil Gibran's work

9.1 Mahabharata

Mahabharata war occurred for 18 days, and the epic *Mahabharata* has over one hundred and fifty references to astronomical events, and a majority of these references are available in the Bhismaparvan. There are references of eclipses, positions of planets, planets, stars, comets and other events.

On the eve of Mahabharata war, Vyasa meets the king Dhrtarastra and describes him a whole set of ill omens he has seen. Vyasa said to Dhrtarastra that there would be an imminent war, the enemies are standing facing each other, ready to fight and that war will bring total destruction. This prediction was based on his observations. Vyasa said that he was observing the Sun every day both at the time of sunrise in the morning and at the time of sunset in the evening. He had noticed the appearance of halos and some comets, which appear to enclose the Sun as if with long arms. Further, the moon had become lustreless during the lunar eclipse on *Kartika purnima*. Such events indicate an imminent war.

Vyasa tells about twelve comets he had seen and the appearance of comets were considered to be harbingers of calamity since *Atharvaveda*. Vyasa described the omens which indicate destruction of the entire population; these include a pair of eclipses, a lunar and a solar eclipse occurring in the same month and within an interval of thirteen days. He said that he knew New Moon coinciding with fourteenth, fifteenth and also on the sixteenth day, but he never knew it coinciding with the thirteenth day. Moreover, in one and the same month, both the Sun and the Moon eclipsed on the thirteenth. Such ill-timed eclipses indicate destruction of the people. During war, a solar eclipse happened on the 14th day of the war when it was nearly the sunset time. It was the same day when Jayadrath was killed by Arjun.

9.2 Ramayana

The *Ramayana* has 7 *Kaandas*. The most common claim of Astronomy in *Ramayana* is that "Bal Kaanda" which mention the exact position of all planets and stars at the time of the birth of Lord Rama. It also mentions eclipses during Rama's exile and the position of planets at the time of other important events.

9.3 Old Testament

In the Hebrew Bible (*Old Testament*), there are various names of stars and planets. Besides the Earth, Venus and Saturn are the only planets expressly stated in the *Old Testament*. The Saturn is represented by the star Kaiwan (or Chiun), worshiped by the Israelites in the desert.

9.4 New Testament

In the *New Testament*, there are 48 Bible verses mentioning Astronomy. Concepts of some verses are given below, as found in the source, and of other verses can be found from the source: (Source: https://bible.knowing-jesus.com/topics/Astronomy).

- Who commands the sun not to shine, and sets a seal upon the stars; Who alone stretches out the heavens. And tramples down the waves of the sea; Who makes the Bear, Orion and the Pleiades, And the chambers of the south.
- Lift up your eyes on high, And see who has created these stars, The One who leads forth their host by number, He calls them all by name; Because of the greatness of His might and the strength of His power, Not one of them is missing.
- Thus the heavens and the earth were completed, and all their hosts.
- Behold, to the Lord your God belong heaven and the highest heavens, the earth and all that is in it.
- For all the gods of the peoples are idols, But the Lord made the heavens.
- You alone are the Lord. You have made the heavens, The heaven of heavens with all their host, The earth and all that is on it, The seas and all that is in them. You give life to all of them, And the heavenly host bows down before You.
- Is not God in the height of heaven? Look also at the distant stars, how high they are!
- If even the moon has no brightness. And the stars are not pure in His sight.

9.5 Puranas

Indian astronomy has a long history stretching from pre-historic to modern times. The earliest roots of Indian astronomy can be dated to the period of Indus Valley Civilization or earlier. Later on, Astronomy developed as a discipline of Vedanga associated with the study of the Vedas. The oldest known text is the could be worked out *Vedanga Jyotisha*.

9.6 Kalidas's Work

Some contributions of Kalidas are:

- *Ritusamhara* describes the six Indian seasons: summer, monsoon/rains, autumn, hemanta (cool), winter, and vasanta (spring). It is considered to be Kalidasa's earliest work.

- In *Kumarasabhavam* (The Birth of Kumara), the style of description of spring set the standard for nature metaphors. It basically talks about the birth of Kumara (Kartikeya), the son of Shiva and Parvati.
- *Meghaduta* (Cloud Messenger) describes how a *yaksa* (or nature spirit), asked a cloud to take a message of love to his wife.

9.7 Khalil Gibran's Work

Some contributions of Khalil Gibran are:

- *The Astronomer*
- *The Earth Gods*
- *What of Time?*
- *Past and Future*

10. Science Fictions

Defining imaginary scenerio of astronomical phenomena/events, science fictions have been written in various languages by many authors. Here, we have mentioned some of them without any kind of preference.

10.1 Isaac Asimov's Science Fiction

Nightfall, 1941

10.2 Ray Bradbury's Science Fiction

Martian Chronicles, 1950

10.3 Narlikar's Science Fictions

Jayant Narlikar has written science fictions in English and Marathi. Some of his science fictions written in English are:

The Return of Vaman, 1990
The Adventure
The Comet

10.4 Premendra Mitra's Science Fictions

Premendra Mitra has written science fictions in Bengali language. Some of them are:

Juddho Jakhan Thamlo (When the War Stopped)
Pinpre Puran (The Story of the Ants)
Prithivir Shatru (The Enemies of the Earth)
Kalapanir Atole
Mangalbairi (The Martian Enemies)
Koral Kit (Horrible Insect)
Akasher Atonko (The Danger from the Sky)
Manusher Protidwondi (The Rival of the Man)
MoyDanober Dweep (The Island of MoyDanob)
Shomaoner Ron(g) Sada (The White Coloured Death)
Shukre Jara Giyechhilo (Those Who Went to Venusi); previously named *Prithibee Chhariye* (Beyond the Earth)

10.5 Satyajit Ray's Science Fictions

Satyajit Ray wrote science fictions in Bengali and English.

Bankubabur Bandhu (Banku Babu's Friend or Mr. Banku's Friend) (Bangali) 1962
The Alien (English) 1967

10.6 Steven Spielberg's Science Fiction

E.T. (The Extra-Terrestrial), 1982

10.7 Fred Hoyle's Science Fictions

The Black Cloud, 1957
A for Andromeda, 1961

11. Contribution of India in Astronomy

11.1 Stars and Constellations in Rig Veda

Hindu mythology has four Vedas: *Rig Veda, Yajur Veda, Sama Veda* and *Atharva Veda*. The *Rig Veda* has an incomplete list of 27 (or 28 stars/asterisms) (nakshatras), which lie along the ecliptic and begin from Pleiades (Krttikas). List of 27 *nakshatras* in sequence is given in Table 1. There has been one more naksatras, Abhijit, lying in between Uttara Ashadha and Shravana.

Table 1. List on nakshatras

No.	Name	No.	Name	No.	Name
1	Krttika (the Pleiades)	10	Uttara phalguni (Denebola)	19	Uttara ashadha
2	Rohini (Aldebaran)	11	Hasta	20	Shravana
3	Mrigashirsha	12	Chitra (Spica)	21	Dhanishta
4	Ardra (Betelgeuse)	13	Svati (Arcturus)	22	Satabhishak (Sadachbia)
5	Punarvasu	14	Vishakha	23	Purva bhadrapada
6	Pushya	15	Anuradha	24	Uttara bhadrapada
7	Aslesh	16	Jyeshth	25	Revati
8	Magha (Regulus)	17	Mula	26	Ashvini
9	Purva phalguni	18	Purva ashadha	27	Bharani

In older tradition, Hindu astronomy had 28 *nakshatras* which were used as celestial markers in the heavens.The angle of 360° was divided into 28 equal parts, and each *nakshatra* measuring $12° \ 51\frac{3'}{7}$. When the *nakshatras* were mapped along the ecliptic, the ecliptic was divided into 27 equal segments, each segment measuring 13° 20′. The *nakshatra* Abhijit, in between Uttara Ashadha and Shravana, was left out without any portion on the ecliptic. (It removed fraction in the arc-minute.) However, while deciding on the timing of an auspicious event, the Abhijit *Nakshatra* becomes important. The coordinates of the twenty-seven *nakshatras* have been specified in the *Surya Siddhantha*. Each *nakshatra* is further divided into four equal *padas*, each *pada* measuring 3° 20′.

11.2 Rasis or Zodiac Signs

As per Indian Astrology, there are 12 *rasis* or zodiac signs. 27 *naksatras* and 12 *rasis* are the basic division to identify particular individual's position of planets at the time of his/her birth. The sky or space is divided into 12 divisions measuring 30° and is known as *rasi*.

Table 2. Rasis or Zodiac signs

No.	Sanskrit name	English name	Ruling Graha (Planet)
1.	Mesha	ARIES	Mangal (Mars)
2.	Rishaba	TAURUS	Shukra (Venus)
3.	Mithuna	GEMINI	Budha (Marcury)
4.	Karkataka	CANCER	Chandra (Moon)
5.	Simha	LEO	Surya (Sun)
6.	Kanya	VIRGO	Budha (Mercury)
7.	Thula	LIBRA	Shukra (Venus)
8.	Vrischika	SCORPIO	Mangal (Mars)
9.	Dhanush	SAGGITARIUS	Brahaspati (Jupiter)
10.	Makara	CAPRICORN	Sani (Saturn)
11.	Kumbha	AQUARIUS	Sani (Saturn)
12.	Meena	PISCES	Brahaspati (Jupiter)

11.3 Aryabhata and his Seminal Contribution to Astronomy

Aryabhata or Aryabhata I (476-550 CE) (Fig. 23) is considered the first of the major mathematician-astronomers from the classical age of Indian mathematics and Indian astronomy. His important contributions are the *Aryabhatiya* and the *Aryasiddhanta*.

Fig. 23: Aryabhata (**476550** CE).
(Courtesy: Google website)

It is believed that he was born in the year 476 in Kusumapura or Pataliputra (present-day Patna, Bihar). Though there are uncertainties about the place of his birth, but it is fairly certain that, at some point of time, he went to Kusumapura for advanced studies and lived there for some time. One version is that Aryabhata was the head of an institution (*kulapa*) at Kusumapura, and, because the university of Nalanda was in Pataliputra at the time and had an astronomical observatory, it is speculated that Aryabhata might have been the head of the Nalanda University as well. Aryabhata is also considered to have set up an observatory at the Sun temple in Taregana, Bihar.

He has written several treatises on Mathematics and Astronomy, some of which are considered lost.

His major work, *Aryabhatiya*, a compendium of Mathematics and Astronomy, was extensively referred to in the Indian mathematical literature and has survived to modern times. The mathematical section of the *Aryabhatiya* covers arithmetic, algebra, plane trigonometry, and spherical trigonometry. It also contains continued fractions, quadratic equations, sums-of-power series, and a table of sines.

The *Aryabhatiya* is written in the style of *sutras*, in which each line is an aid to memory for a complex system. It contains 108 verses and 13 introductory verses, and is divided into four *padas* (chapters):

1. **Gitikapada: (13 verses):** Large units of time *kalpa*, *manvantra*, and *yuga* which present a cosmology different from earlier texts such as Lagadha's *Vedanga Jyotisha* (c. 1st century BCE). There is also a table of sines (*jya*), given in a single verse. The duration of the planetary revolutions during a *mahayuga* is given as 4.32 million years.
2. **Ganitapada (33 verses):** Covering mensuration (*ksetra vyavahara*), arithmetic and geometric progressions, gnomon/shadows (*shanku-chhaya*), simple, quadratic, simultaneous, and indeterminate equations (*kuaka*).
3. **Kalakriyapada (25 verses):** Different units of time and a method for determining the positions of planets for a given day, calculations concerning the intercalary month (*adhikamraa*), *kshaya-tithis*, and a seven-day week with names for the days of week.
4. **Golapada (50 verses):** Geometric/trigonometric aspects of the celestial sphere, features of the ecliptic, celestial equator, node, shape of the Earth, cause of day and night, rising of zodiacal signs on horizon, etc. In addition, some versions cite a few colophons added at the end, extolling the virtues of the work, etc.

Aryabhatiya has described a number of innovations in Mathematics and Astronomy in the verse form, which remained effective for many centuries. The extreme brevity of the text was elaborated in commentaries by his disciple Bhaskara I (c. 600 CE) and later on by Nilakantha Somayaji in his *Aryabhatiya Bhasya*, (1465 CE).

Aryabhatiya's description of relativity of motion is also remarkable. He discussed that just as a man in a boat moving forward sees the stationary objects (on the shore) as moving backward, in the similar manner the stationary stars seen by the people on the Earth as moving exactly towards the west.

Based on the writings of Aryabhata's contemporary, Varahamihira, and later mathematicians and commentators, including Brahmagupta and Bhaskara I, it is believed that Aryabhata's contribution, *Aryasiddhanta*, having astronomical computations, is missing. This work appears to be based on the older *Surya Siddhanta* and uses the midnight-day reckoning, as opposed to the sunrise in *Aryabhatiya*. It also considered to contain a description of several astronomical instruments:

- the gnomon (*shanku-yantra*)
- a shadow instrument (*chhaya-yantra*)
- possibly angle-measuring devices, semicircular and circular (*dhanur-yantra/chakra yantra*)
- a cylindrical stick (*yasti-yantra*)
- an umbrella-shaped device called the (*chhatra-yantra*), and
- water clocks of at least two types, bow-shaped and cylindrical.

A third text, which may have survived in the Arabic translation. It claims that it is a translation done by Aryabhata, but the Sanskrit name of this work is not known.

Contribution in Mathematics

Place value system and zero

The place-value system, first found in the 3rd century, was clearly available in the work of Aryabhata. Though he did not use a symbol for zero, the French mathematician Georges Ifrah argues that in Aryabhata's place-value system, the information about zero was implicit. Aryabhata did not use the Brahmi numerals. Continuing the Sanskritic tradition from Vedic times, Aryabhata used letters of the alphabet to denote numbers, expressing quantities, such as the table of sines.

Approximation of π

Aryabhata worked for getting approximate value of π, and may have come to the conclusion that π is irrational. He obtained $\pi = 3.1416$, which is accurate to three decimal places.

Trigonometry

Aryabhata discussed the concept of sine in his work by the name of *ardha-jya*, which literally means "half-chord". For simplicity, people started calling it *jya*.

Indeterminate equations

Indian mathematicians since ancient times have been greatly interested to find integer solutions to diophantine equations having the form $ax + by = c$.

In general, such diophantine equations may be very difficult. They were discussed extensively in ancient Vedic text *Sulba Sutras*. Aryabhata's method of solving such problems, later on elaborated by Bhaskara, is called the 'Kuaka method'. The meaning of *kuaka* is breaking into small pieces, and a recursive algorithm is involved in the method for writing the original factors in smaller numbers. In Indian mathematics, this algorithm became the standard method for solving first-order diophantine equations. Initially, the whole subject of algebra was called '*kuaka-gaita*' or simply '*kuaka*'.

Algebra

Aryabhata provided nice results for the summation of the series of squares and cubes as the following:

$$1^2 + 2^2 + 3^2 + ... + n^2 = \frac{n(n+1)(2n+1)}{6}$$

and

$$1^3 + 2^3 + 3^3 + ... + n^3 = (1 + 2 + 3 + ... + n)^2$$

Contribution in Astronomy

In the Aryabhata's system of Astronomy, days are reckoned from 'uday', dawn at equator. In some of his later writings on astronomy, which apparently proposed a second model, the days are reckoned from midnight. Though these records are lost but can be partly reconstructed from the discussion in Brahmagupta's *Khandakhadyaka.* In some texts, he seems to assign the apparent motions of heavenly bodies to the Earth's rotation. He may have believed that the planet's orbits as elliptical rather than circular.

Motions of the solar system

Aryabhata insisted that the Earth rotates about its axis daily, and that the apparent movement of the stars is a relative motion due to the rotation of the Earth, contrary to the then-prevailing opinion, that the sky rotates. He provided the number of rotations of the earth in a 'yuga'.

Aryabhata described a geocentric model of solar system, in which the Sun and Moon are each carried by epicycles. They, in turn, revolve around the Earth. In this model, the motions of the planets are each governed by two epicycles, a smaller slow and a fast. In terms of distance from the Earth, the order of the planets was taken as: Moon, Mercury, Venus, Sun, Mars, Jupiter, Saturn, and the asterisms.

The positions and periods of the planets were calculated relative to uniformly moving points. In the case of Mercury and Venus, they move around the Earth at the same mean speed as the Sun. In the case of the Mars, Jupiter, and Saturn, they move around the Earth at specific speeds, representing each planet's motion through the zodiac.

Eclipses

Aryabhata stated that the Moon and planets shine by the reflected sunlight. A scientific approach adopted by Aryabhata to explain the solar and lunar eclipses. He explained eclipses in terms of shadows cast by and falling on the Earth. Thus, the lunar eclipse occurs when the Moon enters into the Earth's shadow. He discussed in detail, the size and extent of the Earth's shadow and then provided the computation and the size of the eclipsed part during an eclipse.

Sidereal periods

Aryabhata calculated the sidereal rotation as 23 hr, 56 min, and 4.1 sec; the modern value is 23 hrs, 56 min, and 4.091 sec. Similarly, his value for the length of the sidereal year at 365 days, 6 hours, 12 minutes, and 30 seconds (365.25858 days) is an error of 3 minutes and 20 seconds over the length of a year (365.25636 days).

Heliocentrism

Some scientists believe that in the work of Aryabhata, there is indication of heliocentric system. It has been however found that Aryabhata's system was not explicitly heliocentric.

11.4 Varahahmihira

Fig. 24: Varahmihira (c. **505**-c. **587**).
(*Courtesy*: Google website)

Varahamihira (c. 505-c. 587), also called Varaha or Mihira (Fig. 24), was a Hindu astronomer and mathematician who lived in Ujjain (Madhya Pradesh). He was born in the Avanti region, roughly corresponding to modern-day Malwa (part of Madhya Pradesh), to Adityadasa. According to one of his own write-ups, he was educated at Kapitthaka.

Varrahamihira's most famous work was the *Brihat Samhita*, an encyclopedic work on architecture, temples, planetary motions, eclipses, timekeeping, astrology, seasons, cloud formation, rainfall, agriculture, mathematics, gemology, perfumes and many other topics. According to Varahamihira, in some verses he was merely summarizing earlier existing literature on Astronomy, Shilpa Sastra and temple architecture, yet his presentation of different theories and models of design are among the earliest texts that have survived. The chapters of *Brihat Samhita* and verses of Varahamihira were quoted by the Persian traveler and scholar Al Biruni.

Varahamihira is also credited with writing several authoritative texts on Astronomy and Astrology. He learned the Greek language, and praised the Greeks (Yavanas) in his text saying them "well trained in the sciences". Some scholars consider him the one who understood and introduced the zodiac signs, predicted calculations for the auspicious ceremonies and made astrological computations.

Pancha-Siddhantika

Varahamihira's main work is the book *Panchasiddhantika* dated about 575 CE, which provides information about older Indian texts which are now lost. The work is a treatise on mathematical Astronomy and it summarises five earlier astronomical treatises by five texts, namely, the *Surya Siddhanta*, *Romaka Siddhanta*, *Paulisa Siddhanta*, *Vasishtha Siddhanta* and *Paitamaha Siddhanta*. It is a compilation of '*Vedanga Jyotisha*' and Hellenistic Astronomy (with Greek, Egyptian and Roman elements). Varahamihira was the first one to state that the ayanamsa, or the shifting of the equinox is 50.32 seconds.

Five Siddhrantas of Indians

- *Surya Siddhanta*, the siddhnta of the Sun, considered to be composed by Laadeva, but actually composed by Mayasura also known as Mamuni Mayan as mentioned in the text itself.
- *Vasishtha Siddhanta*, so called from one of the stars of the Great Bear, composed by Vishnucandra,
- *Paulisa Siddhnata*, so called from Paulisa, the Greek, from the city of Saintra, which is supposed to be Alexandria, composed by Paulisa.
- *Romaka Siddhanta*, so called from the Rum, i.e., the subjects of the Roman Empire, composed by rshea.
- *Paitahama Siddhanta*.

Brihat-Samhita

Another important contribution of Varahamihira is the encyclopedic *Brihat Samhita*. It contains 106 chapters and is considered the "great compilation". Although this text is mostly about divination, it also includes a wide range of subjects other than divination. It covers wide-ranging subjects of human interest, including Astronomy, planetary movements, eclipses, rainfall, clouds, architecture, growth of crops, manufacture of perfume, matrimony and domestic relations. The volume expounds on gemstone evaluation criterion found in the *Garuda Purana*, and elaborates on the sacred Nine Pearls from the same text.

On Astrology

Based on the Vedic Astrology system, Varahamihira has given *Brihajjataka*, which is predictive Astrology, written in Sanskrit. The most famous contribution on Astrology of Varahamihira is considered *Hora Shastra* or *Brihadjathaka*. It is mostly prepared in code language.

His son Prithuyasas also contributed to Hindu Astrology and his book *Hora Sara*, is a famous contribution on horoscopy. Bengali poet astrologer Khana (also named Lilavati elsewhere) is believed to be the daughter-in-law of Varahamihira.

Contributions

Varahamihira has great contributions in several fields. Some of them are as the following.

Trigonometry: Varahamihira improved the accuracy of the sine tables given by Aryabhata.

Combinatorics: He was one of the first mathematicians who discovered a version of what is now called the Pascal's triangle. He used it for the calculation of binomial coefficients. He also records the first known 4×4 magic square.

Physics: Varahamihira has contributed in Physics also. He explained the reflection as the back-scattering of particles and the refraction as the ability of particles to penetrate inner spaces of the material.

11.5 Brahamagupta

According to his own statement, Brahmagupta (Fig. 25) was born in 598 CE. He lived in Bhillamala, Gurjaradesa (modern Bhinmal in Rajasthan) during the reign of the Chavda dynasty ruler, Vyagrahamukha. He was the son of Jishnugupta and was Hindu by religion. He lived and worked there for a good part of his life.

Fig. 25: Brahmagupta (**598** CE-**668** CE). (*Courtesy*: Google website)

Bhillamala was the capital of the Gurjaradesa, the second-largest kingdom of western India, comprising southern Rajasthan and northern Gujarat in modern-day India. It was also a centre of learning for Mathematics and Astronomy. Brahmagupta became an astronomer of the Brahmapaksha school, one of the four major schools of Indian Astronomy during this period. He studied the five traditional *Siddhanthas* on Indian Astronomy as well as the work of other astronomers including Aryabhata I, Latadeva, Pradyumna, Varahamihira, Simha, Srisena, Vijayanandin and Vishnuuser.

In the year 628, at the age of 30, he composed the *Brahmasphuasiddhanta*, which is believed to be a revised version of the revised Siddhanta of the Brahmapaksha school. Scholars state that he incorporated a great deal of originality to his revision, adding a considerable amount of new material. The book has 24 chapters with 1008 verses. The large portion of book deals with Astronomy, but it also contains key chapters on Mathematics, including algebra, geometry, trigonometry and algorithmics, which are believed to contain new insights due to Brahmagupta himself.

Later, Brahmagupta moved to Ujjaini, Avanti, which was also a major centre for Astronomy in central India. At the age of 67, he composed his next well known work *Khanda-khadyaka*, a practical manual of Indian Astronomy for the use of students.

Brahmagupta died in 668 CE, and he is presumed to have died in Ujjain.

Controversy

Brahmagupta is considered involved in the criticism of the work of rival astronomers, and his *Brahmasphutasiddhanta* displays one of the earliest discrepancies among Indian mathematicians. The division was primarily about the use of Mathematics to the physical world, rather than about the Mathematics itself. In Brahmagupta's case, the disagreements developed largely from the choice of astronomical parameters and theories.

Mathematics

Brahmagupta has big contribution in Mathematics also as the following.

Algebra

Brahmagupta has given solution of the general linear equation in *Brahmasphutasiddhanta*. He found solutions of equation $ax^2 + bx = c$ as the following.

$$x = \frac{\pm\sqrt{4ac + b^2} - b}{2a}$$

and

$$x = \frac{\sqrt{ac + \frac{b^2}{4}} - \frac{b}{2}}{a}$$

He further proceeded to solve systems of simultaneous indeterminate equations using the method that the desired variable must first be isolated, and then the equation must be divided by the desired variable's coefficient. In particular, he recommended using the "pulverizer" to solve equations with several unknowns.

Arithmetic

The four fundamental operations (addition, subtraction, multiplication, and divi- sion) were known for long time before Brahmagupta. This current system is based on the Hindu Arabic number system and first appeared in *Brahmasphutasiddhanta*. Brahmagupta describes the multiplication of two positive integers m and n, as addition of m, n tines or vice versa. In Medieval Europe, Indian arithmetic was known as "Modus Indorum" meaning the method of Indians. In *Brahmasphutasiddhanta*, multiplication was named Gomutrika. In *Brahmasphutasiddhnta*, Brahmagupta has provided details of operations on fractions. The reader is expected to know the basic arithmetic operations as far as taking the square root, although he has given explanation for finding the cube and cube-root of an integer and later he provided rules utilized for the computation of squares and square roots. Later on, he has given rules for dealing with five types of combinations of fractions:

$$\frac{a}{c}+\frac{b}{c} \qquad \frac{a}{c}\times\frac{b}{d} \qquad \frac{a}{1}+\frac{b}{d}$$

$$\frac{a}{c}+\frac{b}{d}\times\frac{a}{c}=\frac{a(d+b)}{cd} \qquad \frac{a}{c}-\frac{b}{d}\times\frac{a}{c}=\frac{a(d-b)}{cd}$$

Series

Further Brahmagupta proceeded to give the sum of the squares and cubes of the first n integers as

$$1^2+2^2+3^2+...+n^2=\frac{n(n+1)(2n+1)}{6}$$

and $$1^3+2^3+3^3+...+n^3=\left[\frac{n(n+1)}{2}\right]^2$$

Zero

Brahmagupta's *Brahmasphuasiddhanta* is the first book providing rules for arithmetic manipulations which apply to zero and to negative numbers. The *Brahmasphutasiddhanta* is the earliest known text to treat zero as a number in its own right, rather than as simply a placeholder digit to represent another number as used by the Babylonians or as a symbol for a lack of quantity as used by Ptolemy and the Romans.

It was also included that the sum of two positive numbers is positive, of two negative numbers, negative; of a positive and a negative is their difference; when they are equal, the sum is zero. The sum of a negative and zero is negative, and of a positive and zero positive, and of two zeros is zero.

$$(+m)+(+n)=+(m+n)$$
$$(-m)+(-n)=-(m+n)$$
$$(+m)+(-n)=+(m-n)$$
$$(+m)+(-n)=0$$
$$(-m)+0=-m$$
$$(+m)+0=+m$$
$$0+0=0$$

A negative number minus zero is negative, a positive number minus zero is positive; zero minus zero is zero. When a positive is subtracted from a negative or a negative from a positive, then it is added.

$$(-m)-0=-m$$
$$(+m)-0=+m$$
$$0-0=0$$
$$(-m)-(+n)=-(m+n)$$
$$(+m)-(-n)=+(m+n)$$

He continued his work and described multiplication. The product of a negative and a positive is negative, of two negative numbers is positive, and of positive numbers is positive; the product of zero and a negative, of zero and a positive, or of two zeros is zero.

$$(-m) \times (+n) = -(m \times n)$$
$$(-m) \times (-n) = +(m \times n)$$
$$(+m) \times (+n) = +(m \times n)$$
$$(-m) \times 0 = 0$$
$$(+m) \times 0 = 0$$
$$0 \times 0 = 0$$

But his description of division by zero differs from our modern understanding. A positive divided by a positive or a negative divided by a negative is positive; a zero divided by a zero is zero; a positive divided by a negative is negative; a negative divided by a positive is negative. A negative or a positive divided by zero has that zero as its divisor, or zero divided by a negative or a positive has that negative or positive as its divisor. The square of a negative or of a positive is positive; the square of zero is zero. A number is the square of its square root.

$$\frac{+m}{+n} = +\left(\frac{m}{n}\right) \quad \frac{-m}{-n} = +\left(\frac{m}{n}\right) \quad \frac{0}{0} = 0 \quad \frac{+m}{-n} = -\left(\frac{m}{n}\right)$$

$$\frac{+m}{-n} = -\left(\frac{m}{n}\right) \quad \frac{-m}{+n} = -\left(\frac{m}{n}\right) \quad \frac{+m}{0} = 0 \quad \frac{-m}{0} = 0$$

$$\frac{0}{+m} = +m \qquad \frac{0}{-m} = -m \qquad (-m)^2 = +(m^2)$$

$$(+m)^2 = +(m^2) \qquad (0)^2 = 0 \qquad (\sqrt{m})^2 = m$$

Pythagorean triplets

In *Brahmasphutasiddhanta*, Brahmagupta provides a formula which can be used for generating Pythagorean triples.

Pell's equation

Brahmagupta then gave a recurrence relation for generating solutions to second degree equation such as $Nx^2 + 1 = y^2$ (called Pell's equation) by using the Euclidean algorithm. He knew the Euclidean algorithm as the "pulverizer" as it breaks numbers down into ever smaller pieces. The key to his solution was the identity,

$$(x_1^2 - Ny_1^2)(x_2^2 - Ny_2^2) = (x_1x_2 + Ny_1y_2)^2 - N(x_1y_2 + x_2y_1)^2$$

which is a generalisation of an identity discovered by Diophantus:

$$(x_1^2 - y_1^2)(x_2^2 - y_2^2) = (x_1x_2 + y_1y_2)^2 - (x_1y_2 + x_2y_1)^2$$

Making use of his identity and the fact that if (x_1, y_1) and (x_2, y_2) are solutions to the equations $x^2 - Ny^2 = k_1$ and $x^2 - Ny^2 = k_2$, respectively, then $(x_1x_2 + Ny_1y_2, x_1y_2 + x_2y_1)$ is a solution to $x^2 - Ny^2 = k_1k_2$, he could find integral solutions to Pell's equation through a series of equations of the form $x^2 - Ny^2 = k_i$. However, Brahmagupta could not apply his solution uniformly for all possible values of N, rather he could show that if $x^2 - Ny^2 = k$ has an integer solution for $k = \pm 1, \pm 2$, or ± 4, then $x^2 - Ny^2 = 1$ has a solution. The solution of the general Pell's equation was later on given by Bhaskara II in c.1150 CE.

Geometry

Brahmagupta has made contribution in geometry also.

Brahmagupta's formula

Brahmagupta's most famous contribution in geometry is his formula for cyclic quadrilaterals. Given the lengths of the sides of any cyclic quadrilateral, Brahmagupta gave an approximate and an exact formula for the figure's area. For given lengths *p*, *q*, *r* and *s* (Fig. 26) of a cyclic quadrilateral, the approximate area is:

$$\text{Area} = \left(\frac{p+r}{2}\right)\cdot\left(\frac{q+s}{2}\right)$$

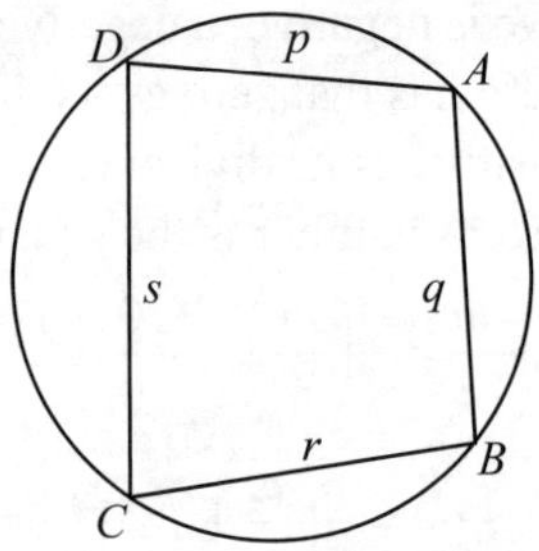

Fig. 26: Brahmagupta's formula.

while, letting $t = (p + q + r + s)/2$, the exact area is $\sqrt{(t-p)(t-q)(t-r)(t-s)}$. Although Brahmagupta did not mention explicitly that these quadrilaterals are cyclic, but it is apparent from his rules that this is the case.

Triangles

A substantial part of the work of Brahmagupta is on geometry. One theorem gives the lengths of two segments a triangle base is divided by its altitude. Thus, the lengths of the two segments are

$$\frac{1}{2}\left(b \pm \frac{c^2 - a^2}{b}\right)$$

He then gave a theorem on rational triangles. A triangle with rational sides *a*, *b*, *c* and rational area is of the form:

$$a = \frac{1}{2}\left(\frac{u^2}{v} + v\right)$$

$$b = \frac{1}{2}\left(\frac{u^2}{w} + w\right)$$

$$c = \frac{1}{2}\left(\frac{u^2}{v} - v + \frac{u^2}{w} - w\right)$$

for some rational numbers *u*, *v*, and *w*.

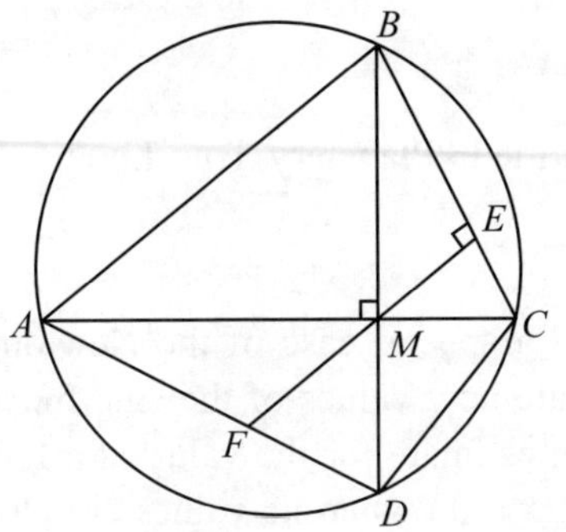

Fig. 27: Brahmagupta's theorem states that $AF = FD$.

Brahmagupta's theorem

For the construction given in Fig. 27, Brahmagupta's theorem states that $AF = FD$.

Brahmagupta continued his work. He said that in a non-unequal cyclic quadrilateral (that is, an isosceles trapezoid), the length of each diagonal is $\sqrt{pr + qs}$.

He continued to give formulas for the lengths and areas of geometric figures, such as the circumradius of an isosceles trapezoid and a scalene quadrilateral, and the lengths of diagonals in a scalene cyclic quadrilateral. This leads up to Brahmagupta's famous theorem.

Imaging two triangles within a cyclic quadrilateral with unequal sides, the two diagonals are the two bases. Their two segments are separately the upper and lower segments (formed) at the intersection of the diagonals. The two (lower segments) of the two diagonals are two sides in a triangle; the base (of the quadrilateral is the base of the triangle). Its perpendicular is the lower portion of the (central) perpendicular; the upper portion of the (central) perpendicular is half of the sum of the (sides) perpendiculars diminished by the lower (portion of the central perpendicular).

π

Brahmagupta gave values of π. As a practical value, used 3 for π, and $\sqrt{10} \approx 3.1622$ as an accurate value of π. The error in this accurate value is less than 1%.

Measurements and constructions

In some text, Brahmagupta gave about the constructions of various figures with arbitrary sides. He essentially manipulated right triangles to produce isosceles triangles, scalene triangles, rectangles, isosceles trapezoids, isosceles trapezoids with three equal sides, and a scalene cyclic quadrilateral.

After giving the value of π, he considered the geometry of plane figures and solids, such as calculating volumes and surface areas (or empty spaces dug out of solids). He calculated the volume of rectangular prisms, pyramids, and the frustum of a square pyramid. He further calculated the average depth of a series of pits. For the volume of a frustum of a pyramid, he found the pragmatic value as the depth times the square of the mean of the edges of the top and bottom faces, and he found the superficial volume as the depth times their mean area.

Trigonometry

Sine table

In *Brahmasphutasiddhanta*, entitled "Planetary True Longitudes", Brahmagupta gave a sine table.

Interpolation formula

Brahmagupta devised and used a special case of the Newton-Stirling interpolation formula of the second-order to interpolate new values of the sine function from other values already tabulated. The formula gives an estimate for the value of a function f at a value $a + xh$ of its argument (with $h > 0$ and $-1 \le x \le 1$) when its values are already known at $a - h$, a and at $a + h$. The formula for the estimate is:

$$f(a+xh) \approx f(a) + x\frac{\Delta f(a) + \Delta f(a-h)}{2} + x^2 \frac{\Delta^2 f(a-h)}{2!}$$

where Δ is the first-order forward-difference operator, expressed as $\Delta f(a) = f(a+h) - f(a)$.

Astronomy

Some of the important contributions made by Brahmagupta in Astronomy are his methods for calculating the position of heavenly bodies over time (ephemerides), their rising and setting, conjunctions, and the calculation of solar and lunar eclipses.

He explained that since the Moon is closer to the Earth than the Sun, the degree of the illuminated part of the Moon depends on the relative positions of the Sun and the Moon, and this can be computed from the size of the angle between the two bodies.

He further discussed in his treatise *Khandakhadyakawas*, the methods to explore the longitudes of the planets, diurnal rotation, lunar and solar eclipses, risings and settings, the moon's crescent and conjunctions of the planets.

11.6 Symbolic Relation between Mathematics and Astronomy

Celestial objects, *Rasis* are represented in terms of mathematical symbols.

Table 3. Mathematical symbols for celestial objects and Rasis

Symbol	Name	Symbol	Name	Symbol	Name
☿	Mercury	♀	Venus	⊕	Earth
♂	Mars	♃	Jupiter	♄	Saturn
♅	Uranus	♆	Neptune	♇	Pluto
☉	Sun	○	Full Moon	●	New Moon
☾	Left Moon	☽	Right Moon		
♈	Aries	♋	Cancer	♎	Libra
♑	Capricorn	♉	Taurus	♌	Leo
♏	Scorpio	♒	Aquarius	♊	Gemini
♍	Virgo	♐	Sagittarius	♓	Pisces

11.7 Evidence of the Precision of Equinox from Vedic Literature

The equinoxes occur at the two intersections of the celestial equator and the ecliptic. There are two equinoxes, vernal equinox in spring and autumnal equinox in fall. At the equinoxes, the days and nights are equal. At the time of Varahamihira (550 CE) the vernal equinox was in Ashvini *nakshatra*. Indian calendar was reckoned according to the constellation at the vernal equinox. During the era of 3400-2210 BCE, the vernal equinox was in the Pleiades *nakshatra*. The vernal equinox lay in Orion around 4500 BCE.

The above is good enough to find the evidence of the precision of equinox from vedic literature.

11.8 Jai Singh and his Jantar Mantar

Jain Singh, full name, Sawai Jai Singh II (Fig. 28), was a Rajput king. He was born on 3 November 1688 in Amber and died on 21 September 1743 in Jaipur (Rajasthan). He founded the fortified city of Jaipur and shifted his capital from Amber to Jaipur.

Fig. 28: Sawai Jai Singh II (**1688-1743**).
(*Courtesy*: Google website)

Jai Singh was a great mathematician, astronomer, and town planner. He set up the famous observatories known as Jantar Mantars. The word 'Jantar' was derived from the Sanskrit word 'Yantra', meaning "instrument, machine", and the word 'Mantar' was derived from the Sanskrit word 'Mantrana', meaning 'consult, calculate'. Thus, the word 'Jantar Mantar' literally means 'calculating instrument'.

The Jantar Mantar is a collection of nineteen architectural astronomical instruments, which was completed in 1734 and is located near City Palace and Hawa Mahal.

Jai Singh noticed that the *Zij*, used for determination of the position of celestial objects, was not in the agreement with the positions calculated on the table. He constructed five new observatories in different cities in order to create a more accurate *Zij*, and created astronomical tables, which were in continuous use in India for a century. Also, it was used to measure time. Thus, there were five Jantar Mantars in New Delhi, Jaipur, Ujjain, Mathura and Varanasi in India. The Jantar Mantar in Jaipur is the largest one. At present only four Jantar Mantars exist, as the Jantar Mantar at Mathura was torn down just before the Revolt of 1857.

The Jantar Mantar consists of nineteen instruments for measuring time, predicting eclipses, tracking location of major stars as the Earth revolved around the Sun, ascertaining the declinations of planets, and determining the celestial altitudes and related ephemerides. The instruments are (alphabetical):

1. **Chakra Yantra** has four semicircular arcs on which a gnomon casts a shadow, and gives the declination of the Sun at four specified times of the day. This data corresponds to the noon at four observatories around the world (Greenwich in UK, Zurich in Switzerland, Notke in Japan and Saitchen in the Pacific). It is equivalent of a wall of clocks giving local times in different parts of the world.
2. **Dakshin Bhitti Yantra** measures meridian, altitude and zenith distances of celestial bodies.
3. **Digamsha Yantra** is a pillar in the middle of two concentric outer circles, used to measure azimuth of the sun and to calculate the time of sunrise and sunset forecasts.
4. **Disha Yantra** used to describe the directions.
5. **Dhruva Darshak Pattika** is used to observe and find the location of pole star with respect to other celestial bodies.
6. **Jai Prakash Yantra** has two hemispherical bowl-based sundials with marked marble slabs that map inverted images of the sky and allow the observer to move inside the instrument. It measures altitudes, azimuths, hour angles, and declinations.
7. **Kapali Yantra** measures coordinates of celestial bodies in azimuth and equatorial systems. Any point in sky can be visually transformed from one coordinate system to another.
8. **Kanali Yantra**
9. **Kranti Vritta** Yantra measures longitude and latitude of celestial bodies.
10. **Laghu Samrat Yantra** is the smaller sundial at the monument, inclined at 27°, to measure time, though less accurately than Vrihat Samrat Yantra.
11. **Misra Yantra** (meaning mixed instrument) is a compilation of five different instruments.
12. **Nadi Valaya Yantra** (two sundials on different faces of the instrument), the two faces represent north and south hemispheres, and measure the time to an accuracy of less than a minute.
13. **Palbha Yantra**
14. **Rama Yantra** is an upright building used to find the altitude and the azimuth of the Sun.
15. **Rashi Valaya Yantra** has 12 gnomon dials that measure ecliptic coordinates of stars, planets and all 12 constellation systems.
16. **Shastansh Yantra** (next to Vrihat Samrat Yantra) has a 60° arc built in the meridian plane within a dark chamber. At noon, the sun's pinhole image falls on a scale below enabling the observer to measure the zenith distance, declination, and the diameter of the Sun.
17. **Unnatamsa Yantra** is a metal ring divided into four segments by horizontal and vertical lines, with a hole in the middle. The position and orientation of the instrument allows to measure the altitude of celestial bodies.
18. **Vrihat Samrat Yantra** (meaning the great king of instruments) is the world's largest (27 m high) gnomon sundial measuring time in intervals of 2 seconds using shadow cast

from the sunlight. Its face is angled at 27 degrees, the latitude of Jaipur. It is a UNESCO World Heritage site. The Hindu *chhatri* (small cupola) on top is used as a platform for announcing eclipses and the arrival of monsoons.

19. **Yantra Raj Yantra** (a 2.43-metre bronze astrolabe, one of the largest in the world) is used only once a year to calculate the Hindu calendar.

The instruments are in most cases huge structures. The scale to which they have been built has been alleged to increase their accuracy.

12. Constellations

On the celestial sphere, a constellation is an area where a group of visible stars forms a perceived outline or pattern, typically representing an animal, mythological person or creature, or an inanimate object. The word "constellation" comes from the Late Latin term constellatio, meaning "set of stars". The word came into use during the 14th century.

The origins of the earliest constellations likely go back to prehistory. People used them to relate stories of their beliefs, experiences, creation, or mythology. Different countries and cultures adopted their own constellations.

The 48 traditional Western constellations are Greek. Constellations in the far southern sky were added from the 15th century until the mid-18th century when European explorers began travelling to the Southern Hemisphere. Twelve ancient constellations belong to the zodiac.

In 1922, the International Astronomical Union (IAU) formally accepted the modern list of 88 constellations, and in 1928 adopted official constellation boundaries that together cover the entire celestial sphere. Each IAU constellation has an official 3-letter abbreviation.

Table 4. Constellations in the IAU list and their abbreviations

Constellation	Abb	Constellation	Abb	Constellation	Abb
Andromeda	And	Antlia	Ant	Apus	Aps
Aquarius	Aqr	Aquila	Aql	Ara	Ara
Aries	Ari	Auriga	Aur	Bootes	Boo
Caelum	Cae	Camelopardalis	Cam	Cancer	Cnc
Canes Venatici	CVn	Canis Major	CMa	Canis Minor	CMi
Capricornus	Cap	Carina	Car	Cassiopeia	Cas
Centaurus	Cen	Cepheus	Cep	Cetus	Cet
Chamaeleon	Cha	Circinus	Cir	Columba	Col
Coma Berenices	Com	Corona Australis	CrA	Corona Borealis	CrB
Corvus	Crv	Crater	Crt	Crux	Cru
Cygnus	Cyg	Delphinus	Del	Dorado	Dor
Draco	Dra	Equuleus	Equ	Eridanus	Eri
Fornax	For	Gemini	Gem	Grus	Gru

Contd...

Hercules	Her	Horologium	Hor	Hydra	Hya
Hydrus	Hyi	Indus	Ind	Lacerta	Lac
Leo	Leo	Leo Minor	LMi	Lepus	Lep
Libra	Lib	Lupus	Lup	Lynx	Lyn
Lyra	Lyr	Mensa	Men	Microscopium	Mic
Monoceros	Mon	Musca	Mus	Norma	Nor
Octans	Oct	Ophiuchus	Oph	Orion	Ori
Pavo	Pav	Pegasus	Peg	Perseus	Per
Phoenix	Phe	Pictor	Pic	Pisces	Psc
Piscis Austrinus	PsA	Puppis	Pup	Pyxis	Pyx
Reticulum	Ret	Sagitta	Sge	Sagittarius	Sgr
Scorpius	Sco	Sculptor	Scl	Scutum	Sct
Serpens	Ser	Sextans	Sex	Taurus	Tau
Telescopium	Tel	Triangulum	Tri	Triangulum Australe	TrA
Tucana	Tuc	Ursa Major	UMa	Ursa Minor	UMi
Vela	Vel	Virgo	Vir	Volans	Vol
Vulpecula	Vul				

Following the formal definition, other star patterns or groups called asterisms are not constellations, but are also used by observers to navigate the night sky. An asterism may have several stars within a constellation, or may share stars in more than one constellation. Examples of asterisms include the Pleiades and Hyades within the constellation Taurus and the False Cross split between the southern constellations Carina and Vela, or Venus' Mirror in the constellation of Orion.

Review Questions

1. Discuss in brief about seven ancient civilizations in Astronomy.
2. Describe phases of the Moon and provide explanation for it. If the Moon has been flat in shape, could we see its phases.
3. How idea was obtained that the distance between the Moon and Earth is smaller than that between the Sun and Earth?
4. How information about the depth in the sky was obtained?
5. How it was decided that the the size of the Sun is larger than that of the Moon?
6. How it was decided that the Earth's surface is not flat, but spherical in shape?
7. How the size of the Earth was estimated first time.
8. Why stars are considered to be fixed in position? What is sidereal period of a celestial body?

9. Describe the geocentric model. Show that this system could not explain the retrograde motion of Mars.
10. Describe the heliocentric model. How it could explain the retrograde motion of the Mars.
11. What is Copernicus' opinion about planetary system? What was reaction on it?
12. Though the Mercury and Venus revolve around the Sun, but they are noticed only when they are east or west of the Sun. Comment on it.
13. Define parallax. How it can be used to distinguish between nearby and far distant stars.
14. Write name of scientist who used the telescope first time for observing outside the Earth. What were consequences of those observations?
15. When seen from the Earth, the Mars shows its retrograde motion. Explain. How the problem of retrograde motion was sorted out.
16. Write opinions of Tycho Brahe and Johannes Kepler about arrangement of planets in our solar system.
17. Write observations of Copernicus about heliocentric system. Write the name of person who propagated this idea of heliocentric system and what have been consequences of that?
18. Describe geocentric system for the arrangement of planets. What is heliocentric system for the arrangement of planets?
19. Write sequence of planets in Tycho Brahe's system.
20. Write sequence of planets in Copernicus' system.
21. Write names of celestial objects which were known up to the time of Copernicus.
22. Write names of two celestial objects which were considered planets in Tycho Brahe's system, but are not now.

    Ans: Sun, Moon
23. Write name of one celestial object which was not considered a planet in Tycho Brahe's system, but is now.
24. Write about the role of Johannes Kepler in Astronomy.
25. Write about the role of Galileo Galilee in Astronomy.
26. Write about the role of Newton in Astronomy.
27. Describe about the Astronomy in classical literature.
28. Describe the contribution of India in Astronomy.
29. Describe the contribution of Aryabhata in Astronomy.
30. Describe the contribution of Varahahmihira in Astronomy.
31. Describe the contribution of Brahamagupta in Astronomy.
32. Describe the contribution of Jai Singh in Astronomy.
33. How many *nakshatras* were in the older tradition in Hindu astronomy? Which *nakshatra* is left out when the *nakshatras* were mapped along the ecliptic?

    Ans: Earth
34. Write notes on the following:
    (i) Ancient Babylonian Astronomy
    (ii) Ancient Indian Astronomy
    (iii) Ancient Greek Astronomy
    (iv) Ancient Mayan Astronomy

(v) Ancient Egyptian Astronomy
(vi) Ancient Chinese Astronomy
(vii) Ancient Persian Astronomy
(viii) Phases of Moon
(ix) Full-Moon night
(x) New-Moon night
(xi) Rising and setting of the Sun
(xii) Horizon
(xiii) Distances of the Sun and the Moon from the Earth
(xiv) Sizes of the Sun and the Moon
(xv) Shape of the Earth
(xvi) Size of the Earth
(xvii) Seasons on the Earth
(xviii) Day and night on the Earth
(xix) Magnitudes of stars
(xx) Geocentric model
(xxi) Ptolemy
(xxii) Pathani Samanta
(xxiii) Tycho Brahe
(xxiv) Retrograde motion of the Mars
(xxv) Heliocentric model
(xxvi) Copernican revolution
(xxvii) Role of Giordano Bruno
(xxviii) Kepler and his laws of planetary motion
(xxix) Galileo Galilee and his contribution in Astronomy
(xxx) Pisa Tower and equivalence principle
(xxxi) Newton and his contribution in Astronomy
(xxxii) Astronomy in Mahabharata
(xxxiii) Astronomy in Ramayana
(xxxiv) Astronomy in Old Testament
(xxxv) Astronomy in New Testament
(xxxvi) Astronomy in Puranas
(xxxvii) Astronomy in Kalidas's work
(xxxviii) Astronomy in Khalil Gibran's work
(xxxix) Science fictions
(xL) Stars and constellations in *Rig Veda*
(xLi) *Rasis* or zodiac signs
(xLii) Aryabhata and his ontribution in Astronomy
(xLiii) Varahahmihira and his ontribution in Astronomy
(xLiv) Brahamagupta and his ontribution in Astronomy
(xLv) Jai Singh and his ontribution in Astronomy
(xLvi) Constellations

Multiple-Choice Questions

1. Which civilization has the first record of the appearances of Halley's comet?
A. Babylonian B. Greek C. Egyptian D. Chinese **Ans. A**
2. Which civilization has first formulated theories and mathematical equations to understand the universe?
A. Babylonian B. Greek C. Egyptian D. Chinese **Ans. B**
3. Anaxagoras belonged to which civilization?
A. Babylonian B. Greek C. Egyptian D. Chinese **Ans. B**
4. Eratosthenes belonged to which civilization?
A. Babylonian B. Greek C. Egyptian D. Chinese **Ans. B**
5. Pythagoras belonged to which civilization?
A. Babylonian B. Greek C. Egyptian D. Chinese **Ans. B**
6. Aryabhata belonged to which civilization?
A. Babylonian B. Greek C. Indian D. Chinese **Ans. C**
7. The Great Pyramid of Giza belonged to which civilization?
A. Babylonian B. Greek C. Egyptian D. Chinese **Ans. C**
8. The first solar calendar was prepared in which civilization?
A. Babylonian B. Greek C. Egyptian D. Indian **Ans. C**
9. Who proposed first explanation about the phases of the Moon?
A. Einstein B. Galileo C. Anaxagoras D. Newton **Ans. C**
10. The idea about the spherical shape of Moon emerged from
A. solar eclipse B. lunar eclipse
C. phases of Moon D. spherical shape of Earth **Ans. C**
11. Out of the Sun, Moon and Earth, which one is between the others two on full-moon night.
A. Sun B. Earth C. Moon D. Any one **Ans. B**
12. Out of the Sun, Moon and Earth, which one is between the others two on new-moon night.
A. Sun B. Earth C. Moon D. Any one **Ans. C**
13. When the Moon is half in shape, the angle between the directions joining the Moon to the Sun and to the Earth is about
A. 30° B. 45° C. 60° D. 90° **Ans. D**
14. Knowing that the size of the Sun is larger than that of the Moon, from the Earth, the distance of the Sun is larger than that of the Moon can be decided with the help of
A. solar eclipse B. lunar eclipse
C. phases of the Moon D. seasons on the Earth **Ans. A**
15. Approximate relation between angular sizes S and M of the Sun and Moon, respectively, is
A. S = 200 M B. S = 2 M C. S = M D. 2 S = M **Ans. C**
16. The size of the Earth was first time estimated with the help of
A. solar eclipse B. lunar eclipse
C. phases of the Moon D. seasons on the Earth **Ans. B**

17. Who was never convinced with heliocentric system?
    A. Copernicus B. Tycho Brahe C. Kepler D. Einstein **Ans. B**
18. Who was burnt alive when he supported heliocentric system?
    A. Copernicus B. Galileo C. Kepler D. Giordano Bruno **Ans. D**
19. Which body shows phases?
    A. Mars B. Jupiter C. Mercury D. Saturn **Ans. C**
20. Which body shows phases?
    A. Mars B. Jupiter C. Venus D. Saturn **Ans. C**
21. Which body does not show phases?
    A. Moon B. Mercury C. Venus D. Mars **Ans. D**
22. Which body does not show phases?
    A. Moon B. Mercury C. Venus D. Jupiter **Ans. D**
23. Which body does not show phases?
    A. Moon B. Mercury C. Venus D. Saturn **Ans. D**
24. The phases are shown by
    A. inner planets only B. both inner and outer planets
    C. outer planets only D. neither inner nor outer planets **Ans. A**
25. Who was author of the book entitled *On the Revolution of the Celestial Spheres*?
    A. Galileo B. Copernicus C. Hipparchus D. Tycho Brahe **Ans. B**
26. Who introduced magnitude system of stars?
    A. Galileo B. Copernicus C. Hipparchus D. Tycho Brahe **Ans. C**
27. Who derived famous laws for planetary motion around the Sun?
    A. Copernicus B. Tycho Brahe C. Kepler D. Einstein **Ans. C**
28. Whose data were used for derivation of Kepler's laws for planetary motion around the Sun?
    A. Copernicus B. Tycho Brahe C. Kepler D. Einstein **Ans. B**
29. The Sun rises in east and sets in west because the
    A. Sun moves from east towards west B. Earth revolves around the Sun
    C. Earth spins about its own axis D. Sun spins about its own axis **Ans. C**
30. Reason for seasons on the Earth is the
    A. Sun moves from east towards west B. Earth revolves around the Sun
    C. Earth spins about its own axis D. Sun spins about its own axis **Ans. B**
31. Which is not a naked eye planet?
    A. Saturn B. Jupiter C. Uranus D. Mars **Ans. C**
32. In geocentric system, at the centre of planetary system is the
    A. Sun B. Earth C. Moon D. Mars **Ans. B**
33. In heliocentric system, at the centre of planetary system is the
    A. Sun B. Earth C. Moon D. Mars **Ans. A**

34. The maximum angular separation of the Mercury from the Sun is about
A. 15° B. 28° C. 45° D. 90° **Ans. B**
35. The maximum angular separation of the Venus from the Sun is about
A. 15° B. 32° C. 47° D. 90° **Ans. C**
36. The conclusion that sidereal periods of planets are indicators of their distances from the Sun, was drawn by
A. Copernicus B. Tycho Brahe C. Kepler D. Einstein **Ans. A**
37. The Venus appears as a morning star, when situated on which side of the Sun?
A. East B. West C. North D. South **Ans. B**
38. The Venus appears as a evening star, when situated on which sidc of the Sun?
A. East B. West C. North D. South **Ans. A**
39. The Mercury appears as a morning star, when situated on which side of the Sun?
A. East B. West C. North D. South **Ans. B**
40. The Mercury appears as a evening star, when situated on which side of the Sun?
A. East B. West C. North D. South **Ans. A**
41. Who modified the preface in book entitled *On the Revolution of the Celestial Spheres of Nicolaus Copernicus*?
A. Osiander B. Galileo C. Kepler D. Bruno **Ans. A**
42. Which one does not show retrograde motion when seen from the Earth?
A. Mercury B. Mars C. Jupiter D. Saturn **Ans. A**
43. Which one does not show retrograde motion when seen from the Earth?
A. Venus B. Mars C. Jupiter D. Saturn **Ans. A**
44. Who agreed to recast his teaching given in favour of heliocentric system?
A. Copernicus B. Tycho Brahe C. Kepler D. Galileo **Ans. D**
45. Who was placed under house arrest as he was in favour of heliocentric system?
A. Copernicus B. Tycho Brahe C. Kepler D. Galileo **Ans. D**
46. Which body was considered a planet in geocentric system, but not in heliocentric system?
A. Mars B. Moon C. Jupiter D. Saturn **Ans. B**
47. Which body was considered a planet in geocentric system, but not in heliocentric system?
A. Sun B. Mars C. Jupiter D. Saturn **Ans. A**
48. Who is author of *Brihat Samhita*?
A. Aryabhata B. Brahmagupta C. Varrahamihira D. Bhaskara **Ans. C**
49. Who is author of *Aryabhatiya*?
A. Aryabhata B. Brahmagupta C. Varrahamihira D. Bhaskara **Ans. A**
50. How many verses are in *Aryabhatiya*?
A. 101 B. 108 C. 121 D. 151 **Ans. C**
51. Who is author of *Pancha-Siddhantika*?
A. Aryabhata B. Brahmagupta C. Varrahamihira D. Bhaskara **Ans. C**
52. Who is author of *Brahmasphuasiddhanta*?
A. Aryabhata B. Brahmagupta C. Varrahamihira D. Bhaskara **Ans. B**

53. Who is author of *Hora Sara*?
    A. Prithuyasas B. Brahmagupta C. Varrahamihira D. Bhaskara **Ans. A**
54. Who is author of *Brihajjataka*?
    A. Prithuyasas B. Brahmagupta C. Varrahamihira D. Bhaskara **Ans. C**
55. Who is author of *Hora Shastra*?
    A. Prithuyasas B. Brahmagupta C. Varrahamihira D. Bhaskara **Ans. C**
56. Who is author of *Brahmasphuasiddhanta*?
    A. Prithuyasas B. Brahmagupta C. Varrahamihira D. Bhaskara **Ans. B**
57. Who is author of *Khanda-khadyaka*?
    A. Prithuyasas B. Brahmagupta C. Varrahamihira D. Bhaskara **Ans. B**

2 Astronomical Coordinates

In a clear, dark night, we see sparkling points of light apparently situated on a vast sphere (called, the celestial sphere) in the sky. All these points are not necessarily the stars. They may be the stars, star clusters, galaxies, planetary nebulae, etc. Because of the large distances of these heavenly bodies (also known as the cosmic objects) from us, it is difficult to resolve them with the naked eyes and therefore, each of them appears like a star. In order to specify the position of an astronomical object on the celestial sphere, one requires a coordinate system. Further, the coordinates of a celestial body observed from the Earth depends on the position of the observer on the surface of the Earth. Thus, it is also essential to know the position of the observer on the surface of the Earth. In this chapter, we have introduced various coordinate systems used to specify position on the surface of the Earth and to specify positions of various celestial bodies in the sky.

The Earth revolves around the Sun and the seasons on the Earth change regularly. After one year, the seasons repeat, and we can count the number of days in one year. How many times the Sun rises (for example) in one year is the number of days in one year. In a period of four years[1], the Sun rises 1461 times. Thus, the number of days in one year is 365.25. Further, it was assumed that the angular displacement of the Earth is 1° per day. In order to make a round figure, it was decided that in one revolution, the angular displacement of the Earth is 360° (instead of 365.25°)[2].

In radian, the angular displacement is

$$\text{angle} = \frac{\text{arc}}{\text{radius}} = \frac{2\pi r}{r} = 2\pi$$

Thus, we have

$$2\pi \text{ rad} = 360°$$

The Earth makes a rotation of 360° in 24 hours. So we have

$$360° \equiv 24 \text{ h or } 15° \equiv 1 \text{ h}$$

Example 1.

Convert $5^h\ 16^m\ 27^s$ into angle unit (degrees, arc minutes and arc seconds).

---

[1]It is interesting to know how a year is calculated. One can easily notice that the sun rises on the horizon each day at different position. This position oscillates between two extremes. Starting from one extreme, when the Sun comes again to that extreme is the duration of one year.

[2]The rounding of number could be 370°. The number 370 can be divided by 2, 5, 10, 37 and 370 only whereas the number 360 can be divided by 2, 3, 4, 5, 6, 8, 10 and several other numbers; scientists preferred to consider 360 and not 370.

Solution:

We have

$$27^s = \frac{27}{60} = 0.45^m$$

Therefore,

$$16^m27^s = 16 + 0.45 = 16.45^m = \frac{16.45}{60} = 0.27417^h$$

and
$$5^h16^m27^s = 5 + 0.27417 = 5.27417^h$$

We know $1^h \equiv 15°$. Therefore, we have

$$5.27417^h = 15 \times 5.27417 = 79.11255°$$

Now, we have

$$0.11255° = 60 \times 0.11255 = 6.753'$$

and
$$0.753' = 60 \times 0.753 = 45.18''$$

Thus, we have $5^h\ 16^m\ 27^s \equiv 79°\ 6'\ 45.18''$.

Example 2.

Convert 15° 26′ 36″ into time unit (hours, minutes and seconds).

Solution:

We have

$$36'' = \frac{36}{60} = 0.6'$$

Therefore,

$$26'36'' = 26 + 0.6 = 26.6' = \frac{26.6}{60} = 0.44333°$$

and
$$15°26'36'' = 15 + 0.44333 = 15.44333°$$

We know $15° \equiv 1^h$. Therefore, we have

$$15.44333° = 15.44333/15 = 1.02956^h$$

Now, we have

$$0.02956^h = 60 \times 0.02956 = 1.7736^m$$

and
$$0.7736^m = 60 \times 0.7736 = 46.42^s$$

Thus, we have $15°\ 26'\ 36'' = 1^h\ 1^m\ 46.42^s$.

1. Point in a Plane

In order to specify position of a point in a plane, a coordinate system is required. When a particle is moving, its coordinates change with time. In order to introduce a coordinate system, also called the reference system, in a plane, let us consider a straight line *AB* (Fig. 1) in the plane and a point *O* on the line. With the help of this reference system comprising the line *AB* and the point *O* (Fig. 1a), position of a point *P* in the plane can be specified. Now, it is up to the observer which coordinate system one wants to use.

For the Cartesian coordinate system, we draw a straight line *PQ* through the point *P* and perpendicular to the line *AB* (Fig. 1b). When the point *O* is treated as the origin of the coordinate system, then the distance *OQ* represents the *x*-coordinate, and the distance *PQ*, is the y-coordinate of the point *P*. It is obvious that when we change the position of, either the line *AB* or the point *O*, the coordinates (*x*, *y*) of the point *P* change. It shows that the coordinates of a point depend on the reference system.

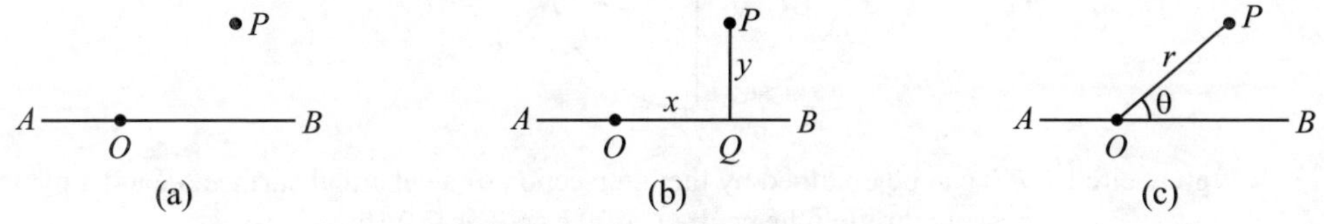

Fig. 1: Coordinates in a plane surface.

For the polar coordinate system, we join the points *O* and *P* through a straight line (Fig. 1c). The distance *OP* represents the radial coordinate *r* and angle *POB* represents the angular coordinate θ. Thus, the polar coordinates of the point *P* are (*r*, θ).

2. Spherical Surface

In the Astronomy and Astrophysics[3], we often deal with the spherical surfaces. For example, the Earth's surface is taken as spherical in shape; celestial sphere, on which heavenly bodies are assumed to be situated, is spherical. Before discussing about the coordinate systems for the spherical surfaces, let us introduce some definitions pertaining to the spherical surfaces.

2.1 Great-circle and Small-circle

We know that intersection of a spherical surface with a plane is a circle. For a given spherical surface, the radius of circle so obtained depends on the position of the intersecting plane. The radius of the circle is obviously the largest when the plane passes through the centre of the circle. This circle of the largest radius is known as the *great-circle*. Any of other circles have radii smaller than that of the great-circle. This circle of smaller radius is known as the *small-circle*. Parallel to a great-circle, there are obviously a large number of small-circles having various radii. By rotating the position of the intersecting plane, we get another great-circle and corresponding small-circles.

2.2 Poles and Equator

For spherical surface SS, shown in Fig. 2, suppose *GC* is the great-circle obtained by the intersection of a plane passing through the centre *C* of the spherical surface. Now, a straight line perpendicular to the plane of the great-circle and passing through *C* intersects the spherical surface at two points *R* and *T*. These points are known as the poles corresponding to the great-circle. This great circle is called the equator. If we consider another great-circle obtained by intersection of another plane passing through the centre *C*, we get another pair of poles.

[3]Some persons have opinion that observational (experimental) aspect is known as Astronomy whereas the theoretical aspect is known as the Astrophysics. Large number of persons consider the Astronomy and the Astrophysics as the synonyms. Quite often, both words are used together as the Astronomy and Astrophysics (abbreviated as A&A).

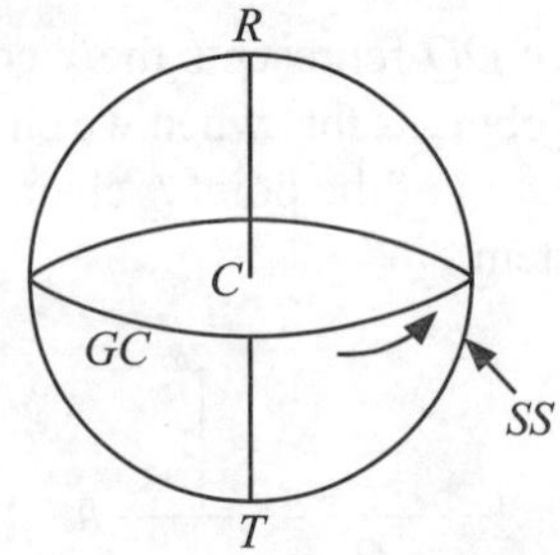

Fig. 2: A great circle (*GC*) can be obtained by the intersection of a spherical surface *SS* and a plane passing through the centre *C* of the spherical surface.

On the other side, when we consider a straight line passing through the centre *C*, then a plane whose normal is this line and passes through the *C* intersects the spherical surface in a great-circle.

2.3 Coordinates of a Point on a Spherical Surface

Having these definitions, we now introduce a coordinate systems for a spherical surface. When we transform from a plane surface (discussed in the preceding section) to a spherical surface, the lines *AB* and *PQ* of the plane surface become the arcs of great-circles (Fig. 3) and the coordinates of a point *P* on the spherical surface are expressed in terms of the angles subtended by the arcs *OQ* and *PQ* at the centre *C* of the sphere.

For an orthogonal coordinate system, the great-circle of arc *PQ* would pass through the poles *R* and *T*, corresponding to the great-circle of the arc *OQ*. Thus, the coordinates of the point *P* are (ψ, θ), where $\psi = \angle OCQ$ and $\theta = \angle PCQ$. With the help of these angles ψ and θ, various coordinate systems are introduced, which are used in the Astronomy and Astrophysics.

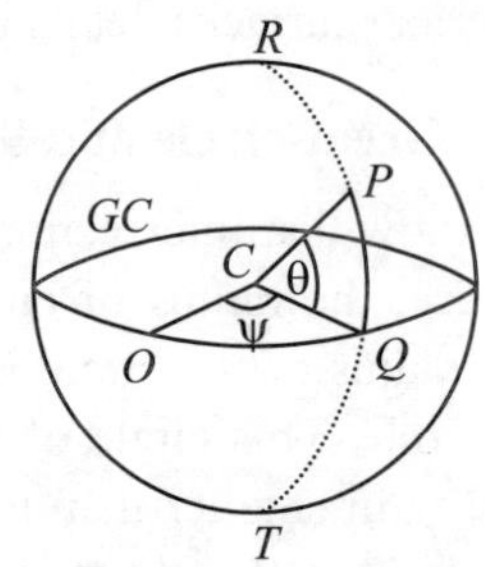

Fig. 3: Coordinates (ψ, θ) of a point *P* on a spherical surface.

3. Spherical Triangle

Suppose, *A*, *B*, *C* are three points on a spherical surface, as given in Fig. 4. Let the arcs *AB*, *BC*, *CA*, lying on the spherical surface, be parts of the great circles. The triangle *ABC* so formed is known as the spherical triangle.

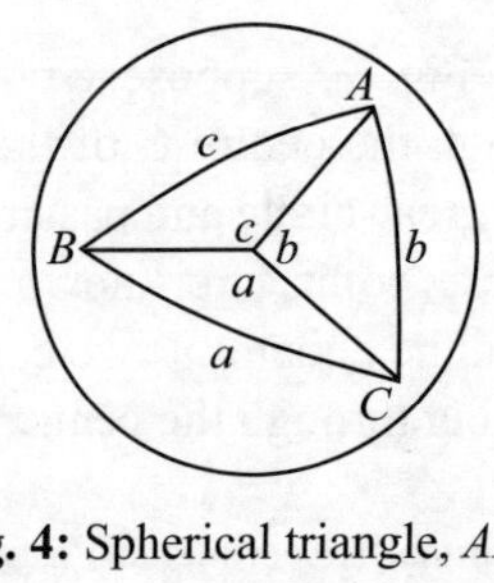

Fig. 4: Spherical triangle, *ABC*.

Here, the vertices and the angles at the vertices both are denoted by the common upper case letters *A*, *B* and *C*. The angles *A*, *B*, *C* of triangle are equal to the angles between the planes which intersect the surface of the sphere or, equivalently, the angles between the tangent vectors of the great-circle arcs where they meet at the vertices. The angles are measured in radians and are such that[4]

$$\pi < A + B + C < 3\pi$$

The lengths of sides opposite the angles A, B, C, are, respectively, denoted by the lower case letters a, b, c. When the radius of the sphere is taken as unity, the lengths of arcs are numerically equal to the angles (in radian) subtend by the arcs at the centre of the sphere.

The radius of the sphere is generally taken as unity. For some specific practical problems on a sphere of radius R, the measured lengths of the sides are divided by R before using the identities (cosine and sine rules) given below. Further, after the calculations on the unit sphere, the sides a, b, c are multiplied by R.

The spherical triangle is quite different from a triangle in a plane surface. The trigonometry dealing with the spherical triangle is known as the spherical trigonometry, and cosine and sine rules are as the following.

3.1 Cosine Rules

The cosine rules are the fundamental identities of the spherical trigonometry. They are as the following.

$$\cos a = \cos b \cos c + \sin b \sin c \cos A$$

$$\cos b = \cos c \cos a + \sin c \sin a \cos B$$

$$\cos c = \cos a \cos b + \sin a \sin b \cos C$$

3.2 Sine Rules

The sine rules, given below, are also the fundamental identities of the spherical trigonometry.

$$\frac{\sin A}{\sin a} = \frac{\sin B}{\sin b} = \frac{\sin C}{\sin c}$$

4. Terrestrial Latitude and Longitude

This is also known as the geographical coordinate system. For the practical purposes, our Earth is taken as a spherical body spinning about its own axis passing through its centre C (Fig. 5) from the west towards the east. The axis of rotation intersects the Earth's surface at two points denoted by N and S, called the north pole and the south pole, respectively. Corresponding to these poles, the great-circle is known as the equator (EQUA). This equator obviously divides the Earth into two hemispheres, called the northern hemisphere and the southern hemisphere. The equator is used as the reference of coordinate system. The reference point O on the equator is specified in the following manner.

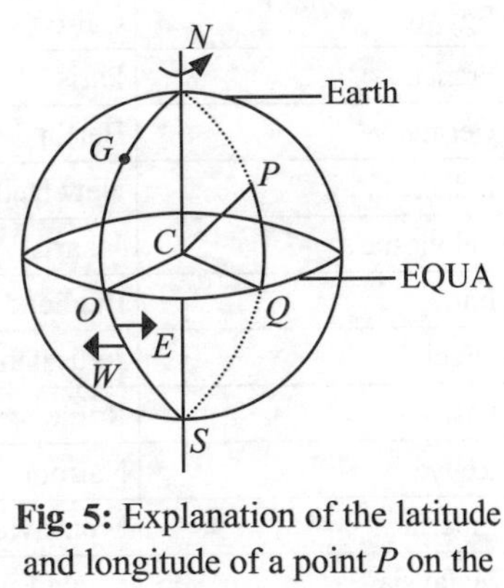

Fig. 5: Explanation of the latitude and longitude of a point P on the

[4]In case of a triangle ABC in a plane, we have $A + B + C = \pi$.

The great-circles passing through the poles *N* and *S* are known as the meridians. The meridian which passes through the fundamental instrument (transit circle) installed in Greenwich observatory (*G*), in England, is known as the principal or standard meridian. The point of intersection of this standard meridian with equator is the reference point *O*. For a point *P* on the Earth's surface, the angle *PCQ* is called the latitude of *P* and the angle *OCQ* is called the longitude of *P*.

When the point *P* is in between the equator and the north pole *N*, the latitude is said to be the north latitude. The north latitude can vary from 0° to 90° north.

The angle *NCP* is known as colatitude of point *P*. Therefore, we have

$$\text{colatitude} = 90° - \text{latitude}$$

When the point *P* is in between the equator and the south pole *S*, the latitude is said to be south latitude. The south latitude can vary from 0° to 90° south. The angle *SCP* is colatitude of the point *P*.

Longitudes are measured from 0° to 180° east and from 0° to 180° west from the Greenwich meridian. Thus, the position of a point on the Earth's surface can be expressed in terms of the latitude and the longitude for that point. The latitude and the longitude of the capitals of some countries are given in Table 1. These are coordinates of a specific point in the city.

Table 1. Coordinates of capitals of some countries

Country	Capital	Latitude	Longitude
Argentina	Buenos Aires	36° 30′ S	60° 00′ W
Austria	Vienna	48° 12′ N	16° 22′ E
Brazil	Brasilia	15° 47′ S	47° 55′ W
Cambodia	Phnom Penh	11° 33′ N	104° 55′ E
Canada	Ottawa	45° 27′ N	75° 42′ W
Colombia	Bogota	04° 34′ N	74° 00′ W
Denmark	Copenhagen	55° 41′ N	12° 34′ E
Egypt	Cairo	30° 01′ N	31° 14′ E
France	Paris	48° 50′ N	02° 20′ E
Germany	Berlin	52° 30′ N	13° 25′ E
India	New Delhi	28° 37′ N	77° 13′ E
Indonesia	Jakarta	06° 09′ S	106° 49′ E
Iraq	Baghdad	33° 20′ N	44° 30′ E
Israel	Jerusalem	31° 47′ N	35° 12′ E
Italy	Rome	41° 54′ N	12° 29′ E
Kenya	Nairobi	01° 30′ N	36° 48′ E
Liberia	Monrovia	06° 18′ N	10° 47′ W
Malaysia	Kuala Lumpur	03° 09′ N	101° 41′ E
Mexico	Mexico	19° 20′ N	99° 10′ W
Russia	Moskua	55° 45′ N	37° 35′ E

5. Celestial Sphere

When we look towards the sky, the cosmic objects appear to be situated on a vast spherical surface of which the individual observer is the centre. This imaginary sphere is known as the celestial sphere. The eye, of course, fails to give any indication about the distances of these objects. However, it allows us to make estimates of the angle subtended by a pair of objects at the observer's eye. This angle is known as the angular separation between two given objects. Hence, for each position of an observer, there is a celestial sphere whose centre is at the position of the observer.

6. Altitude and Azimuth

There are several coordinate systems to specify the position of an astronomical object (e.g., stars) on the celestial sphere. One of them is the altazimuth coordinate system. This is also known as the horizon coordinate system. In this system (Fig. 6), the observer *D* (on the surface of the Earth) is the centre of the celestial sphere. The line joining the observer *D* and the Earth's centre intersects the celestial sphere (CS) at two points *Z* and *K*, called the zenith and the nadir, respectively. The zenith is vertically overhead. The great circle corresponding to the poles *Z* and *K* is known as the celestial horizon or simply horizon. Thus, the horizon is the intersection of the celestial sphere and the plane tangential to the Earth's surface passing through the observer *D*.

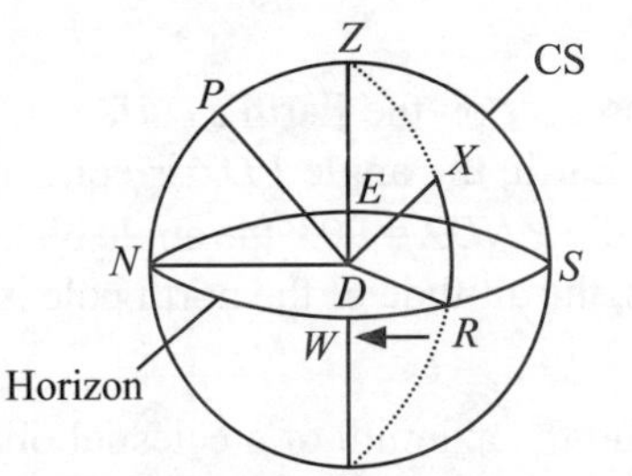

Fig. 6: In the altazimuth coordinate system, the angle *NESR* is known as the azimuth and the angle *XDR* is known as the altitude of an object *X*, situated on the celestial sphere.

The horizon obviously divides the celestial sphere into two hemispheres, of which the upper hemisphere is visible to the observer, whereas the lower one is hidden from the observer by the Earth. The horizon is used as the great circle for the reference of coordinate system. Now, we need to specify the position of the reference point on the horizon. This is done in the following manner.

Let the line *DP* be parallel to the spin-axis of the Earth and passing through the observer. Suppose, the line *DP* interacts the celestial sphere at the point *P*. When the latitude of observer is north, the point *P* is known as the north celestial pole or simply the north pole. The great circles passing through *Z* and *K* are known as the vertical circles. The vertical circle passing through *P*, that is, *ZPNK* is defined as the prime vertical. It intersects the horizon at the point *N*, which is taken as the reference point on the horizon. This point *N* is also known as the north point on the horizon. The point *S* on the horizon, exactly opposite to *N*, is the south point, and the east (*E*) and the west (*W*) points have the directions at right angles to the directions of *N* and *S*. The point *N*, *E*, *S* and *W* are known as cardinal points.

Now, the position of an object *X* on the celestial sphere can be expressed in terms of the $\angle XDR$ (called the altitude) and the $\angle NESR$, measured along the direction of the arrow with *N* as the reference point (called the azimuth). The altitude can assume the value from 0° to 90°. The angle *ZDX* is known as zenith distance (ZD) of the object *X*. Thus, we have

$$\text{ZD} = 90° - \text{altitude}$$

Since the hemisphere below the horizon is hidden from the observer by the Earth, the altitude is negative for the objects below the horizon. The azimuth can assume the values from 0° to 360°.

It can be noticed that the altitude and the azimuth both of an object at a given time depend on the position of the observer on the Earth's surface. Further, as the Earth spins from the west towards the east about its own axis passing through its centre, all the stars appear to travel across the sky along the circular paths from the east towards the west around the spin-axis. Hence, the altitude and the azimuth coordinates of the objects change continuously with time. In the northern hemisphere, there is however a star whose position appears to change very little. This star, named the Polaris, is visible to the naked eyes.

Example 3.

Show that for an observer, the altitude of the north pole is equal to the observer's latitude.

Solution:

Suppose, the latitude of the observer on the Earth's surface is ϕ degree north. Since the *PD* is parallel to the spin-axis of the Earth, the angle *PDZ* is equal to the colatitude of the observer. Thus, the $\angle PDZ = 90° - \phi$. As the $\angle NDZ = 90°$, the angle NDP is equal to ϕ. It is obviously the altitude of the north pole. Thus, the altitude of the north pole is equal to the observer's latitude.

Example 4.

For an observer, the altitude and the azimuth of a celestial object are 43° 20′ 13″ and 52° 14′ 38″, respectively. Calculate the zenith distance of the object.

Solution:

The zenith distance has no consideration of the azimuth. The zenith distance is

$$ZD = 90° - 43°20'13'' = 46°39'47''$$

Example 5.

An observer is standing on a bench 2.5 km due north-west of a mountain that has a height of 1500 m. Moon appears to be directly behind the peak of the mountain. Calculate the altitude and the azimuth of the Moon.

Solution:

The observer is due the north-west of the mountain and thus, the Moon is due the south-east of the observer. Measurement of the azimuth begins at due the north and proceeds in an easterly direction so that due the east has an azimuth of 90° and the south-east a further 45° at (90 + 45 =) 135°. The mountain's angular height is equal to the Moon's altitude. The tangent of this angle is the ratio of the mountain's height to its distance at (1500/2500) = 0.6.

$$\text{Moon's altitude} = \tan^{-1} 0.6 = 31°$$

At this moment, the Moon has the azimuth and the altitude of 135° and 31°, respectively, which uniquely define its position in the sky.

7. Standard Celestial Sphere

It has been mentioned that around each point on the surface of the Earth, there is a celestial sphere. For example, there is a celestial sphere around the observer *D*. For each celestial sphere, the positions of the celestial poles and the celestial equator, etc., depend on the position of the observer. Hereafter, we shall refer to a celestial sphere whose centre is at the centre of the Earth. This celestial sphere is often referred to as the standard celestial sphere or geocentric celestial sphere. Now, the positions of the celestial poles and the celestial equator do not depend on the position of the observer. The celestial poles, or simply the poles, are now defined as the points where the spin-axis of the Earth intersects the standard celestial sphere. The great circle corresponding to these poles is known as the celestial equator. Obviously, the celestial equator is the intersection of the celestial sphere and the plane of the terrestrial equator. With the introduction of the standard celestial sphere, the horizon also would have the centre at the Earth's centre.

7.1 Celestial Poles, Celestial Equator and Celestial Meridians

The spin-axis of the Earth intersects the celestial sphere at two points, called the celestial north pole and the celestial south pole. Thus, the terrestrial and the celestial poles lie on a common line passing through the centre of the Earth (Fig. 7).

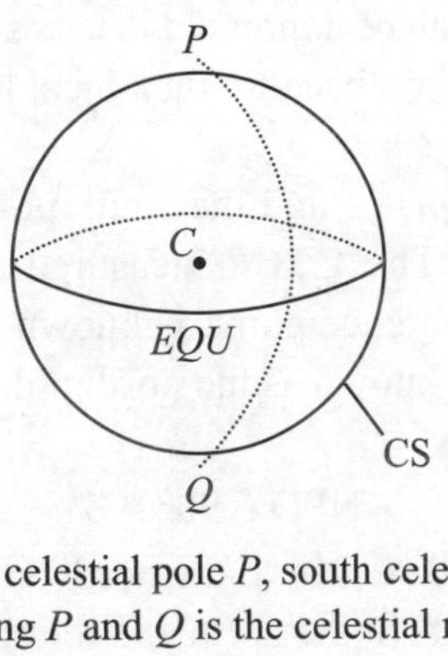

Fig. 7: Celestial sphere (CS), north celestial pole *P*, south celestial pole *Q*, celestial equator *EQA*; line joining *P* and *Q* is the celestial meridian.

The intersection of the plane of the terrestrial equator with the celestial sphere is a circle, called the celestial equator. Thus, the terrestrial equator and the celestial equator both lie in a common plane.

The planes containing the spin axis (i.e., passing through the centre of the Earth, the north and south celestial poles) intersect the the celestial sphere in the form of arcs, called the celestial meridians. The terrestrial meridian and the celestial meridian of a place lie in a common plane.

8. Declination and Hour-angle

It is also known as the equatorial coordinate system. The coordinates of an object can also be expressed in terms of the declination and the hour-angle. In Fig. 8, *P* and *Q* are north and south celestial poles, respectively. These are the points of the intersection of the standard celestial sphere (SCS) and the spin-axis of the Earth. The great circle corresponding to the poles is the equator (EQU). For this coordinate system, the equator is used as the great circle for the reference. Now, the reference point on the equator can be specified in the following manner.

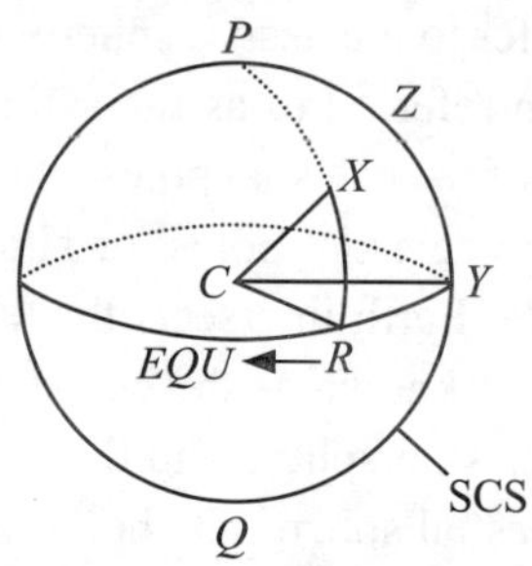

Fig. 8: Declination and hour-angle coordinates system. The angle *YCR* is known as the hour-angle and the angle *XCR* is known as the declination of an object *X*, situated on the celestial sphere.

The great circles passing through the poles *P* and *Q* are known as the celestial meridians. Suppose, *Z* is the zenith for an observer in northern hemisphere. Then, the celestial meridian passing through *Z* is known as the observer's meridian. The observer's meridian intersects the equator at the point *Y*, which is taken as the reference point on the equator. Now, the position of an object *X* on the celestial sphere can be expressed in terms of the $\angle XCR$, called the declination of the object. It is the north declination because the object is between the equator and the north pole.

For an object, between the equator and the south pole, the declination would be south. The declination is denoted by δ. The $\angle YCR$ measured along the direction of arrow (i.e., towards the west) with *Y* as the reference point, is known as the hour-angle of the object and is denoted by *H*. The angle *PCX* is known as the north polar distance (NPD) of the star. Thus, we have

$$\text{NPD} = 90° - \delta$$

Declination is measured from 0° to 90° to the north or the south of the equator. In order to use the same formulas for the objects in the north and the south hemispheres, the declination is treated as an algebraic quantity. The north declination carries a positive sign whereas the south declination carries a negative sign. The value of the hour-angle varies from 0° to 360° or from 0^h to 24^h ($360° \equiv 24^h$).

As the objects appear to travel across the sky along the circular paths around the spin-axis of the Earth, the declination of an object remains unchanged. However, the hour-angle changes with time due to the rotation of the Earth. It also differs from place to place because of the change in the position of the zenith, corresponding to an observer.

Example 6.

The declination and the hour angle of an object are 37° 42′ 16″ and 52° 37′ 42″, respectively. Calculate the north polar distance (NPD) of the object.

Solution:

The north polar distance has no consideration of hour angle. The north polar distance of the object is

$$NPD = 90° - 37°42'16'' = 52°17'44''$$

Example 7.

From where on the Earth's surface are the declination of all the stars equal to their altitudes?

Solution:

The declination is measured relative to celestial equator whereas the altitude is measured relative to the observer's horizon. Thus, when the horizon and the celestial equator are equivalent, the declination and the altitude are equivalent. Remembering that the celestial equator is a projection of the Earth's equator, then standing on the Earth's equator means that the celestial equator runs in an arc overhead. As the observer moves towards either of the pole, the celestial equator forms an arc that has a progressively lower and lower maximum altitude. At the poles, the horizon and the celestial equator follow the same line. Therefore, each star has equal altitudes and declinations when observed from the Earth's poles.

9. Rotating Sky

The celestial sphere is static and the Earth is spinning inside it. The appearance of the sky depends on the time and position of the observer on the Earth's surface. The place in the sky that is invariant when observing from the northern hemisphere is the north celestial pole. In order to understand the path of the stars across the sky, it is only necessary to remember that the Earth spins once a day about its own axis. Hence, an observer's view of the sky moves through 360° each day. The stars appear to circle the sky (Fig. 9). The position of the celestial pole does not change, but the star of interest appears to rotate around the pole.

Fig. 9: Rotating sky.
(*Courtesy*: Google website)

The size and the position of circles in the sky described by stars depend on the position of the star on the celestial sphere and the observer's position on the Earth. It is obvious that the stars close to the celestial pole travel in small circles around the pole. The stars reach their highest position in the sky, known as the upper culmination, when they cross the observer's meridian. At this moment, their altitude is given by the altitude of the celestial pole (equal to the observer's latitude, ϕ) plus the difference in the declination between the star and the north celestial pole. The north celestial pole has a declination of 90°. Hence, the altitude of a high declination star at the upper culmination is expressed as

$$a_{max} = \phi + (90 - \delta) = 90 + \phi - \delta$$

The stars away from the north celestial pole have smaller declination and describe larger circles in the sky. The altitude is measured from the closest point on the horizon. For the southern cardinal point, the maximum altitude is expressed as

$$a_{max} = 180 - (90 + \phi - \delta) = 90 - (\phi - \delta)$$

In both the cases the star's minimum altitude also occurs on the observer's meridian but directly below the north celestial pole. As the motion is circular, the star's minimum latitude is

$$a_{min} = \phi - (90 - \delta) = \phi + \delta - 90$$

Let us see the situation when the observer's latitude plus the star's declination ($\phi + \delta$) is less than 90° so that the star's minimum latitude is negative. This is the case of the setting of a star which dips below the observer's horizon. The path of such a star across the sky is still circular but part of the circle is hidden by the Earth.

The stars which never leave the sky (even though they are not visible during the day time) are said to be circumpolar. The condition for a star to be circumpolar is

$$(\delta + \phi) \leq 90$$

It shows that when the observer is positioned at north pole ($\phi = 90°$) the same stars remain in the sky at all the times. The stars which are visible are those of the north celestial hemisphere and they all circle parallel to the horizon about the north celestial pole which is at the zenith. On movement towards south more stars become visible but some stars are no longer circumpolar and disappear from the view for a period of time.

For an observer on equator, the north and south celestial poles lie at the north and south cardinal points on the horizon. The celestial equator loops overhead and all the stars describe a perfect semi-circle paths across the sky which is parallel to the celestial equator. There are no circumpolar stars, but the whole of the celestial sphere appears in the sky throughout the course of the daytime.

Example 8.

As observed from a city (latitude 50° N, longitude 8° W), calculate the maximum altitude of a star (right ascension 10°, declination +67°) when it crosses the observer's meridian. Calculate if the star disappears from the sky. If not, calculate the minimum altitude of the star.

Solution:

We have $\delta = 67°$ and $\phi = 50°$. Star crosses the observer's meridian at upper culmination. As ($\delta > \phi$), upper culmination takes place north of zenith. The altitude of star at the upper culmination is

$$a_{max} = 90 + \phi - \delta = 90 + 50 - 67 = 73°$$

As δ is larger than $(90 - \phi)$, the star is circumpolar when observed from the city.

The altitude of the star at the lower culmination is

$$a_{min} = \phi + \delta - 90 = 50 + 67 - 90 = 27°$$

Thus, the star remains in the sky at all the times with an altitude that varies between 27° and 73°.

Example 9.

As observed from York (latitude 54° N, longitude 1° W), calculate the maximum altitude of the star Vega (declination +39°) when it crosses the observer's meridian. Whether the star disappears from the sky. If not, calculate the minimum altitude of the star.

Solution:

Star crosses the observer's meridian at upper culmination. As $(\delta < \phi)$, the upper culmination takes place south of the zenith. The altitude of the star at the upper culmination is

$$a_{max} = 90 - \phi + \delta = 90 - 54 + 39 = 75°$$

As the δ (39°) is larger than $(90 - \phi)$ $(90 - 54 = 36°)$, the star is circumpolar when observed from York. The altitude of star at the lower culmination is

$$a_{min} = \phi + \delta - 90 = 54 + 39 - 90 = 3°$$

Thus, the star remains in the York sky at all times with an altitude that varies between 3° and 75°.

Example 10.

As observed from York (latitude 54° N, longitude 1° W), calculate the maximum altitude of the star Spica (declination –11°) when it crosses the observer's meridian. Whether the star disappears from the sky. If not, calculate the minimum altitude of the star.

Solution:

The star crosses the observer's meridian at the upper culmination. As $(\delta < \phi)$, the upper culmination takes place south of the zenith. The altitude of the star at the upper culmination is

$$a_{max} = 90 - \phi + \delta = 90 - 54 - 11 = 25°$$

As δ (–11°) is smaller than $(90 - \phi)$ $(90 - 54 = 36°)$, the star is not circumpolar when observed from York. Thus, the star does not remain in the York sky at all the times.

10. Ecliptic

Besides the spin motion about its own axis, the Earth revolves around the Sun in one year in an elliptical path, where the Sun, denoted by *S*, is situated at a focus of the ellipse (Fig. 10). The points *A*, *B*, *D*, *E*, *F* denote various positions of the Earth revolving around the Sun.

Since the observations are generally made from the Earth, the Sun appears to describe an elliptical orbit around the Earth. In Fig. 11, *C* denotes the centre of the Earth and the ellipse represents the apparent orbit of the Sun relative to the Earth. The sequence of positions of the Sun, namely, *a*, *b*, *c*, *d*, *e*, f (Fig. 11) relative to the Earth corresponds to the sequence of the positions *A*, *B*, *C*, *D*, *E*, *F* (Fig. 10) of the Earth in its orbit around the Sun, respectively. During one year, the Sun thus appears to make a complete revolution around the Earth.

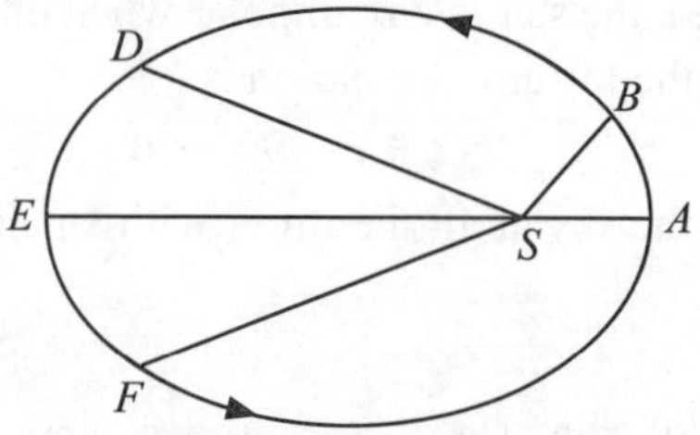

Fig. 10: The ellipse shows the path of the Earth revolving around the Sun, which is situated at the focus *S* of the ellipse.

The plane of the apparent orbit of the Sun relative to the Earth (which is also the plane of the orbit of the Earth revolving around the Sun) is known as the plane of the ecliptic and the great circle obtained with the intersection of this plane and the celestial sphere is known as the ecliptic.

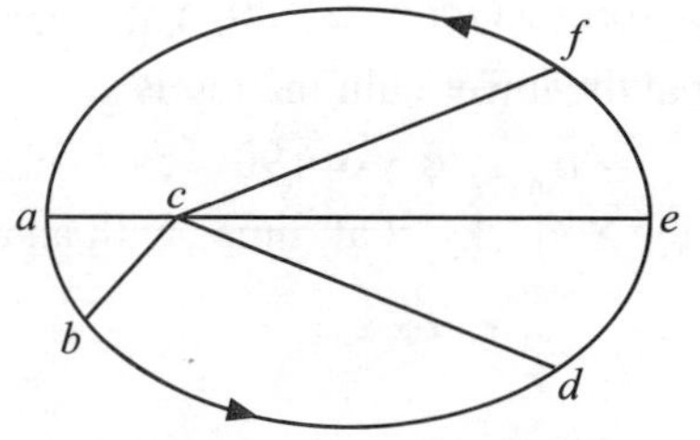

Fig. 11: The ellipse shows the apparent path of the Sun moving around the Earth, which is situated at the focus (*C*) of the ellipse.

From the observations, it is found that the ecliptic is inclined at an angle of about 23.5° to the celestial equator (Fig. 12). This inclination is known as the obliquity of the ecliptic. In Fig. 12, P is the pole of the equator (EQU) and K is pole of the ecliptic (ECL). Relative to the Earth, the Sun appears to move on the celestial sphere along the ecliptic in the direction shown with the help of the arrows.

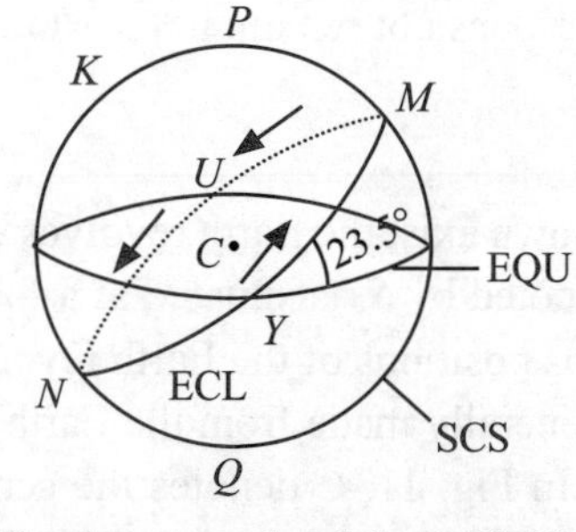

Fig. 12: The ecliptic (ECL) is inclined at an angle of about **23.5°** to the celestial equator (EQU). The ecliptic intersects the celestial equator at the vernal equinox *Y* and the autumnal equinox *U*.

The ecliptic obviously intersects the equator at two points, *Y* and *U*. At these points, the declination of the Sun is zero degree. During the motion from *Y* to *M*, and from *M* to *U*, the Sun is in the northern hemisphere, and therefore its declination is north. On the other side, during the motion from *U* to *N*, and from *N* to *Y*, the Sun is in the southern hemisphere, and therefore its declination is south. The position *Y*, at which the declination of the Sun changes from the south to the north is known as the vernal equinox. The Sun is at this position around March 21 each year. The position *M* at which the north declination of the Sun is the maximum (about 23.5° N) is known as the summer solstice. The Sun is at this position around June 21 each year. The position *U* at which the declination of the Sun changes from the north to the south is known as the autumnal equinox. The Sun is at this position around September 23 each year. The position *N* at which the south declination of the Sun is the maximum (about 23.5° S) is known as the winter solstice. The Sun is at this position around December 21 each year.

11. Right Ascension and Declination

We have seen that the altitude and the azimuth coordinates of a star vary with the rotation of the Earth. Further, the hour-angle coordinate changes with the rotation of the Earth. However, the declination of an object does not depend on the rotation of the Earth. Now, we intend to introduce a coordinate system that does not vary with the rotation of the Earth nor does it vary from place to place. For such requirement, the equator (*EQU*) can be used as a great circle for the reference and the vernal equinox (*Y*) is taken as the reference point on the equator (Fig. 13).

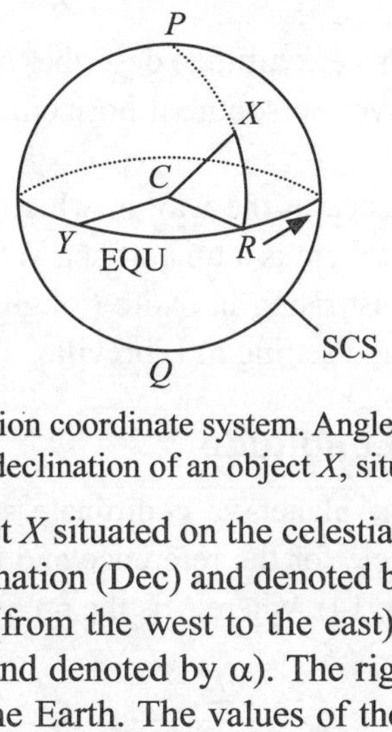

Fig. 13: Right ascension and declination coordinate system. Angle *YCR* is known as right ascension and angle *XCR* is known as declination of an object *X*, situated on celestial sphere.

Now, the position of an object *X* situated on the celestial sphere can be expressed in terms of the angle *XCR* (called the declination (Dec) and denoted by δ) and the angle *YCR*, measured along the direction of the arrow (from the west to the east) with the *Y* as the reference point (called the right ascension (*RA*) and denoted by α). The right ascension also, obviously does not change with the rotation of the Earth. The values of the right ascension are expressed in terms of the hours ($360° \equiv 24^h$). In this coordinate system, four positions of the Sun have the following coordinates.

	δ	α
(i) Vernal equinox	0°	0^h
(ii) Summer solstice	23.5°	6^h
(iii) Autumnal equinox	0°	12^h
(iv) Winter solstice	–23.5°	18^h

Example 11.

How does the Sun's highest position in the sky vary throughout the year in Quito which is on the equator in Ecuador, South Africa, and in the Libreville which is on the equator in Gabon, Africa?

Solution:

On 21 Dec (or close to it, depending on the calendar for that year) the Sun's position on the celestial sphere reaches its farthest position south, at a declination of –23.5°, when the Earth's southern hemisphere is tilted towards the Sun. The observer's latitude (for both the Quito and the Libreville) is 0° and so upper culmination takes place south of the zenith ($\delta < \phi$) at this time and the altitude is

$$a_{max} = 90 - \phi + \delta = 90 - 0 - 23.5 = 66.5°$$

As the year progresses, the Sun's declination increases towards zero which it reaches on (or close to) 20 March. At this time the value of a_{max} is 90°. During April and May the Sun continues to increase its declination until it peaks on the summer solstice on (or close to) 21 June. At this time the Sun's declination is 23.5° and the altitude is

$$a_{max} = 90 + \phi - \delta = 90 + 0 - 23.5 = 66.5°$$

The Sun's declination continues to decrease, reaching 0° on (or close to) 22 September. At this time the value of a_{max} is 90°.

Throughout the year, the Sun's peak altitude describes an oscillation between 66.5° above the northern horizon to 66.5° above the southern horizon, passing through the zenith in the March and in the September.

There is essentially no difference in the way in which the Sun travels across the sky in Quito and in Libreville, except that there is a time offset. When the Sun is directly overhead in Libreville (longitude 10° E) it is just rising in Quito (longitude 79° W). Six months later, the Sun is directly overhead in Quito and setting in Libreville.

12. Celestial Latitude and Longitude

For the study of the motion of the planets, a coordinate system is often used in which the ecliptic (ECL) is used as great circle for the reference and the vernal equinox is taken as the reference point on the ecliptic (Fig. 14), where *K* is the pole of the ecliptic.

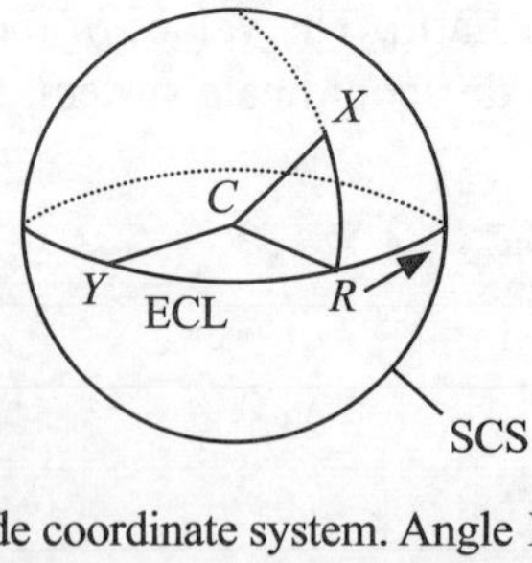

Fig. 14: Celestial latitude and longitude coordinate system. Angle *YCR* is known as the celestial longitude and the angle *XCR* is known as the celestial latitude of a planet *X*, situated on the celestial sphere.

Now, position of a planet *X* on the celestial sphere can be expressed in terms of the angle *XCR* (called the celestial latitude and is denoted by β) and the angle *YCR*, measured along the direction of the arrow (i.e., from the west to the east) with *Y* as a reference point (called the celestial longitude and is denoted by λ). The latitude above the ecliptic is known as the north latitude and is considered positive, whereas the latitude below the ecliptic is known as the south latitude and is considered negative. The value of λ varies from 0° to 360° around the ecliptic.

13. Galactic Latitude and Longitude

For the study of our galaxy, a coordinate system is used in which the galactic equator (GEQU) (Fig. 15) is used as a great circle for the reference. The galactic equator almost coincides the central line of Milky Way in the sky. GP is the galactic north pole. The reference point *O* on the galactic equator is a point in the constellation Sagittarius and is located at the centre of our galaxy. Now, the position of an astronomical object (e.g., star, cluster) *S* on the celestial sphere can be specified in terms of the angle *SCR* (called the galactic latitude and denoted by *b*) and the angle *OCR*, measured along the direction of the arrow with *O* as the reference point (called the galactic longitude and denoted by *l*). It may be mentioned here that the galactic coordinates cannot be determined by the direct observations. They are calculated by using the equatorial coordinates of the star of the galactic pole (GP) and of the point *O* in the direction of the centre of the galaxy.

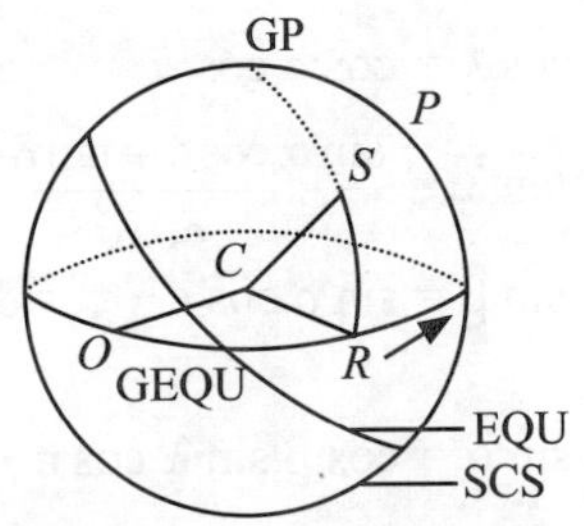

Fig. 15: Galactic latitude and longitude coordinate system. The angle *OCR* is known as the galactic longitude and the angle *SCR* is known as the galactic latitude of an astronomical object *S*, situated on the celestial sphere.

Note:

The vernal equinox itself processes slowly westward relative to the fixed stars and completes one revolution in about 26,000 years. Therefore, all the coordinates where vernal equinox is taken as a reference point are modified from time to time. For example, the present coordinates (e.g, RA and Dec) are with reference to the position of vernal equinox on 1 January 2000. Earlier the coordinates were with reference to the position of the vernal equinox on 1 January 1950. This date is known as epoch.

14. Conversion of Coordinates

Let us first summarize the coordinate systems and notations used there.

Horizontal coordinates

A – azimuth; *a* – altitude

Equatorial coordinates

α – right ascension; δ – declination; h – hour angle

Ecliptic coordinates

λ – ecliptic longitude; β – ecliptic latitude

Galactic coordinates

l – galactic longitude; b – galactic latitude

Miscellaneous

λ_0 – observer's longitude; ϕ_0 – observer's latitude;

ε – obliquity of the ecliptic

θ_L – local sidereal time; θ_G – Greenwich sidereal time

14.1 Hour Angle ⇔ Right Ascension

$$h = \theta_L - \alpha \quad \text{or} \quad h = \theta_G - \lambda_0 - \alpha$$

$$\alpha = \theta_L - h \quad \text{or} \quad \alpha = \theta_G - \lambda_0 - h$$

14.2 Equatorial ⇔ Ecliptic

We have

$$\cos\beta \sin\lambda = \cos\delta \sin\alpha \cos\varepsilon + \sin\delta \sin\varepsilon$$

$$\cos\beta \cos\lambda = \cos\delta \cos\alpha$$

so that

$$\tan\lambda = \frac{\sin\alpha \cos\varepsilon + \tan\delta \sin\varepsilon}{\cos\alpha}$$

$$\sin\beta = \sin\delta \cos\varepsilon - \cos\delta \sin\varepsilon \sin\alpha$$

We have

$$\cos\delta \sin\alpha = \cos\beta \sin\lambda \cos\varepsilon - \sin\beta \sin\varepsilon$$

$$\cos\delta \cos\alpha = \cos\beta \cos\lambda$$

so that

$$\tan\alpha = \frac{\sin\lambda \cos\varepsilon - \tan\beta \sin\varepsilon}{\cos\lambda}$$

$$\sin\delta = \sin\beta \cos\varepsilon + \cos\beta \sin\varepsilon \sin\lambda$$

14.3 Equatorial ⇔ Ecliptic ⇔ Horizontal

Note that the Azimuth (A) is measured from the South point, turning positive to the West. Zenith distance, the angular distance along the great circle from the zenith to a celestial object, is simply the complementary angle of the altitude: $90° - a$. We have

$$\cos a \sin A = \cos\delta \sin h$$

so that

$$\cos a \cos A = \cos\delta \cos h \sin\phi_0 - \sin\delta \cos\phi_0$$

$$\tan A = \frac{\sin h}{\cos h \sin\phi_0 - \tan\delta \cos\phi_0}$$

$$\sin a = \sin\phi_0 \sin\delta + \cos\phi_0 \cos\delta \cos h$$

We have

$$\cos\delta \sin h = \cos a \sin A$$

$$\cos\delta \cos h = \sin a \cos\phi_0 + \cos a \cos A \sin\phi_0$$

so that

$$\tan h = \frac{\sin A}{\cos A \sin\phi_0 + \tan a \cos\phi_0}$$

sin A

$$\sin\delta = \sin\phi_0 \sin a - \cos\phi_0 \cos a \cos A$$

14.4 Equatorial ⇔ Galactic

These equations are for converting equatorial coordinates referred to B1950.0. If the equatorial coordinates are referred to another equinox, they must be precessed to their place at B1950.0 before applying these formulas. We have

$$l = 303° - \arctan\left(\frac{\sin(192.25° - \alpha)}{\cos(192.25° - \alpha)\sin 27.4° - \tan\delta \cos 27.4°}\right)$$

$$\sin b = \sin\delta \sin 27.4° + \cos\delta \cos 27.4° \cos(192.25° - \alpha)$$

These equations convert to the equatorial coordinates referred to B1950.0.

$$\alpha = \arctan\left(\frac{\sin(l - 123°)}{\cos(l - 123°)\sin 27.4° - \tan b \cos 27.4°}\right) + 12.25°$$

$$\sin\delta = \sin b \sin 27.4° + \cos b \cos 27.4° \cos(1 - 123°)$$

Review Questions

1. Convert $4^h\ 23^m\ 25^s$ into degrees, arc minutes and arc seconds.
2. Convert 45° 37′ 45″ into hours, minutes and seconds.
3. Describe a reference system for the position of a point in a plane. How it can be modified for a spherical surface?
4. How, in general, coordinates of a point on a spherical surface are described?
5. What are the great-circle and the small-circle? Is meridian a part of a great-circle?
6. On the surface of the Earth, define the poles, equator, meridian, standard meridian.
7. Describe the terrestrial latitude and longitude system of coordinates.
8. For a celestial sphere, describe the poles, equator and meridian.
9. Describe the altazimuth coordinate system.
10. For an observer, the altitude and the azimuth of a celestial object are 33° 25′ 18″ and 56° 24′ 48″, respectively. Calculate the zenith distance of the object.
11. Describe the declination and hour-angle coordinate system.
12. The declination and hour angle of a object are 39° 38′ 17″ and 62° 45′ 35″, respectively. Calculate the north polar distance of the object.
13. Discuss the rotating sky, and describe the circumpolar and non-circumpolar stars. How it is decided if a star is circumpolar or not?

14. Define the ecliptic. Describe the right ascension and declination coordinate system. Write the right ascension and declination of:
    (i) Vernal equinox,
    (ii) Summer solstice,
    (iii) Autumnal equinox, and
    (iv) Winter solstice.
15. Describe the celestial latitude and longitude coordinate system.
16. Describe the galactic latitude and longitude coordinate system.
17. Write notes on the following:
    (i) Great-circle, small-circle
    (ii) Spherical triangle
    (iii) Equator and poles on the surface of the Earth
    (iv) Terrestrial meridians and principal meridian
    (v) Terrestrial latitude and longitude
    (vi) Celestial sphere and standard celestial sphere
    (vii) Altazimuth coordinate system
    (viii) Declination and hour-angle
    (ix) Rotating sky
    (x) Circumpolar and non-circumpolar stars
    (xi) Ecliptic
    (xii) Right ascension and declination
    (xiii) Celestial latitude and longitude
    (xiv) Galactic latitude and longitude

Multiple-Choice Questions

1. $4^h\,26^m\,30^s$ is equivalent to
   A. 60° 26′ 30″ B. 74° 0′ 0″ C. 66° 37′ 30″ D. 67° 36′ 30″
   Ans. C
2. 60° 26′ 30″ is equivalent to
   A. $4^h\,2^m\,1.7^s$ B. $4^h\,1^m\,46^s$ C. $4^h\,26^m\,30^s$ D. $4^h\,30^m\,26^s$ **Ans. B**
3. How many dimensions are in a plane surface?
   A. 0 B. 1 C. 2 D. 3 **Ans. C**
4. Number of great circles for a spherical surface is
   A. zero B. one C. two D. infinite **Ans. D**
5. Intersection of a spherical surface with a plane is
   A. ellipse B. hyperbola C. circle D. parabola
   Ans. C
6. Intersection of a spherical surface with a plane passing through the centre of sphere is
   A. ellipse B. parabola C. great-circle D. small-circle
   Ans. C
7. Intersection of a spherical surface with a plane not passing through the centre of sphere is
   A. ellipse B. parabola C. great-circle D. small-circle
   Ans. D

8. Equator is
   A. elliptical B. spherical C. great circle D. small circle
   Ans. C
9. For a spherical triangle *ABC*, the sum of angles *A*, *B*, *C* satisfy the condition
   A. $A + B + C = \pi$ B. $\pi < A + B + C < 3\pi$
   C. $\pi \leq A + B + C \leq 3\pi$ D. $A + B + C < \pi$ **Ans. B**
10. Terrestrial latitude has the range
    A. 0°-90° N B. –90°-90° C. 0°-180° E D. –180°-180° **Ans. A**
11. Latitude of a place on the Earth is 55°. Colatitude of the place is
    A. 55° B. 35° C. 125° D. 235° **Ans. B**
12. A great circle passing through the poles of the Earth is known as
    A. Equator B. Meridian C. Equinox D. Solstice **Ans. B**
13. For altitude A, the zenith distance is defined as
    A. 90° + A B. 90° – A C. 180° – A D. 180° + A **Ans. B**
14. Suppose, latitude of an observer is ϕ. Colatitude of observer is
    A. $90° + \phi$ B. $90° - \phi$ C. $180° - \varphi$ D. $180° + \phi$ **Ans. B**
15. Which coordinate does not change with time.
    A. Declination B. Hour angle C. Altitude D. Azimuth **Ans. A**
16. Which coordinate does not change with time.
    A. Right ascension B. Hour angle C. Altitude D. Azimuth **Ans. A**
17. Through which country, the standard meridian of the Earth passes.
    A. India B. China C. England D. America **Ans. C**
18. Great circles joining zenith and nadir are known as
    A. Meridians B. Vertical circles C. Equator D. Ecliptic **Ans. B**
19. The points N and S on the observer's horizon are known as
    A. north and south poles B. meridians
    C. cardinal points D. equinoxes **Ans. C**
20. Altitude of a celestial object is 40°. The zenith distance of object is
    A. 40° B. 50° C. 140° D. 220° **Ans. B**
21. Angle between celestial equator and ecliptic is
    A. 15.5° B. 23.5° C. 45° D. 90° **Ans. B**
22. Azimuth of a celestial objects can assume the values from
    A. 0° to 90° B. 0° to 180° C. 0° to 360° D. –90° to 90° **Ans. C**
23. Spin-axis of the Earth intersects the the Earth's surface at
    A. north and south poles B. celestial poles
    C. equinoxes D. cardinal points **Ans. A**
24. For declination 35°, the north polar distance is
    A. 145° B. 215° C. 35° D. 55° **Ans. D**
25. Spin-axis of the Earth intersects the celestial sphere at
    A. terrestrial poles B. celestial poles C. equinoxes D. cardinal points
    Ans. B

26. Sun is at summer solstice around
    A. 21 March B. 21 June C. 22 September D. 21 December
    Ans. B
27. Sun is at vernal equinox around
    A. 21 March B. 21 June C. 22 September D. 21 December
    Ans. A
28. Sun is at autumnal equinox around
    A. 21 March B. 21 June C. 22 September D. 21 December
    Ans. C
29. The Sun is at winter solstice around
    A. 21 March B. 21 June C. 22 September D. 21 December
    Ans. D
30. For Sun, the declination $\delta = 0°$ and right ascension $\alpha = 0^h$ at
    A. Vernal equinox B. Summer solstice C. Autumnal equinox D. Winter solstice
    Ans. A
31. For the Sun, the declination $\delta = 23.5°$ and right ascension $\alpha = 6^h$ at
    A. Vernal equinox B. Summer solstice C. Autumnal equinox D. Winter solstice
    Ans. B
32. For the Sun, the declination $\delta = 0°$ and right ascension $\alpha = 12^h$ at
    A. Vernal equinox B. Summer solstice C. Autumnal equinox D. Winter solstice
    Ans. C
33. For the Sun, the declination $\delta = -23.5°$ and right ascension $\alpha = 18^h$ at
    A. Vernal equinox B. Summer solstice C. Autumnal equinox D. Winter solstice
    Ans. D
34. Which coordinate system is preferably used for the study of the motion of planets in solar system?
    A. Right ascension and declination B. Celestial latitude and longitude
    C. Terrestrial latitude and longitude D. Altitude and azimuth **Ans. B**
35. For the study of a distant cosmic object, preferably used coordinate system is
    A. Right ascension and declination B. Celestial latitude and longitude
    C. Terrestrial latitude and longitude D. Altitude and azimuth **Ans. A**
36. For a spherical triangle *ABC*, the sum of the angles $A + B + C$ is
    A. $\pi/2$ B. less than π C. π D. more than π
    Ans. D
37. An observer is situated at a place on the Earth where latitude is 50° N. The declination of non-circumpolar star is
    A. 77° B. 65° C. 42° D. 36° **Ans. D**
38. An observer is situated at a place on the Earth where latitude is 32° N. The declination of circumpolar star is
    A. 77° B. 55° C. 44° D. 18° **Ans. A**

3 Time

Time is a very important parameter in Astronomy and Astrophysics (A&A). A periodic motion repeats after an interval of time, called the time-period of the motion. Revolution of a planet around the Sun, spin motion of a planet about its own axis, revolution of a satellite around a planet, spin motion of a satellite, revolution of binary stars about their centre-of-mass, are a few examples of the periodic motions in A&A. It is worth mentioning that in spite of the best efforts, the measurement of time could not be transformed to a decimal system, unlike the case of the length and the mass. It is because the measurement of the time is based on the natural phenomena, rather than on the man-made phenomena. However, the apparent solar day varies regularly and no watch can be fabricated for the apparent solar day. The watches are designed for the motion of a fictitious Sun whose motion is parallel to the equator of the Earth. The mid-day (12.00 noon) is defined when the fictitious Sun is situated at the standard meridian assigned to a country. Thus, the local time varies from one country to another. Because of the difference between the sidereal time and the mean solar time, the stars appear early each day, and consequently, the stars in the winter nights are different from those in the summer nights. In this chapter, we have discussed about all these aspects related to the time.

1. Apparent Solar Day

Let us perform a simple experiment. Suppose, we put a tall pole *AB* vertically in a plane ground *PQ* (Fig. 1). On a sunny day, the shadow of the pole rotates around the pole and its size changes with the apparent motion of the Sun around the Earth. Relative to the Earth, the Sun appears to move from the east towards the west. In fact, the Earth spins from the west towards the east and the Sun is stationary.

In the morning hours, the shadow of the pole is quite long and besides the rotation of the shadow, its size decreases with the motion of the Sun. The size becomes the minimum (may be zero) and then again increases. When the Sun is directly overhead the meridian of the pole, the shadow is the minimum. The shadow may disappear (zero size) when the Sun is overhead the pole. That instant is known as the noon at that place. That is, the Sun is at the local meridian. The time interval between the two successive returns of the Sun to a given meridian of a place is defined as the apparent solar day. After an apparent solar day, the event repeats and the Sun appears to have covered an angle of 360° around the Earth.

It is however noticed that the time taken by the Sun to complete one circle around the Earth varies from day to day. That is, the difference between two successive appearances of the Sun on a local meridian varies from day to day. The reason for this variation may be assigned to two causes:

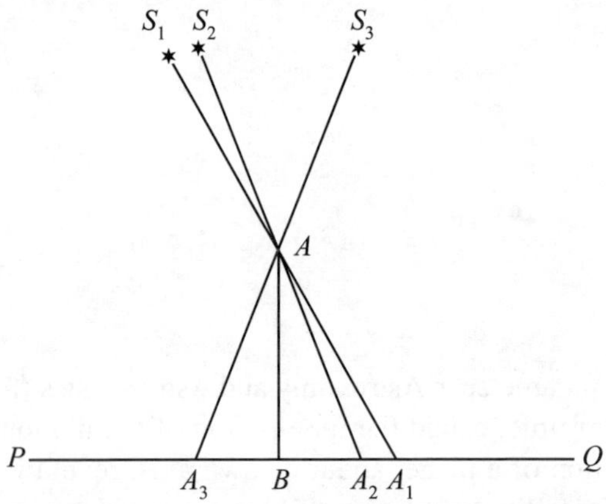

Fig. 1: Pole AB is erected vertically upward on a plane ground PQ. For different positions S_1, S_2, S_3 of the Sun, the positions of shadow of the upper most point A of the pole are A_1, A_2, A_3, respectively.

(i) The speed of the Earth revolving in the elliptical orbit around the Sun is not constant. It varies with its position. Speed is the largest when the Sun is nearest to the Earth (perihelion) and the smallest when the Sun is farthest from the Earth (aphelion). The variation follows the Kepler's second law of planetary motion, which states that the areal velocity is constant.

(ii) The ecliptic is inclined at an angle of 23.5° with the celestial equator. Owing to the spin motion of the Earth, the apparent motion of the Sun is parallel to the celestial equator, whereas its actual motion is along the ecliptic. Therefore, one has to take the projection of the motion of the Sun, along ecliptic, on the celestial equator. When the Sun crosses the celestial equator (at both the equinoxes), the projection of Sun's daily shift (relative to the background stars) on the celestial equator is small. On the other side, when the Sun is farthest from the celestial equator (at both the solstices), the projection of Sun's daily shift on the celestial equator is large.

Thus, in the months of June and December, the Sun is farthest from the celestial equator and therefore, the Earth's daily shift along the ecliptic corresponds to a large shift on the celestial equator. Further, in the months of March and September, the Sun is closest to the celestial equator and therefore, the Earth's daily shift along the ecliptic corresponds to a small shift on the celestial equator. Consequently, the length of a day is short in March and September as compared to that in June and December. (A day includes the bright part and the dark both.)

1.1 Sundial

A sundial (Fig. 2) is a device that tells about the time of day with the help of the apparent position of the Sun in the sky. It was in common use in early days. Thus, a sundial gives the apparent solar time. It can obviously be used when the Sun is visible.

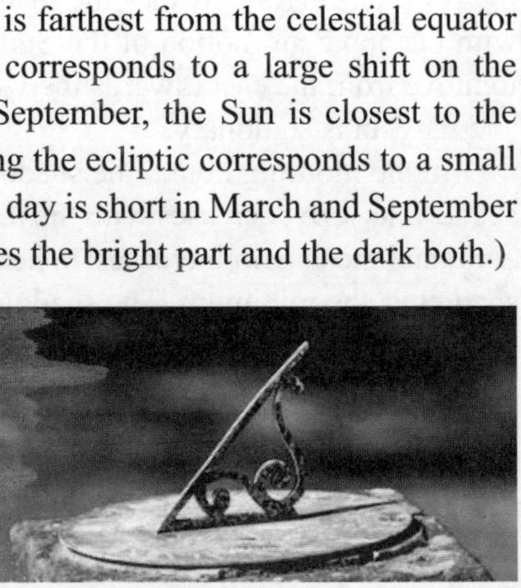

Fig. 2: Sundial.
(*Courtesy*: Google website).

2. Mean Solar Day

Since the apparent solar day varies from day to day, it is not possible to design (manufacture) a watch which can work according to the apparent solar day. Therefore, it has been decided by the international scientific community to take the average of apparent solar days over a period of one year and to call it as the mean solar day, or simply a day.

This duration of one day (one mean solar day) is divided into 24 equal parts and each part is known as one hour. Thus, a day has 24 hours. Each hour is divided into 60 equal parts and each part is known as one minute. Thus, one hour has 60 minutes. One minute is further divided into 60 equal parts and each part is known as one second. Hence, one minute has 60 seconds.

A watch which runs at a constant rate is designed for mean solar day[1]. In order to use full bright-time (when the Sun is visible) in a day for the working purposes, the beginning of day was decided to be at the mid-night. Thus, the mid of day, 12 hours, was named as the noon. Obviously, the mean Sun is at the local meridian at noon (12 hours).

Thus, we have introduced a fictitious body, called the mean Sun, which moves along the celestial equator at a constant speed. This mean Sun may be ahead or behind the real Sun, or at the same position as the real Sun.

It is obvious that there is difference between the apparent solar day and the mean solar day. The difference varies smoothly from day to day. For example, in the year 1998, the length of the apparent solar day on some dates has been found as given in Table 1.

Table 1. Length of apparent solar day in 1998

February 11	24 hours
March 26	24 hours − 18.1 seconds
May 14	24 hours
June 19	24 hours + 13.1 seconds
July 26	24 hours
September 16	24 hours − 21.3 seconds
November 3	24 hours
December 22	24 hours + 29.9 seconds

Notice that , for example, apparent solar time on 26 March 1998 has been shorter than the mean solar day by 18.1 seconds and on 19 June 1998 has been longer than the mean solar day by 13.1 seconds.

The mean solar time is still not perfectly constant from one century to the next. But, it may be considered constant for the most purposes.

Note:

The length of the mean solar day is slowly increasing due to tidal acceleration of the Moon by the Earth. In this tidal process, the Earth loses its energy and therefore its rotational motion is

[1]Notice that decimal system could not be introduced in the system of time. It is not possible to introduce decimal system, as the time depends on the natural phenomena.

slowing down at very small rate. Consequently, from time to time, the time shown by a watch is decreased by one second. Such second is known as the leap second. This adjustment of time was done first time on 30 June 1972 when the watch was slowed down by 11 seconds. Later on the adjustment is done by 1 second at the beginning of day of month. In the night at 23h 59m 59s the time is changed to 24h. Some dates when the time was modified are given in Table 2.

Table 2. Some dates of modification of time

31 December 1972	31 December 1973	31 December 1974
31 December 1975	31 December 1976	31 December 1977
31 December 1978	31 December 1979	30 June 1981
30 June 1982	30 June 1983	30 June 1985
31 December 1987	31 December 1989	31 December 1990
30 June 1992	30 June 1993	30 June 1994
31 December 1995	30 June 1997	31 December 1998
31 December 2005	31 December 2008	30 June 2012
31 December 2016		

3. Equation of Time

The equation of time describes the difference between the apparent solar time and the mean solar time. It is because the projection of the speed of the real Sun on the equator is not constant, whereas that of the mean Sun (fictitious Sun) is steady. There is no universally accepted definition of the sign of the equation of time. Some scientists show it positive when the sundial is ahead of a clock, and others show when the clock is ahead of the sundial. Thus, we have

$$\text{Equation of time} = \text{Apparent solar time} - \text{Mean solar time}$$

or

$$\text{Equation of time} = \text{Mean solar time} - \text{Apparent solar time}$$

The equation of time is zero when the mean solar time and the apparent solar time are equal.

4. Local Time

After introduction of the mean Sun, the local time is 12 hours (noon) when the mean Sun is overhead the local meridian for that location. Thus, in a country, every place has its own time, as different places have different meridians. Obviously, it is difficult to manage with the time in a country. Therefore, it has been decided to have one meridian passing over the given country as the standard meridian for that country. Hence, all places in that country have a common meridian, called the standard meridian for that country[2]. The longitude of the standard meridian for India is 82.5° E.

Thus, when the mean Sun is over the standard meridian, then it is 12 hours (noon) for that country. The time defined in this manner is known as the local time. The time corresponding

[2]This standard meridian not necessarily passes over the capital of the country.

to the meridian over the Greenwich is known as the Greenwich Meridian Time (GMT). As the rotation of 360° is equivalent to 24 hours, we have

$$24 \text{ hours} \equiv 360° \qquad \text{and therefore} \qquad 1 \text{ hour} \equiv 15°$$

Obviously, the difference between the local time in India and the GMT is 82.5/15 = 5.5 hr (5 hr 30 min). As the longitude of the standard meridian is East, the time in India is large by 5 hr 30 min as compared to the GMT. That is, if it is 1.00 hr in England, the corresponding time in India is 6 hr 30 min. The relation between the local time and the GMT can be expressed as the following.

Table 3. Standard meridian and time-difference for some countries

Country	Longitude	Difference from GMT
United States (Hawai)	150° W	–10 Hr
United States (Alaska)	135° W	–9 Hr
United States (Colorado)	105° W	–7 Hr
United States (Illinois)	90° W	–6 Hr
Canada (British Columbia)	105° W	–7 Hr
Canada (Saskatchewan)	90° W	–6 Hr
Canada (Ontario)	75° W	–5 Hr
Canada (Nova Scotia)	60° W	–4 Hr
Argentina	45° W	–3 Hr
Brazil	30° W	–2 Hr
England	0°	0 Hr
Spain	0°	0 Hr
Ireland	0°	0 Hr
Portugal	0°	0 Hr
Germany	15° E	1 Hr
France	15° E	1 Hr
Italy	15° E	1 Hr
Iran	52° 30′ E	3 Hr 30 min
Pakistan	75° E	5 Hr
India	82° 30′ E	5 Hr 30 min
Sri Lanka	82° 30′ E	5 Hr 30 min
Nepal	86° 15′ E	5 Hr 45 min
Bangladesh	90° E	6 Hr
South Korea	135° E	9 Hr

(i) Suppose, the terrestrial longitude of the standard meridian of a country is a°E. Then, the local time in hours is

$$\text{local time} = \text{GMT} + \frac{a}{15}$$

(ii) Suppose, the terrestrial longitude of the standard meridian of a country is a°W. Then, the local time in hours is

$$\text{local time} = \text{GMT} - \frac{a}{15}$$

The above two relations can be combined as the following.

$$\text{local time} = \text{GMT} \pm \frac{a}{15}$$

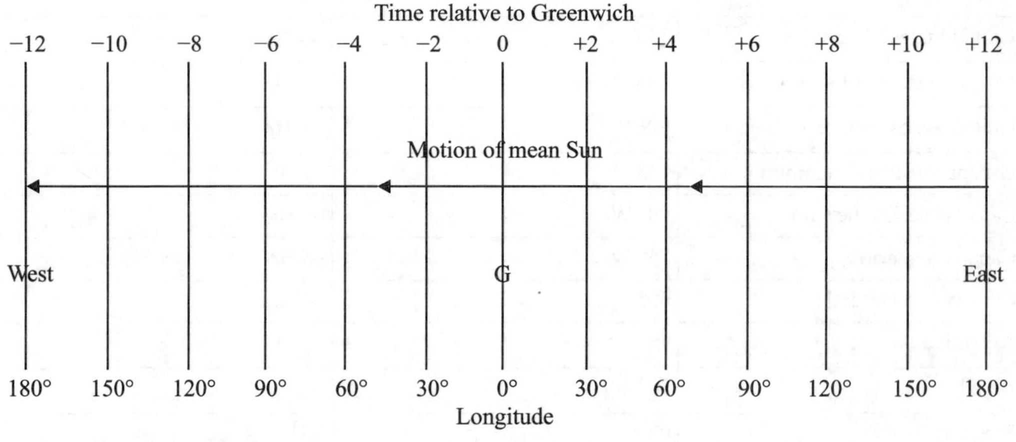

Fig. 3: Time relative to the GMT corresponding to the longitude of the local meridian.

The plus sign is taken when the longitude of place is the East, and the minus sign when the longitude is the West. For a given local time and the GMT (both in hours), the longitude (in degrees) of the standard meridian of country is

$$\text{longitude} = 15(\text{local time} - \text{GMT})$$

When the longitude is positive, the longitude is the East and when negative, the longitude is the West. For some countries, the longitude of the standard meridian and the difference of time with respect to the GMT is given in Table 3. Notice that the countries of large size have two or more standard meridians. The plus sign with time shows that the standard meridian for the country is on the east to the Greenwich meridian and the negative sign shows that the standard meridian for the country is on the west to the Greenwich meridian.

Note:

The mean Sun revolves parallel to the celestial equator which is parallel to the terrestrial equator. Thus, the local time depends on the terrestrial longitude and has no dependence on the terrestrial latitude. The longitudes of some meridians and the corresponding times relative to the GMT are shown in Fig. 3.

Example 1.

Two countries *A* and *B* have the standard meridians at 37° 30′ E and 22° 30′ W longitudes, respectively. Calculate the local times in the countries when the GMT is 6 hr 30 min.

Solution:

Given, the GMT = 6 hr 30 min = 6.5 hr. For the country *A*, the longitude is 37° 30′ E = 37.5° E. As the longitude is the East, local time in *A* is larger than the GMT. Thus, the local time in *A* is a

$$\begin{aligned}\text{local time} &= \text{GMT} + \frac{a}{15}\\ &= 6.5 + \frac{37.5}{15} = 9 \text{ hr}\end{aligned}$$

For the country *B*, the longitude is 22° 30′ W = 22.5° W. As the longitude is the West, local time in *B* is smaller than the GMT. Thus, the local time in *B* is

$$\begin{aligned}\text{local time} &= \text{GMT} - \frac{b}{15}\\ &= 6.5 - \frac{22.5}{15} = 5 \text{ hr}\end{aligned}$$

Example 2.

Local times in two countries *A* and *B* are 2 hr 15 min and 8 hr 45 min, respectively. If the GMT is 5 hr 15 min, calculate the longitudes of the standard meridians of the countries.

Solution:

Given, the GMT = 5 hr 15 min = 5.25 hr. The local time in the country *A* is 2 hr 15 min = 2.25 hr. As the time in *A* is smaller than the GMT, the longitude of *A* is the West. The longitude of the standard meridian of *A* is

$$\begin{aligned}\text{longitude} &= 15(\text{local time} - \text{GMT})\\ &= 15(2.25 - 5.25) = -45°\end{aligned}$$

As negative sign is there, the longitude of the standard meridian of *A* is 45° W.

Local time in the country *B* is 8 hr 45 min = 8.75 hr. As the time in *B* is larger than the GMT, the longitude of *B* is the East. So, the longitude of the standard meridian of *B* is

$$\begin{aligned}\text{longitude} &= 15(\text{local time} - \text{GMT})\\ &= 15(8.75 - 5.25) = 52.5°\end{aligned}$$

As positive sign is there, the longitude of standard meridian of *B* is 52° 30′ E.

Example 3.

Longitudes of two countries *A* and *B* are at 97° 30′ E and 120° E, respectively. Corresponding to the time $6^h\ 45^m$ in *A*, calculate the time in *B*.

Solution:

The difference between the longitudes of the countries *A* and *B* is

$$120° - 97°\ 30' = 22°\ 30'$$

which is equivalent to $1^h\ 30^m$.

Since the longitude is the East and the longitude of B is larger than that of A, the time in B is more than that in A. The time in B is

$$6^h\,45^m + 1^h\,30^m = 8^h\,15^m.$$

Example 4.

Local times in two countries A and B are $5^h\,45^m$ and $7^h\,15^m$, respectively. If longitude of A is at 97° 30′ E, calculate the longitude of B.

Solution:

The difference between the times in two countries is

$$7^h\,15^m - 5^h\,45^m = 1^h\,30^m$$

which is equivalent to 22° 30″. Since the country A is on the East to the Greenwich and the longitude of A is larger than 22° 30′, the country B also is on the East. As the time in B is larger than that in A, therefore the longitude of B is larger than that of A. So, the longitude of B is

$$97°\,30' + 22°\,30' = 120°\text{ E}$$

5. Sidereal Time

To understand the sidereal time, let us consider an arrangement shown in Fig. 4, where the revolution of the Earth around the Sun S is shown. The Earth is spinning about its own axis. In figure, E_1 and E_2 are two positions of the Earth on two successive days. In the background of the Sun, there are the stars situated at very large distances. They are also shown in the figure. When the Earth is at the position E_1, suppose the Sun as well as the distant stars are on the meridian of a person at position P on the Earth.

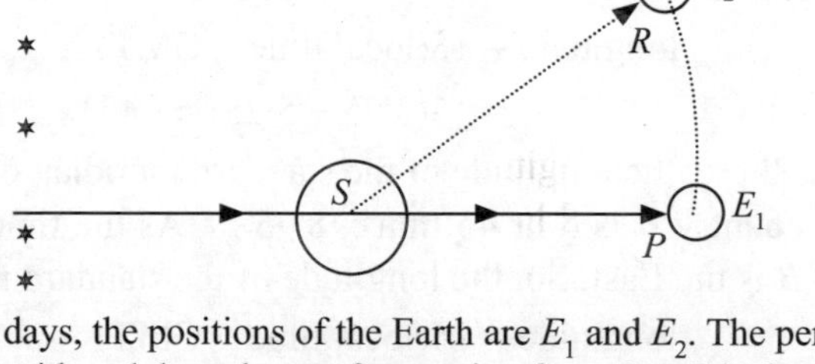

Fig. 4: On two successive days, the positions of the Earth are E_1 and E_2. The period of the rotation from P to Q is the sidereal day whereas for rotation from P to R is the solar day.

Now, when the Earth completes one rotation about its axis, the Earth reaches at the position E_2, as it revolves around the Sun. After completion of one rotation, the position of the person on the Earth is Q. In this position, the distant stars are again on the meridian of the person, but the Sun has not yet arrived on the meridian of the person. When the Earth rotates a bit more and the position of person is at R, the Sun is on the meridian of the person. The time taken for rotation from P to Q is known as the sidereal day whereas the time taken for the rotation from P to R is known as the apparent solar day. Thus, the sidereal time is measured relative to the fixed stars, whereas the apparent solar time is measured relative to the Sun. Obviously, the solar day is longer than the sidereal day. Our watch measures the mean solar day. That is, the mean solar day is equal to 24 hours, and thus, the sidereal day is smaller than 24 hours.

Because of this difference, the stars rise a bit early each day. Interesting to note that after one year, the rising time of stars repeats, showing that the number of sidereal days in one year is large by 1 as compared to the number of solar days. We know that in one year the Earth makes 365.2425 rotations[3] relative to the Sun. Therefore, in one year the Earth makes 365.2425 + 1 rotations relative to the fixed stars. Suppose, d is the length of the sidereal day. Thus, we have

$$d\,(365.2425 + 1) = 24 \times 365.2425$$

Therefore, we have

$$d = \frac{24 \times 365.2425}{366.2425} = 23.9344697 \text{ hr}$$

$$= 23 \text{ hr } 56 \text{ min } 4.0907 \text{ s}$$

The difference between a mean solar day and a sidereal day is

$$24 \text{ hr} - (23 \text{ hr } 56 \text{ min } 4.0907 \text{ s}) = 3 \text{ m } 55.9093 \text{ s}$$

Therefore, the stars appear $3^m\ 56^s$ early each day. In the time of 6 months, the shift is about 12^h. That is, the stars which are visible at mid-night today appear at noon after six months. Still there are some small corrections to be accounted for. The correct value of the sidereal day is about 23^h, 56^m, 4.0916^s (23.9344699 hours or 0.99726958 mean solar day). It is the time taken by the Earth for making one rotation relative to the vernal equinox. The vernal equinox itself processes slowly westward relative to the fixed stars and completes one revolution in about 26,000 years. So, the sidereal day is about 0.0084 seconds shorter than the Earth's period of rotation relative to the fixed stars[4].

Example 5.

Show that the stars which appear at mid-night today appear at noon after six months.

Solution:

We know that daily shift of a star is $3^m\ 56^s = 3.9333^m$. The shift in 6 months (182.5 days) is

$$3.9333^m \times 182.5 = 717.8273^m = 11.96^h$$

Hence, the shift of distant stars observed in 6 months is 11.96^h. So, the stars appear at noon.

Example 6.

Calculate the angle of rotation by the Earth in one solar day.

---

[3]A common person knows that a year (number) which can be divided by 4 is a leap year. For example, the year 1964 is divisible by 4. A leap year has 366 days. A year, which is not a leap year, has 365 days. It is however not complete story about the leap years. Though each century-year can be divided by 4, but it is not leap year. In case of century-years, the year which can be divided by 400 is a leap year. For example, the years 1700, 1800, 1900 are not the leap years, but the years 1600 and 2000 are the leap years. Consequently, in the span of 400 years, there are 97 leap years. Thus, the average number of days in a year is

$$= 365 + \frac{97}{400} = 365.2425$$

[4]The difference of 1 day in 26,000 years corresponds to

$$\frac{24 \times 60 \times 60}{26000 \times 365.2425} = 0.0091 \text{ sec/day}$$

Solution:

It is obvious that the Earth rotates about its axis through an angle of 360° in one sidereal day. It shows that in one mean solar day (or in one solar day) the Earth rotates through an angle larger than 360°. As the Earth makes one additional rotation in the time period of one year. Therefore, the additional rotation of the Earth per day is

$$\frac{360}{366.2425} = 58'59'' \approx 59'$$

Thus, in the time-period of one mean solar day, the Earth rotates about its axis through an angle of 360° 59′.

5.1 Diurnal Motion of Stars

Diurnal motion (Fig. 5) refers to the apparent daily motion of the stars around the Earth, or more precisely around the two celestial poles. It is caused by the Earth's rotation about its own axis. So, every star apparently moves on a circle, that is called the diurnal circle. The time for one complete rotation is 23 hours, 56 minutes and 4.09 seconds (1 sidereal day).

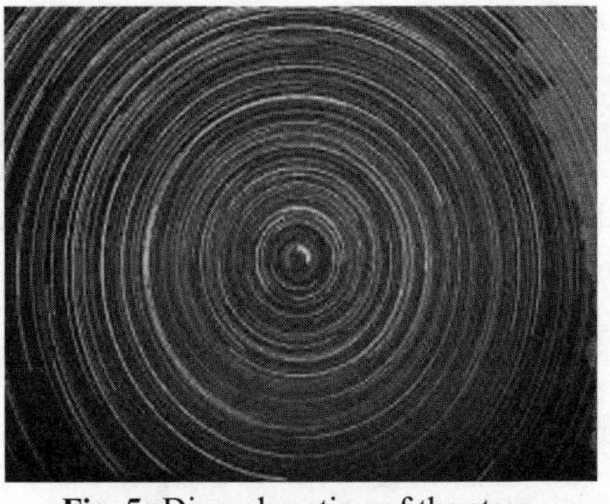

Fig. 5: Diurnal motion of the stars.
(*Courtesy*: Google website)

Example 7.

Suppose, the sidereal and solar clocks show the same time at the midnight on 31 December then what time the sidereal clock shows at midnight on 18 August.

Solution:

The time period of interest is 230 solar days. If there are 366.26 sidereal days in 365.24 solar days then the number of sidereal days in 230 solar days is

$$\frac{366.26}{365.24} \times 230 = 230.624 \text{ sidereal days}$$

It is equivalent to 230 sidereal days, 15 sidereal hours, 25 sidereal minutes. The sidereal clock therefore reads 15:25 at midnight on 18 August.

6. Calendar

Generally, 365 days are taken equal to one year, called the civil year. The interval between two successive passages of the Sun through the vernal equinox is defined as the solar year or tropical year. In the time of Roman Emperor Julius Caesar, the astronomers regarded 365.25 days as equal to one tropical year. Hence, one day accumulated in every four (civil) years [(365.25 − 365) × 4 = 1]. In order to accommodate this extra day, in the year 46 BC, the Emperor made it a rule that every fourth year, which can be completely divided by 4, should be regarded as consist of 366 days instead of 365 days, so that the civil and the tropical years could correspond with each other. Thus, the leap year was introduced and extra day was added to the month of February.

The astronomers of 16th century, however, found that actual tropical year was equal to about 365^d, 5^h, 48^m and 46^s. That is, 1 solar year = 365.242199 days. Thus, according to the above convention, 365.25 – 365.242199 = 0.007801 day accumulates each year. Hence, in 400 years, this excess amounts to 0.007801 × 400 = 3.1204 days. Therefore, in 1582, Pope Gregory of Rome introduced a correction that the century years, which are not completely divisible by 400, would have 365 days instead of 366 days. For example, the years 2000 and 2400 are the leap years (as 2000 and 2400 are divisible by 400), but the years 2100, 2200, 2300 are not the leap years (as 2100, 2200, 2300 are not divisible by 400). Thus, a correction of 3 days in 400 years was introduced in the Julian calendar. This correction was, however, adopted in England much later, in 1752. The present calendar is based on this correction and is known as the Gregorian calendar (or Julian calendar with the Gregorian correction). Obviously, 0.1204 day per four hundred years is still left unaccounted in the present calendar. This is, however, not troublesome for a long time to come. It amounts to 1 day in 400/0.1204 = 3322.26 years.

6.1 Systematics in Gregorian Calendar

Some systematics which can be observed in the present calendar are as the following.

1. After a normal year of 365 days, the next year's weekdays are shifted by one day, since 365 = (7 × 52) + 1.
2. After a leap year of 366 days, the next year's weekdays are shifted by two days, since 366 = (7 × 52) + 2.
3. If we define odd days as the days remained after dividing the days by 7, then the normal year of 365 days has 1 odd day and the leap year of 366 days has 2 odd days. A century has 24 leap years and 76 ordinary years. So, the total number of odd days is 2 × 24 + 76 = 124 = (7 × 17) + 5 = 5.
4. Once in 400 years one century has 25 leap years and 75 ordinary years, resulting in 6 odd days. Hence, in 400 years, we have (5 × 3) + 6 = 21 = (7 × 3) = 0 odd day. That is, the Gregorian calendar repeats in every 400 years.
5. Within a century, in 28 years, we have 7 leap years and 21 ordinary years. Thus, we have 21 + (7 × 2) = 35 = (7 × 5) = 0 odd day. That is, within a century, Gregorian calendar repeats in every 28 years.
6. The number of odd days and shift M' for various months are given in Table 4. The shift M' for a month is obtained as the sum of the shift of the previous year plus the number of odd days in the previous year. When the shift is larger than 6, we subtract 7 from the shift. In case of a leap year, an asterisk is placed on the number.

Table 4. Number of odd days and shift M'

Month	Days	Odd days	Shift M'
January	31	3	0
February	28 29*	0 1*	3
March	31	3	3 4*
April	30	2	6 0*
May	31	3	1 2*
June	30	2	4 5*
July	31	3	6 0*
August	31	3	2 3*
September	30	2	5 6*
October	31	3	0 1*
November	30	2	3 4*
December	31	3	5 6*

*For leap year.

Using the above systematics, a day D falling on the particular date, month and year may be expressed as

$$D = R_7[5\{\text{Rem}(X/4) + \text{Int}(Y/4)\} + \text{Rem}(Y/4) + M' + \delta + \text{DATE}]$$

where DATE is the date, M' is parameter (Table 4) corresponding to the month. X and Y are parts of the year. For example, for the year 1985, $X = 19$ and $Y = 85$. Rem(X/4) is remainder after dividing X by 4, Int(Y/4) is integral part of Y/4, and Rem (Y/4) is remainder after dividing Y by 4. R_7 represents the remainder after dividing by 7. δ is the parameter to be determined according to the days assigned to the value of D. When the values of D are assigned to days as given in Table 5, then δ comes out to be zero for the non-leap years and 1 for the leap years.

Defining $M = M' + \delta$, we get the values of M as given in Table 6 and the present calendar may be represented by the relation

$$D = R_7[5\{\text{Rem}(X/4) + \text{Int}(Y/4)\} + \text{Rem}(Y/4) + M + \text{DATE}]$$

However, 0.1204 day per 400 years is left unaccounted. It accumulates and gives a difference of one day after about 3322 years, i.e., around the year (1752 + 3322) = 5074.

Example 8.

What day of week was on 2 October 1869?

Solution:

For 1869, we have $X = 18$, $Y = 69$.

$$\text{Rem}(X/4) = 2, \quad \text{Int}(Y/4) = 17, \quad \text{Rem}(Y/4) = 1$$

For October $\quad M = 0$ (Table 6), $\quad$ DATE = 2

$$D = R_7[5\{2 + 17\} + 1 + 0 + 2] = R_7[98] = 0$$

$D = 0$ corresponds to Saturday (Table 5). That is, 2 October 1869 was on Saturday.

Table 5. *D* versus Day

D	Day
1	Sunday
2	Monday
3	Tuesday
4	Wednesday
5	Thursday
6	Friday
0	Saturday

Table 6. Values of *M*

Month	Shift M	
January	0	6*
February	3	2*
March	3	
April	6	
May	1	
June	4	
July	6	
August	2	
September	5	
October	0	
November	3	
December	5	

*For the leap year.

6.2 Finding Day Without Calculations

Using above systematics along with the Tables 3 and 4, a chart shown in Table 7 may be obtained. Rem(*X*/4) is denoted by *n*. For the given year, find *X* and *Y*. Now find the letter (*A*, *B*, *C*, etc.) in the row *Y* and column *n* (Table 7). Now look for that letter opposite in the given month (if the letter is written with an asterisk, then see the month with asterisk, e.g., Jan* and Jan are different; Feb* and Feb are different. Then, move down to a column of days. This column of days along with numbers written on the left to it is the calendar for given month and year.

Table 7. A chart for calendar

Y				*n*			
				0	1	2	3
00				G*	E	C	A
01	10	19	28	A	F	D	B
02	11	20	29	B	G	E	C
03	12	21	30	C	A	F	D
04	13	22	31	E*	C*	A*	F*
05	14	23	32	F	D	B	G
06	15	24	33	G	E	C	A
07	16	25	34	A	F	D	B
08	17	26	35	C*	A*	F*	D*
09	18	27	36	D	B	G	E

Contd...

37	56	75	94	E	C	A	F
38	57	76	95	F	D	B	G
39	58	77	96	A*	F*	D*	B*
40	59	78	97	B	G	E	C
41	60	79	98	C	A	F	D
42	61	80	99	D	B	G	E
43	62	81		F*	D*	B*	G*
44	63	82		G	E	C	A
45	64	83		A	F	D	B
46	65	84		B	G	E	C
47	66	85		D*	B*	G*	E*
48	67	86		E	C	A	F
49	68	87		F	D	B	G
50	69	88		G	E	C	A
51	70	89		B*	G*	E*	C*
52	71	90		C	A	F	D
53	72	91		D	B	G	E
54	73	92		E	C	A	F
55	74	93		G*	E*	C*	A*

For example, for the year 1869, $n = \text{Rem}(18/4) = 2$ and $Y = 69$. The letter below $n = 2$ and opposite to $Y = 69$ is E. For October, the letter E is in sixth column, and hence, the calendar for October 1869 is as the following.

1	8	15	22	29	F
2	9	16	23	30	St
3	10	17	24	31	S
4	11	18	25		M
5	12	19	26		T
6	13	20	27		W
7	14	21	28		Th

Table 8. Week-days for the letter D and month

Jan*		Apr		Jul	A	B	C	D	E	F	G
Jan		–		Oct	G	A	B	C	D	E	F
–		May		–	F	G	A	B	C	D	E
Feb*		–		Aug	E	F	G	A	B	C	D
Feb		Mar		Nov	D	E	F	G	A	B	C
–		Jun		–	C	D	E	F	G	A	B
Sep		–		Dec	B	C	D	E	F	G	A

1	8	15	22	29	S	M	T	W	Th	F	St
2	9	16	23	30	M	T	W	Th	F	St	S
3	10	17	24	31	T	W	Th	F	St	S	M
4	11	18	25		W	Th	F	St	S	M	T
5	12	19	26		Th	F	St	S	M	T	W
6	13	20	27		F	St	S	M	T	W	Th
7	14	21	28		St	S	M	T	W	Th	F

7. Julian Day

The Julian day is the continuous count of days since the beginning of the Julian period. It is used primarily by the astronomers, and in software for easily calculating the elapsed days between the two given events. The Julian day number (JDN) is the integer assigned to a whole solar day in the Julian day count starting from the noon universal time (UT), where the Julian day number 0 is assigned to the day starting at noon on Monday, January 1, 4713 BC. For a given day D, month M and year Y, JDN can be calculated with the help of the following relation.

$$\text{JDN} = (1461 \times (Y + 4800 + (M - 14)/12))/4 + (367 \times (M - 2 - 12 \times ((M - 14)/12)))/12 - (3 \times ((Y + 4900 + (M - 14)/12)/100))/4 + D - 32075$$

The months from January to December are assigned the digits from 1 to 12, respectively. The Julian day number for the day starting at 12:00 UT (noon) on January 12000, was 2451545.

The above formula can be calculated as the following:

(i) Drop the fractional part of all results of all multiplications and divisions. (Integer division rounding towards zero is used exclusively, that is, the positive values are rounded down and the negative values are rounded up.)
Follow the following steps:

$$A = (M - 14)/12$$

$$B = 1461 \times (Y + 4800 + A)/4$$

$$C = 367 \times (M - 2 - 12 \times A)/12$$

$$F = (Y + 4900 + A)/100$$

$$E = 3 \times F/4$$

$$\text{JDN} = B + C - E + D - 32075$$

For 12:00 UT (noon) on January 1, 2000, we have $D = 1$, $M = 1$, $Y = 2000$, and we have

$$A = -1$$

$$B = 2483334$$

$$C = 336$$

$$F = 68$$

$$E = 51$$
$$\text{JDN} = 2451545$$

For the time other than 12:00 UT (noon), the Julian day (JD) can be calculated with the help of the following expression, where JDN is calculated from the above expression.

$$\text{JD} = \text{JDN} - 1 + \frac{\text{hour} + 12}{24} + \frac{\text{minute}}{1440} + \frac{\text{second}}{86400}$$

The Julian Date for 00:30:00.0 UT January 1, 2013, is 2456293.520833.

Finding Day of Week on given Julian Day Number

The day of the week W1 (for an afternoon or evening UT) can be determined from the Julian Day Number J using the following expression:

$$W_1 = \text{mod}(J + 1, 7)$$

W_1	0	1	2	3	4	5	6
Day of the week	Sun	Mon	Tue	Wed	Thu	Fri	Sat

If the moment in time is after midnight UT (and before 12:00 UT), then one is already in the next day of the week.

Table below shows some JD and day of the week for given date and UT.

Date	UT	JD	Day
1 Jan 2000	12: 0: 0.00	2451545.000000	Sat
1 Jan 2013	0:30: 0.00	2456293.520833	Tue
5 Jul 2010	11:12: 0.00	2455382.966667	Mon
4 Dec 2009	5:10:20.00	2455169.715509	Fri
8 Jul 1949	18:15: 0.00	2433106.260417	Fri
7 Aug 1987	2:15: 0.00	2447014.593750	Fri

7.1 Julian or Gregorian Calendar from Julian Day Number

This is an algorithm to convert a Julian Day Number, J, to a date in the Julian or Gregorian calendar. The algorithm is valid for Julian day numbers greater than or equal to 0. All the variables have integer values, and the notation "a div b" indicates integer division, and "mod(a, b)" denotes the modulus operator.

For Gregorian calendar:

$$A = (4 \times J + 274277)/146097$$
$$B = J + 1401 + (A \times 3)/4 - 38$$
$$C = 4 \times B + 3$$
$$E = C - (C/1461) \times 1461$$
$$F = E/4$$

$$G = 5 \times F + 2$$
$$H = G - (G/153) \times 153$$
$$I = G/153 + 2$$
$$D = H/5 + 1$$
$$M = I - (I/12) \times 12 + 1$$
$$Y = C/1461 - 4716 + (14 - M)/12$$

D, M, and Y are the numbers of the day, month, and year respectively for the afternoon at the beginning of the given Julian day.

For example, for Julian date number $J = 2451545$, we have

$$A = 68$$
$$B = 2452959$$
$$C = 9811839$$
$$E = 1224$$
$$F = 306$$
$$G = 1532$$
$$H = 2$$
$$I = 12$$
$$D = 1$$
$$M = 1$$
$$Y = 2000$$

Thus, on the Julian day number 2451545, the Gregorian date was 1 Jan 2000. Table below shows Gregorian date and weekday for some given Julian day numbers.

JD	Date	Day
2451545	1 Jan 2000	Sat
2456294	1 Jan 2013	Tue
2455383	5 Jul 2010	Mon
2455170	4 Dec 2009	Fri
2433106	8 Jul 1949	Fri
2447015	7 Aug 1987	Fri

Review Questions

1. Describe the apparent solar day and the mean solar day. What is the equation of time? Which one of these is followed in our watches?
2. What is the difference between the real Sun and the mean Sun?
3. Two countries A and B have standard meridians at 35° 27″ E and 21° 43″ W longitudes, respectively. Calculate the local times in the countries when the GMT is 6 hr 43 min.

4. Local times in two countries *A* and *B* are 2 hr 18 min and 9 hr 37 min, respectively. If the GMT is 4 hr 25 min, calculate the longitudes of the standard meridians of the countries.
5. What is the sidereal day for the Earth? Explain the reason that the apparent solar day is longer than the sidereal day.
6. Discuss the sidereal time. Estimate the length of a sidereal day.
7. Show that the Gregorian calendar repeats in every 400 years.
8. Show that the Gregorian calendar repeats in every 28 years.
9. Define a leap year. Which of the following years are leap years?
   1960, 2100, 2123, 2200, 2275, 2300, 2356, 2400, 2496, 2500

   Ans. 1960, 2356, 2400, 2496
10. Calculate the Julian day number and week day for the UT 5:45:00 on 10 June 1953.
11. What was the Gregorian date and week day for for the Julian day number 2455269.
12. Write notes on the following:
    (i) Apparent solar day (ii) Sundial
    (iii) Mean Sun (iv) Mean solar day
    (v) Equation of time (vi) Local mean time
    (vii) GMT (viii) Sidereal time
    (ix) Diurnal motion of stars (x) Leap year
    (xi) Julian day

Multiple-Choice Questions

1. Solar time varies from day to day because the
   A. Earth is not spherical in shape
   B. Sun is moving in our galaxy Milky Way
   C. Earth revolves around the Sun in an elliptical orbit
   D. Earth spins about its own axis **Ans. C**
2. The mean solar day is equal to
   A. 24 hr B. 23 hr 56 m C. 24 hr 3 m D. 12 hr **Ans. A**
3. The time based on the apparent motion of the actual Sun is
   A. sidereal time B. mean solar time C. solar time D. none of them
   Ans. C
4. The apparent motion of the real Sun is along the
   A. celestial equator B. ecliptic
   C. local meridian D. spin axis of the Earth **Ans. B**
5. The motion of the mean Sun is along
   A. celestial equator B. ecliptic
   C. local meridian D. spin axis of the Earth **Ans. A**
6. Longitude of a country is East. The local time in the country is
   A. larger than GMT B. equal to GMT C. smaller than GMT D. none of them
   Ans. A
7. Longitude of a country is West. The local time in the country is
   A. larger than GMT B. equal to GMT C. smaller than GMT D. none of them
   Ans. C

8. Local time depends on
   A. latitude B. both latitude and longitude
   C. longitude D. none of latitude and longitude **Ans. C**
9. When the local time is larger than the GMT, the longitude of the country is
   A. East B. West C. North D. South **Ans. A**
10. When the local time is smaller than the GMT, the longitude of the country is
    A. East B. West C. North D. South **Ans. B**
11. The time measured relative to the distant stars is known as the
    A. solar time B. mean solar time C. local time D. sidereal time
    Ans. D
12. The time measured relative to the Sun is known as
    A. solar time B. mean solar time C. local time D. sidereal time
    Ans. A
13. Standard meridians of two countries *A* and *B* correspond to 5h and 6h 30m, respectively. Which of the followings is correct?
    A. *B* is east to *A* B. *B* is west to *A* C. *B* is north to *A* D. *B* is south to *A*
    Ans. A
14. Earth rotates 360° in one
    A. solar day B. mean solar day C. sidereal day D. year **Ans. C**
15. In one solar day, Earth rotates about its own axis through an angle approximately
    A. 359° 29′ B. 360° C. 360° 59′ D. 180° **Ans. C**
16. Orbit of fictitious Sun relative to the Earth is
    A. ellipse B. circle C. parabola D. hyperbola **Ans. B**
17. Angle between the orbit of the fictitious the Sun and the celestial equator is
    A. 0° B. 23.5° C. 45° D. 60° **Ans. A**
18. A sundial gives the
    A. mean solar time B. apparent solar timeC. sidereal time D. local time **Ans. B**
19. The weekday on the Julian day number 2451545 is
    A. Fri B. Sat C. Sun D. Mon **Ans. B**
20. The weekday on the Julian day number 2433106 is
    A. Fri B. Sat C. Sun D. Mon **Ans. A**

4 Atomic Physics

Balmer in 1885 found that observed lines from the hydrogen atom could be fitted through the expression

$$\lambda = 3646\frac{n^2}{n^2 - 4}$$

$$n = 3, 4, 5, \ldots$$

Bohr provided for this expression.

1. Bohr Model

Niels Bohr in 1913 proposed a model for the hydrogen atom which could explain successfully the origin of the observed spectral lines. This theory is applicable to the hydrogen-like atoms also, such as He^+, Li^{++}, etc. Therefore, we discuss here the Bohr model for a hydrogen-like atom. The model is based on the following three postulates proposed by Bohr.

(i) The atom consists of a heavy and positively charged nucleus situated at its centre and the electron is revolving in a circular path round the nucleus without radiating energy (Fig. 1). Such orbitals are known as the stationary orbits. It shows that the classical electrodynamics theory is not applicable in the present scenario.

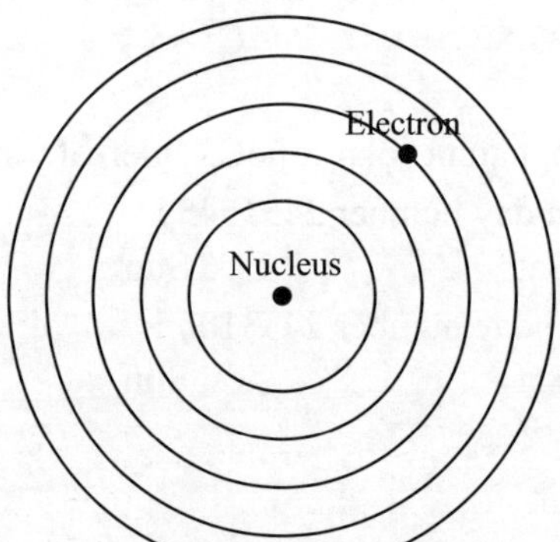

Fig. 1: In the Bohr model, an electron revolves around the nucleus in one of the stationary and circular orbits without radiating energy. Nucleus is shown by a small solid circle at the centre and the electron is shown by a rather big solid circle.

Thus, two forces:

(i) the centrifugal force in the outward direction and (ii) the electrostatic attraction in the inward direction, acting on the electron are balancing to each other, and we have

$$\frac{Ze^2}{r^2} = \frac{mv^2}{r} \tag{4.1}$$

where m, e, v are, respectively, the mass, charge and velocity of electron revolving in a circular orbit of radius r whose centre coincides with the centre of the nucleus[1]. Ze is the charge on the nucleus (for hydrogen atom, the atomic number Z is 1).

(ii) Out of an infinite number of circular orbits, only possible orbits for the revolution of an electron are those for which orbital angular momentum is an integer multiple of $\hbar$, where $\hbar = h/2\pi$. That is,

$$mvr = n\hbar \tag{4.2}$$

where n is a non-zero positive integer, called the **principal quantum number**. Hence, the circular orbits in which the electron can revolve are not continuous, but discrete.

(iii) Since the direction of motion of electron is changing continuously, the electron is accelerated and according to the classical electrodynamics, the accelerated electron should radiate its energy. Bohr assumed that in spite of the accelerated motion, the electron does not radiate its energy and such orbits are called the **stationary orbits**.

(iv) Whenever an electron jumps from an outer orbit of energy E_i to an inner orbit of energy E_f, an electromagnetic radiation (i.e., a photon) of energy $h\nu$ is emitted, so that

$$E_i - E_f = h\nu \tag{4.3}$$

where ν is the frequency of emitted photon and h is the Planck constant. In the reverse process, an electron in an inner orbit can jump to an outer one after absorbing an energy $h\nu$. From equations (4.1) through (4.3), one can derive various parameters as below.

1.1 Radius of Orbit

Using the value of v from equation (4.2) in (4.1), we get

$$\frac{Ze^2}{r^2} = \frac{m}{r}\left(\frac{nh}{2\pi mr}\right)^2$$

or

$$r = \frac{h^2}{4\pi^2 me^2}\frac{n^2}{Z} \tag{4.4}$$

Thus, the radius of the n-th orbit is

$$r_n = r_1\frac{n^2}{Z}$$

where $r_1 (\equiv h^2/4\pi^2 me^2) = 5.29 \times 10^{-9}$ cm $= 0.529$Å is the radius of the first Bohr orbit of hydrogen atom, also called the **Bohr radius**. Thus, radius r_n of an stationary orbit is proportional to n^2 and inversely proportional to Z.

1.2 Speed of Electron

Using equation (4.4) in (4.2), we get

[1]Since the mass of the nucleus is much larger than that of the electron. Therefore, here we have taken it to be infinite, so that the centre of the circular orbit of the electron coincides with the centre of the nucleus. Further, the reduced mass of electron and nucleus is equal to the mass of electron.

$$mv\frac{h^2}{4\pi^2 me^2}\frac{n^2}{Z} = n\hbar$$

or $$v = \frac{2\pi e^2}{h}\frac{Z}{n}$$

Thus, the velocity in the *n*-th orbit is

$$v_n = v_1\frac{Z}{n} \tag{4.5}$$

where $v_1(\equiv 2\pi e^2/h) = 2.187 \times 10^8$ cm/s is the speed of electron in first Bohr orbit of hydrogen atom. Thus, v_n is inversely proportional to *n* and proportional to *Z*.

1.3 Energy of Electron

Total energy of the electron in an orbit is the sum of the kinetic energy and the potential energy. The kinetic energy (K_n) of electron in the *n*-th orbit is (using equation 4.5)

$$K_n = \frac{1}{2}mv_n^2 = \frac{2\pi^2 me^4}{h^2}\frac{Z^2}{n^2}$$

The force of attraction (*F*) between the electron and the nucleus,

$$F = -\frac{Ze^2}{r^2}$$

The potential energy (*U*) of the electron,

$$U = \int_r^\infty F\,dr = -\int_r^\infty \frac{Ze^2}{r^2}$$

$$dr = -Ze^2\int_r^\infty \frac{dr}{r^2} = -\frac{Ze^2}{r}$$

Using equation (4.4) here, we have the potential energy of the electron in the *n*-th orbit,

$$U_n = -Ze^2\frac{4\pi^2 me^2 Z}{h^2 n^2} = -\frac{4\pi^2 me^4}{h^2}\frac{Z^2}{n^2}$$

Total energy (E_n) of the electron in the *n*-th orbit,

$$E_n = K_n + U_n = \frac{2\pi^2 me^4}{h^2}\frac{Z^2}{n^2} - \frac{4\pi^2 me^4}{h^2}\frac{Z^2}{n^2}$$

$$= \frac{2\pi^2 me^4}{h^2}\frac{Z^2}{n^2} = -E_1\frac{Z^2}{n^2}$$

where $E_1(\equiv 2\pi^2 me^4/h^2) = 2.178 \times 10^{-11}$ erg = 13.61 eV is the amount of energy of electron in the first Bohr-orbit of the hydrogen atom. Thus, the total energy is inversely proportional to n^2 and proportional to Z^2. The negative sign in the energy indicates that the electron is bound to the nucleus. Total energy of the electron is the lowest when it is in the state with the principal quantum number $n = 1$. This state is known as the ground state of hydrogen atom. Other states with $n > 1$ are called the excited states. The energy of a level is obviously discrete, and not continuous.

Radius, velocity and energy for the first five orbits of hydrogen atom are as the following.

n	r_n(Å)	v_n(m/s)	E_n(eV)
1	0.529	2.188×10^6	13.606
2	2.117	1.094×10^6	3.401
3.	4.766	7.292×10^5	1.512
4	8.467	5.469×10^5	0.850
5	13.229	4.375×10^5	0.544

1.4 Line Spectrum

In the hydrogen atom ($Z = 1$), when the electron in the initial state (other than ground state) with energy E_i jumps to a lower state with energy E_f, a photon of energy $h\nu$ is emitted. Thus, we have (using equation 4.3)

$$h\nu = E_i - E_f = -\frac{E_1}{n_i^2} + \frac{E_1}{n_f^2}$$

or

$$\nu = \frac{E_1}{h}\left(\frac{1}{n_f^2} - \frac{1}{n_i^2}\right)$$

Using the relation $\nu = c/\lambda$, where c is the speed of light and λ the wavelength of photon, we get

$$\frac{1}{\lambda} = \frac{E_1}{hc}\left(\frac{1}{n_f^2} - \frac{1}{n_i^2}\right)$$

or

$$\bar{\nu} = R_\infty\left(\frac{1}{n_f^2} - \frac{1}{n_i^2}\right)$$

where $\bar{\nu}$ ($\equiv 1/\lambda$) is known the wavenumber of photon and R_∞($\equiv E_1/hc$) = 1.09678×10^5 cm^{-1} is called the **Rydberg constant**. (The suffix ∞ is used because the mass of the nucleus of atom is taken as infinite.) Now, various series observed in the spectrum of hydrogen atom can be expressed as below:

(i) **Lyman series:** $n_f = 1$ and $n_i = 2, 3, 4$, so on. The lines in this series are denoted by L_α, L_β, L_γ, ... corresponding to $n_i = 2, 3, 4, ...$, respectively.

(ii) **Balmer series:** $n_f = 2$ and $n_i = 3, 4, 5$, so on. The lines in this series are denoted by H_α, H_β, H_γ, so on, corresponding to $n_i = 3, 4, 5$, so on, respectively.

(iii) **Paschen series:** $n_f = 3$ and $n_i = 4, 5, 6$, so on.

(iv) **Brackett series:** $n_f = 4$ and $n_i = 5, 6, 7$, so on.

(v) **Pfund series:** $n_f = 5$ and $n_i = 6, 7, 8$, so on.

Wavelengths in Å of the first four lines and the limit of each series are as below:

Line	Lyman	Balmer	Paschen	Brackett	Pfund
1	1215.7	6564.7	18756.2	40522.7	74598.5
2	1025.7	4862.7	12821.6	26258.7	46537.7

Contd...

3	972.5	4341.7	10941.1	21661.2	37405.5
4	949.7	4102.9	10052.2	19450.9	32969.9
∞	911.8	3647.0	8205.8	14588.2	22794.0

Wavenumbers in cm^{-1} of the first four lines and the limit of each series are as below:

Line	Lyman	Balmer	Paschen	Brackett	Pfund
1	82258.50	15233.06	5331.57	2467.76	1340.51
2	97491.56	20564.63	7799.32	3808.26	2148.79
3	102823.13	23032.38	9139.83	4616.55	2673.40
4	105290.88	24372.89	9948.12	5141.16	3033.07
∞	109678.00	27419.50	12186.44	6854.88	4387.12

Similar series can be obtained for hydrogen-like atoms.

The energy of a stationary state with $n = \infty$ is zero. Above this state, electron is free and always has a positive energy. This region of the positive energy is called the **continuum**. The amount of energy required for a transition of an electron from the ground state to zero-energy state ($n = \infty$), is called the **ionization potential** of atom. The energy levels with $n = 1, 2, 3, ...$, are also called K, L, M, ... levels, respectively. Some of the emission lines (not to the scale, in the emission) of these series are shown in Fig. 2.

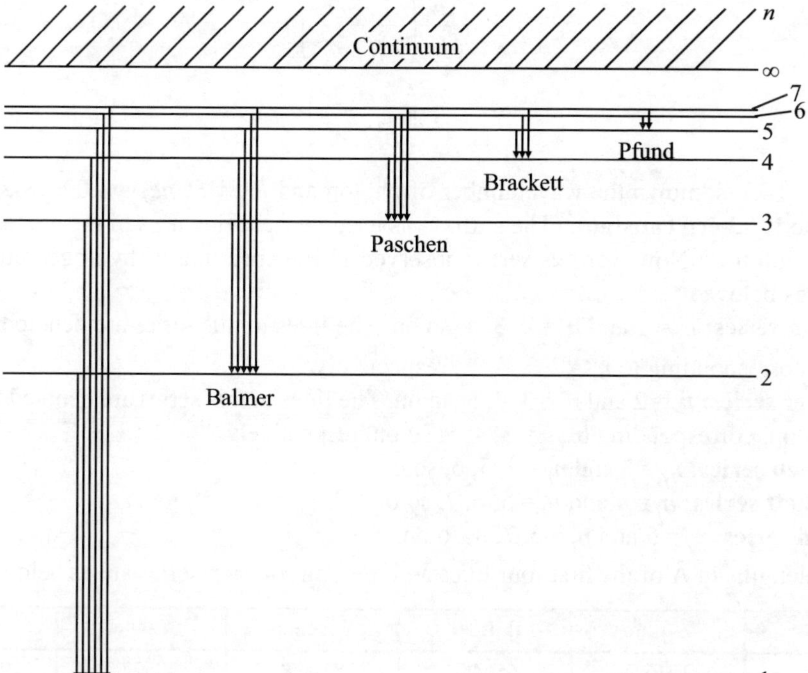

Fig. 2: Some emission lines of the Lyman, Balmer, Paschen, Brackett and Pfund series.

Example 1.

Lifetime of an excited state of hydrogen atom is 10^{-8} sec. How many revolutions does an electron in $n = 2$ state of the hydrogen atom make before falling to $n = 1$ state?

Solution:

In a hydrogen atom, an electron revolves in a stationary orbit of radius r with the velocity v is expressed as

$$r = \frac{h^2 n^2}{4\pi^2 me^2}$$

and
$$v = \frac{2\pi e^2}{hn}$$

Number of revolutions of electron in the orbit per second is

$$f = \frac{v}{2\pi r} = \frac{4\pi^2 me^4}{h^3 n^3}$$

$$= \frac{2\times\pi^2\times 9.1\times 10^{-28}\times(4.8\times 10^{-10})^4}{(6.63\times 10^{-27})^2 n^3}$$

$$= \frac{3.27\times 10^{15}}{n^3}\text{Hz}$$

For the orbit with $n = 2$, we have $f = 3.27 \times 10^{15}/8 = 4.09 \times 10^{14}$ Hz.

Thus, the number of revolutions made in 10^{-8} sec is $4.09 \times 10^{14} \times 10^{-8} = 4.09 \times 10^6$.

Example 2.

The series limit wavelength of Balmer series in the hydrogen spectrum is 3647.04 Å. Calculate the Rydberg constant for hydrogen atom.

Solution:

For the Balmer series, we have

$$\frac{1}{\lambda} = R_\infty\left(\frac{1}{2^2} - \frac{1}{n_i^2}\right)$$

where
$$n_i = 3, 4, 5, ...$$

where R_∞ is the Rydberg constant for hydrogen atom. For the series limit $n_i = \infty$ and we have

$$\frac{1}{3647.04\times 10^{-8}} = R_\infty\left(\frac{1}{2^2} - \frac{1}{\infty^2}\right)$$

Thus, $R_\infty = 1.0968 \times 10^5$ cm^{-1}.

Example 3.

Wavelength of the first line of Lyman series is 1215.7 Å. Calculate the wavelength of the second line of Balmer series.

Solution:

For the Lyman series, we have

$$\frac{1}{\lambda} = R_\infty\left(\frac{1}{1^2} - \frac{1}{n_i^2}\right)$$

where $$n_i = 2, 3, 4, 5, ...$$

and for the Balmer series, we have

$$\frac{1}{\lambda} = R_\infty\left(\frac{1}{2^2} - \frac{1}{n_i^2}\right)$$

where $$n_i = 3, 4, 5, ...$$

For the first line of Lyman series, we have

$$\frac{1}{1215.7\times10^{-8}} = R_\infty\left(\frac{1}{1} - \frac{1}{4}\right)$$

For the second line of Balmer series, we have

$$\frac{1}{\lambda} = R_\infty\left(\frac{1}{4} - \frac{1}{16}\right)$$

Thus, we have

$$\frac{\lambda}{1215.7\times10^{-8}} = \frac{3/4}{3/16}$$

or $$\lambda = 4862.8 \text{ Å}$$

Example 4.

How much the minimum energy is required to remove an electron from the hydrogen atom in the ground state, and also in $n = 6$ state?

Solution:

Energy of electron in a state with quantum number n is

$$E_n = -\frac{2\pi^2 me^4}{h^2 n^2} = -\frac{2\times\pi^2\times9.1\times10^{-28}\times(4.8\times10^{-10})^4}{(6.63\times10^{-27})^2 n^2}$$

$$= -\frac{2.169\times10^{-11}}{n^2}\text{ erg} = -\frac{2.169\times10^{-11}}{1.60\times10^{-12}\times n^2}$$

$$= -\frac{13.556}{n^2}\text{ eV}$$

Thus, the minimum energy required to remove an electron from the ground ($n = 1$) state is 13.556 eV and from the $n = 6$ state is $13.556/6^2 = 0.377$ eV.

1.5 Effect of Finite Mass of Nucleus

In the Bohr theory of hydrogen-like atom, the centre of the circular orbit of electron was assumed to coincide with the centre of the nucleus. This is possible when the mass of nucleus is taken as infinite. But, the mass of nucleus is finite and therefore, both the nucleus of mass M and the electron of mass m revolve with an angular velocity ω about their common centre-of-mass c.m. (Fig. 3). Following the definition of the centre-of-mass, we have

$$Mx = m(r - x)$$

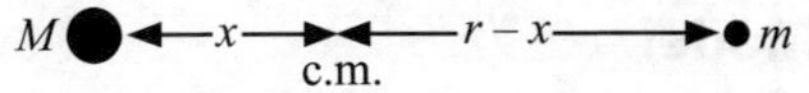

Fig. 3: Nucleus of mass M is at a distance x from the centre-of-mass c.m. and the electron of mass m is at a distance $r - x$ from the centre-of-mass on the opposite side.

Thus,

$$x = \frac{mr}{M+m}$$

and

$$(r - x) = \frac{Mr}{M+m}$$

Total angular momentum of the system revolving about the centre-of-mass with the angular velocity ω is

$$L = Mx^2\omega + m(r-x)^2\omega$$

$$= M\left(\frac{mr}{M+m}\right)^2\omega + m\left(\frac{Mr}{M+m}\right)^2\omega$$

$$= \frac{Mm}{M+m}r^2\omega = \mu r^2\omega$$

where $\mu \equiv Mm/(M+m)$ is the reduced mass of the system. Hence, the total angular momentum of the system is equal to the angular momentum of a fictitious body of mass μ revolving around the nucleus of infinite mass at a distance r. Therefore, to account for the finite mass of the nucleus, the mass of electron m should be replaced by μ. Hence, the radius of the n-th orbit is

$$r_n = \frac{h^2}{4\pi^2\mu e^2}\frac{n^2}{Z}$$

Further, the energy of electron in the n-th orbit is

$$E_n = -\frac{4\pi^2\mu e^4}{h^2}\frac{Z^2}{n^2} = -\frac{Mm}{M+m}\frac{4\pi^2 e^4}{h^2}\frac{Z^2}{n^2}$$

$$= -\frac{M}{M+m}E_1\frac{Z^2}{n^2}$$

where $E_1 (\equiv 2\pi^2 me^4/h^2)$ is the energy of the first Bohr-orbit of hydrogen atom. Hence, the wave number $\bar{\nu}$ of the photon due to a transition between the levels with quantum numbers n_i and n_f is

$$\bar{\nu} = \frac{M}{M+m}\frac{E_1}{hc}\left(\frac{1}{n_f^2} - \frac{1}{n_i^2}\right) = R_M\left(\frac{1}{n_f^2} - \frac{1}{n_i^2}\right)$$

where

$$R_M = \frac{M}{M+m}\frac{E_1}{hc} = \frac{M}{M+m}R_\infty$$

Here, R_M and R_∞ are Rydberg constants for the finite and infinite masses of the nucleus, respectively. This modification due to the finite mass of the nucleus, led to the discovery of deuterium − an isotope of the hydrogen. Nucleus of deuterium has one proton and one neutron. The hydrogen is found to have one more isotope, called the tritium whose nucleus has one proton and two neutrons.

2. Bohr-Sommerfeld Model

Bohr theory of quantized circular orbits could explain successively about the generation of line spectrum of hydrogen-like atoms. However, with the help of the instruments of good resolution, each line consists of a number of closely packed lines. For example, the H_α line was found to have five components. Such arrangement of lines is known as the **fine structure** of spectral line. The Bohr model was obviously incapable to explain the fine structure. In order to explain the fine structure of spectral lines, Sommerfeld extended the Bohr model and presented a model called the **Bohr-Sommerfeld model**.

2.1 Orbit of Electron-

According to the Bohr-Sommerfeld model, the electron in the hydrogen-like atom revolves in an elliptical path around the nucleus which is situated at one of the foci, *S*, of the ellipse (Fig. 4). Using the polar coordinates *r* and θ in a plane with origin at the focus *S* of the ellipse, the radial and angular components of the angular momentum are

$$p_r = m\frac{dr}{dt}$$

and

$$p_\theta = mr^2\frac{d\theta}{dt}$$

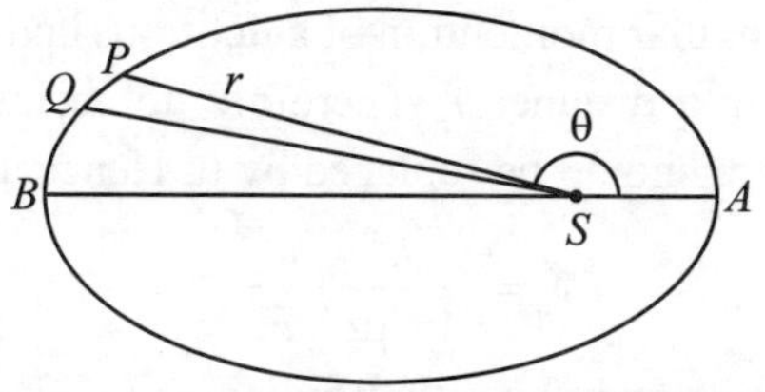

Fig. 4: In the Bohr-Sommerfeld model, an electron revolves around the nucleus in an elliptical path. Nucleus is situated at the focus *S* of the ellipse.

Since the *r* and θ both are periodic functions of the time *t*, the space quantization conditions in the present case are as the following[2].

$$\oint p_r dr = n_r h$$

and

$$\oint p_\theta d\theta = n_\theta h \tag{4.6}$$

Here, n_r and n_θ are non-zero positive integers. Now, the electron can move in only those orbits for which both the quantization conditions are satisfied simultaneously. Here, also the centrifugal force in the outward direction on the electron is balanced by the electrostatic attraction in the inward direction. Thus, we have

[2]For the Bohr model, *r* was not varying (circular path). So the only quantization condition was

$$\oint p_\theta d\theta = nh$$

Here, the angular momentum $p_\theta = mvr$ is constant. Thus, we have

$$\int_0^{2\pi} mvr d\theta = nh \qquad mvr2\pi = nh \qquad mvr = n\frac{h}{2\pi}$$

It is the quantization condition of the Bohr-model.

$$\frac{Ze^2}{r^2} = \frac{mv^2}{r}$$

where m, e, v are, respectively, the mass, charge and velocity of electron. Equation of ellipse in polar coordinates (r, θ) with origin at the focus S is

$$\frac{1}{r} = C + D \cos \theta \tag{4.7}$$

The constants C and D can be determined in the following manner. Let a, b and e be the semi-major axis, semi-minor axis and eccentricity of the ellipse. The coordinates of point A are $[a(1 - e), 0]$ whereas the coordinates of the point B are $[a(1 + e), \pi]$. Thus, we have

$$\frac{1}{a(1-e)} = C + D$$

and

$$\frac{1}{a(1+e)} = C - D$$

On solving these equations, we get

$$C = \frac{1}{a(1-e^2)}$$

and

$$D = \frac{e}{a(1-e^2)}$$

Using the values of C and D in equation (4.7), the equation of ellipse is

$$\frac{a(1-e^2)}{r} = 1 + e \cos \theta \tag{4.8}$$

Now, suppose in time dt, the electron moves from the position P to the position Q, so that $\angle PSQ = d\theta$. The area of the triangle PSQ is

$$\text{Area of triangle} = \frac{1}{2}\text{Base} \times \text{Height}$$

$$= \frac{1}{2} r \times r\,d\theta = \frac{1}{2} r^2 d\theta$$

Hence, the areal velocity is

$$\text{Areal velocity} = \frac{\text{Area}}{\text{Time}} = \frac{1}{2} r^2 \frac{d\theta}{dt}$$

According to the Kepler's second law of planetary motion, the areal velocity of a particle moving in an elliptical path is constant. That is,

$$\frac{1}{2} r^2 \frac{d\theta}{dt} = \frac{k}{2}$$

so that

$$r^2 \frac{d\theta}{dt} = k \tag{4.9}$$

where k is a constant. It shows that the θ-component of the angular momentum

$$p_\theta = mr^2 \frac{d\theta}{dt} = mk$$

is constant throughout the motion. Thus,

$$n_\theta h = \oint p_\theta \, d\theta = \oint mk \, d\theta = mk2\pi \tag{4.10}$$

To evaluate the first integral in equation (4.6), we take logarithm of (4.8) and then differentiate to get

$$\log[a(1 - e^2)] - \log r = \log (1 + e \cos \theta)$$

Differentiation of this gives

$$-\frac{dr}{r} = -\frac{e\sin\theta}{1+e\cos\theta} d\theta$$

or

$$\frac{1}{r}\frac{dr}{d\theta} = \frac{e\sin\theta}{1+e\cos\theta} \tag{4.11}$$

From the first equation of (4.6), we have

$$n_r h = \oint p_r dr = \oint m\frac{dr}{dt} dr = \oint m \frac{dr}{d\theta}\frac{d\theta}{dt}\frac{dr}{d\theta} d\theta$$

$$= \oint m\left(\frac{1}{r}\frac{dr}{d\theta}\right)^2 r^2 \frac{d\theta}{dt} d\theta = \oint m\left(\frac{e\sin\theta}{1+e\cos\theta}\right)^2 k \, d\theta$$

$$= mk\int_0^{2\pi} \frac{e^2 \sin^2\theta}{(1+e\cos\theta)^2} d\theta$$

Here, we have used equations (4.9) and (4.11). Using the solution of the integral (Appendix *C*), we have

$$n_r h = mk2\pi\left(\frac{1}{\sqrt{1-e^2}} - 1\right)$$

$$= n_\theta h\left(\frac{1}{\sqrt{1-e^2}} - 1\right)$$

Thus, we have

$$\frac{1}{\sqrt{1-e^2}} = \frac{n_r}{n_\theta} + 1$$

or

$$1 - e^2 = \frac{n_\theta^2}{(n_r + n_\theta)^2} = \frac{n_\theta^2}{n^2} \tag{4.12}$$

Here, we have used equation (4.10). Using relation $b^2 = a^2(1 - e^2)$, we have

$$\frac{b^2}{a^2} = \frac{n_\theta^2}{n^2}$$

or

$$\frac{b}{a} = \frac{n_\theta}{n}$$

where $n = n_r + n_\theta$ is known as the total quantum number. Since n_r as well as n_θ both are integer, n also is an integer. Here, n_θ can assume integer values ranging from 1 to n with increment of unity. Thus, there can be n number of ellipses for a given value of n.

2.2 Energy of Electron in an Orbit

Total energy E of electron is sum of potential energy, radial kinetic energy and angular kinetic energy. The force of attraction F between electron and nucleus is

$$F = -\frac{Ze^2}{r^2}$$

The potential energy U is

$$U = \int_r^\infty F\,\mathrm{d}r = -\int_r^\infty \frac{Ze^2}{r^2}\mathrm{d}r = -Ze^2\int_r^\infty \frac{\mathrm{d}r}{r^2} = -\frac{Ze^2}{r}$$

Radial kinetic energy K_r is[3]

$$K_r = \frac{1}{2}m\left(\frac{\mathrm{d}r}{\mathrm{d}t}\right)^2 = \frac{1}{2}m\left(\frac{\mathrm{d}r}{\mathrm{d}\theta}\frac{\mathrm{d}\theta}{\mathrm{d}t}\right)^2$$

$$= \frac{m}{2r^2}\left(\frac{1}{r}\frac{\mathrm{d}r}{\mathrm{d}\theta}\right)^2\left(r^2\frac{\mathrm{d}\theta}{\mathrm{d}t}\right)^2$$

$$= \frac{mk^2}{2r^2}\left(\frac{e\sin\theta}{1+e\cos\theta}\right)^2$$

Here, we have used equations (4.9) and (4.11). Angular kinetic energy,

$$k_\theta = \frac{1}{2}mr^2\left(\frac{\mathrm{d}\theta}{\mathrm{d}t}\right)^2 = \frac{m}{2r^2}\left(r^2\frac{\mathrm{d}\theta}{\mathrm{d}t}\right)^2 = \frac{mk^2}{2r^2}$$

Here, we have used equation (4.9). Therefore, the total energy,

$$E = -\frac{Ze^2}{r} + \frac{mk^2}{2r^2}\left(\frac{e\sin\theta}{1+e\cos\theta}\right)^2 + \frac{mk^2}{2r^2}$$

On substituting the value of r from equation (4.8) here, we have

$$E = -Ze^2\frac{1+e\cos\theta}{a(1-e^2)} + \frac{mk^2}{2}\left(\frac{e\sin\theta}{1+e\cos\theta}\right)^2\left(\frac{1+e\cos\theta}{a(1-e^2)}\right)^2 + \frac{mk^2}{2}\left(\frac{1+e\cos\theta}{a(1-e^2)}\right)^2.$$

$$= -Ze^2\frac{1+e\cos\theta}{a(1-e^2)} + \frac{mk^2(1+e^2+2e\cos\theta)}{2a^2(1-e^2)^2} \tag{4.13}$$

The energy E should not depend on the angle θ. We have freedom to choose the parameter k. Thus, the parameter k should be such that the coefficient of $\cos\theta$ is zero. On putting the coefficient of $\cos\theta$ equal to zero, we get

$$-\frac{Ze^2}{a(1-e^2)} + \frac{mk^2}{a^2(1-e^2)^2} = 0$$

or

$$k = \left[\frac{Ze^2a(1-e^2)}{m}\right]^{1/2} \tag{4.14}$$

[3]In a plane with polar coordinates, components of the velocity are

$$v_r = \frac{\mathrm{d}r}{\mathrm{d}t} \text{ and } v_\theta = r\frac{\mathrm{d}\theta}{\mathrm{d}t}$$

On substituting the value of k in equation (4.13), the total energy

$$E = -\frac{Ze^2}{a(1-e^2)} + \frac{mk^2(1+e^2)}{2a^2(1-e^2)^2}$$

$$= -\frac{Ze^2}{a(1-e^2)} + \frac{m(1+e^2)}{2a^2(1-e^2)^2}\frac{Ze^2a(1-e^2)}{m}$$

$$= \frac{Ze^2}{2a}$$

The total energy is inversely proportional to the semi-major axis a of the ellipse. From equation (4.14), the semi-major axis is

$$a = \frac{k^2m}{Ze^2(1-e^2)} = \left(\frac{n_\theta h}{2\pi m}\right)^2 \frac{m}{Ze^2(n_\theta/n)^2}$$

$$= r_1\frac{n^2}{Z} \tag{4.15}$$

where we have used equations (4.10) and (4.12). Here, $r_1(\equiv h^2/4\pi^2me^2)$ is the Bohr radius. The semi-minor axis b of the ellipse is calculated with the help of the relation

$$b = \frac{an_\theta}{n} = \frac{r_1nn_\theta}{Z}$$

Using equation (4.15), the total energy,

$$E = -\frac{Ze^2}{2}\frac{4\pi^2me^2}{h^2}\frac{Z}{n^2} = -E_1\frac{Z^2}{n^2}$$

where $E_1(\equiv 2\pi^2me^4/h^2)$ is the amount of energy of electron in the first Bohr-orbit of hydrogen atom. Thus, the total energy of the Sommerfeld elliptical orbit is the same as that of the Bohr circular orbit.

Also in the Sommerfeld model n_θ can assume the values from 1 to n and n_r can assume the values from $(n - 1)$ to 0. The $n_\theta = 0$ is not possible as semi-minor axis cannot be zero because in that case the ellipse reduces to a straight line. Hence, for a given value of n, there can be n ellipses having equal semi-major axis, but different semi-minor axis. Though in the Bohr-Sommerfeld model, for a given value of n, we have n different elliptical orbits, but their energies are the same. Such energy levels are referred to as the **degenerate energy levels**. Since no new energy levels are obtained, there are no new spectral lines.

However, in an elliptical path, the velocity of electron varies from point to point. As the velocity of electron is quite large, so the mass of the electron should vary with the velocity according to the relation

$$m = \frac{m_0}{\sqrt{1-(v/c)^2}}$$

where m_0 is the rest-mass of electron and c the speed of light. With the variation of mass of electron in elliptical orbit, the path around the nucleus no longer remains a close and simple ellipse, but it becomes a complicated curve showing a precessing ellipse. After accounting for the relativistic correction for the variation of mass of electron, the total energy of the electron was found as:

$$E = -E_1 \frac{Z^2}{n^2}\left[1 + \frac{Z^2\alpha^2}{n^2}\left(\frac{n}{n_\theta} - \frac{3}{4}\right)\right] \tag{4.16}$$

where $\alpha(\equiv 2\pi e^2/hc) = 1/137$ is the fine-structure constant. In equation (4.17), higher order terms, which are not significant, are dropped. Now, the total energy not only depends on n, but also on n_θ. So, the energy level with a given value of n breaks into n components with $n_\theta = 1, 2, ..., n$. Selection rules for the radiative transitions between the level are $\Delta n_\theta = \pm 1$.

2.3 Bohr-Sommerfeld Model for H_α Line

First line of Balmer series, H_α, corresponds to a transition between the levels $n = 2$ and $n = 3$. According to the Bohr-Sommerfeld model, $n = 2$ level splits into two levels with $n_\theta = 1$ and 2, and $n = 3$ level splits into three levels with $n_\theta = 1$, 2 and 3. The wavenumbers of levels are

$$\bar{\nu} = \frac{R}{n^2}\left[1 + \frac{\alpha^2}{n^2}\left(\frac{n}{n_\theta} - \frac{3}{4}\right)\right] \tag{4.17}$$

where R is the Rydberg constant. The shift is given as

$$\Delta\bar{\nu} = -\frac{R\alpha^2}{n^4}\left(\frac{n}{n_\theta} - \frac{3}{4}\right)$$

Here, $R\alpha^2 = 5.8435718\ \text{cm}^{-1}$. Since $n_\theta \le n$, all shifts are negative. The shifts are given in the following table.

n	n_θ	$\Delta\bar{\nu}/R\alpha^2$	$\Delta\bar{\nu}(\text{cm}^{-1})$
3	3	−1/324	−0.0180
	2	−1/108	−0.0541
	1	−1/36	−0.1623
2	2	−1/64	−0.0913
	1	−5/64	−0.4565

The levels and splitting of levels is shown in figure. below, where the energies of the levels are not to the scale. Following the selection rule $\Delta n_\theta = \pm 1$, H_α line is found to have three components packed together.

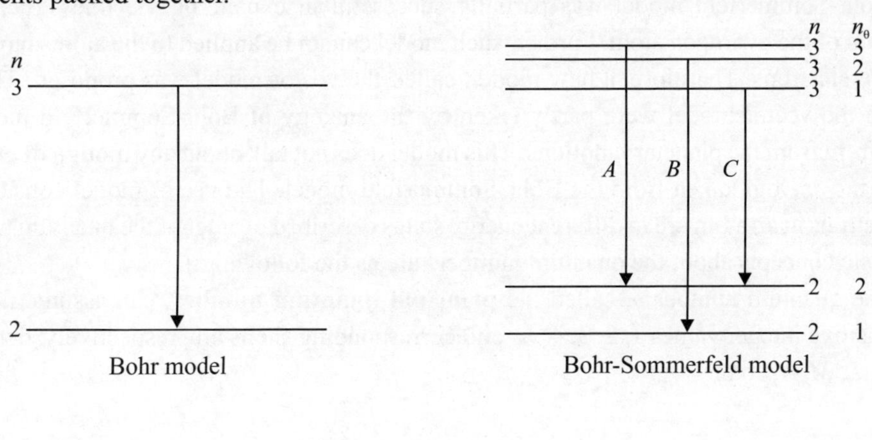

Since high resolution instruments found five components of H_α line, success of Bohr-Sommerfeld model was therefore partial.

Example 5.

Following the Bohr-Sommerfeld model, find out the number of components of first line of Paschen series.

Solution:

First line of Paschen series involves $n = 3$ and $n = 4$ principal quantum numbers. According to the Bohr-Sommerfeld model, $n = 3$ level splits into three levels with $n_\theta = 1, 2$ and 3, and $n = 4$ level splits into four levels with $n_\theta = 1, 2, 3$ and 4. This splitting is shown in the figure below, where the energy of the levels is not to the scale. Following the selection rule $\Delta n_\theta = \pm 1$, the first line of the Paschen series is found to have five components packed together.

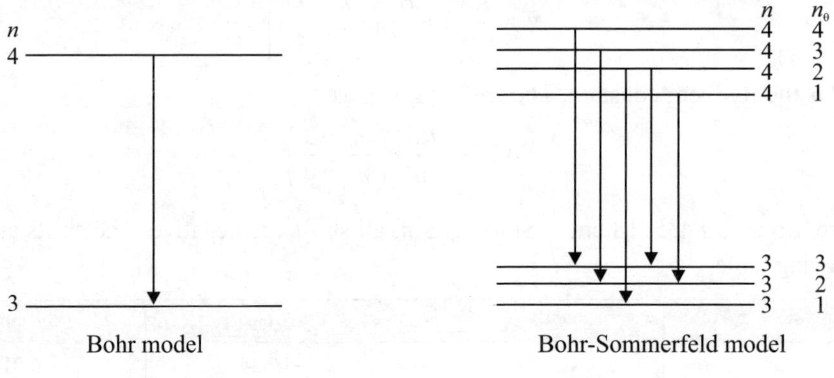

2.4 Effect of Finite Mass of Nucleus

In the Bohr-Sommerfeld model, effect of finite mass of nucleus can be accounted for by replacing the mass of electron m by reduced mass μ. Total energy of electron is given by equation (4.17).

3. Vector Model

The Bohr-Sommerfeld model was partially successful in explaining the fine structure line spectrum of the hydrogen atom. Further, such model cannot be applied to the atoms having two or more electrons. Therefore, a new model, called the vector model was proposed. The ideas used in the vector model were partly taken by the analogy of Bohr-Sommerfeld model and partly from from the planetary motions. This model does not talk about any motion of electrons, except the analogy taken from the Bohr-Sommerfeld model. The vector model considers that the electrons in an atom are in different energy states described by a set of the quantum numbers. The basic concepts about the quantum numbers are as the following:

(i) The quantum number n, called the **principal quantum number**, can assume non-zero, positive integer values 1, 2, 3, 4 .... and corresponding shells are, respectively, denoted by K, L, M, N, so on.

(ii) In analogy with the revolution of electron around a nucleus, an electron has an angular quantum number l, called the **azimuthal quantum number**, which can assume the values, ranging from zero to $(n - 1)$ with the increment of unity. The orbitals for $l = 0, 1, 2, 3, 4, 5, 6, \ldots$ are, respectively, denoted by $s, p, d, f, g, h, i, \ldots$.

(iii) Though the idea of space quantization was introduced in Bohr-Sommerfeld model, but in the vector model one needs a certain preferential direction about which the orbital angular momentum may get its orientation. Conventionally, preferential direction is taken as z-axis around which angular momentum l^* is assumed to rotate. The projection of l^* on the z-axis is denoted by m_l, called the **magnetic quantum number**.
The magnetic quantum number m_l is quantized and can assume the integer values ranging from $-l$ to l with the increment of unity. For example, for $l = 2$, m_l can assume the values $-2, -1, 0, 1, 2$. Hence, the m_l can assume $(2l + 1)$ values.

(iv) These three quantum numbers n, l and m_l could not suffice to explain observed spectra of atoms. Therefore, from the analogy of the planetary motion, spin of electron was introduced. Electron was assumed earlier to revolve around the nucleus and now electron is also assumed to spin about its own axis. Hence, the electron is now assumed to have spin angular momentum, denoted by s^*. Similar to l^*, according to quantum mechanics, the magnitude of s is $s^* = \sqrt{s(s+1)}$, where s is called the **spin quantum number**.
The projection of s^* on the z-axis is denoted by m_s, called the **spin magnetic quantum number**. The spin magnetic quantum number m_s is quantized and can assume the values ranging from $-s$ to s with the increment of unity. Hence, m_s can assume $(2s + 1)$ values. In order to explain the observed spectra of atoms, requirement was however to assign a value one-half to s. This non-integer value for s was not consistent with the normal quantization rules of integer values. But, this half-integer value for s was retained as it could explain the observed spectra of atoms. Hence, for $s = 1/2$, m_s can assume the values, $-1/2$ and $1/2$. The experimental proof for the spin of electron and its value one-half was later on given by Stern and Gerlach.

3.1 Stern-Gerlach Experiment

The concept of this experiment was initiated by Otto Stern in 1921 and the experiment was first successfully conducted by Walther Gerlach in 1922. The experiment demonstrated that the spatial orientation of the angular momentum is quantized. Thus, an atomic-scale system was shown to have intrinsically quantum properties.

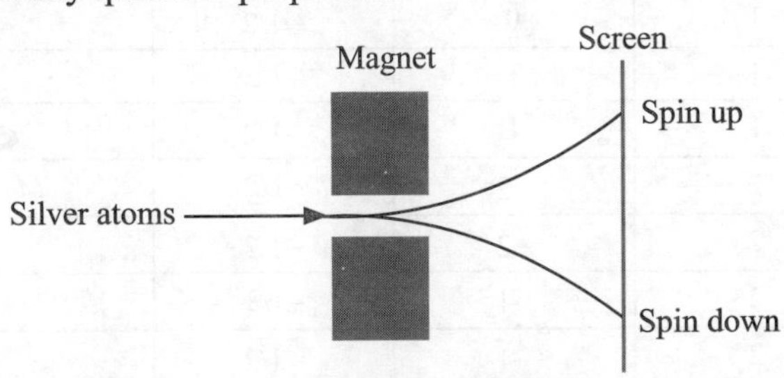

Fig. 5: Outline of the Stern-Gerlach experiment where the spatially varying magnetic field is applied.

In the original experiment, silver atoms were allowed to pass through a spatially varying magnetic field. Due to the magnetic field gradient, the particles having non-zero magnetic moment from a straight path. These deflected particles are finally incident on a detector screen, such as a glass slide.

The screen reveals discrete points of accumulation, rather than a continuous distribution, owing to their quantized spin. Historically, this experiment was decisive in convincing the physicists about the angular-momentum quantization in all the atomic-scale systems.

The experiment is normally conducted using electrically neutral particles such as the silver atoms. Electron configuration of the silver is $1s^2 2s^2 2p^6 3s^2 3p^6 4s^2 3d^{10} 4p^6 5s^1 4d^{10}$. There is only one unpaired electron $5s^1$. The spin of this electron can be up or down.

3.2 Pauli's Exclusion Principle

Table 1. Possible quantum states

Quantum states				Shell	Sub-shell
n	l	m_l	m_s		
1	0	0	+1/2	*K*	*s*
1	0	0	−1/2		
2	0	0	+1/2	*L*	*s*
2	0	0	−1/2		
2	1	−1	+1/2	*L*	*p*
2	1	−1	−1/2		
2	1	0	+1/2		
2	1	0	−1/2		
2	1	1	+1/2		
2	1	1	−1/2		
3	0	0	+1/2	*M*	*s*
3	0	0	−1/2		
3	1	−1	+1/2	*M*	*p*
3	1	−1	−1/2		
3	1	0	+1/2		
3	1	0	−1/2		
3	1	1	+1/2		
3	1	1	−1/2		
3	2	−2	+1/2	*M*	*d*
3	2	−2	−1/2		
3	2	−1	+1/2		
3	2	−1	−1/2		
3	2	0	+1/2		

3	2	0	−1/2		
3	2	1	+1/2		
3	2	1	−1/2		
3	2	2	+1/2		
3	2	2	−1/2		

All the electrons in an atom are identical in mass, charge and spin. According to the vector model, quantum state of an electron in an atom is specified by a set of four quantum numbers, n, l, m_l and m_s. On the basis of the experimental observations, Wolfgang Pauli in 1925 proposed a principle, called the Pauli's exclusion principle, regarding the occupancy of the quantum states by the electrons. This principle states that a given quantum state cannot be occupied by two or more electrons. In other words, the principle can be described as the following.

In an atom, no two electrons can be in the same quantum state.

It is equivalent to the following statement.

No two electrons can have the same set of four quantum numbers, n, l, m_l and m_s.

With the help of Pauli's exclusion principle, in a given atom, different quantum states can be assigned to electrons. The possible quantum states in different sub-shells and shells for n = 1, 2 and 3 are given in Table 1.

3.3 Aufbau Principle

Now, the point under consideration is about the configuration of electrons; whether all the sub-shells of a shell are filled first before going to the next shell or not. Answer to this question was given in the form of the 'aufbau principle', represented in following diagram. According to this principle, the sub-shells are filled following the arrows, starting from the top. For example, one can notice that the sub-shell 4*s* is filled before the sub-shell 3*d*.[4]

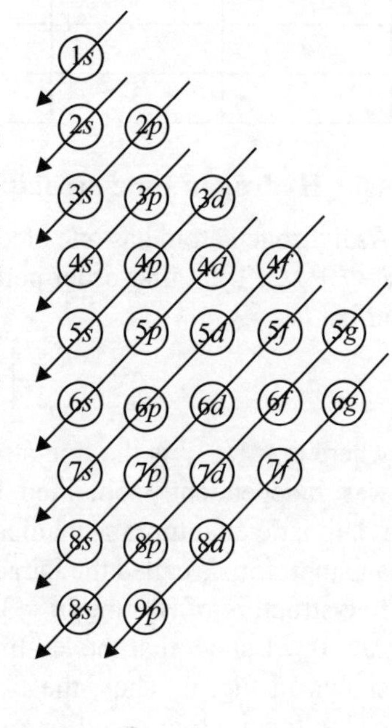

Following the aufbau principle, electron configuration in some atoms is given in Table 2.

However, there are some exceptions to the aufbau principle. Some examples of the exceptions are Chromium (Cr), Copper (Cu), Palladium (Pd), Silver (Ag), Gold (Au).

4. One Valence Electron Atoms

For one valence electron atom, its orbital and terms, due to spin-orbit interaction, are given Table 3.

[4]For these subshells if we take values of $(n + l)$, we have
1, 2. 3, 3, 4, 4, 5, 5, 5, 6, 6, 6, 7, 7, ...
That is the value of $(n + l)$ goes on increasing.

Table 2. Electron configuration

Atom	Symbol	Number of electrons	Electron configuration
Hydrogen	H	1	$1s^1$
Helium	He	2	$1s^2$
Carbon	C	6	$1s^2 2s^2 2p^2$
Nitrogen	N	7	$1s^2 2s^2 2p^3$
Oxygen	O	8	$1s^2 2s^2 2p^4$
Sodium	Na	11	$1s^2 2s^2 2p^6 3s^1$
Chlorine	Cl	17	$1s^2 2s^2 2p^6 3s^2 3p^5$
Iron	Fe	26	$1s^2 2s^2 2p^6 3s^2 3p^6 4s^2 3d^6$
Iodine	I	53	$1s^2 2s^2 2p^6 3s^2 3p^6 4s^2 3d^{10} 4p^6 5s^2 4d^{10} 5d^5$

Table 3. Terms for various orbitals

Orbital	l	s	L	S	J	Terms
s	0	1/2	0	1/2	1/2	$^2S_{1/2}$
p	1	1/2	1	1/2	1/2, 3/2	$^2P^0_{1/2}$, $^2P^0_{3/2}$
d	2	1/2	2	1/2	3/2, 5/2	$^2D_{3/2}$, $^2D_{5/2}$
f	3	1/2	3	1/2	5/2, 7/2	$2F^0_{5/2}$, $^2F^0_{7/2}$

4.1 Hydrogen Fine-structure

As hydrogen atom has one electron, we have $s = 1/2$ and j can assume the values $j = l - 1/2$ and $j = l + 1/2$. Total first-order perturbation correction due to spin-orbit interaction and relativistic effect is

$$\Delta\bar{\nu} = \frac{R\alpha^2}{n^3}\left[-\frac{1}{j+1/2}+\frac{3}{4n}\right]$$

where $\alpha = 1/137$ is the fine-structure constant. This equation was independently obtained by Dirac from a completely relativistic quantum mechanical treatment of hydrogen atom and therefore is called the Dirac shift. Let us now consider the fine-structure of H_α-line ($n = 3 \rightarrow n = 2$). The selection rules $\Delta j = 0, \pm 1$ show that the H_α-line splits into five components, shown in Fig. 6. Thus, the five components of the Hα-line of hydrogen atom could be explained after inclusion of the electron-spin.

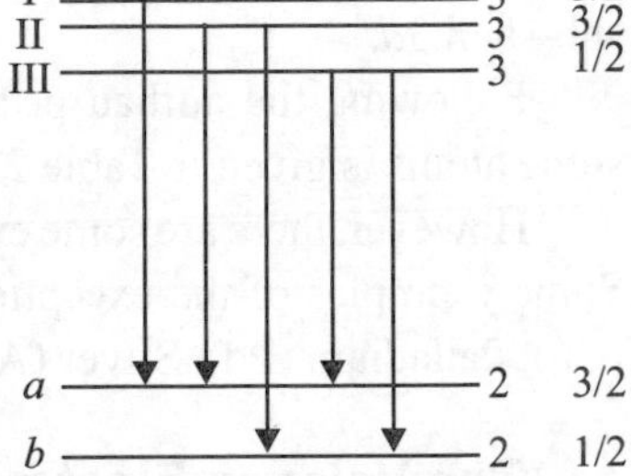

Fig. 6: Dirac transitions for H_α line (not to the scale).

5. Two Valence Electrons Atoms

We have to remind that the relativistic effect is important for hydrogen atom only. For two valence electrons in an atom, there are four angular momenta l_1^*, l_2^*, s_1^*, s_2^* which can have the following six possible interactions with the energies $\Delta\bar{v}_1$, $\Delta\bar{v}_2$, $\Delta\bar{v}_3$, $\Delta\bar{v}_4$, $\Delta\bar{v}_5$, $\Delta\bar{v}_6$, respectively.

s_1^* with s_2^* $\quad$ l_1^* with l_2^*, $\quad$ l_1^*, with s_1^* $\quad$ l_2^* with s_2^* $\quad$ l_1^* with s_2^* $\quad$ l_2^* with s_1^*

Depending on the electron configurations, in an atom, some of these energies are large as compared to others, whereas in other atoms, the different energies predominate. Based on the relative values of these interaction energies, two rather simple, ideal models have been developed. One or other of them gives fair agreement with the experiment in most cases. One of them is called the **LS-coupling** and other is called the **jj-coupling**.

In the *LS*-coupling, the interaction energies $\Delta\bar{v}_1$ and $\Delta\bar{v}_2$ are assumed to predominate over $\Delta\bar{v}_3$ and $\Delta\bar{v}_4$, while $\Delta\bar{v}_5$ and $\Delta\bar{v}_6$ are considered negligibly small. In the *jj*-coupling, the interaction energies $\Delta\bar{v}_3$ and $\Delta\bar{v}_4$ are assumed to predominate over $\Delta\bar{v}_1$ and $\Delta\bar{v}_2$, while $\Delta\bar{v}_5$ and $\Delta\bar{v}_6$ are again considered negligibly small. About these two coupling schemes, we have discussed here.

5.1 LS Coupling

This coupling is also known as the Russell-Saunders coupling. The resultant orbital quantum number *L*, obtained from the *ll*-coupling, and the resultant spin quantum number *S*, obtained from the *ss*-coupling, of two valence electrons in an atom combine in various possible relative orientations to give *J* whose value varies from $|L - S|$ to $(L + S)$ in the steps of unity. For the given values of *L*, *S* and *J*, the term is written as

$$(2S+1)_{L\,J}$$

Thus, we have

$$\Delta\bar{v}_3 + \Delta\bar{v}_4 = \frac{A}{2}[J(J+1) - L(L+1) - S(S+1)]$$

where
$$A = \frac{L(L+1)+l_1(l_1+1)-l_2(l_2+1)}{4L(L+1)}\,\frac{R\alpha^2 Z^4}{n_1^3 l_1(l_1+1)(l_1+1/2)} + \frac{L(L+1)+l_2(l_2+1)-l_1(l_1+1)}{4L(L+1)}\,\frac{R\alpha^2 Z^4}{n_2^3 l_2(l_2+1)(l_2+1/2)}$$

Two valence electrons have spin quantum numbers $s_1 = 1/2$ and $s_2 = 1/2$. The resultant spin quantum number *S* can have two values 0 and 1.

Two valence electrons with orbital quantum numbers l_1 and l_2, combine in various possible relative orientations. Resultant orbital quantum number *L* varies from $|l_1 - l_2|$ to $(l_1 + l_2)$ with the increment of unity. The terms corresponding to $L = 0, 1, 2, 3, 4, \ldots$ are, respectively, denoted by *S*, *P*, *D*, *F*, *G*, ...

5.2 *jj*-coupling

In the *jj*-coupling, for one electron, l_1^* combines with s_1^* to form j_1^*. For the second electron, l_2^* combines with 2_1^* to form j_2^*. For two valence electrons with the angular quantum numbers j_1 and j_2, these angular quantum numbers combine in various possible relative orientations. The resultant angular quantum number *J* which varies from $|j_1 - j_2|$ to $(j_1 + j_2)$. For the given values of j_1, j_2 and *J*, the term is written as

$$(j_1, j_2)_J$$

The energy shift $\Delta\bar{\nu}$ is expressed as

$$\Delta\bar{\nu} = \frac{b_1}{2}[j_1(j_1+1) - l_1(l_1+1) - 3/4] + \frac{b_2}{2}[j_2(j_2+1) - l_2(l_2+1) - 3/4] + b_3 j_1 j_2$$

with $b_1 = \frac{R\alpha^2 Z^4}{n_1^3 l_1(l_1+1)(l_1+1/2)}$ $b_2 = \frac{R\alpha^2 Z^4}{n_2^3 l_2(l_2+1)(l_2+1/2)}$

The expression for b_3 is not known.

Under the *ls*-coupling, for each electron separately, the orbital quantum number l and the spin quantum number s, combine together to give angular quantum number j. Since for an electron, $s = 1/2$, the angular quantum number j can have two values, $|(l - 1/2)|$ and $(l + 1/2)$.

6. Hyperfine Structure

After accounting for the spin of valence electrons, scientists could describe the fine-structure observed in an atomic spectrum. When the fine-structure lines were detected with an instrument having further good resolving power, each of them was found to have some components. These components are referred to as the **hyperfine-structure** (HFS) of the atomic spectrum. For explanation of the hyperfine-structure, scientists accounted for the nucleus of the atom. A nucleus consists of protons and neutrons, each of them having spin quantum number 1/2. Hence, a nucleus can have either half-integer or an integer value for its nuclear spin quantum number I.

For explanation of the HFS of an atomic spectrum, scientists introduced an additional vector $I^*[=\sqrt{I(I+1)}]$ in the vector model. Here, I is the nuclear spin quantum number when the nucleus is in its ground state. The magnetic HFS can be found only when $I \neq 0$. A nucleus consists of protons and neutrons, each of them having the spin 1/2, as they are fermions.

Hence, the quantum number I can assume either half integer or an integer value. Now, the quantum vector J^* is coupled with the quantum vector I^*, representing the total mechanical moment F^* of the atom. For a single valence electron, j^* and J^* are the same. For an atom having two or more valence electrons, the J^* is obtained either through the LS coupling or through the jj coupling.

6.1 21-cm Line of Hydrogen Atom

A radio hiss was noticed in 1930. Since it was found to vary on a daily cycle, it was concluded that the source of the radio hiss was extraterrestrial. It was noticed that the radio waves appeared to come from the centre of the galaxy. These discoveries were noted and published by Jan Oort in 1940. In 1944, Hendrik van de Hulst predicted that the neutral hydrogen could produce radiation at a frequency of 1420·4058 MHz due to two closely spaced energy levels in the ground state of hydrogen atom.

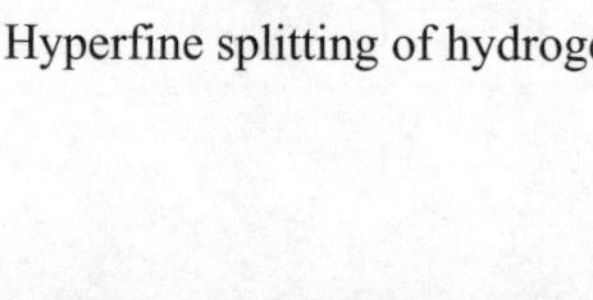

Fig. 7: Hyperfine splitting of hydrogen atom.

Hydrogen atom has one electron. Its ground state 1*s* has $l = 0$. Because of spin 1/2 of electron, the value of J is 1/2. The nucleus of hydrogen atom has one proton and thus, the nuclear spin $I = 1/2$. Owing to hyperfine interaction with the nuclear spin, the ground state level of hydrogen atom split into two components with hyperfine quantum number $F = 0$ and 1. In the free space, the wavelength, corresponding to the separation between these two levels, is 21.1061140542 cm which corresponds to the frequency of 1,420,405,751.7667 Hz. This wavelength falls within the microwave region of the electromagnetic spectrum, and it is observed frequently in the radio astronomy because as the Earth's atmosphere is transparent in this region.

This transition is highly forbidden with an extremely small transition rate (Einstein *A*-coefficient) of 2.9×10^{-15} s^1, and consequently, the mean radiative lifetime of the excited state is about 10 million years. A spontaneous occurrence of the transition cannot be seen in a laboratory on the Earth. But, it can be artificially induced using a hydrogen MASER. It is commonly observed in the astronomical settings, such as the hydrogen clouds in our galaxy and others.

The 21 cm line (1420.4 MHz) was first detected in 1951 by Ewen and Purcell at Harvard University. Then, the first maps of the neutral hydrogen in the galaxy were prepared. If we assume that the hydrogen atoms are uniformly distributed throughout the galaxy, each direction towards the galaxy would produce 21-cm hydrogen line. However, the difference between each of these lines would be the Doppler shift in each of these lines. Hence, one can calculate the relative speed of each arm of our galaxy. The rotation curve of our galaxy has been calculated using the 21 cm hydrogen line. It is therefore possible to use the plot of the rotation curve and the velocity to determine the distance to a certain point within the galaxy.

7. Emission and Absorption Lines

A spectral line is the result of interaction between a quantum system (atom, molecule or atomic nucleus) and a single photon. In the emission process, the system goes from an upper state to a lower state, according to the selection rules, a photon having energy equal to the energy-difference between the two states is emitted. In the absorption process, the system goes from a lower state to an upper state, according to the selection rules, after absorption of a photon having energy equal to the energy-difference between the two states. In the absorption, the system goes to higher energy state. Then, the energy is spontaneously re-emitted, either in the same frequency as the original or in a cascade, where the sum of the energies of the photons emitted is equal to the energy of the one absorbed photon (assuming that the system returns to its original state). A spectral line is thus a dark (absorption) or bright (emission) line in an otherwise uniform and continuous spectrum. The spectral lines are often used to identify the atoms and molecules. These observed lines can be compared with the previously collected lines of atoms and molecules, and are thus used to identify the atomic and molecular components of the stars and the planets.

The type of line (emission or absorption) depends on the kind of the material and its temperature relative to another emission source. The intensity of light, over a narrow frequency range, is reduced due to the absorption by the material and re-emission in the random directions. In emission, the intensity of light, over a narrow frequency range, increases due to emission by the material.

8. Line Widths and Doppler Shifts

It is known that no spectral line is strictly sharp, but has a finite width. Many lines are found to have observable magnitude of the breath. In general, a spectral line appears densest at its centre and it fades towards the edges. For convenience, the width of a spectral line is defined as the distance between the two points, one on each side of the centre, at which the intensity is half of that at the centre. It is therefore often called the **half intensity width**. The known mechanisms responsible for the line-broadening may be classified into two categories: (i) Homogeneous mechanisms and (ii) Inhomogeneous mechanisms.

(i) In the **homogeneous mechanisms**, the central frequency as well as the shape of a line generated by each atom in the gas is the same. That is, all atoms have equal contribution in the formation of line. Examples of such broadening are the natural broadening and collision broadening.

(ii) In the **inhomogeneous mechanisms**, the central frequency due to a group of atoms is shifted. Though each atom in the groups has the same central frequency as well as the same shape of the line, various groups of atoms have different central frequencies. Example of such broadening is the Doppler broadening.

8.1 Natural Broadening

According to the classical electromagnetic theory, motion of a vibrating charge is continuously damped due to the loss of energy. The damped harmonic motion of an electron can be expressed as

$$\frac{d^2y}{dt^2} + b\frac{dy}{dt} + \omega_0^2 y = 0$$

where $\nu_0 = \omega_0/2\pi$ is the characteristic frequency of vibration and b is damping coefficient. The electric field associated with such a damped oscillator is expressed as

$$E(t) = E_0\, e^{-(b/2)t}\, e^{-i\omega_0 t}$$

$$= E_0\, e^{-(i\omega_0 + b/2)t}$$

The frequency spectrum of amplitude is given by the Fourier transform

$$E(\omega) = \frac{1}{2\pi}\int_0^{\infty} E_0 \exp\left[-\frac{b}{2} - i\omega_0 + i\omega\right] t\, dt$$

$$= \frac{E_0}{2\pi}\frac{1}{i(\omega_0 - \omega) + b/2}$$

The intensity distribution is therefore,

$$I(\omega) = |E(\omega)|^2 = \frac{E_0^2}{4\pi^2}\frac{1}{i(\omega_0 - \omega)^2 + b^2/4}$$

Hence, the line shape function is

$$g(\omega) = \frac{K}{(\omega_0 - \omega)^2 + b^2/4}$$

where K is a constant which can be determined by applying the condition of normalization as the following.

$$1 = \int_{-\infty}^{\infty} g(\omega)\mathrm{d}\omega = K\int_{-\infty}^{\infty} \frac{\mathrm{d}\omega}{(\omega_0 - \omega)^2 + b^2/4}$$

Therefore,

$$1 = \frac{2K}{b}\left[\tan^{-1}\frac{2(\omega_0 - \omega)}{b}\right]_{-\infty}^{\infty} = \frac{2K\pi}{b}$$

and thus we have $K = b/2\pi$. The line shape function is therefore

$$g(\omega) = \frac{b}{2\pi}\frac{1}{(\omega_0 - \omega)^2 + b^2/4}$$

or for the frequency, we have

$$g(\nu) = \frac{b}{2\pi}\frac{1}{4\pi^2(\nu_0 - \nu)^2 + b^2/4} \tag{4.18}$$

The line-shape of such kind is called the **Lorentzian line shape** and is shown in Fig. 8.

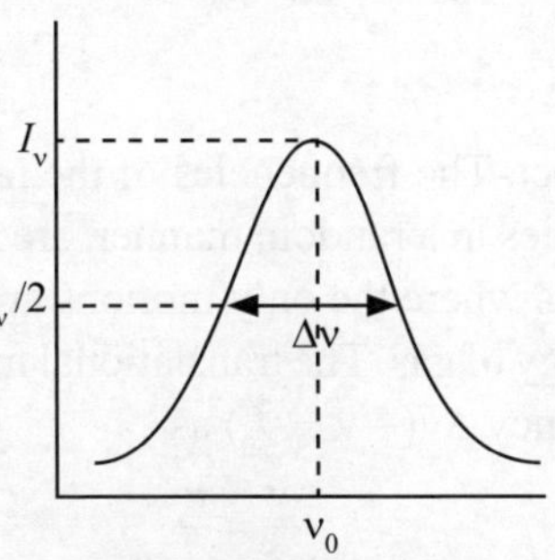

Fig. 8: Lorentzian line shape

Equation (4.18) suggests that the maximum intensity is $2/\pi b$ and the intensity drops to half of the maximum value when the two terms in the denominator are equal. That is,

$$|\nu - \nu_0| = \frac{b}{4\pi}$$

Thus, the natural half-intensity width is

$$\Delta\nu = 2|\nu - \nu_0| = \frac{b}{2\pi}$$

The natural width, which is classically attributed to the damping of radiation, is quantum mechanically regarded as due to the finite lifetimes of energy levels and consequent uncertainties in their energy values. When ΔE denotes the spread in energy value and Δt is the mean lifetime, then we have the uncertainty relation

$$\Delta E\ \Delta t = \hbar$$

For the ground state, Δt is very large and therefore, the uncertainty in energy is small. Also, for a metastable level, Δt is large and these states are also very sharp. For an ordinary level, $\Delta t \approx 10^{-8}$ s and the states have the finite widths. For a transition between two levels at energies E_1 and E_2, the uncertainties ΔE_1 and ΔE_2 can be calculated from their lifetimes and with the help of the uncertainty principle. The line width $\Delta \nu$ is therefore

$$\Delta_\nu = \frac{\Delta E_1 + \Delta E_2}{\hbar}$$

Using $\Delta E \Delta t = \Delta E \tau = \hbar$, where τ represents the lifetime of level, we have

$$\Delta \nu = \frac{1}{\tau_1} + \frac{1}{\tau_2}$$

Here, τ_1 and τ_2 are the radiative lifetimes of the two levels[5]. Using $\nu = c/\lambda$ so that $\Delta \nu = -c\Delta\lambda/\lambda^2$, we get

$$\Delta \lambda = \frac{\lambda^2}{c}\left(\frac{1}{\tau_1} + \frac{1}{\tau_2}\right)$$

Radiative lifetime of a level is related to the Einstein A-coefficients of the radiative transitions in the downward direction from that level as

$$\tau_j = 1\Big/\sum_i A_{ji} \qquad i < j$$

8.2 Doppler Broadening

It arises due to the Doppler effect. The frequencies of the radiation from the atoms, which are moving with the thermal velocities in a random manner, are modified. The Doppler broadening can be calculated easily for a gas where the only motions are the random motions of the atoms associated with the thermal energy of gas. The translational motion of an atom emitting radiation produces a change in the frequency $\Delta\nu(= \nu - \nu_0)$ as

$$\frac{\Delta \nu}{\nu_0} = \frac{u}{c} \tag{4.19}$$

where u is the velocity component for motion of atom along the line-of-sight, ν_0 the rest frequency and ν the observed frequency. According to the Maxwell-Boltzmann distribution, the number of particles having velocity between u and $u + du$ is

$$\mathrm{d}n = n_0\left[\frac{M}{2\pi kT}\right]^{1/2} \exp\left[-\frac{Mu^2}{2kT}\right]\mathrm{d}u \tag{4.20}$$

where M is the molecular weight, k the Boltzmann constant and T the kinetic temperature of gas. The intensity I of the line obtained as a function of ν is

[5]There can be other kinds of lifetimes, such as the collisional lifetime, which depends on the collisional rates and the density of the colliding partners.

$$I(\nu) = I_0 \exp\left[-\frac{Mc^2}{2kT\nu_0^2}(\nu - \nu_0)^2\right]$$

Here, we have used equation (4.19). The shape of such a frequency distribution is called **Gaussian line shape** and is shown in Fig. 9. A comp`1arison of the Lorentzian and the Gaussian shapes give that the Gaussian shape is sharper than the Lorentzian shape.

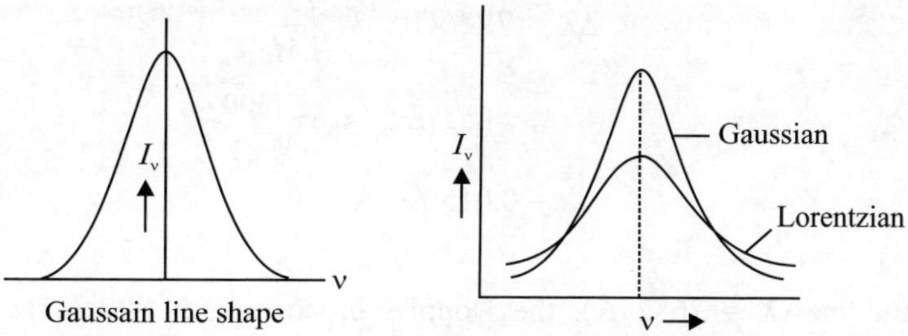

Fig. 9: Gaussian line shape. Comparison of the Gaussian and the Lorentzian line shapes

Now, we have

$$\frac{I_0}{I} = \exp\left[\frac{Mc^2}{2kT\nu_0^2}(\nu - \nu_0)^2\right]$$

According to the half-intensity width, we have

$$\exp\left[\frac{Mc^2}{2kT\nu_0^2}(\Delta\nu)^2\right] = 2$$

We have expressed $\Delta\nu = |\nu - \nu_0|$. Thus, we have

$$\Delta\nu = \left(\frac{\ln 2}{Mc^2/2kT\nu_0^2}\right)^{1/2}$$

The half-intensity width is

$$\Delta\nu = 2\left(\frac{\ln 2}{Mc^2/2kT\nu_0^2}\right)^{1/2} = 1.67\frac{\nu_0}{c}\left(\frac{2kT}{M}\right)^{1/2}$$

Using $\nu = c/\lambda$ so that $\Delta\nu = -c\Delta\lambda/\lambda^2$ and thus, the half-intensity width in terms of the wavelength is

$$\Delta\lambda = 1.67\frac{\lambda_0}{c}\left(\frac{2kT}{M}\right)^{1/2}$$

Using $c = 3 \times 10^{10}$ cm/s, $k = 1.38 \times 10^{-16}$ erg/K and 1 amu = 1.67×10^{-24} gm (or $c = 3 \times 10^8$ m/s, $k = 1.38 \times 10^{-23}$ J/K and 1 amu = 1.67×10^{-27} kg), we get

$$\Delta\lambda = 7.16 \times 10^{-7}\lambda_0\sqrt{\frac{T}{M}}$$

and

$$\Delta\nu = 7.16 \times 10^{-7}\nu_0\sqrt{\frac{T}{M}}$$

where T is the temperature in kelvin and M is the molecular weight in amu.

Example 6.

For a sodium line (λ = 5893 Å), calculate the Doppler broadening at room temperature of 300 K.

Solution:

For the sodium $M = 23$. The line broadening is

$$\Delta\lambda = 7.16 \times 10^{-7} \lambda_0 \sqrt{\frac{T}{M}}$$

$$= 7.16 \times 10^{-7} \times 5893 \sqrt{\frac{300}{23}}$$

$$= 0.015 \text{ Å}$$

Example 7.

For a potassium line (λ = 7682 Å), the Doppler broadening is 800 MHz. Calculate the temperature of potassium vapour. Atomic mass of potassium is 39.

Solution:

We have

$$\Delta\nu = 7.16 \times 10^{-7} \frac{c}{\lambda_0} \sqrt{\frac{T}{M}}$$

Therefore,

$$800 \times 10^6 = 7.16 \times 10^{-7} \frac{3 \times 10^{10}}{7682 \times 10^{-8}} \sqrt{\frac{T}{39}}$$

Hence,

$$\sqrt{\frac{T}{39}} = 2.861$$

and

$$T = 319.2 \text{ K}$$

8.3 Collision Broadening

One of the important external causes in line broadening is the collision damping. This effect is produced due to the collisions of two atoms, one of which is in the process of emitting or absorbing the radiation. This phenomenon of the collision broadening is based on the assumption that when during the emission or absorption of radiation of frequency ν_0 by an atom, the atom collides elastically with another atom, there is probability that the phase and amplitude of radiation may change considerably.

In the classical treatment of the collision damping, we assume that the radiation is an undamped oscillator which is expressed as

$$x = A_0 \cos(\omega_0 t + \phi)$$

Suppose, τ is interval between the time when the atom begins radiating and the time when a collision takes place. Now, for each atom of gas, τ has a different value, so that for the gas as a whole there is a certain probability distribution of τ's.

If we assume that the distribution of τ's is the same as taken generally for the free path of atoms of the same gas, the probability of τ lying between τ and $\tau + d\tau$ is

$$dw = \frac{1}{\tau_0} e^{-\tau/\tau_0} d\tau$$

where τ_0 is mean interval, i.e., the average time between the two successive collisions. When the Fourier analysis of the set of wave trains is made, the relative intensity I as a function of frequency ν is obtained as

$$I(\nu) = \text{constant} \frac{1}{4\pi^2(\nu_0 - \nu)^2 + 1/\tau_0^2}$$

The intensity drops to its half-value when

$$|\nu_0 - \nu| = \frac{1}{2\pi\tau_0}$$

It gives the collision damping half intensity width as

$$\delta = 2|\nu_0 - \nu| = \frac{1}{\pi\tau_0}$$

We can now find the order of magnitude of the line broadening due to collision damping in the following manner. We need to find a value of τ_0. According to the kinetic theory of gases, the average velocity of an atom or molecule is

$$v_0 = \sqrt{\frac{8RT}{\pi M}}$$

where R is the gas constant, T the kinetic temperature of gas and M the molecular weight. The mean free length l_0 is

$$l_0 = v_0 \tau_0$$

Thus, the half-intensity breath is

$$\delta = \frac{1}{\pi\tau_0} = \frac{1}{\pi l_0}\sqrt{\frac{8RT}{\pi M}}$$

Review Questions

1. Describe the Bohr model for hydrogen-like atom. Discuss the effect of the finite mass of nucleus.
2. Describe the Bohr-Sommerfeld model for hydrogen atom. Show that it could explain partly the fine structure of spectral line.
3. Describe the vector model of atoms.
4. Discuss about the spectrum of one valence electron atoms.
5. Discuss about the spectrum of two valence electrons atoms.
6. Discuss about the 21 cm line of hydrogen atom.
7. Discuss about the line widths and Doppler shifts.
8. Write notes on the following:
   (i) Vector model (ii) Pauli's exclusion principle
   (iii) 21 cm line of hydrogen atom (iv) Stern-Gerlach experiment

(v) Aufbau principle
(vi) LS coupling
(vii) *jj*-coupling
(viii) Hyperfine structure
(ix) Emission and absorption lines
(x) Natural broadening
(xi) Doppler broadening
(xii) Collision broadening

Multiple-Choice Questions

1. H_α line of hydrogen atom was found to have how many fine-structure components?
   A. 3 B. 4 C. 5 D. 6 **Ans. C**
2. 21 cm spectral line is produced by
   A. H_2 molecule B. CO C. H_2O D. H atom **Ans. D**

5 Molecules in Interstellar Medium

Molecule is an entity having two or more atoms. CO, H_2O, CO_2 are examples of molecules. We are aware that the universe is expanding continuously. Though the boundaries of the universe are not known, but scientists believe that it contains several hundred billion (10^{11}) galaxies and our galaxy, Milky Way, is one of them. Estimation is that the Milky Way has about hundred billion (10^{11}) stars, and our Sun is one of them. It shows that our solar system is very small as compared to the universe. The medium between the stars is known as the interstellar medium (ISM). In some parts of the ISM, the molecules have been found. In this chapter, we have discussed about the molecules in the ISM.

1. Opinion of Scientists

Because of the γ-rays, x-rays present in the ISM, majority of scientists was against the existence of molecules in the ISM. A section of scientists, however, had opinion that in some parts of the ISM, the temperature may be so low that the molecules may exist there. As the molecules do not emit radiations lying in the optical window, their presence in the ISM could not be noticed directly. In 1932, Karl Jansky discovered that the radiations of long wavelengths also reach the Earth's surface from the ISM.

Fig. 1: Antenna of the Karl Jansky (left) and Transparency of Earth's atmosphere (right).

(*Courtesy*: Google website)

This discovery opened a new astronomical world, called the **radio window** (RW), having a range from few millimetres to few decameters in wavelength. As the transitions between the rotational levels in the molecules produce radiations lying in the RW, scientists opined that if the molecules exist in the ISM, they could be detected now through their rotational transitions. But, for three decades after the discovery of RW, no molecules were detected. It supported the scientists who were against the existence of the molecules in the ISM.

Table 1. Molecules identified in the interstellar space

Molecules of two atoms:
AlCl, CO, CP, CS, H_2, HCl, N_2, NH, AlF, C_2, CH, CH^+, CN, KCl, NO, NaCl, NP, SO^+, SiN, NS, OH, SO, SiC, SiO, SiS, CO^+, HF, PO, FeO, SH, TiO, SH^+, CF^+, FeH, OH^+, CN–, AlO, O_2, H_2^+, MgH, ND, NaH, O_2^+, HCl^+, ArH^+, CeO
Molecules of three atoms:
C_2H, C_2S, H_3^+, HCN, HNC, HCO, DCO^+, HNO, H_2O, N_2H^+, OCS, SO_2, HCS^+, *c*-SiC_2, NaCN, NaOH, HDO, C_3, CO_2, CH_2, N_2O, CNO, MgCN, HOC^+, MgNC, D_2H^+, HCP, CCD, DCN, DNC, AlNC, SiCN, SiNC, N_2D^+, H_2O^+, H_2Cl^+, FeCN, KCN, HO_2, HCO^+, D_2S, CCP, D_2O, OCN–, NH_2, CCO, CO_2^+, TiO_2, AlOH, HNSi
Molecules of four atoms:
l-C_3H, *c*-C_3H, C_3O, $HCNH^+$, H_2CO, C_3N–, C_3S, H_2CS, HNCS, H_3O^+, NH_3, HNCO, D_2CO, HC_2N, H_2CN, NH_2D C_2H_2, CH_3, C_4, HDCO, HDCS, ND_2H, NCCN, HSCN, C_2HD, HOOH, $HOCO^+$, HCNO, ND_3, PH_3, *l*-C_3H^+, D_2CS, HCO_2^+, H_2SiO, HMgNC, C_3N
Molecules of five atoms:
C_4Si, CH_2CN, HCOOH, HC_3N, H_2CCO, H_2NCN, CH_4, *c*-C_3H_2, SiH_4, C_3HD, C_4D, C_5, H_2COH^+, HCCNC, HNCCC, DC_3N, CH_3D, HCCCN, CH_2NH, H_2CCC, *l*-C_3H_2, CH_3O, HNCNH, CNCHO, C_4H–, GeH_4, NH_3D^+
Molecules of six atoms:
C_2H_4, C_5H, CH_3CN, CH_3OH, CH_3SH, HC_2CHO, H_2C_4, NH_2CHO, HC_3NH^+, CH_3OD, C_5N, H_2CCCC, HC_4N, E-HNCHCN, *c*-H_2C_3O, CH_2CNH, CD_3OH, CH_2DOH, CHD_2OH, CH_3NC, C_5N–
Molecules of seven atoms:
HC_5N, C_6H, CH_3NH_2, *c*-C_2H_4O, CH_3C_2H, CH_3CHO, CH_3CCD, HC_4OH, C_6H–, HC_4CN, C_3H_4, HCO_2NH_2, CH_2DCCH, DC_5N, HCCCCCN, CH_2CHCN, H_2C_5, H_2CHOH, *c*-C_2H_5
Molecules of eight atoms:
CH_3C_3N, $HCOOCH_3$, CH_2CHCHO, CH_3COOH, $(NH_2)_2CO$, C_7H, C_3H_3NO, H_2C_6, CH_2OHCHO, $HCOCH_2OH$, CH_3CHNH, C_4H_4, C_2H_6, NH_2CH_2OH, CH_2CCHCN, H_2NCH_2CN, H_2CCHCN, CH_3C_2NC
Molecules of nine atoms:
CH_3C_4H, $(CH_3)_2O$, HC_7N, C_2H_5OH, C_8H, CH_3CH_2CN, $H(C{\equiv}C)_3CN$, C_8H–, CH_3CONH_2, CH_3CH_2OH, CH_3CHCH_2, CH_3CH_2SH
Molecules of ten and more atoms:
$HC_{11}N$, $(CH_3)_2CO$, CH_3C_5N, C_{60}, $C_{14}H_{10}$, $C_{16}H_{10}$, CH_3C_6H, CH_3COCH_2OH, C_{60}^+, C_{24}, $(CH_2OH)_2$, C_{70}, C_6H_6, HC_8CN, CH_3COOCH_3, C_3H_7CN, HC_9N, C_3H_8, CH_3CH_2CHO, $C_{14}H_{10}^+$, C_2H_5OCHO, *c*-C_4H_5N, NH_2CH_2COOH

Weinreb et al., Nature 200, 829-831 (1963) discovered the first molecule OH in ISM through its 18 cm radiation. With the detection of OH molecule in ISM, scientists were astonished as their opinion shattered down. However, they got interested in search of more molecules in the ISM and at least 252 molecules, given in Table 1, have been identified in the ISM.

2. Cosmic Molecular Laboratories

In the ISM, some regions have been found to have varieties of molecules. These regions may be called the cosmic molecular laboratories. Names and positions of some of them are given in Table 2.

Table 2. Cosmic molecular laboratories

Object	Right ascension	Declination
Orion Nebula	$05^h\ 35^m\ 17.3^s$	−05 23′ 28″
IRC+10216	$09^h\ 47^m\ 57.406^s$	13° 16′ 43.56″
Epsilon Persei	$03^h\ 57^m\ 51.232^s$	40° 00′ 36.7752″
HD 124314	$14^h\ 15^m\ 1.606^s$	−16° 42′ 24.38″
Zeta Persei	$03^h\ 54^m\ 7.922^s$	31° 53′ 1.0812″
Zeta Ophichuhi	$16^h\ 37^m\ 9.539^s$	−10° 34′ 1.52″
Rho Ophiuchi	$16^h\ 28^m\ 06^s$	−24° 32′ 30″
NGC 7822	$0^h\ 3^m\ 00^s$	68° 37′ 00″
TMC 1	$4^h\ 38^m\ 38.6^s$	25° 36′ 18″
B1-b Cloud	$3^h\ 33^m\ 20.8^s$	31° 07′ 34″
L134 (Dark Cloud)	$15^h\ 51^m\ 26^s$	−2° 43′ 31″
IC 443G Nebula	$06^h\ 13^m\ 42^s$	22° 33′ 30″
Sgr B2	$17^h\ 44^m\ 9.5^s$	−28° 21′ 20″
W51	$19^h\ 21^m\ 27^s$	14° 24′ 30″
NGC 7027	$21^h\ 07^m\ 1.7^s$	42° 14′ 11″
V Y Canis Majoris	$7^h\ 22^m\ 58.32^s$	−25° 46′ 03.23″
16 Cygni B	$19^h\ 41^m\ 48.95^s$	50° 31′ 30.2″
NGC 6210	$16^h\ 44^m\ 29.5^s$	23° 47′ 59.7″
IRAS 16293-2422	$16^h\ 32^m\ 22.6^s$	−24° 28′ 33.0″
W 49	$19^h\ 10^m\ 17^s$	9° 06′ 00″
NGC 1952	$5^h\ 34^m\ 30^s$	22° 01′ 00″
NGC 2024	$5^h\ 41^m\ 54^s$	−01° 51′ 00″
NGC 2264	$06^h\ 41^m\ 06^s$	09° 53′ 00″
Rosette Nebula	$06^h\ 33^m\ 45^s$	04° 59′ 54″
SK 66-169	$05^h\ 36^m\ 54.50^s$	−66° 38′ 25″
NGC 1333	$3^h\ 29^m\ 02^s$	31° 20′ 54″
NGC 1333 IRAS 4A	$3^h\ 29^m\ 10.49^s$	31° 13′ 31″
NGC 1333 IRAS 2	$3^h\ 28^m\ 49.7^s$	31° 14′ 47″
NGC 4945	$13^h\ 05^m\ 27.5^s$	−49° 28′ 06″
NGC 7538	$23^h\ 13^m\ 45.7^s$	61° 28′ 21″
W 33A	$18^h\ 14^m\ 14.4^s$	−17° 55′ 50″
CRL 618	$4^h\ 42^m\ 53.64^s$	36° 00′ 53.5″
GL 2591	$20^h\ 27^m\ 35.8^s$	40° 01′ 14″
CLD 2	$4^h\ 38^m\ 21^s$	25° 41′ 00″
Horse head Nebula	$5^h\ 40^m\ 59^s$	−02° 27′ 30″

Benard 1	$3^h\ 33^m\ 20.8^s$	31° 7′ 34″
IRAS 04368+2557	$4^h\ 39^m\ 53.59^s$	26° 03′ 5.5″
Cernis 52	$3^h\ 43^m\ 0.3^s$	31° 58′ 26″
NGC 7023	$21^h\ 00^m\ 05^s$	68° 10′ 00″
DR21(OH)	$20^h\ 39^m\ 02^s$	42° 19′ 38″
Red rectangle nebula	$6^h\ 19^m\ 58.21^s$	−10° 38′ 14.69″
HD 27778	$4^h\ 23^m\ 59.76^s$	24° 18′ 3.563″
Cassiopeia A	$23^h\ 23^m\ 26^s$	58° 48′ 00″
NGC 6334	$17^h\ 19^m\ 58^s$	−35° 57′ 47″
55 Cygni	$20^h\ 48^m\ 56.29^s$	46° 06′ 50.88″
W3 (OH)	$2^h\ 27^m\ 4.1^s$	61° 52′ 22″
Cygni OB2-12	$20^h\ 32^m\ 40.95^s$	41° 14′ 29.28″

3. Classification of Molecules

Rotation of a molecule, in general, can be described in terms of three moments of inertia, I_a, I_b, I_c (such that $I_a \le I_b \le I_c$), about three orthogonal inertial axes, *a*, *b*, *c*, respectively, passing through the centre-of-mass of the molecule[1]. Based on the principal moments of inertia, the molecules may be classified into three main categories:

(i) Spherical top molecules (ii) Symmetric top molecules

(iii) Asymmetric top molecules

Molecules may also be classified in terms of the polar and non-polar molecules.

Corresponding to the moments of inertia, I_a, I_b, I_c, rotational constants *A*, *B*, *C* are expressed as

$$A = \frac{h}{8\pi^2 I_a c}$$

$$B = \frac{h}{8\pi^2 I_b c}$$

$$C = \frac{h}{8\pi^2 I_c c}$$

such that $A \ge B \ge C$.

3.1 Spherical Top Molecules

When all the three moments of inertia of molecule are equal, $I_a = I_b = I_c$, it is known as the spherical top molecule. Examples of such molecules are CH_4, SF_6, CCl_4, etc.

3.2 Symmetric Top Molecules

When two of the three moments of inertia of molecule are equal, it is known as the symmetric top molecule. Depending on which two moments of inertia are equal, the symmetric top molecules are of two types: the oblate symmetric top molecules, and the prolate symmetric top molecules.

[1]About three principal inertial axis, moments of inertia of molecule are calculated. The axis corresponding to the lowest moment of inertia is designated as the *a*-axis; the axis corresponding to the largest moment of inertia is designated as the *c*-axis, and the remaining axis corresponding to the intermediate moment of inertia is designated as the *b*-axis.

Oblate symmetric top molecules

When $I_a = I_b < I_c$, the molecule is known as the oblate symmetric top. Examples of such molecules are BF_3, BCl_3, etc. The *c*-axis is known as the axis of symmetry and it is used as the preferential direction in quantum mechanics. The projection of rotational quantum number *J* on the axis of symmetry is denoted by k_c. The k_c can assume the values ranging from $-J$ to $+J$ with the increment of unity. The rotational energy levels are denoted by J_{k_c}. For example, for $J = 2$, we have $k_c = -2, -1, 0, 1, 2$, and the levels are denoted as $2_{-2}, 2_{-1}, 2_0, 2_1, 2_2$.

Prolate symmetric top molecules

When $I_a < I_b = I_c$, the molecule is known as the prolate symmetric top. Examples of such molecules are CH_3Cl, CH_3F, CH_3CH, NH_3, etc. The *a*-axis is known as the axis of symmetry and it is used as the preferential direction in quantum mechanics. The projection of the rotational quantum number *J* on the axis of symmetry is denoted by k_a. The k_a can assume the values ranging from $-J$ to $+J$ with the increment of unity. The rotational energy levels are denoted by J_{ka}. For example, for $J = 2$, we have $k_a = -2, -1, 0, 1, 2$, and levels are denoted as $2_{-2}, 2_{-1}, 20, 2_1, 2_2$.

A special category of the prolate symmetric top molecules is the linear molecules. When all the atoms of a given molecule are aligned along a straight line, the molecule is known as the linear molecule. The moment of inertia about the molecular-axis is obviously zero, whereas the moments of inertia about other two mutually perpendic- ular axes, and perpendicular to the molecular-axis are equal. Thus, we have $I_a = 0$, and $I_b = I_c \neq 0$. Examples of linear molecules are CO, H_2, CO_2, OCS, HCN, C_2H_2, etc.

All the diatomic molecules obviously belong to the category of the linear molecules. The diatomic molecules are further classified into two categories:

(i) Homonuclear molecules (ii) Heteronuclear molecules

In a homonuclear diatomic molecule, both the atoms are the same. Examples of such molecules are, H_2, O_2, N_2, etc. In these molecules, the centre-of-mass, centre of positive charges, and centre of negative charges coincide with each other and lie at the mid-point on molecular bond. These molecules obviously have no permanent electric dipole moment.

In a heteronuclear diatomic molecule, both the atoms are different from each other. Examples of such molecules are, CO, OH, SiO, etc. In these molecules, the centre of positive charges does not coincide with the centre of negative charges, and therefore, they have permanent electric dipole moment.

3.3 Asymmetric Top Molecules

When all the three moments of inertia of molecule are unequal, $I_a < I_b < I_c$, it is known as the asymmetric top molecule. In these molecules, there is no preferential direction and any of *a*, *b*, *c* can be used as a preferential direction. Therefore, for the treatment of asymmetric top molecules, there are six representations (discussed later on). Examples of such molecules are H_2O, H_2CO, CH_3OH, CH_2CHCl, etc. Large number of molecules are the asymmetric top molecules.

3.4 Polar and Non-polar Molecules

Depending on whether a molecule has permanent electric dipole moment or not, the molecules can also be divided into two categories:

(i) polar molecules (ii) non-polar molecules

A large number of molecules are polar in nature. Examples of non-polar molecules are homonuclear diatomic molecules (H_2, O_2, etc.), C_6H_6, CH_4, etc. In general, only polar molecules interact with the electromagnetic radiation. The electric dipole moment of a molecule interacts with the electric vector of the electromagnetic radiation.

Depending on the direction of the electric dipole moment, asymmetric top molecules are called the *a*-type, *b*-type, etc. For example, H_2CO is *a*-type asymmetric top molecule, as its electric dipole moment lies along the *a*-axis of inertia; H_2O is *b*-type asymmetric top molecule, as its electric dipole moment lies along the *b*-axis of inertia.

Example 1.

Using the coordinates given in Å of atoms in H_2CC molecule, show that H_2CC is *a*-type asymmetric top molecule.

Atom	*x*	*y*	*z*
C	0.0	0.8163	0.0
C	0.0	–0.4747	0.0
H	0.9366	–1.0251	0.0
H	–0.9366	–1.0251	0.0

Solution:

All the atoms of H_2CC molecule lie in the *xy*-plane. The moment of inertia of molecule (comprising of four atoms H, H, C, C) about the *x*-axis is

$$I_x = \Sigma m_i y_i^2 = 12(0.8163)^2 + 12(-0.4747)^2 + 2 \times 1(-1.0251)^2$$
$$= 12.802 \text{ amu Å}^2$$

The moment of inertia of the molecule about the *y*-axis is

$$I_y = \Sigma m_i x_i^2 = 1(0.9366)^2 + 1(-0.9366)^2 = 1.754 \text{ amu Å}^2$$

Following the theorem of the perpendicular axis for the moment of inertia, we have moment of inertia of molecule about the *z*-axis as

$$I_z = I_x + I_y = 12.802 + 1.754 = 14.556 \text{ amu Å}^2$$

Thus, the *y*-axis is the *a*-axis of inertia, the *x*-axis is the *b*-axis of inertia and the *z*-axis is the *c*-axis of inertia. As the $I_a \neq I_b \neq I_c$, the H_2CC is asymmetric top molecule. The electric dipole moment would be along the axis of symmetry. Since electric dipole moment of molecule is along the *a*-axis of inertia, it is *a*-type asymmetric top molecule.

4. Rigid Diatomic Molecule

As an approximation, a diatomic molecule may be considered as a rigid rotator consisting of two atoms with masses m_1 and m_2 connected by a rigid bond of length *r* (Fig. 2).

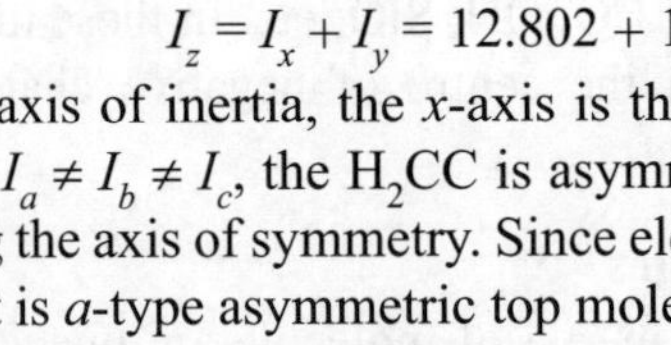

Fig. 2: A rigid diatomic molecule having two atoms of masses m_1 and m_2, separated by a distance *r*.

Let us consider rotation of this rigid rotator about an axis perpendicular to the bond of the molecule and passing through the centre-of-mass. The solution of Schrödinger equation for this rigid rotator gives the energy

$$E = \frac{P^2}{2I} = \frac{\hbar^2}{2I} J(J+1)$$

where P is the angular momentum, J the rotational quantum number and I the moment of inertia of molecule about the axis of rotation. Here, each energy level has $(2J + 1)$-fold degeneracy. When r_1 and r_2 are the distances of atoms from the axis of rotation, we have

$$I = m_1 r_1^2 + m_2 r_2^2$$

$$m_1 r_1 = m_2 r_2$$

and
$$r = r_1 + r_2$$

These relations give

$$r_1 = \frac{m_2 r}{m_1 + m_2}$$

$$r_2 = \frac{m_1 r}{m_1 + m_2}$$

and
$$I = \mu r^2$$

where the reduced mass μ is expressed as

$$\mu = \frac{m_1 m_2}{m_1 + m_2}$$

or
$$\frac{1}{\mu} = \frac{1}{m_1} + \frac{1}{m_2}$$

Spectoscopists prefer to express the energy in terms of wavenumber (which is inverse of the wavelength). Thus, we have

$$F(J) = \frac{E}{hc} = \frac{h}{8\pi^2 Ic}(J(J+1)$$

$$= BJ(J+1)$$

where the rotational constant $B = h/8\pi^2 Ic$. The selection rules for the radiative transitions are $\Delta J = \pm 1$. Denoting the lower level by J and the upper level by J', the separation between the two levels is

$$\bar{\nu} = F(J') - F(J) = B[J'(J'+1) - J(J+1)]$$

Applying the selection rule $\Delta J = +1$, i.e., $J' - J = 1$, the wavenumber for the transition $J \rightarrow (J + 1)$ is

$$\bar{\nu} = B[(J+1)(J+2) - J(J+1)] = 2B(J+1)$$

The rotational quantum number J can assume integer values 0, 1, 2, so on. The energies of the rotational levels are 0, $2B$, $6B$, $12B$, $20B$, $30B$, so on. On applying the selection rules $\Delta J = +1$, we get the spectral lines with wavenumbers $2B$, $4B$, $6B$, $8B$, $10B$ so on.

4.1 Non-rigid Diatomic Rotator

It has seen that for a rigid rotator, the separation between two adjacent spectral lines remains constant, which is equal to $2B$. But, the experimental observations show that this separation goes on decreasing with the increase of J. Hence, the assumption to consider a molecule as a rigid rotator is not correct. The bond between the atoms is elastic and therefore its length changes. Consideration of the elastic nature of bond leads to an additional term for the wavenumber

$$F(J) = BJ(J+1) - DJ^2(J+1)^2 \tag{5.1}$$

where D is a small, positive quantity known as the centrifugal distortion constant. Denoting the lower level by J and the upper level by J', the separation between two levels is

$$\begin{aligned} \bar{\nu} &= F(J') - F(J) \\ &= B[J'(J'+1) - J(J+1)] - D[J'^2(J'+1)^2 - J^2(J+1)^2] \end{aligned}$$

Applying the selection rule $\Delta J = +1$, i.e., $J' = J + 1$ gives the wavenumber $\bar{\nu}$ for the transition $J \rightarrow (J+1)$ as

$$\begin{aligned} \bar{\nu} &= B[(J+1)(J+2) - J(J+1)] - D[(J+1)^2(J+2)^2 - J^2(J+1)^2] \\ &= 2B(J+1) - 4D(J+1)^3 \end{aligned}$$

The first term is the same as for the rigid molecule. The second term gives the shifts of the lines from that of the rigid molecule which increases as $(J+1)^3$.

5. Non-rigid Linear Polyatomic Molecules

In a linear polyatomic molecule, all the atoms are aligned along a line, called the molecular axis. A linear molecule consisting of n atoms has $n - 1$ bonds between adjacent atoms. A linear polyatomic molecule can also be treated similar to a diatomic molecule. For these molecules also we have $I_a = 0, I_b = I_c$. The rotational motion is described by the rotational quantum number J and energy of rotational level is

$$F(J) = BJ(J+1) - DJ^2(J+1)^2$$

where B is the rotational constant, and centrifugal distortion constant D is very small. The selection rules for radiative transitions are $\Delta J = \pm 1$. The wavenumber for the transition $J \rightarrow J + 1$ is

$$\begin{aligned} \bar{\nu} &= F(J+1) - F(J) \\ &= [B(J+1)(J+2) - D(J+1)^2(J+2)^2] - [BJ(J+1) - DJ^2(J+1)^2] \\ &= 2B(J+1) - 4D(J+1)^3 \end{aligned}$$

where $J = 0, 1, 2, 3$, so on.

6. Symmetric Top Molecules

For a symmetric top molecule, we have two categories:

(i) the prolate symmetric top where $I_a < I_b = I_c$, and
(ii) the oblate symmetric top where $I_a = I_b < I_c$.

6.1 Rigid Prolate Top Molecules

For a rigid prolate symmetric top, $I_a < I_b = I_c$; $A > B = C$, and a is the symmetry axis, and the energies of the levels are

$$E(A, B, C) = \frac{P_a^2}{2I_a} + \frac{P_b^2}{2I_b} + \frac{P_c^2}{2I_c}$$

$$= \frac{P_a^2}{2I_a} + \frac{P_b^2 + P_c^2}{2I_b}$$

$$= \frac{P_a^2 + P_b^2 + P_c^2}{2I_b} + \left(\frac{1}{2I_a} - \frac{1}{2I_b}\right)P_a^2$$

where P_a, P_b, P_c are the components of angular momentum. We have

$$P_a^2 + P_b^2 + P_c^2 = J(J+1)\hbar^2$$

and $$P_a^2 = K^2\hbar^2$$

where K is the projection of J on the symmetry-axis. It is often denoted as k_a. Thus, the energy is

$$E_{JK} = \frac{h^2}{8\pi^2 I_b} J(J+1) + \frac{h^2}{8\pi^2}\left(\frac{1}{I_a} - \frac{1}{I_b}\right)K^2$$

The wavenumber

$$F(J, K) = \frac{E_{JK}}{hc} = BJ(J+1) + (A - B)K^2$$

For a prolate top, the levels are denoted as J_{k_a}. For a given J, the energy increases with the increase of K.

6.2 Rigid Oblate Top Molecules

For a rigid oblate symmetric top, $I_a = I_b < I_c$; $A = B > C$, and c is the symmetry-axis, and the energies of the levels are

$$E(A, B, C) = \frac{P_a^2}{2I_a} + \frac{P_b^2}{2I_b} + \frac{P_c^2}{2I_c}$$

$$= \frac{P_a^2 + P_b^2}{2I_b} + \frac{P_c^2}{2I_c}$$

$$= \frac{P_a^2 + P_b^2 + P_c^2}{2I_b} - \left(\frac{1}{2I_b} - \frac{1}{2I_c}\right)P_c^2$$

where P_a, P_b, P_c are the components of angular momentum. We have

$$P_a^2 + P_b^2 + P_c^2 = J(J+1)\hbar^2$$

and $$P_c^2 = K^2\hbar^2$$

where K is the projection of J on the symmetry-axis. It is often denoted as k_c. Thus, the energy is

$$E_{JK} = \frac{h^2}{8\pi^2 I_b} J(J+1) - \frac{h^2}{8\pi^2}\left(\frac{1}{I_b} - \frac{1}{I_c}\right)K^2$$

The wavenumber

$$F(J, K) = \frac{E_{JK}}{hc} = BJ(J+1) - (B - C)K^2$$

For a oblate top, the levels are denoted as J_{k_c}. For a given J, the energy decreases with the increase of K.

For both the prolate and oblate symmetric tops, the angular momentum quantum number J can assume the values, $J = 0, 1, 2, 3, \ldots$ and its projection K on the axis of symmetry can assume the values $K = 0, \pm1, \pm2, \ldots, \pm J$. One can assume that the positive and negative values of K

correspond to the clockwise and anticlockwise rotations about the axis of symmetry. All levels with $K > 0$ are doubly degenerate, as we have square-term K^2 in the expression of the energy. The energy levels for the rigid prolate top and rigid oblate top are shown in Fig. 3. The selection rules for transitions are

$$\Delta J = \pm 1$$

and

$$\Delta K = 0$$

The wavenumber of the transition $(J, K) \rightarrow (J + 1, K)$ is

$$\bar{\nu} = F(J + 1, K) - F(J, K) = 2B(J + 1)$$

where $J = 0, 1, 2, \ldots$. This expression is the same for both the prolate and oblate symmetric top molecules. The allowed transitions are also shown in Fig. 3. Thus, the spectrum is independent of K. However, this treatment is for a rigid molecule.

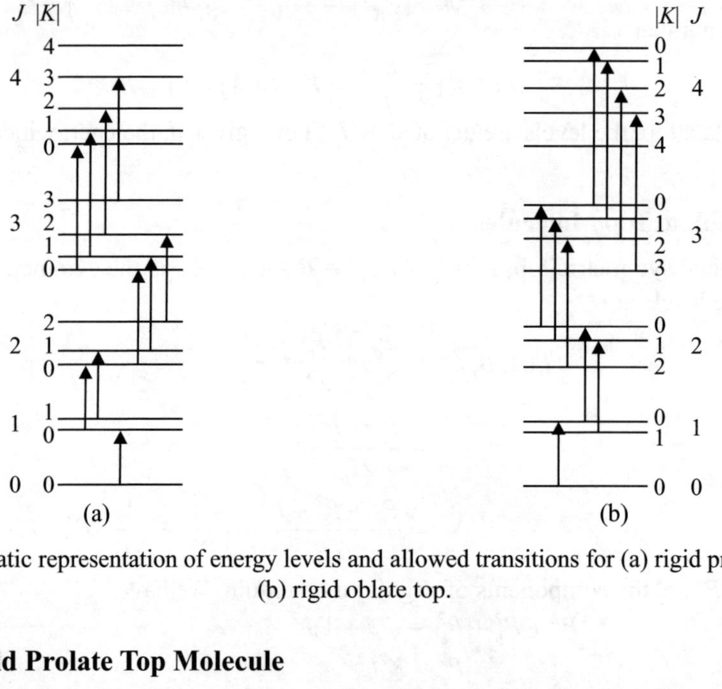

Fig. 3: Schematic representation of energy levels and allowed transitions for (a) rigid prolate top and (b) rigid oblate top.

6.3 Non-rigid Prolate Top Molecule

A symmetric top molecule is not rigid and owing to the rotational motion, the centrifugal stretching appears in the picture. On accounting for the stretching effects for prolate top molecule, we get

$$F(J, K) = BJ(J + 1) + (A - B)K^2 - D_J J^2(J + 1)^2 - D_{JK} J(J + 1)K^2 - D_K K^4$$

where D_J, D_{JK}, D_K, called the distortional constants, have small values. The selection rules are still the same as stated above. The wavenumber of the transition $(J, K) \rightarrow (J + 1, K)$ is

$$\begin{aligned}\bar{\nu} = F(J + 1, K) - F(J, K) &= [B(J + 1)(J + 2) + (A - B)K^2 - DJ\,(J + 1)^2\,(J + 2)^2 \\ &\quad - D_{JK}(J + 1)(J + 2)K^2 - D_K K^4] - [BJ(J + 1) + (A - B)K^2 - D_J J^2(J + 1)^2 \\ &\quad - D_{JK} J(J + 1)K^2 - D_K K^4] \\ &= 2B(J + 1) - 4D_J(J + 1)^3 - 2D_{JK}\,(J + 1)K^2\end{aligned}$$

Now, the spectrum depends on K. For a given value of J, we have $(2J + 1)$ values of K. Since K is appearing as K^2, there are $(J + 1)$ levels for the rotational quantum number J.

6.4 Non-rigid Oblate Top Molecule

A symmetric top molecule is not rigid and owing to the rotational motion, the centrifugal stretching appears in the picture. On accounting for the stretching effects for the prolate top molecule, we get

$$F(J, K) = BJ(J+1) - (B-C)K^2 - D_J J^2(J+1)^2 - D_{JK} J(J+1)K^2 - D_K K^4$$

where D_J, D_{JK}, D_K, called the distortional constants, have small values. The selection rules are still the same as stated above. The wavenumber of the transition $(J, K) \rightarrow (J + 1, K)$ is

$$\begin{aligned}\bar{\nu} &= F(J+1, K) - F(J, K)\\ &= [B(J+1)(J+2) - (B-C)K^2 - D_J (J+1)^2 (J+2)^2 - D_{JK}(J+1)(J+2)K^2 - D_K K^4]\\ &\quad - [BJ(J+1) - (B-C)K^2 - D_J J^2(J+1)^2 - D_{JK} J(J+1)K^2 - D_K K^4]\\ &= 2B(J+1) - 4D_J (J+1)^3 - 2D_{JK} (J+1)K^2\end{aligned}$$

Now, the spectrum depends on K. This expression is the same as for prolate top molecules. For a given value of J, we have $(2J + 1)$ values of K. Since K is appearing as K^2, there are $(J + 1)$ levels for the rotational quantum number J.

7. Asymmetric Top Molecules

For an asymmetric top molecule all three moments of inertia are different, $I_a < I_b < I_c$; $A > B > C$, and the energy of asymmetric top molecule is

$$E(A, B, C) = \frac{P_a^2}{2I_a} + \frac{P_b^2}{2I_b} + \frac{P_c^2}{2I_c}$$

The wavenumber is

$$\begin{aligned}F(A, B, C) &= \frac{E(A,B,C)}{hc}\\ &= \frac{AP_a^2 + BP_b^2 + CP_c^2}{\hbar^2} \qquad (5.2)\end{aligned}$$

where P_a, P_b and P_c are the components of angular momentum. For the asymmetric top molecule, K is not a good quantum number, the energy levels may be specified by giving a value of J, and the value k_a for the limiting prolate and k_c for the limiting oblate symmetric top. Thus, the rotational levels may be designated as $J_{k_a k_c}$ or J_τ, where a pseudo quantum number τ is expressed as $\tau = k_a - k_c$, which can assume the values from $-J$ to $+J$ with the increment of one.

For an asymmetric top molecule, there is no preferential axis and the molecular inertial axes, a, b and c may be identified with the space fixed axes, x, y and z in six ways, shown in Table 3. For two scalars σ and ρ, Equation (5.2) can be expressed as

$$F(\sigma A + \rho, \sigma B + \rho, \sigma C + \rho) = \sigma F(A, B, C) + \rho J(J+1) \qquad (5.3)$$

Table 3. Six representations

	I^r	I^l	II^r	II^l	III^r	III^l
x	b	c	c	a	a	b
y	c	b	a	c	b	a
z	a	a	b	b	c	c

where we have used $P_a^2 + P_b^2 + P_c^2 = J(J+1)\hbar^2$. Now, we choose

$$\sigma = \frac{2}{A-C}$$

and
$$\rho = -\frac{A+C}{A-C}$$

so that
$$\sigma A + \rho = 1$$

and
$$\sigma B + \rho = \frac{2B-A-C}{A-C} = \kappa$$

$$\sigma C + \rho = -1$$

Here, κ is known as the Ray asymmetry parameter and its value varies between –1 and +1. κ = –1 for the prolate top molecule and κ = +1 for the oblate top molecule.

Table 4. Expressions for $E(\kappa)$

J	κ_a	κ_c	$E(\kappa)$
0	0	0	0
1	0	1	$\kappa - 1$
	1	1	0
	1	0	$\kappa + 1$
2	0	2	$2[\kappa - (\kappa^2+3)^{1/2}]$
	1	2	$\kappa - 3$
	1	1	4κ
	2	1	$\kappa + 3$
	2	0	$2[\kappa + (\kappa^2+3)^{1/2}]$
3	0	3	$5\kappa - 3 - 2[4\kappa^2 + 6\kappa + 6]^{1/2}$
	1	3	$2[\kappa - (\kappa^2+15)^{1/2}]$
	1	2	$5\kappa + 3 - 2[4\kappa^2 - 6\kappa + 6]^{1/2}$
	2	2	4κ
	2	1	$5\kappa - 3 + 2[4\kappa^2 + 6\kappa + 6]^{1/2}$
	3	1	$2[\kappa + (\kappa^2 + 15)1/2]$
	3	0	$5\kappa + 3 + 2[4\kappa^2 - 6\kappa + 6]^{1/2}$

From equation (5.3) we have

$$F(1, \kappa, -1) = F(\kappa)$$
$$= \frac{2}{A-C}F(A,B,C) - \frac{A+C}{A-C}J(J+1)$$

and thus the wavenumber of an asymmetric rotor is

$$F(A, B, C) = F(J_\tau)$$
$$= \frac{A+C}{2}J(J+1) + \frac{A-C}{2}F(\kappa)$$

Computation of $E(\kappa)$ is quite complicated and beyond the scope of this book. However, some expressions for $E(\kappa)$ are given in Table 4. Figure 4 shows the variation of $E(\kappa)$ versus κ up to $J = 5$. Here we have considered a rigid asymmetric top molecule.

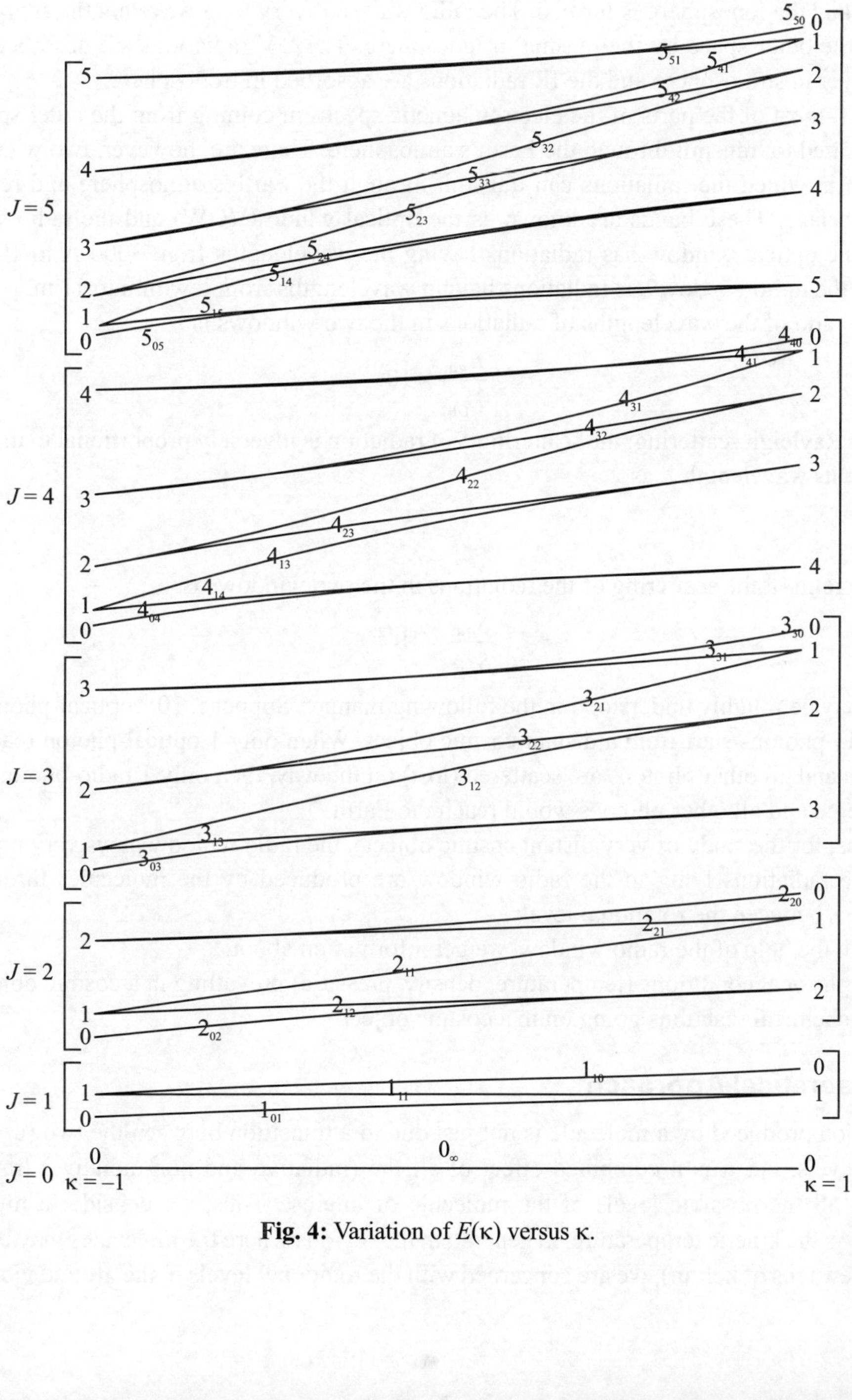

Fig. 4: Variation of $E(\kappa)$ versus κ.

8. Importance of Molecules in the ISM

Radiations coming from the outer space have full electromagnetic spectrum consists of the γ-rays, x-rays, UV-radiations, visible light, IR-radiations, microwaves and radio waves. The γ-rays and x-rays are absorbed in the uppermost part of the Earth's atmosphere where the gas is ionized, and the ionosphere is formed. The radio waves of very long wavelengths are reflected back to the outer space by the plasma in ionosphere. The UV radiations are absorbed in the ozone layer in stratosphere and the IR radiations are absorbed in troposphere.

Thus, most of the parts of the electromagnetic spectrum coming from the outer space are not permitted to transmit through the Earth's atmosphere. There are, however, two wide bands (regions) in which the radiations can transmit through the Earth's atmosphere and reach the earth's surface. These bands are known as the optical window (OW) and the radio window (RW). The optical window has radiations having the wavelengths from 3000 Å to 10,000 Å whereas the radio window has radiations having wavelengths from few mm to 11 m.

The ratio of the wavelengths of radiations in the two windows is

$$\frac{\lambda_{RW}}{\lambda_{OW}} \approx 10^5$$

For Rayleigh scattering, the scattering S of radiation is inversely proportional to the fourth power of its wavelength λ as

$$S \propto \frac{1}{\lambda^4}$$

The ratio of the scattering of the radiations in the two windows is

$$\frac{S_{RW}}{S_{OW}} \approx 10^{-20}$$

It may be roughly understood in the following manner. Suppose, 10^{20} optical-photons and 10^{20} radio-photons start from a distant cosmic object. When only 1 optical-photon reaches on the Earth and all other photons are scattered (lost) on the way, then only 1 radio-photon would be scattered and all other photons would reach the Earth.

Thus, for the study of very distant cosmic objects, the radio window plays very important role. The radiations lying in the radio window are produced by the molecules through the transitions between the rotational levels.

With the help of the radio window, we get information about:

(i) the physical conditions (temperature, density, pressure) prevailing in a cosmic object.

(ii) the chemical reactions going on in a cosmic object.

9. Theoretical Approach

A radiation produced by a molecule is not just due to a transition between the two (upper and lower) levels, but it is a combined effect of all the (radiative and non-radiative) transitions between all the possible levels of the molecule of interest. Thus, we consider a multilevel system. As the kinetic temperature, in general, in the regions where the molecules may be found is low (few tens of kelvin), we are concerned with the rotational levels in the ground vibrational state and the ground electronic state. The number of rotational levels considered depends on the kinetic temperature in the region, so that populations of further higher levels are negligibly small. Rotational levels of asymmetric top molecules are classified into the groups, according to the direction of electric dipole moment of the molecule.

For *a*-type molecule (molecule having electric dipole moment along the inertial axes for the lowest moment of inertia), the levels are classified according to the selection rules:

$$J:\ \Delta J = 0, \pm 1$$

$$k_a, k_c:\ \text{odd, even} \longleftrightarrow \text{odd, odd} \qquad \text{(set I)}$$

$$\text{even, even} \longleftrightarrow \text{even, odd} \qquad \text{(set II)}$$

For *b*-type molecule (molecule having electric dipole moment along the inertial axes for middle moment of inertia), the levels are classified according to the selection rules:

$$J:\ \Delta J = 0, \pm 1$$

$$k_a, k_c:\ \text{odd, even} \longleftrightarrow \text{even, odd} \qquad \text{(set I)}$$

$$\text{even, even} \longleftrightarrow \text{odd, odd} \qquad \text{(set II)}$$

For *c*-type molecule (molecule having electric dipole moment along the inertial axes for highest moment of inertia), the levels are classified according to the selection rules:

$$J:\ \Delta J = 0, \pm 1$$

$$k_a, k_c:\ \text{even, odd} \longleftrightarrow \text{odd, odd} \qquad \text{(set I)}$$

$$\text{even, even} \longleftrightarrow \text{odd, even} \qquad \text{(set II)}$$

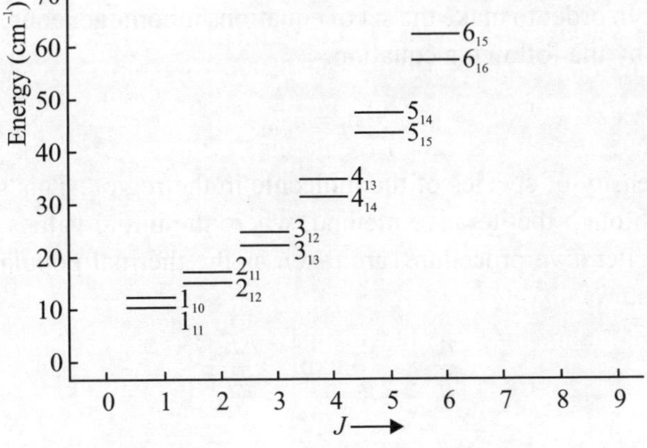

Fig. 5: 23 Rotational levels of ortho-H$_2$CO.

When the electric dipole moment is inclined to the inertial axes, radiative transitions correspond to various components of electric dipole moment.

H_2CO, the first organic molecule detected in the ISM, is *a*-type asymmetric top molecule and its rotational levels are classified into ortho (set I) and para (set II) species. Energy level diagram of 23 lower ortho levels of H_2CO shown in Fig. 5. These levels are connected through radiative and non-radiative (collisional) transitions. For these 23 levels, a set of statistical equilibrium equations coupled with equations of radiative transfer is expressed as (Sobolev large velocity gradient analysis)

$$n_i \sum_{\substack{j=1 \\ j\neq i}}^{23} P_{ij} = \sum_{\substack{j=1 \\ j\neq i}}^{23} n_j P_{ji} \qquad i = 1, 2, ..., 23 \tag{5.4}$$

where n denotes the population density of energy level and parameter P is as the following.

(i) For radiative transition

$$P_{ij} = \begin{cases} (A_{ij} + B_{ij} I_{v,bg})\beta_{ij} + n_{H_2} C_{ij} & i > j \\ B_{ij} I_{v,bg} \beta_{ij} + n_{H_2} C_{ij} & i < j \end{cases}$$

(ii) For collisional transition

$$P_{ij} = n_{H_2} C_{ij}$$

Here, A and B are the Einstein coefficients, C the rate coefficient for the collisional transition and n_{H_2} the density of molecular hydrogen. The escape probability β for the transition is

$$\beta_{lu} = \beta_{ul}$$

$$= \frac{1 - \exp(-\tau_v)}{\tau_v}$$

where optical thickness τ_v is expressed as

$$\tau_v = \frac{hc}{4\pi(\mathrm{d}v_r/\mathrm{d}r)}[B_{lu} n_l - B_{ul} n_u]$$

where $(\mathrm{d}v_r/\mathrm{d}r)$ denotes the velocity-gradient in the cloud. This is non-linear, homogeneous set of equations. Here, all possible radiative and collisional (non-radiative) transitions between 23 rotational levels are considered. Equation (5.4), being homogeneous set of equations, cannot have unique solution. In order to make the set of equations inhomogeneous, one of the equations in the set is replaced by the following equation.

$$\sum_{i=1}^{23} n_i = n_{\text{total}}$$

where n_{total} is total density of species of the molecule in the region. This system of non-linear equations is solved through the iterative method, where the initial values of level population-densities, to start the iterative procedure, are taken as the thermal population-densities at the given kinetic temperature.

$$\frac{n_u^t}{n_l^t} = \frac{g_u}{g_l} \exp\left[-\frac{\Delta E_{ul}}{kT_k}\right]$$

and

$$\sum_{i=1}^{23} n_i^t = n_{\text{total}}$$

Here, the prefix *t* denotes the thermal values. g_u and g_l are the statistical weights of the upper and lower levels, respectively; ΔE_{ul} is the energy difference between the two levels and T_k is kinetic temperature in the region. For the given kinetic temperature, the thermal population densities of the levels are obtained, and are used as the initial values for the iterative procedure. The input parameters for solving the above set of equations are the radiative transition probabilities (Einstein *A* and *B* coefficients) and the collisional rate coefficients for the transitions between the levels. After solving the above set of equations, we get non-thermal population densities, for which excitation temperature T_{ex} is expressed as

$$\frac{n_u^{nt}}{n_l^{nt}} = \frac{g_u}{g_l}\exp\left[-\frac{\Delta E_{ul}}{kT_{ex}}\right]$$

and

$$\sum_{i=1}^{23} n_i^{nt} = n_{\text{total}}$$

Here, the prefix *nt* denotes the non-thermal values.

10. Radiative Transitions

A system having discrete energy levels can be transferred from one stage to another, after absorption or emission of energy, which is equal to the energy difference between the two levels. The required energy can be exchanged through a number of ways. The exchange of energy can be due to collision with some particle, or due to the absorption or emission of a photon of proper frequency. After the absorption of a photon, whose energy ($h\nu$) is exactly equal to energy difference between two levels, the system can be transferred from a lower level to an upper one. In the reverse process, a photon of proper frequency may be emitted when the system is transferred from an upper level to a lower one. In addition to the fact that energy of photon is exactly equal to the energy difference between the two levels, the transition must be radiatively allowed according to the selection rules. There are no selection rules for the collisional transitions, and therefore, between all the pairs of levels. In the following subsections, we have dealt with radiative transitions.

10.1 Absorption

Out of various energy levels of a system, consider two levels, the lower level *l* and the upper level *u*, having energies E_l and E_u, respectively. After absorption of a photon of proper frequency ν (such that $h\nu = E_u - E_l$), a system can be transferred from the lower level to the upper level, provided the transition is radiatively allowed. Such absorption of a photon is termed the absorption or stimulated absorption. Because of the absorption of photon, the population density of the lower level decreases and the population density of the upper level increases.

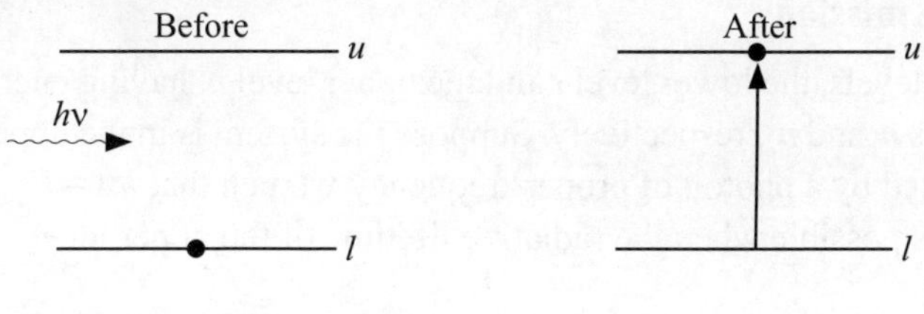

The rate of change of the population of upper level depends on the population density n_l of the lower level and the density of radiations having frequency ν. That is,

$$\frac{dn_u}{dt} = B_{lu} n_l \rho(\nu)$$

where B_{lu} is the Einstein *B*-coefficient for absorption and represents the probability of absorption of photon of frequency ν. The ρ(ν) is the density of photons having frequency ν.

10.2 Spontaneous Emission

Let us consider two levels, the lower level *l* and the upper level *u*, having energies E_l and E_u, and the population densities n_l and n_u, respectively. Suppose that the system, after staying in the upper (excited) state *u* for a period, called the radiative lifetime of the level, can deexcite to the lower level *l*. It is possible when the transition is radiatively allowed. As there is no interaction with the system from outside, the probability of such deexcitation is proportional to the population density n_u of the upper level. In this process, a photon of proper frequency ν (such that $h\nu = E_u - E_l$) is emitted. The probability of the emission of photon is the same in all directions. That is, there is no preferential direction.

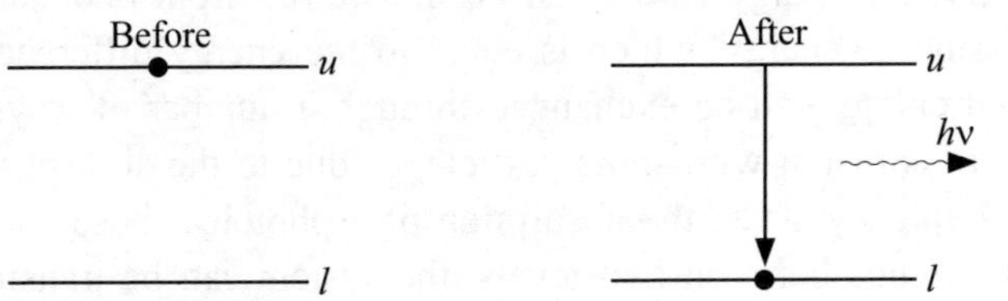

The rate of change of the population of upper level depends on the population density n_u of the upper level as

$$-\frac{dn_u}{dt} = A_{ul} n_u$$

where A_{ul} is the Einstein *A*-coefficient for the spontaneous emission. It gives the probability of emission of photon in absence of any external interaction.

Radiative Lifetime

For known values of the Einstein *A*-coefficients, the radiative lifetime t_j of the level *j* can be calculated as

$$t_j = 1 \Big/ \sum_i A_{ji}$$

Here, the summation is carried over all lower levels *i*.

10.3 Stimulated Emission

Now, consider two levels, the lower level *l* and the upper level *u*, having energies E_l and E_u, and population densities n_l and n_u, respectively. Suppose the system is in the upper (excited) state *u*, and can be stimulated by a photon of proper frequency ν (such that $h\nu = E_u - E_l$) to deexcite to the lower level. It is possible when the radiative lifetime of the upper level is sufficiently large, so that the system is in the upper level at the time of the interaction. In this process, we get two photons of the same frequency. The population density of the upper level decreases and that of the lower level increases.

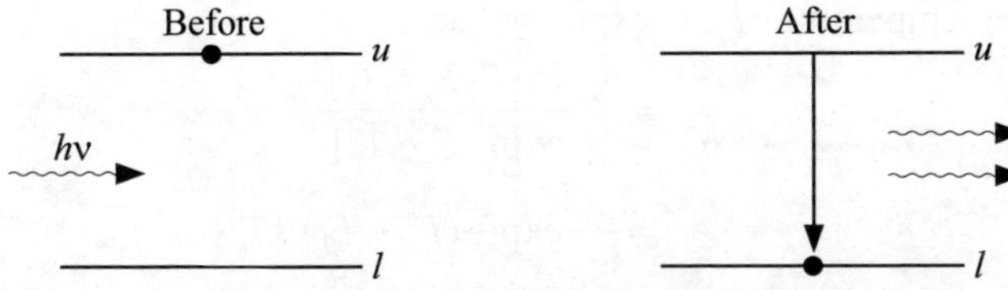

The rate of change of the population of upper level depends on the population density n_u of the upper level and the density of radiations having frequency ν. That is,

$$-\frac{\mathrm{d}n_u}{\mathrm{d}t} = B_{ul}n_u\rho(\nu)$$

where B_{ul} denotes the Einstein *B*-coefficient for the stimulated emission and ρ(ν) the density of photons having frequency ν. It shows the probability of emission of a photon in the presence of incident photons of the same frequency.

10.4 Relation between the Einstein *A* and *B* coefficients

We have seen that for a pair of levels, between which a transition is radiatively allowed, there is one Einstein *A* coefficient and two Einstein *B* coefficients, one for the absorption and the other for the emission of photon. Let us now consider all three types of radiative transitions together. The rate of change of population density of upper level *u* can be expressed as the following.

$$\frac{\mathrm{d}n_u}{\mathrm{d}t} = B_{lu}n_l\rho(\nu) - B_{ul}n_u\rho(\nu) - A_{ul}n_u \tag{5.5}$$

Consider a cavity at kinetic temperature *T*. The energy density of radiations having frequency ν is given by the Planck's black-body radiation formula

$$\rho(\nu) = B\nu(T) = \frac{8\pi h\nu^3}{c^3}\frac{1}{\exp(h\nu/kT)-1} \tag{5.6}$$

Suppose, we have a system having two levels only, the upper level *u* and the lower level *l*. Then, we have the rate of decrease of the population density of upper level equal to the rate of increase of the population density of the lower level, and vice versa. Therefore, we have

$$-\frac{\mathrm{d}n_u}{\mathrm{d}t} = \frac{\mathrm{d}n_l}{\mathrm{d}t}$$

In the equilibrium state, we have

$$\frac{\mathrm{d}n_u}{\mathrm{d}t} = 0$$

and

$$\frac{\mathrm{d}n_l}{\mathrm{d}t} = 0$$

and from equation (5.5), we get

$$\rho(\nu)\left[\frac{n_l}{n_u}B_{lu} - B_{ul}\right] = A_{ul}$$

or $$\rho(\nu) = \frac{A_{ul}}{n_l B_{lu}/n_u - B_{ul}} \tag{5.7}$$

According to the Boltzmann's law, the ratio of population densities n_u and n_l, in the thermal equilibrium, at kinetic temperature T_k, is expressed as

$$\frac{n_u}{n_l} = \frac{g_u \exp[-E_u/kT_k]}{g_l \exp[-E_l/kT_k]}$$

$$= \frac{g_u}{g_l} \exp[-(E_u - E_l)/kT_k]$$

where g_u and g_l are the statistical weights, and E_u and E_l the energies, of the upper and lower levels, respectively. Thus, we have

$$\frac{n_u}{n_l} = \frac{g_u}{g_l} \exp(-\Delta E_{ul}/kT_k) \tag{5.8}$$

where $\Delta E_{ul} = (E_u - E_l)$ denotes the energy difference between the two levels. Einstein considered the relation

$$g_l B_{lu} = g_u B_{ul}$$

Using equation (5.8) in (5.7), we have

$$\rho(\nu) = \frac{A_{ul}}{\frac{g_l}{g_u} \exp(h\nu/kT_k) \frac{g_u}{g_l} B_{ul} - B_{ul}}$$

$$= \frac{A_{ul}}{B_{ul}} \frac{1}{\exp(h\nu/kT_k) - 1} \tag{5.9}$$

Comparison of equations (5.6) and (5.9) gives

$$\frac{8\pi h\nu^3}{c^3} \frac{1}{\exp(h\nu/kT_k) - 1} = \frac{A_{ul}}{B_{ul}} \frac{1}{\exp(h\nu/kT_k) - 1}$$

Thus, we have

$$A_{ul} = \frac{8\pi h\nu^3}{c^3} B_{ul}$$

It obviously shows that by knowing the Einstein *A*-coefficient for an optically allowed transition, the both Einstein *B*-coefficients can be calculated and vice versa. Therefore, one has to calculate the Einstein *A*-coefficients.

In a terrestrial laboratory, spectrum of a molecule of interest is analyzed, and accurate rotational and distortional constants are derived. Electric dipole moment also of the molecules is obtained experimentally.

For known spectroscopic parameters (rotational and centrifugal distortion coefficients, and electric dipole moment), the radiative transition probabilities for the radiative transitions are calculated with the help of a software. The software provides energies of the rotational levels and the line-strengths *S* for the radiative transitions between the levels. The line-strength *S* is related to the Einstein *A*-coefficient, A_{ul}, between the upper level *u* and the lower level *l* as

$$A_{ul} = \frac{64\pi^4 \nu^3 \mu^2}{3 g_u h c^3} S$$

where g_u is the statistical weight of upper level u, μ is the component of electric dipole moment of the molecule and ν the frequency of the transition.

11. Collisional Transitions

Though the collisional transitions between the levels do not follow any selection rules, but computation of the collisional rate coefficients is a tedious job. As an example, suppose, one wants to calculate the collisional rate coefficients for the rotational transitions in ortho-H_2CO colliding with He atom[2]. The H_2CO is first optimized with the help of a software to get the coordinates of constituent atoms of H_2CO with origin as the centre-of-mass of H_2CO. The H_2CO is a planar molecule and its structure is shown in Fig. 6.

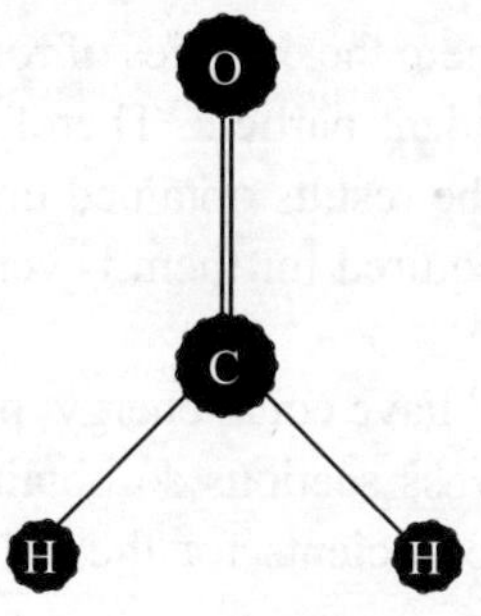

Structure of H_2CO molecule.

Atom	Coordinates (Å)		
	x	*y*	*z*
H	0.000000	0.937086	−1.114961
H	0.000000	−0.937086	−1.114961
C	0.000000	0.000000	−0.525916

Fig. 6: Structure of H_2CO molecule.

Next, the potential energy surface (PES) of the H_2CO molecule is obtained. For calculation of the interaction potential between, H_2CO and colliding partner He atom, positions of constituent atoms of H_2CO are kept fix and the position of He atom is varied in terms of spherical polar coordinates (R, θ, ϕ) with the origin at the centre-of-mass of H_2CO.

In order to account for the Basis Set Superposition Errors (BSSE), scientists include the concept of the ghost atom and perform the calculations as the following:

(i) Energy $E_1(R, \theta, \phi)$ of H_2CO + He.

(ii) Energy $E_2(R, \theta, \phi)$ of H_2CO while the He is present as the ghost atom.

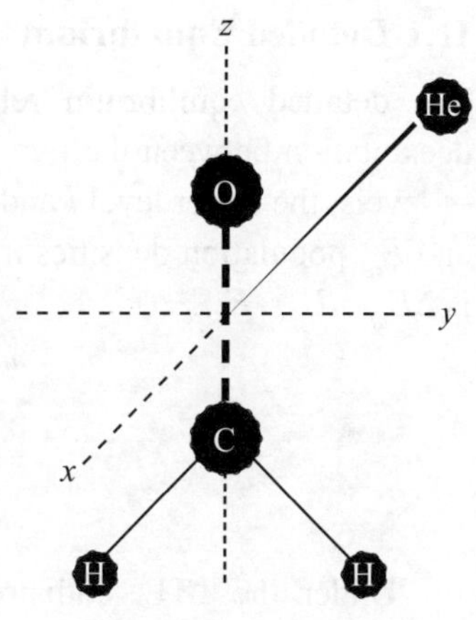

Fig. 7: Interaction of H_2CO molecule with the He atom.

[2]In a molecular cloud, the colliding partner is H_2 molecule. The H_2 has ortho and para species and the calculations become too complicated. For convenience, the H_2 is considered structureless particle and is often replaced by the He atom, as both of them (He atom and H_2 molecule) have two electrons and two protons. In order to account for the difference in their mass, the collisional rate coefficients are scaled in terms of the reduced masses of the He and H_2.

(iii) Energy $E_3(R, \theta, \phi)$ of He while all the atoms of H_2CO are present as the ghost atoms.

The calculations may be done with the help of a software, say MOLPRO. The interaction potential $V(R, \theta, \phi)$ between the H_2CO and He atom is

$$V(R, \theta, \phi) = E_2(R, \theta, \phi) + E_3(R, \theta, \phi) - E_1(R, \theta, \phi)$$

The interaction potential $V(R, \theta, \phi)$ is calculated for a large number of grid-points, so that the interaction potential is quite accurate.

For the calculation of the scattering cross sections for the rotational transitions, this interaction potential is used as input to the computer code MOLSCAT.

The computer code MOLSCAT has provision for calculation of collisional cross sections under three approximations:

(i) Infinite Order Sudden (IOS) approximation.
(ii) Coupled States (CS) approximation.
(iii) Close Coupling (CC) approximation.

In the IOS approximation is a high energy approximation, where the energies of rotational levels are neglegible in comparison with the energy of the incident particle. Therefore, the results obtained in the IOS approximation are not so accurate. The results obtained under the CS or CC approximation are accurate. But, the computer-time required for them is very large and extremely large, respectively.

At a given temperature in a cosmic object, all particles do not have equal energy; particles follow the Maxwellian distribution of energy. The collisional cross sections so obtained are averaged over the Maxwellian distribution to obtain the rate coefficients for the collisional transitions between the levels.

11.1 Detailed Equilibrium

The detailed equilibrium relates the rate coefficients for the collisional excitation and deexcitation between the two levels. Suppose, in the local thermal equilibrium (LTE), a pair of levels, the lower level l and the upper level u, have statistical weights g_l and g_u, energies E_l and E_u, population densities n_l and n_u, respectively. The energy difference between the levels is $\Delta E_{ul} = E_u - E_l$.

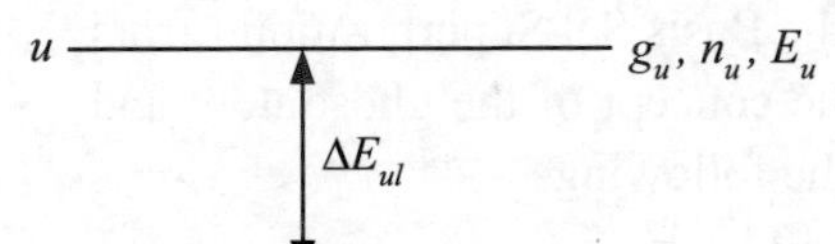

Under the LTE, collisional rate coefficients, C_{lu} for the excitation and C_{ul} for the deexcitation, are related as

$$n_u C_{ul} = n_l C_{lu}$$

or

$$C_{lu} = \frac{n_u}{n_l} C_{ul}$$

Following the Maxwellian distribution, the population densities of the energy levels at the kinetic temperature T_k are expressed as

$$\frac{n_u}{n_l} = \frac{g_u}{g_l}\exp\left(-\frac{\Delta E_{ul}}{kT_k}\right)$$

Using this relation, we get

$$C_{lu} = \frac{g_u}{g_l}\exp\left(-\frac{\Delta E_{ul}}{kT_k}\right)C_{ul} \tag{5.10}$$

It shows that for the calculation of collisional rate coefficients, one requires to calculate the rate coefficients for one direction (excitation or deexcitation) and the rate coefficients for the reverse direction can be calculated with the help of the detailed equilibrium (equation 5.10).

12. Anomalous Phenomena in Cosmic Molecules

Besides the normal emission and absorption, some molecules in the ISM have been found to show anomalous phenomena. There are two anomalous phenomena shown by some molecules:
(i) MASER action, and (ii) Anomalous absorption.

12.1 Natural MASERs

After the discovery of OH in 1963 in the ISM, its 18 cm radiation was found to show the kinetic temperature (measured from the half-width of the line-profile) around 100 K, and the brightness temperature (measured from height of the line-profile) around 1000 K (Fig. 8). Some scientists tried to interpret this discrepancy between these two temperatures in terms of the stimulated emission (phenomena responsible for the LASER and MASER action). LASERs and MASERs on the Earth are the artificial devices and much efforts are required for their construction. Therefore, the majority of scientists was not ready to accept the idea of natural MASERs.

For the improvement of the resolution of radio-telescope, Very Large Baseline Interferometry (VLBI) experiments were performed. After using the VLBI, for 18 cm radiation, the kinetic temperature was still around 100 K, but the brightness temperature was found around 10^{12} K. It shows that the angular size of object was much smaller than the resolution of single-dish radio telescope. So high value cannot be of kinetic temperature, as at this temperature, the molecules would break into its constituent atoms; atoms would be stripped off their electrons; and even existence of the nucleus may not be possible. Then, there was no option left, but to accept the existence of the natural MASER.

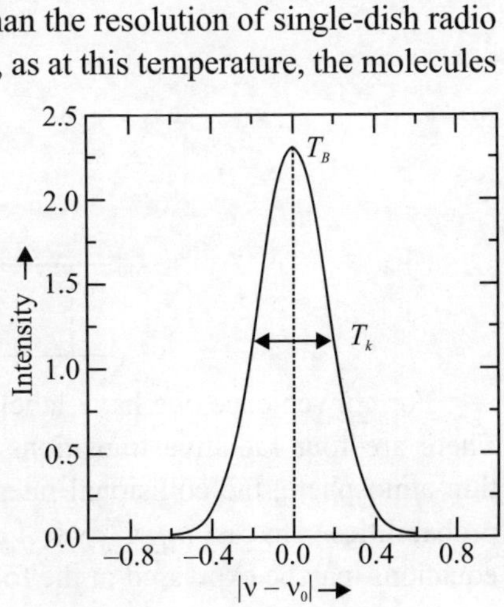

Fig. 8: Profile of a line.

The H_2O was second molecule to show MASER action in the transition $6_{16} - 5_{23}$ at 22.2 GHz. Some more molecules have been found to show weak MASER action. The Einstein *A*-coefficients are used for the computation of radiative lifetimes (t) of rotational levels. In general, the radiative lifetime of upper level is smaller than that of the lower one. But, the reverse is found for a MASER transition. However, having larger lifetime of upper level is not sufficient criterion of the

MASER action. For the MASER action, the population inversion ($n_u g_l / n_l g_u > 1$) between the upper level u and the lower level l is also required. Here, n and g are, respectively, the population density and the statistical weight of the level. For a radiative transition to be a MASER line, following two criteria are required:

$$t_u > t_l$$

and
$$\frac{n_u g_l}{n_l g_u} > 1 \tag{5.11}$$

12.2 Anomalous Absorption

Our universe is expanding and we are not at the boundary of the universe, thus we have a cosmic microwave background (CMB) around us, and it has been found to correspond to the background temperature of 2.73 K, discovered by *A*. Penzias and R. Wilson in 1964.

Accordingly, the brightness temperature of all radiations should be larger than 2.73 K. However, some molecular transitions have shown the brightness temperature less than 2.73 K, showing the absorption against the CMB (anomalous absorption). The first transition found showing the anomalous absorption is $1_{11} - 1_{10}$ transition of ortho-H_2CO at 4.83 GHz. The second radiation found showing anomalous absorption is $2_{20} - 2_{11}$ transition of para-*c*-C_3H_2 at 21.6 GHz. Later on some more lines have been found to show the anomalous absorption.

Anomalous absorption of a spectral line is a unique phenomenon, as the brightness temperature (corresponding to the intensity) of radiation becomes less than the background temperature 2.73 K (cosmic microwave background). The phenomenon of the anomalous absorption has not been yet produced in any terrestrial laboratory.

Anomalous absorption in $1_{11} - 1_{10}$ transition

In order to understand the anomalous absorption of $1_{11} - 1_{10}$ transition, let us consider two doublets of $J = 1$ and $J = 2$, as shown in figure.

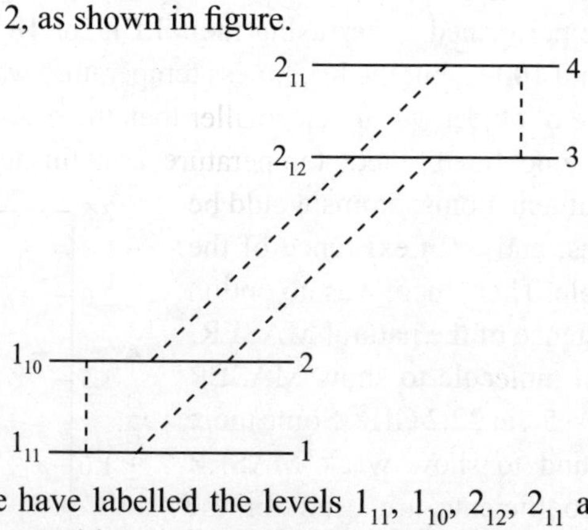

For convenience, we have labelled the levels 1_{11}, 1_{10}, 2_{12}, 2_{11} as 1, 2, 3, 4, respectively. There are four radiative transitions, shown by the dotted lines in figure. For an optically thin atmosphere, the collisional rates are negligible in comparison to the radiative transition probabilities. That is, $n_{H_2} C_{ul} << A_{ul}$. Then, for these four levels, the statistical equilibrium equations may be expressed in the following form:

$$n_1(n_{H_2} C_{13} + n_{H_2} C_{14}) = n_3 A_{31}$$

$$n_2(n_{H_2} C_{23} + n_{H_2} C_{24}) = n_4 A_{42}$$

$$n_3 A_{31} = n_1 n_{H_2} C_{13} + n_2 n_{H_2} C_{23}$$

$$n_4 A_{42} = n_1 n_{H_2} C_{14} + n_2 n_{H_2} C_{24}$$

Here, n's are the population densities of the levels. These equations can be rearranged as

$$\frac{n_2}{n_1} = \frac{C_{14}}{C_{23}}$$

For an anomalous absorption, the requirement is that $n_2 < n_1$. It shows that for the anomalous absorption, we should have $C_{14} < C_{23}$. This is, when $C_{14} < C_{23}$ or in turn $C_{41} < C_{32}$, the transition between the levels 1_{11} and 1_{10} has absorption against the CMB. It obviously shows that the rate coefficients for collisional transitions play an important role.

We have

$$\frac{n_u^{nt}/n_l^{nt}}{n_u^t/n_l^t} = \exp\left[-\frac{\Delta E_{ul}}{k}\left(\frac{1}{T_{ex}} - \frac{1}{T_k}\right)\right]$$

or

$$\ln\left[\frac{n_u^{nt}/n_l^{nt}}{n_u^t/n_l^t}\right] = -\frac{\Delta E_{ul}}{k}\left(\frac{1}{T_{ex}} - \frac{1}{T_k}\right)$$

For 4.83 GHz line of H_2CO at kinetic temperature $T_k = 100$ K, we have plotted $\ln\left[\frac{n_u^{nt}/n_l^{nt}}{n_u^t/n_l^t}\right]$ versus excitation temperature T_{ex} in Fig. 9. Negative value of the excitation temperature corresponds to the MASER action. The region between the lines $T_{ex} = 0$ and $T_{ex} = 2.73$ corresponds to the anomalous absorption.

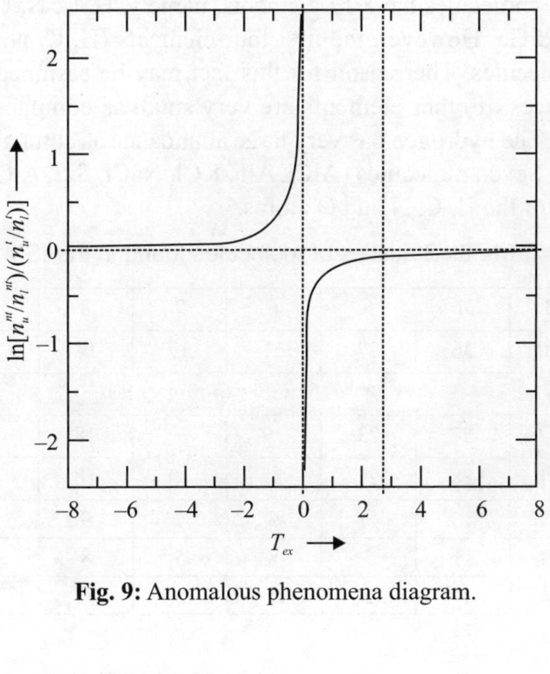

Fig. 9: Anomalous phenomena diagram.

13. Radio Molecule Lines

Electromagnetic radiations having frequency in the range from around 20 kHz to around 300 GHz are known as the radio lines. Such frequencies are produced through the transitions between the rotational levels in a vibrational state of the molecule. The wavelngth of radio lines range from about 0.1 cm to about 150 km. The first detection of molecule in interstellar medium was 18 cm (1665 MHz) radiation of the OH molecule. Information about the first detection some early detected molecules is goven in Table 5, where we have given the name of molecule, year of first detection, and the frequency through which the molecule was detected.

Table 5. First detection of some molecules in cosmic objects

Molecule	Year	Freq (MHz)
OH	1963	1665
NH_3	1968	23694
H_2O	1969	22235
H_2CO	1969	4830
CO	1970	115271
HCN	1970	88632
HC_3N	1970	9098

14. Statistical Analysis

Table 6 has information about various elements present in molecules, given in Table 1. Four molecules (C_{60}, C^+_{60}, C_{70}, C_{24}) are not included here, as they have Carbon element only. Table 6 shows that the molecules have 17 elements, namely, H, C, N, O, D, Si, S, Al, Mg Cl, P, F, K, Na, Fe, Ti and Ge. However, the first four elements (H, C, N, O) has been found in a large number of molecules. The reason for this fact may be assigned to their large cosmic abundances. Abundances of other elements are very small as compared to those of the four (H, C, N, O) elements. The hydrogen has very large abundance as compared to those of Carbon, Nitrogen and Oxygen. Seven molecules (AlCl, AlF, KCl, NaCl, SiS, AlO, D_2S), out of 252, are found not to have any of the H, C, N and O atoms.

Table 6. Statistics of molecules found in the ISM

Atoms	2	3	4	5	6	7	8	9	> 9	Total
Molecules	46	50	36	27	21	19	18	12	19	248
Atom	Number of molecules									
H	16	22	27	23	19	19	18	12	19	175
C	11	31	28	24	21	19	18	12	19	183
N	9	19	15	11	9	7	9	4	7	90
O	15	21	11	5	8	5	8	4	7	84
D	1	9	8	5	4	3				30

Si	4	3	1	1					9
S	7	5	6		1			1	20
Al	3	2							5
Mg	1	2	1						4
Cl	5	1							6
P	3	2	1						6
F	2								2
K	1	1							2
Na	2	2							4
Fe	2	1							3
Ti	1	1							2
Ge				1					

The first row in Table 6 shows the number of atoms in a molecule. The second row gives the number of molecules having the respective number of atoms. For example, out of 248 molecules, 46 molecules are found to have 2 atoms; 50 molecules to have 3 atoms. The columns 2-10 below the third row are the number of molecules where the element (atom) given in the first column (below third row) has been found. For example, there are 16 molecules having 2 atoms where H is present; 31 molecules having 3 atoms where C is present. The data given in the rows 4-7 (for H, C, N, O, respectively) versus the number of atoms (row 1) have been plotted in Fig. 10, where the number of atoms 10 is used for 10 or more atoms in a molecule. In the figure, four molecules (C_{60}, C^+_{60}, C_{70}, C_{24}) are not included, as they have carbon element only.

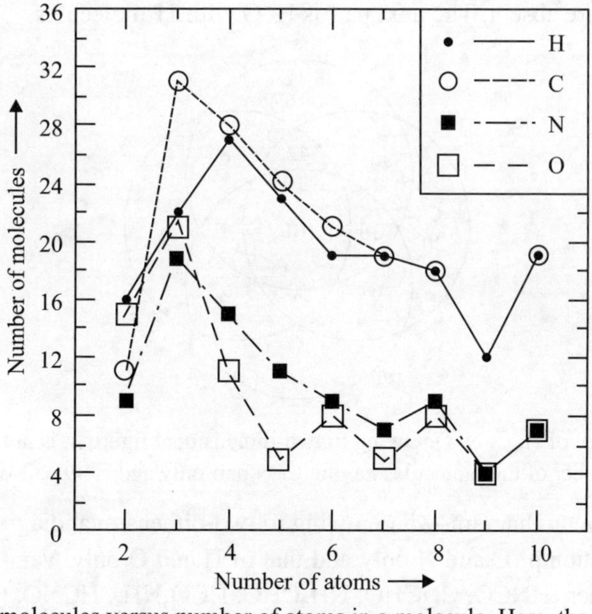

Fig. 10: Number of molecules versus number of atoms in a molecule. Here, the number of atoms 10 denotes 10 or more atoms.

Figure 10 shows the probability of finding C in a molecule (except diatomic molecules) is the largest, in spite of the fact that abundance of C is three thousand times less than that of H, and two times less than that of O. It is also obvious that in a molecule having 7 or more atoms, the elements C and H both are present simultaneously. Further, it is obvious that when besides the C and H, a third element is to be found in a molecule, the probability of existence of N is larger than that of O, in spite of the fact that the cosmic abundance of N is seven times less than that of O.

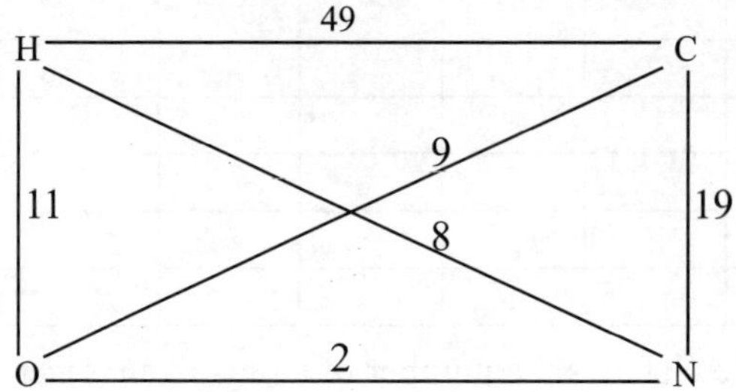

Fig. 11: Number of molecules where various pairs of atoms have been found.

From the curves for N and O in Fig. 10, it may be predicted that the ISM may have more molecules having more than 9 atoms where the probability of existence of H, C and N atoms would be large. Fig. 11 shows the number of molecules where a pair of two elements out of H, C, N, O has been found. For example, there are 9 molecules where C and O have been found, and H and N are absent; 19 molecules where C and N have been found, and H and O are absent. Fig. 11 shows that the pair C–H is found in the largest number (49) of molecules. In these molecules, N and O both are absent. Second pair found in a large number of molecules is C-N; here, H and O both are absent. The next pair is H–O with 11 molecules.

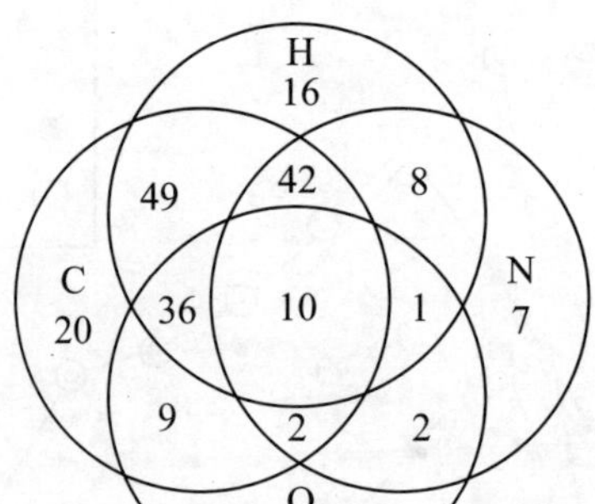

Fig. 12: Venn diagram of H, C, N, O. Owing to two-dimensional figure, it is not possible to show the number of the molecules having C–N pair only and H–O pair only.

Figure 12 is a Venn diagram, where owing to two-dimensional diagram, it is not possible to show the intersection of C and N only and that of H and O only. Venn diagram shows that there are 10 molecules (HNCO, CNCHO, NH_2CHO, HCO_2NH_2, HCNO, $(NH_2)_2CO$, C_3H_3NO, NH_2CH_2OH, CH_3CONH_2, NH_2CH_2COOH) where all the four elements, H, C, N and O, are present simultaneously. There are 42 molecules having H, C and N, and no O; 36 molecules having H, C and O, and no N; 2 molecules having C, N and O, and no H; 1 molecule having H, N and O, and no C. There are 49 molecules having H and C, and no N and O both; 9 molecules having C and O, and no N and H both; 2 molecules having N and O, and no C and H both; 8 molecules having H and N, and no C and O both. There are 20 molecules having C only, and no H, N and O; 16 molecules having H only, and no C, N and O; 7 molecules having N only, and no C, H and O; 13 molecules having O only, and no C, H and N.

From the statistical analysis of 248 molecules, it is found that the probability of finding C in a molecule (except diatomic molecules) is the largest, in spite of the fact that abundance of C is three thousand times less than that of H, and two times less than that of O.

Review Questions

1. Why scientists were against the existence of molecules in the interstellar medium? How their idea changed?
2. Who discovered radio window? What was opinion of scientists after the discovery of radio window?
3. Which was the first molecule detected in the interstellar medium?
4. What is a cosmic molecular laboratory? Write name of 4 cosmic molecular laboratories.
5. Describe how the inertial axes *a*, *b*, *c* for a molecule are assigned. Describe the classification of molecules.
6. What are spherical top, symmetric top and asymmetric top molecules?
7. What are diatomic molecules? Are they prolate or oblate symmetric top molecules?
8. Using following coordinates given in Å of the atoms in H_2O molecule, show that the H_2O is *b*-type asymmetric top molecule.

Atom	*x*	*y*	*z*
H	0.000000	0.762516	–0.467370
H	0.000000	–0.762516	–0.467370
O	0.000000	0.000000	0.116842

9. Using following coordinates given in Å of the atoms in H_2CS molecule, show that the H_2CS is *a*-type asymmetric top molecule.

Structure of H_2CS molecule.

Atom	Coordinates (Å)		
	x	*y*	*z*
H	0.000000	0.920440	–1.603146
H	0.000000	–0.920440	–1.603146
C	0.000000	0.000000	–1.025356
S	0.000000	0.000000	0.584902

10. Using following coordinates given in Å of the atoms in H_2CO molecule, show that the H_2CO is *a*-type asymmetric top molecule.

Atom	Coordinates (Å)		
	x	y	z
H	0.000000	0.937086	−1.114961
H	0.000000	−0.937086	−1.114961
C	0.000000	0.000000	−0.525916
O	0.000000	0.000000	0.673177

11. Describe the rotational spectrum of a rigid diatomic molecule.
12. Describe the rotational spectrum of a non-rigid diatomic molecule.
13. Describe the rotational spectrum of a non-rigid linear polyatomic molecule.
14. Describe the rotational spectrum of a symmetric top molecule.
15. Describe the rotational spectrum of a non-rigid prolate top molecule.
16. Describe the rotational spectrum of a non-rigid oblate top molecule.
17. Describe the rotational spectrum of an asymmetric top molecule.
18. Discuss about the importance of molecules in the interstellar medium.
19. Describe the method of Sobolev large velocity gradient analysis for an asymmetric top molecule.
20. What is radiative transition? Describe spontaneous emission, stimulated absorption and stimulated emission. Derive the relations between the Einstein A and B coefficients.
21. Describe the collisional transitions between rotational levels of an asymmetric top molecule.
22. What is the detailed equilibrium equation for the collisional transitions? Derive it.
23. Write about anomalous phenomena found in the cosmic molecules.
24. Describe how the idea of natural MASERS was obtained.
25. Describe the idea of anomalous absorption.
26. Describe statistical analysis of molecules in the interstellar medium.
27. Write notes on the following:
    (i) Opinion of scientists about molecules in interstellar medium
    (ii) Cosmic molecular laboratories
    (iii) Spherical top molecules
    (iv) Symmetric top molecules
    (v) Asymmetric top molecules
    (vi) Polar and non-polar molecules
    (vii) Rotational spectrum of rigid diatomic molecule
    (viii) Non-rigid diatomic rotator
    (ix) Non-rigid linear polyatomic molecule
    (x) Radiative transitions
    (xi) Collisional transitions
    (xii) Detailed equilibrium for collisional transitions
    (xiii) Anomalous phenomena in cosmic molecules
    (xiv) Natural MASERs
    (xv) Anomalous absorption

(xvi) Statistical analysis of molecules
(xvii) Spontaneous emission
(xviii) Absorption
(xix) Stimulated emission
(xx) Radio molecule lines

Multiple-Choice Questions

1. Scientists were against the existence of molecules in interstellar medium because
   A. γ-rays and x-rays exist in ISM
   B. γ-rays and x-rays are absorbed in ionosphere
   C. UV radians are absorbed by ozone in stratosphere
   D. IR radiations are absorbed in troposphere **Ans. A**
2. Which is the first molecule identified in the ISM?
   A. H_2O B. OH C. CO D. CO_2 **Ans. B**
3. Suppose, I_a, I_b, I_c denote three moments of inertia of molecule about the three principal inertial axes, which relation is correct?
   A. $I_a \geq I_b \geq I_c$ B. $I_a \leq I_b \leq I_c$ C. $I_a = I_b \geq I_c$ D. $I_a \geq I_b = I_c$ **Ans. B**
4. Suppose, I_a, I_b, I_c denote three moments of inertia of molecule about the three principal inertial axes, which relation is impossible?
   A. $I_a \geq I_b \geq I_c$ B. $I_a \leq I_b \leq I_c$ C. $I_a = I_b \leq I_c$ D. $I_a \leq I_b = I_c$ **Ans. A**
5. Suppose, I_a, I_b, I_c denote three moments of inertia of molecule about the three principal inertial axes. A molecule having $I_a = I_b = I_c$ is known as
   A. linear molecule B. spherical top molecule
   C. symmetric top molecule D. asymmetric top molecule **Ans. B**
6. Suppose, I_a, I_b, I_c denote three moments of inertia of molecule about the three principal inertial axes. A molecule having $I_a = 0, I_b = I_c$ is known as
   A. linear molecule B. spherical top molecule
   C. symmetric top molecule D. asymmetric top molecule **Ans. A**
7. Suppose, I_a, I_b, I_c denote three moments of inertia of molecule about the three principal inertial axes. A molecule having $I_a < I_b = I_c$ is known as
   A. linear molecule B. spherical top molecule
   C. symmetric top molecule D. asymmetric top molecule **Ans. C**
8. Suppose, I_a, I_b, I_c denote three moments of inertia of molecule about the three principal inertial axes. A molecule having $I_a = I_b < I_c$ is known as
   A. linear molecule B. spherical top molecule
   C. symmetric top molecule D. asymmetric top molecule **Ans. C**
9. Suppose, I_a, I_b, I_c denote three moments of inertia of molecule about the three principal inertial axes. A molecule having $I_a \neq I_b \neq I_c$ is known as
   A. linear molecule B. spherical top molecule
   C. symmetric top molecule D. asymmetric top molecule **Ans. D**

10. For the rotational constants *A*, *B*, *C*, the ray asymmetric parameter (κ) is
A. $(2B - A - C)/(A - C)$ B. $(2B - A - C)/(C - A)$
C. $(2A - C - B)/(C - B)$ D. $(2C - B - A)/(B - A)$ **Ans. A**
11. Which window of the electromagnetic radiations has the largest scattering?
A. Radio window B. Both radio and optical windows
C. Optical window D. None of them **Ans. C**
12. Which window of the electromagnetic radiations has the lowest scattering?
A. Radio window B. Both radio and optical windows
C. Optical window D. None of them **Ans. A**
13. Which of Einstein coefficients have the same units?
A. A_{ul} and B_{ul} B. A_{ul} and B_{lu} C. B_{lu} and B_{lu} D. A_{ul}, B_{ul} and B_{lu}
Ans. C
14. Einstein coefficient B_{ul} between upper level *u* and lower level *l*, and coefficient B_{lu} are related as
A. $g_l B_{lu} = g_u B_{ul}$ B. $B_{lu} = B_{ul}$ C. $g_u B_{lu} = g_l B_{ul}$ D. $g_l B_{lu} \neq g_u B_{ul}$
Ans. A
15. Which transitions follow the selection rules derived from the quantum mechanics?
A. Radiative B. Radiative and collisional both
C. Collisional D. Neither radiative nor collisional **Ans. A**
16. Which transitions do not follow the selection rules derived from the quantum mechanics?
A. Radiative B. Radiative and collisional both
C. Collisional D. Neither radiative nor collisional **Ans. C**
17. Detailed equilibrium relates
A. Einstein *A*-coefficients B. Einstein *B*-coefficients
C. Collisional rate coefficients D. Einstein *A* and *B* coefficients **Ans. C**
18. Which is the first molecule identified as the natural MASER?
A. H_2O B. OH C. H_2CO D. CO_2 **Ans. B**
19. Which is the first molecule found to show the anomalous absorption?
A. H_2O B. OH C. H_2CO D. *c*-C_3H_2 **Ans. C**
20. Who discovered the cosmic microwave background?
A. Bose and Einstein B. Fermi and Dirac
C. Penzias and Wilson D. Rayleigh and Jeans **Ans. C**

6 Fundamentals

The universe has a large number of objects, generally called the cosmic objects. Examples of them are the stars, planets, satellites, comets, meteors, star clusters, galaxies, nebulae, etc. For the study of a cosmic object, one needs to understand the basic things required for its investigation. This chapter deals with the fundamentals used in the Astronomy and Astrophysics.

1. Parallax

It is also known as the trigonometric parallax. The parallax is a well known phenomenon from the experiments in optics where the position of a real image is coincided with the position of a pin. When the two objects are at different distances, the closer one appears to move relative to the distant one (Fig. 1). The same situation is applicable in case of the closer versus distant stars.

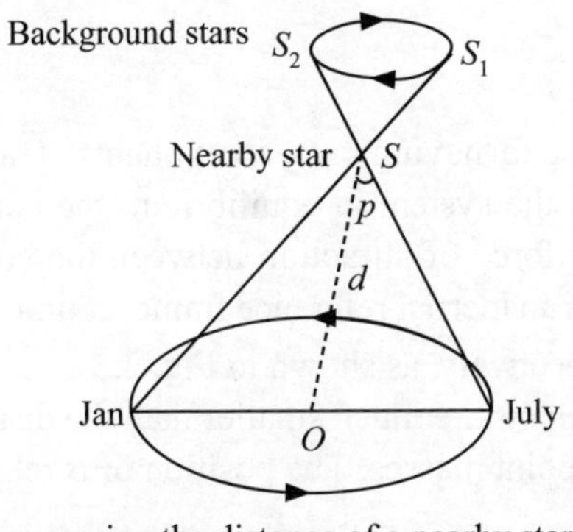

Fig. 1: The parallax method for measuring the distance of a nearby star S. S_1 and S_2 are projections of S on the background of distant stars.

When the observations are made from two opposite extremes in the Earth's orbit around the Sun, a nearby star appears displaced relative to distant stars available in the same part of the celestial sphere. The parallax p is defined as half the apparent angular displacement measured in radian in this way. The distance d to the star is then[1]

$$p = \frac{\text{Astronomical unit}}{d}$$

or
$$d = \frac{1.5 \times 10^{11}\,\text{m}}{p}$$

---

[1]Since the angle is small, we may take p in radians instead of $\tan p$.

Example 1.

The angular separation in the apparent positions of a star relative to the distant stars is 20° 15′ 18″. Calculate the distance of the star from the Sun.

Solution:

1 AU = 1.5×10^{11} m. Half of the angular separation is p = 10° 7′ 39″ = 10.1275° = 0.17667 rad. The distance of the star from the Sun is

$$d = \frac{1.5 \times 10^{11}}{0.17667}$$
$$= 8.49 \times 10^{11} \text{ m}$$

Example 2.

Distance of a star from the Sun is 7.2×10^{11} m. Calculate the parallax shown by the star.

Solution:

We have parallax p

$$p = \frac{\text{Astronomical unit}}{\text{Distance}}$$
$$= \frac{1.5 \times 10^{11} \text{ m}}{7.2 \times 10^{11} \text{ m}}$$
$$= 0.2083 \text{ rad} = 11° \, 56' \, 33.6''$$

2. Two-body System

Let us consider a binary system having two components A and B of masses m_1 and m_2, respectively. In order to keep the system in equilibrium, the components revolve around the centre-of-mass and hence the force of attraction between the components is balanced by the centrifugal forces. Suppose, in an inertial reference frame, at time t, the components are situated at the positions $\vec{r}_1$ and $\vec{r}_2$, respectively, as shown in Fig. 2.

Since the sizes of components are much smaller than the distance under consideration, the masses may be treated as the point masses. The position of B relative to A is

$$\vec{r} = \vec{r}_2 - \vec{r}_1 \tag{6.1}$$

The position vector $\vec{R}$ of the centre-of-mass C is

$$\vec{R} = \frac{m_1\vec{r}_1 + m_2\vec{r}_2}{m_1 + m_2} \tag{6.2}$$

The treatment of a binary system becomes convenient when we consider a coordinate system whose origin is at the centre-of-mass of binary system. In the centre-of-mass system, we have $\vec{R} = 0$ and equation (6.2) gives

$$m_1\vec{r}_1 + m_2\vec{r}_2 = 0 \tag{6.3}$$

Now, $\vec{r}_1$ and $\vec{r}_2$ are the position vectors of the components of the binary relative to the centre-of-mass. Using the expression for $\vec{r}_1$ from equation (6.1) in (6.3), we get

$$m_1(\vec{r}_2 - \vec{r}_1) + m_2\vec{r}_2 = 0$$

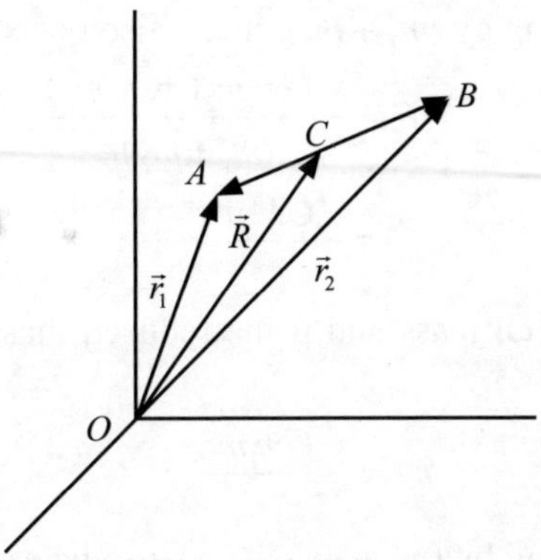

Fig. 2: In an inertial reference frame, at time t, two components A and B of masses m_1 and m_2 of a binary system are situated at the positions $\vec{r}_1$ and $\vec{r}_2$, respectively. $\vec{R}$ position vector of the centre-of-mass C.

On rearranging this relation, we get

$$\vec{r}_2 = \frac{m_1 \vec{r}}{m_1 + m_2} = \frac{\mu}{m_2}\vec{r} \tag{6.4}$$

where $\mu = m_1 m_2/(m_1 + m_2)$ is the reduced mass of the binary system. Using the value of $\vec{r}_2$ in equation (6.1), we get

$$\vec{r} = \frac{m_1 \vec{r}}{m_1 + m_2} - \vec{r}_1$$

and

$$\vec{r}_1 = -\frac{\mu}{m_1}\vec{r}$$

It shows that with respect to the centre-of-mass, the distances of the bodies are inversely proportional to their masses.

2.1 Velocities of Components

Velocity $\vec{v}_1$ of the component A can be obtained by differentiating $\vec{r}_1$ with respect to the time t, as

$$\vec{v}_1 = \frac{d\vec{r}_1}{dt} = -\frac{\mu}{m_1}\frac{d\vec{r}}{dt} = -\frac{\mu}{m_1}\vec{v} \tag{6.5}$$

where $\vec{v}$ is the velocity of B relative to that of A. Velocity $\vec{v}_2$ of the component B can be obtained by differentiating $\vec{r}_2$ with respect to the time t, as

$$\vec{v}_2 = \frac{d\vec{r}_2}{dt} = -\frac{\mu}{m_2}\frac{d\vec{r}}{dt} = -\frac{\mu}{m_2}\vec{v} \tag{6.6}$$

It shows that with respect to the centre-of-mass, the velocities of the bodies are inversely proportional to their masses.

2.2 Total Energy

The gravitational force between the components is

$$F = -\frac{Gm_1 m_2}{r^2}$$

By multiplying and dividing by $(m_1 + m_2)$, it can also be expressed as

$$F = -\frac{Gm_1m_2(m_1 + m_2)}{(m_1 + m_2)r^2}$$

$$= -\frac{GM\mu}{r^2}$$

where $M = (m_1 + m_2)$ is the total mass and μ the reduced mass of the system. The potential energy is

$$U = -\frac{Gm_1m_2}{r}$$

By multiplying and dividing by $(m_1 + m_2)$, it can also be expressed as

$$U = -\frac{Gm_1m_2(m_1 + m_2)}{(m_1 + m_2)r}$$

$$= -\frac{GM\mu}{r}$$

The kinetic energy of binary system is

$$K = \frac{1}{2}m_1v_1^2 + \frac{1}{2}m_2v_2^2$$

Using expressions for v_1 and v_2, we have

$$K = \frac{1}{2}m_1\left(-\frac{\mu}{m_1}v\right)^2 + \frac{1}{2}m_2\left(\frac{\mu}{m_2}v\right)^2$$

$$= \frac{1}{2}\mu^2v^2\left(\frac{1}{m_1} + \frac{1}{m_2}\right)$$

$$= \frac{1}{2}\mu v^2$$

Thus, the total energy E of the system is

$$E = K + U$$

$$= \frac{1}{2}\mu v^2 - \frac{GM\mu}{r}$$

2.3 Angular Momentum

The total orbital angular momentum $\vec{L}$ of the system is

$$\vec{L} = \vec{L}_1 + \vec{L}_2$$

$$= m_1\vec{r}_1 \times \vec{v}_1 + m_2\vec{r}_2 \times \vec{v}_2$$

Using the expressions for $\vec{r}_1$, $\vec{r}_2$, $\vec{v}_1$, $\vec{v}_2$, we get

$$\vec{L} = m_1\left(-\frac{\mu}{m_1}\vec{r}\right) \times \left(-\frac{\mu}{m_1}\vec{v}\right) + m_2\left(\frac{\mu}{m_2}\vec{r}\right) \times \left(\frac{\mu}{m_2}\vec{v}\right)$$

$$= \mu^2 \vec{r} \times \vec{v} \left(\frac{1}{m_1} + \frac{1}{m_2} \right)$$

$$= \mu \vec{r} \times \vec{v}$$

The expressions for total energy and total angular momentum show that in the centre-of-mass reference frame, a two-body problem may be treated as equivalent to one-body problem where a body of mass μ moves at a distance r around a stationary body of mass M.

Case I. System of the Sun and planet

Suppose, the masses of the Sun and a planet are m_s and m_p, respectively. Here, we have $M = m_s + m_p$ and $\mu = m_s m_p/(m_s + m_p)$. Since the mass of planet m_p is negligibly small as compared to the mass of the Sun m_s, we can neglect m_p as compared to m_s and therefore $M = m_s$ and $\mu = m_p$. Thus, we can take the Sun as stationary and planet revolving around the Sun.

Case II. System of nucleus and electron

Suppose, the masses of nucleus and electron are m_n and m_e, respectively. Here, we have $M = m_n + m_e$ and $\mu = m_n m_e/(m_n + m_e)$. Since the mass of electron m_e is negligibly small as compared to the mass of the nucleus m_n, we can neglect m_e as compared to m_n and therefore $M = m_n$ and $\mu = m_e$. Thus, we can take the nucleus as stationary and the electron revolving around the nucleus.

Case III. Binary system of stars

Suppose, the masses of two stars A and B in the binary system are m_a and m_b, respectively. Here, we have $M = m_a + m_b$ and $\mu = m_a m_b/(m_a + m_b)$. Since the masses of the stars are comparable, so the binary system is equivalent to the revolution of a star of mass μ around another star of mass M. If $m_a = m_b = m$, we have $M = 2m$ and $\mu = m/2$.

2.4 Revolution Period

The mass μ is revolving around the mass M at a distance r. The gravitational force between them is balanced by the centrifugal force and we have

$$\frac{GM\mu}{r^2} = \frac{\mu v^2}{r}$$

or
$$v = \sqrt{\frac{GM}{r}}$$

The time-period of revolution is

$$T = \frac{2\pi r}{v} = \frac{2\pi r^{3/2}}{\sqrt{GM}}$$

Example 3.

In a binary system, two stars having masses 5×10^{26} kg and 2×10^{26} kg are at the positions $\vec{r}_1 = 6\hat{i} + 2\hat{j} + 5\hat{k}$ and $\vec{r}_2 = 3\hat{i} - 4\hat{j} + 2\hat{k}$, respectively. Calculate the position vector of the centre-of-mass.

Solution:

The position vector $\vec{R}$ of the centre-of-mass is

$$\vec{R} = \frac{m_1\vec{r}_1 + m_2\vec{r}_2}{m_1 + m_2}$$

On using the values, we get

$$\vec{R} = \frac{5\times10^{26}(6\hat{i} + 2\hat{j} + 5\hat{k}) + 2\times10^{26}(3\hat{i} - 4\hat{j} + 2\hat{k})}{5\times10^{26} + 2\times10^{26}}$$

$$= \frac{36\hat{i} + 2\hat{j} + 29\hat{k}}{7}$$

Example 4.

Show that reduced mass μ of two masses m_1 and m_2 is smaller than both the m_1 and m_2.

Solution:

The reduced mass μ of masses m_1 and m_2 is

$$\frac{1}{\mu} = \frac{1}{m_1} + \frac{1}{m_2}$$

or
$$\mu = \frac{m_1 m_2}{m_1 + m_2}$$

Suppose, $m_1 < m_2$. We have

$$\mu = m_1 \frac{m_2}{m_1 + m_2}$$

Since, $m_2/(m_1 + m_2)$ is smaller than 1, we have $\mu < m_1$. Thus, μ is smaller than both the m_1 and m_2.

Example 5.

Show that the reduced mass of the Sun and a planet is nearly equal to the mass of the planet.

Solution:

Suppose, the mass of the Sun is M and that of the planet m. The reduced mass μ of the combination is

$$\frac{1}{\mu} = \frac{1}{M} + \frac{1}{m}$$

or
$$\mu = \frac{Mm}{M + m}$$

Since $M >> m$, we have

$$\mu \approx \frac{Mm}{m} = m$$

3. Kepler's Laws of Planetary Motion

After analyzing the data about the position and velocity of the planet Mars on various dates, Kepler derived three fundamental laws, which have been found applicable to all the planets revolving around the Sun. These laws known as the Kepler's laws of planetary motion may be stated as the following. (The data used by Kepler were obtained by Tycho Brahe.)

1. A planet revolves around the Sun in an elliptical orbit. This is known as the **law of orbit**. Each of the planets has its elliptical orbit, which is distinct from those for other planets.
2. The line joining one of the foci of the ellipse of orbit and the planet sweeps equal area in equal time. That is, the areal velocity of a planet is constant. This is known as the **law of areal velocity**.
   As shown in Fig. 3, a planet moves in equal times from P to P_1 and from P_2 to P_3, and the areas PSP_1 and P_2SP_3 are equal. Obviously, the speed of the planet during its motion from P to P_1 is faster as compared to that during the motion from P_2 to P_3.
3. For a planet, the square of the time-period for one revolution around the Sun is proportional to the cube of the semi-major axis of the ellipse of the path. This is known as the **law of time-period**.

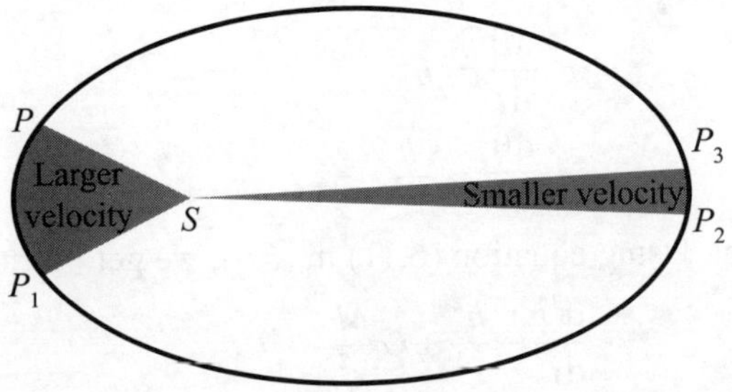

Fig. 3: Motion of a planet around the Sun in elliptical orbit.

It is interesting to note that though these three laws were first derived from the data recorded for the planet Mars, but they are found valid for the revolution of all planets around the Sun. Moreover, the laws are found valid for the revolution of satellites around its planet. These laws can be derived in the following manner.

As we have discussed that the Sun of mass M can be considered stationary and a planet of mass m is revolving around the Sun. Let at a given time, the position vector and the velocity of the planet be $\vec{r}$ and $\vec{v}$, respectively. Then, the total energy E and the total orbital angular momentum $\vec{L}$ of the planet are

$$E = \frac{1}{2}mv^2 - G\frac{mM}{r}$$

and

$$\vec{L} = m\vec{r} \times \vec{v}$$

3.1 Law of Orbit

The force of attraction between the Sun and the planet $\vec{F}$ is

$$\vec{F} = -G\frac{Mm}{r^2}\hat{r}$$

where $\hat{r}$ is unit vector along the radial distance. This is the case of central force. A planet around the Sun revolves under the central force. Thus, the planet has an acceleration

$$\vec{a} = -G\frac{M}{r^2}\hat{r} \tag{6.7}$$

When we use plane polar coordinates (r, θ) with the origin at the mass M, this acceleration can be resolved into the radial and transverse components as:

$$\vec{a} = \left[\frac{d^2r}{dt^2} - r\left(\frac{d\theta}{dt}\right)^2\right]\hat{r} + \frac{1}{r}\frac{d}{dt}\left(r^2\frac{d\theta}{dt}\right)\hat{\theta} \tag{6.8}$$

Comparison of equations (6.7) and (6.8) gives

$$\frac{d^2r}{dt^2} - r\left(\frac{d\theta}{dt}\right)^2 = -G\frac{M}{r^2} \tag{6.9}$$

and

$$\frac{1}{r}\frac{d}{dt}\left(r^2\frac{d\theta}{dt}\right) = 0 \tag{6.10}$$

This equation (6.10) gives

$$r^2\frac{d\theta}{dt} = h$$

or

$$\frac{d\theta}{dt} = \frac{h}{r^2} \tag{6.11}$$

where h is some constant. Using equation (6.11) in (6.9), we get

$$\frac{d^2r}{dt^2} - \frac{h^2}{r^3} + G\frac{M}{r^2} = 0$$

or

$$\frac{d^2r}{dt^2} + \frac{d}{dr}\left(-G\frac{M}{r} + \frac{h^2}{2r^2}\right) = 0$$

Multiplying this equation by dr/dt we get

$$\frac{d^2r}{dt^2}\frac{dr}{dt} + \frac{d}{dr}\left(-G\frac{M}{r} + \frac{h^2}{2r^2}\right)\frac{dr}{dt} = 0$$

$$\frac{d}{dt}\left[\frac{1}{2}\left(\frac{dr}{dt}\right)^2\right] + \frac{d}{dt}\left(-G\frac{M}{r} + \frac{h^2}{2r^2}\right) = 0$$

$$\frac{dt}{dt}\left[\frac{1}{2}\left(\frac{dr}{dt}\right)^2 - G\frac{M}{r} + \frac{h^2}{2r^2}\right] = 0$$

This equation therefore gives

$$\frac{1}{2}\left(\frac{dr}{dt}\right)^2 - G\frac{M}{r} + \frac{h^2}{2r^2} = \frac{E}{m} \tag{6.12}$$

where E/m is a constant. Here, E is the total (kinetic + potential) energy of the system[2]. Equation (6.12) can be rearranged as

---

[2]In the polar coordinates, radial and transverse components of velocity are, respectively,

$$v_r = \frac{dr}{dt} \quad \text{and} \quad v_\theta = r\frac{d\theta}{dt}$$

Thus, the total (kinetic + potential) energy is $E = \frac{1}{2}m\left[\left(\frac{dr}{dt}\right)^2 + r^2\left(\frac{d\theta}{dt}\right)^2\right] + \left(-G\frac{Mm}{r}\right)$

$$\frac{dr}{dt} = \sqrt{2\left(\frac{E}{m} + \frac{GM}{r} - \frac{h^2}{2r^2}\right)} \tag{6.13}$$

Dividing equation (6.13) by (6.11), we get

$$\frac{dr}{d\theta} = \frac{r^2}{h}\sqrt{2\left(\frac{E}{m} + \frac{GM}{r} - \frac{h^2}{2r^2}\right)}$$

$$d\theta = \frac{h}{r^2}\frac{dr}{\sqrt{2\left(\frac{E}{m} + \frac{GM}{r} - \frac{h^2}{2r^2}\right)}} \tag{6.14}$$

This equation can be solved by using the relation

$$r = \frac{1}{u}$$

so that

$$dr = -\frac{1}{u^2}du$$

and equation (6.14) gives

$$d\theta = -\frac{du}{\sqrt{\frac{2E}{mh^2} + \frac{2GMu}{h^2} - u^2}}$$

$$= -\frac{du}{\sqrt{\left(\frac{2E}{mh^2} + \frac{G^2M^2}{h^4}\right) - \left(u - \frac{GM}{h^2}\right)^2}}$$

On integration, we get

$$\theta = \cos^{-1}\left(\frac{u - \frac{GM}{h^2}}{\sqrt{\frac{2E}{mh^2} + \frac{G^2M^2}{h^4}}}\right) + \theta_0$$

where θ_0 is the constant of integration. This equation can be rearranged as

$$u - \frac{GM}{h^2} = \left[\sqrt{\frac{2E}{mh^2} + \frac{G^2M^2}{h^4}}\right]\cos(\theta - \theta_0)$$

Therefore,

$$\frac{1}{r} = \frac{GM}{h^2} + \frac{GM}{h^2}\left[\sqrt{1 + \frac{2h^2E}{mG^2M^2}}\right]\cos(\theta - \theta_0)$$

Using equation (6.11) we get

$$\frac{E}{m} = \frac{1}{2}\left(\frac{dr}{dt}\right)^2 + \frac{h^2}{2r^2} - G\frac{M}{r}$$

$$= \frac{GM}{h^2}\left\{1+\left[\sqrt{1+\frac{2h^2E}{mG^2M^2}}\right]\cos(\theta-\theta_0)\right\}$$

Thus, we have

$$r = \frac{h^2/GM}{\left\{1+\left[\sqrt{1+\frac{2h^2E}{mG^2M^2}}\right]\cos(\theta-\theta_0)\right\}}$$

$$= \frac{a(1-e^2)}{1+e\cos(\theta-\theta_0)}$$

This is equation of ellipse with semi-major axis a and eccentricity e, showing that the path of the motion under a central force is elliptical. This is the Kepler's first law of motion of a planet revolving under the gravitational force. Here, θ_0 is value of θ when r is the minimum, i.e., at the periastron position and

$$e = \sqrt{1+\frac{2h^2E}{mG^2M^2}} \tag{6.15}$$

and

$$a(1-e^2) = \frac{h^2}{GM} \tag{6.16}$$

Using equation (6.15) in (6.16), we get

$$a = -\frac{GMm}{2E} \tag{6.17}$$

It shows that the semi-major axis of the orbit is inversely proportional to the total energy of the system. As *G, M, m, a* all are positive quantities, the negative sign shows that the total energy of the planet is negative.

The total energy is the sum of the kinetic energy and the potential energy. As the kinetic energy can never be negative, it shows that the potential energy is negative and the magnitude of the potential energy is larger than that of the kinetic energy. The negative value of the potential energy is consistent with the attractive force between the Sun and a planet.

3.2 Law of Areal Velocity

Suppose, a body of mass *m* moving in the elliptical path is at a position *P* at time *t* (Fig. 4). After a time interval d*t*, the body reaches at *Q*. Thus, the area swept by the line, joining the focus *O* of the ellipse to the body, in time *dt* is

$$\text{Area} = \frac{1}{2}OP \times PQ$$

$$= \frac{1}{2}r(r\,d\theta)$$

$$= \frac{1}{2}r^2(d\theta)$$

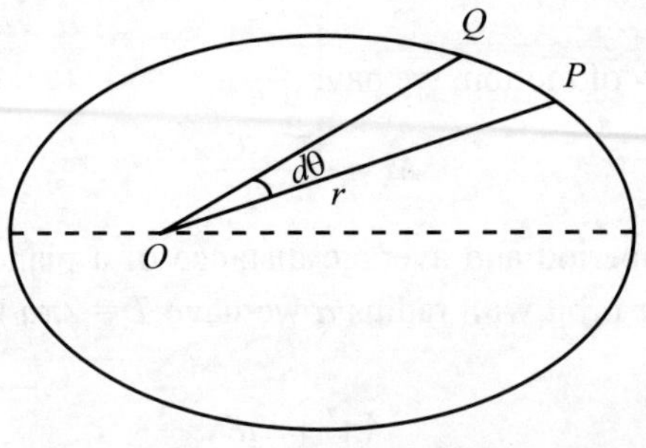

Fig. 4: Body of mass *m* revolving in the elliptical path around a stationary body of mass *M* (situated at *O*), is at a position *P* at time *t*. After a time interval d*t* the body reaches at the position *Q*.

Hence, the areal velocity is

$$\text{Areal velocity} = \frac{\text{Area traversed}}{\text{Time interval}}$$

$$= \frac{1}{2} r^2 \frac{d\theta}{dt} = \frac{h}{2} \tag{6.18}$$

where equation (6.11) is used. This shows that for a particle moving in an elliptical path, the areal velocity is constant. This is the Kepler's second law of motion.

3.3 Law of Time-period

Area of an ellipse with semi-major axis *a* and semi-minor axis *b* is πab. If *T* is the time period of motion, we have

$$\frac{h}{2} = \frac{\pi ab}{T} = \frac{\pi a^2 \sqrt{1-e^2}}{T} \tag{6.19}$$

Here, we have used the relation $b = a\sqrt{1-e^2}$. Using equations (6.17) and (6.15) in (6.19), we have

$$\frac{h}{2} = \frac{\pi a^2}{T} \sqrt{-\frac{2Eh^2}{mG^2M^2}}$$

$$= \frac{\pi h}{T} \sqrt{\frac{a^3}{GM}}$$

Thus, we have

$$1 = \frac{2\pi}{T} \sqrt{\frac{a^3}{GM}}$$

or

$$\frac{a^3}{T^2} = \frac{GM}{4\pi^2} \tag{6.20}$$

This shows that square of the time-period is proportional to the cube of the semi-major axis *a*. This is the Kepler's third law of motion.

3.3.1 Mass of the Sun

From the Kepler's third law of motion, we have

$$M = \frac{4\pi^2}{G}\frac{a^3}{T^2}$$

By knowing the time-period and average distance of a planet, the mass of the Sun can be estimated. For a circular orbit with radius *a* we have $T = 2\pi a/v$, where *v* is the velocity of revolution and thus, we get

$$M = \frac{4\pi^2}{G}\frac{a^3}{(2\pi a/v)^2}$$

$$= \frac{av^2}{G}$$

It shows that with the increase of the radius of the orbit, the orbital velocity of the planet decreases in such a manner so that av^2 remains constant.

Example 6.

The Earth completes one revolution around the Sun in 365.25 days at an average distance of 1.496×10^8 km. Calculate the mass of the Sun.

Solution:

For the Earth revolving around the Sun, we have $a = 1.496 \times 10^8$ km and $T = 365.256$ days and thus

$$M = \frac{4\times(3.14159)^2}{6.67\times10^{-11}}\frac{(1.496\times10^{11})^3}{(365.25\times24\times60\times60)^2}$$

$$= 1.99 \times 10^{30} \text{ kg}$$

Example 7.

A planet is revolving around the Sun in an elliptical orbit. If the minimum and maximum distances of the planet from the Sun are r_{min} and r_{max}, respectively, show that the eccentricity *e* of the orbit is

$$e = \frac{r_{max} - r_{min}}{r_{max} + r_{min}}$$

Solution:

Using the plane polar coordinates (r, θ), the equation of the elliptical orbit is

$$r = \frac{a(1-e^2)}{1+e\cos\theta}$$

where *e* and *a* denote the eccentricity and semi-major axis, respectively, of the ellipse. The distance of planet from the Sun is the minimum when $\theta = 0$ and we have

$$r_{min} = \frac{a(1-e^2)}{1+e}$$

$$= a(1 - e)$$

The distance of planet from the Sun is the maximum when $\theta = \pi$ and we have

$$r_{\max} = \frac{a(1-e^2)}{1-e}$$
$$= a(1+e)$$

Thus, we have

$$r_{\max} + r_{\min} = 2a$$

and
$$r_{\max} - r_{\min} = 2ae$$

It gives

$$e = \frac{r_{\max} - r_{\min}}{r_{\max} + r_{\min}}$$

Example 8.

A planet is revolving around the Sun in an elliptical orbit with eccentricity $e = \sqrt{1+\alpha E}$ and semi-major axis $a = -\beta/E$, where E is the total energy, and α and β are positive constants. Using the Kepler's second law of planetary motion, derive the Kepler's third law of planetary motion.

Solution:

For an ellipse with semi-major axis a, semi-minor axis b and eccentricity e, we have

$$b = a\sqrt{1-e^2}$$

and the area of the ellipse is πab. According to the Kepler's second law of planetary motion, the areal velocity is constant, say v_A. If T denotes the time-period of planetary motion, we have

$$v_A = \frac{\pi ab}{T}$$
$$= \frac{\pi a^2\sqrt{1-e^2}}{T}$$

Using the value of e here, we get

$$v_A T = \pi a^2\sqrt{1-(1+\alpha E)}$$

or
$$v_A^2 T^2 = -\pi^2 a^4 \alpha E$$

Using $E = -\beta/\alpha$, we get

$$v_A^2 T^2 = \pi^2 a^3 \alpha\beta$$

or
$$T^2 \propto a^3$$

It gives the Kepler's third law of planetary motion.

Example 9.

The mean distance of the planet Mars from the Sun is 1.533 times the mean distance between the Earth and the Sun. Calculate the period of revolution of the Mars around the Sun.

Solution:

Following Kepler's third law of planetary motion, the time-periods T_M and T_E of Mars and Earth for revolution around the Sun and their average distances a_M and a_E of Mars and Earth, respectively, are related as the following.

$$\frac{T_M^2}{T_E^2} = \frac{a_M^3}{a_E^3}$$

or
$$T_M = TE\left(\frac{a_M}{a_E}\right)^{3/2}$$

Given, $a_M/a_E = 1.533$. We know, $T_E = 365$ days. Using the values, we get

$$T_M = 365\ (1.533)^{3/2}$$
$$= 692.8 \text{ days}$$

Thus, the time-period of revolution of the Mars around the Sun is 692.8 days.

Note:

Similar exercise can be considered for any pair of planets revolving around the Sun or for the motion of a pair of satellites revolving around a common planet.

Example 10.

The eccentricity of the Earth's orbit for revolution around the Sun is 0.0167. Calculate the ratio of the maximum and the minimum speeds of the Earth in the orbit.

Solution:

The angular momentum of the Earth's motion around the Earth is constant. If at any instant, the velocity is v and the distance from the Sun is r, then we have

$$mvr = \text{constant}$$

or
$$vr = \text{constant}$$

This relation shows that the distance is the minimum when the velocity is the maximum, and vice versa. Thus, we have

$$v_{max}\, r_{min} = v_{min}\, r_{max}$$

or
$$\frac{v_{max}}{v_{min}} = \frac{r_{max}}{r_{min}}$$

For the motion in the elliptical path, the maximum distance r_{max} and minimum distance r_{min} are related to the eccentricity e as

$$\frac{r_{max}}{r_{min}} = \frac{1+e}{1-e}$$

Therefore, we have

$$\frac{v_{max}}{v_{min}} = \frac{1+e}{1-e}$$
$$= \frac{1+0.0167}{1-0.0167} = 1.033$$

4. Newton's Law of Gravitation

About the Newton's law of gravitation, you might have heard a story about the falling of apple from a tree on the head of Newton, from which he got an idea for the law of gravitation between two point-masses separated by a distance. The story may not be correct, but it sounds interesting.

On the scientific ground, it may not be ruled out that the law of gravitation would have been derived from the Kepler's laws for the planetary motion, as discussed in the following.

According to Kepler, a planet (mass m) revolves around the Sun (mass M) in an elliptical orbit. When we use plane polar coordinates (r, θ) with the origin at the Sun, the areal velocity is

$$\text{Areal velocity} = \frac{1}{2} r^2 \frac{d\theta}{dt}$$

According to the Kepler's second law of planetary motion, the areal velocity is constant and therefore,

$$\frac{1}{2} r^2 \frac{d\theta}{dt} = \frac{h}{2}$$

or

$$r^2 \frac{d\theta}{dt} = h \tag{6.21}$$

where h is constant and factor 2 is introduced for convenience. For a motion along a curved path, the radial component a_r and transverse component a_θ of acceleration are expressed as

$$a_r = \frac{d^2 r}{dt^2} - r\left(\frac{d\theta}{dt}\right)^2$$

and

$$a_\theta = \frac{1}{r}\frac{d}{dt}\left(r^2 \frac{d\theta}{dt}\right)$$

Use of equation (6.21) here gives $a_\theta = 0$, showing that the transverse component of acceleration is zero and there is acceleration only along the line joining the planet and the Sun. Now, we want to calculate a_r. From equation (6.21), we have

$$\frac{d\theta}{dt} = hu^2 \tag{6.22}$$

where $r = 1/u$. On differentiation of r with respect to t, we have

$$\frac{dr}{dt} = -\frac{1}{u^2}\frac{du}{dt}$$

$$= -\frac{1}{u^2}\frac{du}{d\theta}\frac{d\theta}{dt}$$

Using equation (6.22) here, we have

$$\frac{dr}{dt} = -h\frac{du}{d\theta}$$

Differentiation of this equation with respect to t gives

$$\frac{d^2 r}{dt^2} = -h\frac{d}{dt}\left(\frac{du}{d\theta}\right)$$

$$= -h\frac{d}{d\theta}\left(\frac{du}{d\theta}\right)\frac{d\theta}{dt}$$

$$= -h^2u^2 \frac{d^2u}{d\theta^2} \tag{6.23}$$

Here, equation (6.22) is used. Using equations (6.22) and (6.23), the expression for radial component of acceleration a_r is

$$a_r = -h^2u^2 \frac{d^2u}{d\theta^2} - \frac{1}{u}h^2u^4$$

$$= -h^2u^2 \left[\frac{d^2u}{d\theta^2} + u\right] \tag{6.24}$$

The elliptical orbit of planet can be expressed as

$$\frac{\ell}{r} = 1 + e \cos \theta$$

or
$$\ell u = 1 + e \cos \theta$$

where e is eccentricity and $\ell = h^2/GM$. Differentiation of this equation with respect to θ twice gives,

$$\ell \frac{d^2u}{d\theta^2} = -e\cos\theta$$

Therefore, we have

$$\ell \frac{d^2u}{d\theta^2} + \ell u = 1$$

or
$$\frac{d^2u}{d\theta^2} + u = \frac{GM}{h^2} \tag{6.25}$$

Using equation (6.25) in (6.24), we have

$$a_r = -GMu^2 = -\frac{GM}{r^2}$$

Thus, the force F on the planet is

$$F = ma_r = -G\frac{mM}{r^2}$$

This shows how the Newton's law of gravitation can be derived from the Kepler's law of planetary motion.

5. Virial Theorem

This theorem helps us in understanding the dynamic behaviour of a system having a large number of bodies. It, however, does not tell about the precise behaviour of any individual body in the system. The theorem states that for a gravitationally bound system of bodies in equilibrium, the total energy of the system is always one-half of the time averaged potential energy. If $\langle U \rangle$ and E denote the time-averaged potential energy and the total energy, respectively, of a system of bodies, then according to the Virial theorem, $E = \langle U \rangle/2$. Since the system of bodies is necessarily bound, and therefore, the E and $\langle U \rangle$ both are negative.

In order to prove the Virial theorem, let us consider a system of large number of bodies of masses m_i, situated at the positions $\vec{r}_i$ and acted upon by the forces $\vec{F}_i$. The linear momentum $\vec{p}_i$ of a particle is expressed as $\vec{p}_i = m_i\,(\mathrm{d}\vec{r}_i/\mathrm{d}t)$, and the force $\vec{F}_i$ is expressed as $\vec{F}_i = (\mathrm{d}\vec{p}_i/\mathrm{d}t)$. Let us consider a quantity

$$Q = \sum_i \vec{p}_i \cdot \vec{r}_i$$

so that

$$\frac{\mathrm{d}Q}{\mathrm{d}t} = \sum_i \left(\frac{\mathrm{d}\vec{p}_i}{\mathrm{d}t} \cdot \vec{r}_i + \vec{p}_i \cdot \frac{\mathrm{d}\vec{r}_i}{\mathrm{d}t} \right)$$

Now, we have

$$\begin{aligned} Q &= \sum_i \vec{p}_i \cdot \vec{r}_i \\ &= \sum_i m_i \frac{\mathrm{d}\vec{r}_i}{\mathrm{d}t} \cdot \vec{r}_i \\ &= \sum_i \frac{m_i}{2} \frac{\mathrm{d}(\vec{r}_i \cdot \vec{r}_i)}{\mathrm{d}t} \\ &= \frac{1}{2} \frac{\mathrm{d}\sum_i m_i r_i^2}{\mathrm{d}t} = \frac{1}{2} \frac{\mathrm{d}I}{\mathrm{d}t} \end{aligned}$$

where, $I = \Sigma m_i r_i^2$ is the moment of inertia of the system about some axis. Thus, we have

$$\frac{\mathrm{d}Q}{\mathrm{d}t} = \frac{1}{2} \frac{\mathrm{d}^2 I}{\mathrm{d}t^2}$$

Now, we have

$$\vec{p}_i \cdot \frac{\mathrm{d}\vec{r}_i}{\mathrm{d}t} = m_i \vec{v}_i \cdot \vec{v}_i = m_i v_i^2$$

Therefore, we get

$$\sum_i \vec{p}_i \cdot \frac{\mathrm{d}\vec{r}_i}{\mathrm{d}t} = \sum_i m_i v_i^2 = 2K$$

where, $K = \frac{1}{2} \sum m_i v_i^2$ is the kinetic energy of the system. Thus, we have

$$\frac{1}{2} \frac{\mathrm{d}^2 I}{\mathrm{d}t^2} = \sum_i \vec{F}_i \cdot \vec{r}_i + 2K$$

or

$$\sum_i \vec{F}_i \cdot \vec{r}_i = \frac{1}{2} \frac{\mathrm{d}^2 I}{\mathrm{d}t^2} - 2K \tag{6.26}$$

Let $\vec{F}_{ij}$ represent the force of interaction on the particle *i* due to the particle *j* in the system, then from the Newton's third law of motion we have $\vec{F}_{ij} = -\vec{F}_{ji}$, and the total force acting on the particle *i* due to all other particles in the system is

$$\vec{F}_i = \sum_j{}' \vec{F}_{ij}$$

where the prime over the summation sign denotes that the sum is taken over all values of *j* except that $j = i$. This convention is followed later on also. Then,

$$\sum_i \vec{F}_i \cdot \vec{r}_i = \sum_i \Big(\sum_{j\neq i} \vec{F}_{ij}\Big) \cdot \vec{r}_i$$

$$= \frac{1}{2}\sum_i \Big[\sum_j{}' \big(\vec{F}_{ij} - \vec{F}_{ji}\big)\Big] \cdot \vec{r}_i$$

After some manipulation, we can write[3]

$$\sum_i \vec{F} \cdot \vec{r}_i = \frac{1}{2}\sum_i \sum_j{}' \vec{F}_{ij} \cdot (\vec{r}_i - \vec{r}_j) \tag{6.27}$$

The force $\vec{F}_{ij}$ on the particle *i* due to the particle *j* is

$$\vec{F}_{ij} = G\frac{m_i m_j}{r_{ij}^2}\hat{r}_{ij} \tag{6.28}$$

where $\vec{r}_i$ and $\vec{r}_j$ are position vectors of the particles *i* and *j*, respectively, and $r_{ij} = |\vec{r}_j - \vec{r}_i|$ is the separation between the particles *i* and *j*, and $\hat{r}_{ij}$ is the unit vector in the direction from *i* to *j*:

$$\hat{r}_{ij} = \frac{\vec{r}_j - \vec{r}_i}{r_{ij}} \tag{6.29}$$

Using equations (6.28) and (6.29) in (6.27), we get

$$\sum_i \vec{F}_i \cdot \vec{r}_i = \frac{1}{2}\sum_i \sum_j{}' G\frac{m_i m_j}{r_{ij}^2}\frac{(\vec{r}_j - \vec{r}_j)}{r_{ij}} \cdot (\vec{r}_i - \vec{r}_j)$$

$$= -\frac{1}{2}\sum_i \sum_j{}' G\frac{m_i m_j}{r_{ij}^3}(\vec{r}_i - \vec{r}_j)^2$$

$$= -\frac{1}{2}\sum_i \sum_j{}' G\frac{m_i m_j}{r_{ij}}$$

$$= -\frac{1}{2}\sum_i \sum_j{}' U_{ij} = U \tag{6.30}$$

where $$U_{ij} = -G\frac{m_i m_j}{r_{ij}}$$

is the potential between the bodies *i* and *j*, and *U* is the total potential energy of the system. Thus, from equations (6.26) and (6.30), we have

$$\frac{1}{2}\left\langle \frac{\mathrm{d}^2 I}{\mathrm{d}t^2} \right\rangle - 2\langle K\rangle = \langle U\rangle$$

[3]In order to understand it, let us consider a case of 3 bodies in a system

$$\sum_{i=1}^{3}\Bigg[\sum_{\substack{j=1\\ j\neq i}}^{3}(\vec{F}_{ij} - \vec{F}_{ji})\Bigg] \cdot \vec{r}_i = (\vec{F}_{12} - \vec{F}_{21}) \cdot \vec{r}_1 + (\vec{F}_{13} - \vec{F}_{31}) \cdot \vec{r}_1 + (\vec{F}_{21} - \vec{F}_{12}) \cdot \vec{r}_2 + (\vec{F}_{23} - \vec{F}_{32}) \cdot \vec{r}_2$$

$$+ (\vec{F}_{31} - \vec{F}_{13}) \cdot \vec{r}_3 + (\vec{F}_{32} - \vec{F}_{23}) \cdot \vec{r}_3 = \vec{F}_{12}(\vec{r}_1 - \vec{r}_2) + \vec{F}_{13} \cdot (\vec{r}_1 - \vec{r}_3) + \vec{F}_{21} \cdot (\vec{r}_2 - \vec{r}_1)$$

$$+ \vec{F}_{23} \cdot (\rightarrow\vec{r}_2 - \vec{r}_3) + \vec{F}_{31} \cdot (\vec{r}_3 - \vec{r}_1) + \vec{F}_{32} \cdot (\vec{r}_3 - \vec{r}_2)$$

$$= \sum_{i=1}^{3}\sum_{\substack{j=1\\ j\neq i}}^{3} \vec{F}_{ij} \cdot (\vec{r}_i - \vec{r}_j)$$

Now, the average may be expressed as

$$\left\langle \frac{d^2I}{dt^2} \right\rangle = \frac{1}{\tau}\int_0^{\tau} \frac{d^2I}{dt^2}$$

$$= \frac{1}{\tau}\left(\left.\frac{dI}{dt}\right|_{\tau} - \left.\frac{dI}{dt}\right|_{0} \right)$$

If the system is periodic with a period τ, we have

$$\left.\frac{dI}{dt}\right|_{\tau} = \left.\frac{dI}{dt}\right|_{0}$$

and thus,

$$\left\langle \frac{d^2I}{dt^2} \right\rangle = 0$$

Even if the system is not strictly periodic, the average $\langle d^2I/dt^2 \rangle$ is zero when it is evaluated over a sufficiently long period of time (i.e., $\tau \to \infty$) assuming that dI/dt is bound. This would describe, for example, a systsem that has attained steady state configuration (an equilibrium). In either case, we have $\langle d^2I/dt^2 \rangle = 0$. Thus, we get

$$2\langle K \rangle + \langle U \rangle = 0$$

so that

$$\langle K \rangle = -\frac{1}{2}\langle U \rangle$$

Now, we have

$$\langle E \rangle = \langle K \rangle + \langle U \rangle = -\frac{1}{2}\langle U \rangle + \langle U \rangle = \frac{1}{2}\langle U \rangle$$

This proves the Virial theorem, which applies to a wide variety of systems ranging from the particles in an ideal gas to a cluster of galaxies.

6. Cosmic Abundances

Abundance of a chemical element (common or rare) is a measure of relative presence of the element as compared to all other elements in a given environment. The abundance may be measured in terms of the mass fraction (the same as weight fraction), mole fraction (fraction of atoms by numerical count), or volume fraction. A common way for representation of abundance is in terms of the volume fraction in mixed gases, such as the planetary atmospheres, and is close to the molecular mole fraction for ideal gas mixtures (i.e., the gas mixtures at relatively low densities and pressures).

For example, in pure water (H_2O), the mass fraction abundance of oxygen is about 89% because it is the fraction of water's mass which is oxygen. However, the mole fraction abundance of oxygen atoms in water is 33%, because out of 3 atoms in water, there only 1 atom of oxygen.

In the universe as a whole, and in the atmospheres of giant planets having gas, such as Jupiter, the mass fraction abundances of hydrogen and helium are about 74% and 23-25%, respectively, whereas their (atomic) mole fractions are closer to 92% and 8%, respectively. However, for example, in the outer atmosphere of Jupiter, hydrogen is in the molecular form whereas the helium is in the atomic form. Hence, the molecular mole fraction (fraction of total gas molecules), as well as the fraction of atmosphere by volume, of the hydrogen is about 86%, and for the helium is about 13%.

The mass fraction of ten most common elements in the Milky Way Galaxy estimated spectroscopically are given in Table 1.

Table 1. Ten most common elements in Milky Way Galaxy

Z	Element	Mass fraction
1	Hydrogen	739,000
2	Helium	240,000
8	Oxygen	10,400
6	Carbon	4,600
10	Neon	1,340
26	Iron	1,090
7	Nitrogen	960
14	Silicon	650
12	Magnesium	580
16	Sulphur	440

Following standard cosmological model, the lightest elements were mostly produced shortly (i.e., within a few hundred seconds) after the Big Bang through the nucleosynthesis. The heavier elements were mostly produced much later inside the stars.

Most of the baryonic mass in the universe is estimated to be in the form of hydrogen (74%) and helium (24%). Though the remaining heavy elements are small fraction, but they can greatly influence the astronomical phenomena. Other than the hydrogen or helium, in Astronomy, all the elements are put under the category, named as the metal. This distinction is significant because hydrogen and helium are the only elements that were produced in the significant quantities in the Big Bang, whereas the others were produced later on. Hence, the metalicity of a galaxy or other object is an indication of the past stellar activity. In the disk of the Milky Way galaxy, the mass of heavy elements is only about 2%.

Hydrogen and helium are most common elements, from the Big Bang. The next three elements (Li, Be, B) are rare because they were poorly synthesized in the Big Bang and also in the stars. The two general trends in the remaining elements produced in the stars are as the following.

(i) The abundances follow the Oddo-Harkins rule which states that an element with an even atomic number is more abundant than both elements with the adjacently smaller and larger atomic numbers. For example, the carbon (atomic number 6) is more abundant than both the boron (atomic number 5) and nitrogen (atomic number 7).

(ii) The abundance of element decreases in general with the increase of atomic number.

Iron is especially common because it represents the minimum energy nuclide that can be made by the fusion of helium in the supernovae.

7. Henry-Draper Star Catalogue

Henry Draper, an amateur astronomer, worked on the spectra of stars. Between 1918 and 1924, an astronomical star catalogue, named Henry Draper Catalogue (HD) was published where the spectroscopic classifications for 225,300 stars was given. Extended version of this catalogue was published between 1925 and 1936 with the name, Henry Draper Extension (HDE) where 46,850 additional stars were included. From 1937 to 1949 the Henry Draper Extension Charts (HDEC) were published in the form of charts, where 86,933 additional stars were included. As of August 2017, 359,083 stars in all were classified. The numbers given in the catalogue are commonly used as a way of identifying stars. The HD catalogue covers the entire sky almost completely down to an apparent photographic magnitude of about 9; the extensions added fainter stars in certain areas of the sky. The construction of the Henry Draper Catalogue was a part of a pioneering effort to classify the stellar spectra.

8. Stellar Classification

All the stars are not similar in their characteristics. Based on their characteristics, the stars are classified into different categories. Basically, there are two classification schemes: the MK system and the Harvard system.

8.1 M-K System

The Morgan-Keenan (MK) system of stellar classification is based on the star's luminosity. Originally, the MK luminosity class was containing roman numerals between I (supergiant star) and V (main sequence). These days, class I stars have been subdivided into Ia-O, Ia and Ib, and classes VI (sub-dwarf) and D (white dwarf) have been added. The following table summarises the MK Luminosity Classes:

Class	Star
Ia-O	extremely luminous supergiants
Ia	luminous supergiants
Ib	less luminous supergiants
II	bright giants
III	normal giants
IV	subgiants
V	main sequence dwarf stars
VI or sd	subdwarfs
D	white dwarfs

8.2 Harvard Scheme

The Harvard system of stellar classification is based on the star's surface temperature. The basic idea behind the spectral classification is that for a given chemical composition, the pattern of absorption lines which are formed in the photosphere of a star depends on the temperature and pressure in the photosphere. Indeed, a great variety of stellar spectra can be observed for different stars. This great variety arises from the variation of the effective temperature of star than they do from the variation of the chemical composition or surface pressure. Thus, to a first approximation, the spectral type of a star yields an estimate of its effective temperature T_e. That is, the spectral type is a function of T_e:

$$\text{Spectral type} = f(T_e)$$

Historically, the spectral types were assigned alphabetically in an order determined primarily by the decreasing strength of hydrogen absorption lines. The stars with the strongest hydrogen lines were called A type stars; the stars with the next strongest hydrogen lines were called B type stars; and so on. It was soon realized that this assignment did not give sequence to the stars according to their effective temperatures.

Through the work of Saha, astronomers learned that O stars have weak hydrogen lines because their photospheres are so hot that the hydrogen is almost ionized, whereas M stars, which are alphabetically close to O stars, have weak hydrogen lines for a completely different reason. The M stars have such cool photospheres that there is little atomic hydrogen in the first excited ($n = 2$) state from where an absorption of a stellar photon can produce a hydrogen line in the Balmer series, which is visible at the optical wavelengths.

At a relatively low temperature of an M star's photosphere, hydrogen atoms prefer to be in their ground electronic state. At the relatively high temperature of an O star's photosphere, they prefer to be ionized. When the spectral types are ordered in a sequence of decreasing effective temperature, they appear in the sequence[4]:

OBAFGKM

Table 2 shows the effective temperature for various classes. The Spectral classes are divided into subclasses denoted by the numbers 0, 1, ..., 9; sometimes decimals are used, e.g., B0.4.

The original classification scheme compiled in the Henry Draper (HD) Catalog included other spectral types, such as N, R, and S. It is now known that such stars have different-looking spectra, not because they are cooler than M stars, but because they have peculiar heavy-element abundances. The R and N stars are rich in carbon compounds. These carbon-stars have effective temperatures comparable to those of the K and M stars. The S stars have photospheres with enhanced abundances of s-process elements. The additional classes R, N, and S are often appended in parentheses to the main spectral types:

OBAFGKM (RNS)

To these classes finer subdivisions are introduced by appending the numbers, ranging from 0 to 9 to each spectral class. For example, G2 is the class of the Sun.

To completely describe a star, the MK luminosity class is appended to the original Harvard classification for the star. For example, the Sun is a main sequence G2 star, therefore its full classification is G2V.

[4]Some students memorize via the mnemonic: Oh Be A Fine Girl Kiss Me!

Table 2. Temperature versus class

Class	Effective temperature
O	≥ 30,000 K
B	10,000 - 30,000 K
A	7,500 - 10,000 K
F	6,000 - 7,500 K
G	5,200 - 6,000 K
K	3,700 - 5,200 K

9. Black-body Spectrum

A body which can absorb radiations of all wavelengths falling on it is known as the black-body. According to the Kirchhoff's law, a good absorber of radiations of particular wavelengths is also a good emitter of the radiations of the same wavelengths. Thus, a black-body can emit radiations of all wavelengths. When the temperature of a black-body is larger than that of the atmosphere around it, it emits the radiations, whereas it absorbs the radiations when the temperature of black-body is less than that of the atmosphere around it.

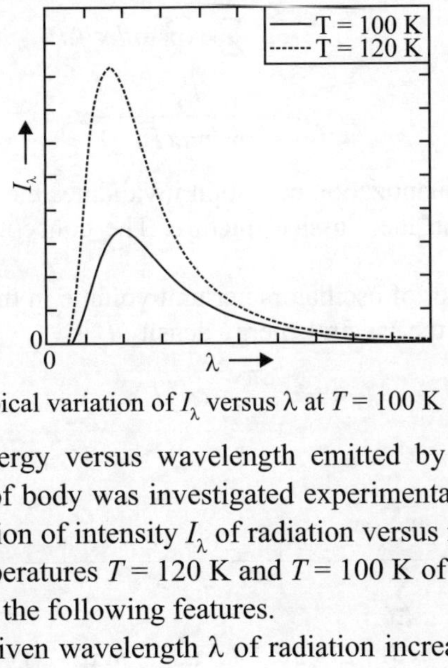

Fig. 5: Typical variation of I_λ versus λ at $T = 100$ K and 120 K.

The distribution of energy versus wavelength emitted by a black-body for different values for the temperature of body was investigated experimentally in 1899 by Lummer and Pringsheim. A typical variation of intensity I_λ of radiation versus its wavelength λ is shown in Fig. 5 for two different temperatures $T = 120$ K and $T = 100$ K of black-body. The black-body spectrum was found to have the following features.

(i) The intensity I_λ for a given wavelength λ of radiation increases with the increase of the temperature T of the body.

(ii) For a given value of temperature T, the intensity I_λ becomes the maximum at some wavelength, denoted by λ_m.

(iii) With the increase of the temperature T of the body, the value of λ_m is found to decrease in a manner such that $\lambda_m T =$ constant.

(iv) The area under the curve was found proportional to T^4.

In 1900, Max Planck made a theoretical attempt to explain the black-body spectrum. He modified the Wien's formula in such a way that the experimental curves could be fitted precisely. Then, he started looking for a sound basis for his modifications. Planck proposed a hypothesis where he considered that the atoms of the wall of a cavity behave like the oscillators, each of them oscillating with a characteristic frequency. Further, he assumed that these oscillators exchange electromagnetic energy with the cavity and for these atomic oscillators, he made the following assumptions:

(i) An oscillator with a characteristic frequency ν, can have energies $E_n = nh\nu$, where h is the Planck constant and the integer n can have integer values 0, 1, 2, 3, ...

(ii) An oscillator can exchange energy with the cavity only in the discrete units. That is, $\Delta E_n = \Delta n\, h\nu$, where $\Delta n = 0, 1, 2, 3, ...$ Thus, the exchange of energy is in the form of the packets of energy, known as the quanta. Thesc quanta were later on named as the photons.

In the Planck's hypothesis, the average energy $\langle E \rangle$ of an oscillator is[5]

$$\langle E \rangle = \frac{\sum_{n=0}^{\infty} E_n \exp(-E_n/kT)}{\sum_{n=0}^{\infty} \exp(-E_n/kT)}$$

$$= \frac{\sum_{n=0}^{\infty} nh\nu \exp(-nh\nu/kT)}{\sum_{n=0}^{\infty} \exp(-nh\nu/kT)}$$

$$= \frac{h\nu}{\exp(h\nu/kT) - 1}$$

In this hypothesis, the quantization condition invalidates the theorem of the equipartition of energy, which is based on the classical picture. The concept of the quantum theory are introduced the first time.

We know that the number of oscillators per unit volume in the frequency range from ν to ν + dν is $(8\pi\nu^2/c^3)$ dν. Thus, the spectral energy density I_ν is

$$I_\nu \, d\nu = \frac{8\pi\nu^2}{c^3} d\nu \langle E \rangle$$

[5]Using $x = \exp(-h\nu/kT)$, we have

$$\langle E \rangle = \frac{h\nu \sum_{n=0}^{\infty} nx^n}{\sum_{n=0}^{\infty} x^n} = \frac{h\nu[x + 2x^2 + 3x^3 + 4x^4 + ...]}{1 + x + x^2 + x^3 + x^4 + ...}$$

$$= \frac{h\nu x[1 + 2x + 3x^2 + 4x^3 + ...]}{1 + x + x^2 + x^3 + x^4 + ...} = \frac{h\nu x(1-x)^{-2}}{(1-x)^{-1}}$$

$$= \frac{h\nu x}{1 - x}$$

Using the expression for x, we have

$$\langle E \rangle = \frac{h\nu \exp(-h\nu/kT)}{1 - \exp(-h\nu/kT)} = \frac{h\nu}{\exp(h\nu/kT) - 1}$$

$$= \frac{8\pi h\nu^3}{c^3} \frac{1}{\exp(h\nu/kT) - 1} d\nu \qquad (6.31)$$

Here, we have used the value of $\langle E \rangle$. This relation is known as the Planck's radiation formula, and it explains successfully the black-body spectrum for all the wavelengths. Thus, the energy density I_ν

$$I_\nu = \frac{8\pi h\nu^3}{c^3} \frac{1}{\exp(h\nu/kT) - 1}$$

This is Planck formula for the black-body spectrum.

Special Cases

It can be easily shown that the Wien's formula, and the Rayleigh and Jeans formula are special cases of this expression.

(i) **Wien's formula:** When $h\nu >> kT$, one can neglect 1 in comparison to the exponential term in denominator and we have the Wien's formula.

$$I_\nu d\nu = \frac{8\pi h\nu^3}{c^3} \exp(-h\nu/kT)\, d\nu$$

(ii) **Rayleigh and Jeans formula:** When $h\nu << kT$, the exponential term can be expanded and we have

$$I_\nu\, d\nu = \frac{8\pi h\nu^3}{c^3}\left[1 + \frac{h\nu}{kT} + \frac{1}{2!}\left(\frac{h\nu}{kT}\right)^2 + \ldots - 1\right]^{-1} d\nu$$

$$= \frac{8\pi kT\nu^2}{c^3}\left[1 + \frac{1}{2!}\frac{h\nu}{kT} + \ldots\right]^{-1} d\nu$$

On neglecting the small terms, we get the Rayleigh and Jeans formula.

$$I_\nu d\nu = \frac{8\pi kT\nu^2}{c^3} d\nu$$

Planck's Radiation Formula in Terms of Wavelengths

Planck's radiation formula in terms of wavelengths can be obtained by using the relation $\nu = c/\lambda$, so that $d\nu = -c\, d\lambda/\lambda^2$, in equation (6.31) as the following.

$$I_\lambda d\lambda = \frac{8\pi ch}{\lambda^5} \frac{1}{\exp(hc/\lambda kT) - 1} d\lambda$$

Since we are concerned with the intervals in frequency and wavelength, the negative sign is not considered. Thus, the energy density I_λ is

$$I_\lambda = \frac{8\pi ch}{\lambda^5} \frac{1}{\exp(hc/\lambda kT) - 1} \qquad (6.32)$$

Special Cases

It can be easily shown that the Wien's formula, and the Rayleigh and Jeans formula are special cases of this expression.

(i) Wien's formula: When $(hc/\lambda) >> kT$, one can neglect 1 in comparison to the exponential term in the denominator and we have the Wien's formula.

$$I_\lambda d\lambda = \frac{8\pi ch}{\lambda^5} \exp(-hc/\lambda kT)\, d\lambda$$

(ii) Rayleigh and Jeans formula: When $(hc/\lambda) << kT$, the exponential term can be expanded and we have

$$I_\lambda\, d\lambda = \frac{8\pi ch}{\lambda^5}\left[1 + \frac{hc}{\lambda kT} + \frac{1}{2!}\left(\frac{hc}{\lambda kT}\right)^2 + \ldots - 1\right]^{-1} d\lambda$$

$$= \frac{8\pi kT}{\lambda^4}\left[1 + \frac{1}{2!}\frac{hc}{\lambda kT} + \ldots\right]^{-1} d\lambda$$

On neglecting the small terms, we get the Rayleigh and Jeans formula.

$$I_\lambda d\lambda = \frac{8\pi kT}{\lambda^4} d\lambda$$

9.1 Wien's Displacement Law

For a given temperature T, if λ_m is the wavelength corresponding to the maximum value of I_λ, then we have

$$\left(\frac{\partial I_\lambda}{\partial \lambda}\right)_{\lambda=\lambda_m} = 0 \qquad (6.33)$$

For convenience, we take logarithm of equation (6.32), and get

$$\ln I_\lambda = \ln(8\pi ch) - 5\ln\lambda - \ln[\exp(hc/\lambda kT) - 1]$$

Differentiation of this equation with respect to λ gives

$$\frac{1}{I_\lambda}\frac{\partial I_\lambda}{\partial \lambda} = -\frac{5}{\lambda} + \frac{hc}{\lambda^2 kT}\frac{\exp(hc/\lambda kT)}{\exp(hc/\lambda kT) - 1}$$

On applying the condition (6.33), we get

$$-5 + \frac{hc}{\lambda_m kT}\frac{\exp(hc/\lambda_m kT)}{\exp(hc/\lambda_m kT) - 1} = 0$$

Putting $x = hc/\lambda_m kT$, we have

$$\frac{xe^x}{e^x - 1} = 5$$

or

$$\frac{x}{1 - e^{-x}} = 5$$

This equation can be solved numerically to get $x = 4.9651$. Thus, we have

$$\frac{hc}{\lambda_m kT} = 4.9651$$

or

$$\lambda_m T = 2.898 \times 10^{-3} \text{ m K}$$

where we have used $h = 6.62 \times 10^{-34}$ J s, $k = 1.38 \times 10^{-23}$ J/K and $c = 3 \times 10^8$ m/s. This relation is known as the Wien's displacement law.

Example 11.

At what wavelength the Sun emits maximum number of photons. Surface temperature of the Sun is 6000 K.

Solution:

According to the Wien's displacement law, we have

$$\lambda_m T = 2.898 \times 10^{-3}$$

or

$$\lambda_m 6000 = 2.898 \times 10^{-3}$$

On solving, we have $\lambda_m = 4.83 \times 10^{-7}$ m = 4830 Å. Thus, the largest number of photons emitted from the Sun is of wavelength λ = 4830 Å.

Example 12.

The spectrum of the Moon gives two maxima at the wavelengths λ = 4800 Å and 14 micron. What conclusions can be drawn from this.

Solution:

Two maxima in the spectrum show that the radiations are mixture from two sources at different temperatures. From the Wien's displacement law, we have

$$\lambda_m T = 2.898 \times 10^{-3}$$

or

$$T = \frac{2.898 \times 10^{-3}}{\lambda_m}$$

Thus, we have

$$T_1 = \frac{2.898 \times 10^{-3}}{4800 \times 10^{-10}} = 6037.5 \text{ K}$$

and

$$T_2 = \frac{2.898 \times 10^{-3}}{14 \times 10^{-6}} = 207 \text{ K}$$

T_1 = 6037.5 K is the temperature of the Sun's surface and T_2 = 207 K is the temperature of the Moon's surface. Thus, the spectrum contains the radiations emitted by the Moon itself and those emitted by the Sun and reflected by the Moon's surface.

Example 13.

Calculate the wavelength of radiations emitted from the body of a person whose temperature is 36.89°C.

Solution:

Temperature of the body of person = 36.89 + 273 = 309.89 K. The wavelength of radiations emitted from the body of person, using the Wien's displacement law is

$$\lambda_m = \frac{2.898 \times 10^{-3}}{T}$$

$$= \frac{2.898 \times 10^{-3}}{309.89} = 9.3517 \times 10 \text{ m}$$

$$= 93517 \text{ Å}$$

This wavelength lies in the far infrared region. Therefore, the radiations from the body of a person cannot be used to see the person. However, a person can be detected with the help of a detector that can detect the radiations of wavelength 93517 Å.

9.2 Stefan-Boltzmann Law

The energy density I of all possible radiations for various wavelengths in a black-body enclosure at temperature T is

$$I = \int_0^\infty I_\lambda \mathrm{d}\lambda = \int_0^\infty \frac{8\pi ch}{\lambda^5} \frac{1}{\exp(hc/\lambda kT) - 1} \mathrm{d}\lambda$$

Putting $x = hc/\lambda kT$, we have

$$I = \frac{8\pi k^4 T^4}{h^3 c^3} \int_0^\infty \frac{x^3}{e^x - 1} \mathrm{d}x$$

We know that

$$\int_0^\infty \frac{x^3}{e^x - 1} \mathrm{d}x = \frac{\pi^4}{15}$$

Therefore, we have

$$I = \frac{8\pi k^4 T^4}{h^3 c^3} \frac{\pi^4}{15}$$

The total emissive power E is related to I at the same temperature T as $E = cI/4$ and therefore, we have

$$E = \frac{2\pi^5 k^4}{15 h^3 c^2} T^4$$

or

$$E = \sigma T^4$$

where $\sigma = 2\pi^5 k^4/15h^3c^2 = 5.67 \times 10^{-8}$ J/m^2 s K^4 is the Stefan-Boltzmann constant and the relation is known as the Stefan-Boltzmann law.

Note:

Planck's explanation of black-body radiation provided a crucial step in the development of quantum concept in Physics. That the energy of an electromagnetic radiation is not continuous, but discrete in the form of packets. This hypothesis of quantization of energy was put on a firm basis by the pioneering work of Einstein, Bohr, and others later on. The concept of quantization of energy was however so radical that even Planck himself was reluctant to accept it.

Example 14.

Total amount of energy radiated by a star per unit surface area per unit time is 16×10^4 times that radiated by the Sun. Estimate the surface temperature of the star when the surface temperature of the Sun is 6000 K.

Solution:

Total amount of energy E radiated per unit surface area per second by a body having surface temperature T is

$$E = \sigma T^4$$

Thus, for the star having surface temperature T_1, we have

$$E_1 = \sigma T_1^4$$

and for the Sun having surface temperature T_2, we have

$$E_2 = \sigma T_2^4$$

Therefore,

$$\frac{E_1}{E_2} = \frac{T_1^4}{T_2^4}$$

or

$$16 \times 10^4 = \frac{T_1^4}{(6000)^4}$$

On solving, we have $T_1 = 1.2 \times 10^5$ K. Thus, the temperature of the surface of star is 1.2×10^5 K.

Example 15.

Two bodies at the temperatures 227°C and 427°C are placed in a vessel maintained at the temperature 27°C. What is the ratio of losses of energy by the bodies.

Solution:

The temperatures of the bodies are

$$T_1 = 227 + 273 = 500 \text{ K}$$

and

$$T_2 = 427 + 273 = 700 \text{ K}$$

The temperatures of environment inside the vessel

$$T_0 = 27 + 273 = 300 \text{ K}$$

The rates of energy losses are

$$E_1 = \sigma(T_1^4 - T_0^4) = \sigma(500^4 - 300^4)$$

and

$$E_2 = \sigma(T_2^4 - T_0^4) = \sigma(700^4 - 300^4)$$

Thus, the ratio of the energy losses is

$$\frac{E_1}{E_2} = \frac{(500^4 - 300^4)}{(700^4 - 300^4)} = 0.234$$

Example 16.

An infrared telescope with an aperture area of 1.0 m^2 is placed in an orbit about the Earth, where the temperature of the telescope is −73°C. At what wavelength does the telescope emit most radiation, how much radiation does it emit and how much radiation does it collect from a Sun-like star situated at 4 ly away. Given, $\sigma = 5.7 \times 10^{-8}$ J/m^2 s K$^4 i$, luminosity of the Sun = 4×10^{26} J/s.

Solution:

Consider a telescope to behave in a similar way as a black-body. The temperature of the telescope is 200 K and so the peak output wavelength is

$$\lambda m = \frac{2.9 \times 10^{-3}}{T}$$

$$= \frac{2.9 \times 10^{-3}}{200} = 1.5 \times 10^{-5} \text{ m}$$

The power radiated per unit area is

$$E = \sigma T^4$$
$$= (5.7 \times 10^{-8}) \times 200^4 \approx 100 \text{ W/m}^2$$

The telescope emits about 100 W/m^2 at a peak wavelength of 15 μm in the middle of the infrared part of the electromagnetic spectrum. The star's power (4×10^{26} W) is radiated equally in all directions and so, at the telescope star separation, the power is spread over a sphere with a diameter of 8 ly ($\approx 8 \times 10^{16}$ m) and therefore,

$$\text{Power per unit area} = \frac{\text{Total power}}{\text{Total area}}$$
$$= \frac{4\times10^{26}}{\pi(8\times10^{16})^2} = 20\times10^{-8}\text{ W/m}^2$$

The telescope collects 2.0×10^{-8} W/m^2 from the star as it has an aperture area of 1 m^2.

Note:

The difficulty of observing in the infrared radiations is shown in this example. The telescope itself is radiating more power than that can be collected from a nearby star.

10. Doppler Redshift

Following the Big Bang theory, the universe is expanding, and each of the cosmic objects is receding away from us and from other objects. Consequently, spectrum from a cosmic object shows the Doppler redshift. Suppose, λ_0 is the wavelength of a particular transition when measured in a laboratory on the Earth's surface and λ the wavelength of the same transition when received from a cosmic object, receding away from us with velocity v. Following the Doppler effect, we have

$$\frac{\lambda}{\lambda_0} = \left(\frac{1+v/c}{1-v/c}\right)^{1/2}$$

The redshift z is described as

$$z = \frac{\lambda - \lambda_0}{\lambda_0}$$
$$1 + z = \frac{\lambda}{\lambda_0}$$

The wavelength of radiation from a cosmic object is

$$\lambda = \lambda_0(1 + z)$$

Thus, the relation between the redshift z and the receding velocity v is

$$1 + z = \left(\frac{1+v/c}{1-v/c}\right)^{1/2} \tag{6.34}$$

10.1 For Non-relativistic Velocity

For a non-relativistic velocity, we have $v << c$, and equation (6.34) gives

$$1 + z = \left(1+\frac{v}{c}\right)^{1/2}\left(1+\frac{v}{c}\right)^{-1/2}$$

$$=\left(1+\frac{v}{2c}\right)\left(1+\frac{v}{2c}\right)$$

Here, we have neglected smaller terms. Now, we get

$$1+z=1+\frac{v}{c}+\frac{v^2}{4c^2}$$

$$=1+\frac{v}{c}$$

Here, also, we have neglected smaller term. Thus, we have

$$z=\frac{v}{c}$$

For a non-relativistic motion, the redshift is proportional to the velocity of the object.

Example 17.

A distant galaxy is receding away with velocity 0.5c, where c denotes the speed of light. Calculate the redshift. What is the wavelength of 21 cm radiation of hydrogen atom, when received from the galaxy?

Solution:

The redshift z is

$$1+z=\left(\frac{1+v/c}{1-v/c}\right)^{1/2}$$

Therefore,

$$1+z=\left(\frac{1+0.5}{1-0.5}\right)^{1/2}$$

or

$$z=\sqrt{3}-1=0.732$$

The wavelength of 21 cm radiation received from the cosmic object is

$$\lambda=\lambda_0(1+z)$$

$$=21(1+0.732)=36.37 \text{ cm}$$

Example 18.

Observations of Sirius show it to be A1 star with apparent magnitude of –1.46 and a parallax of 0.37″. Calculate the size of the Sirius.

Solution:

A parallax of 0.37″ implies that the Sirius lies at a distance of (1/0.37 =) 2.7 parsec. The absolute magnitude of the Sirius is

$$M=m-5\log_{10}d+5$$

$$=-1.46-5\log_{10}2.7+5=1.38$$

The absolute magnitude of the Sirius is expressed using the standard relative scale but an absolute scale is required when the actual size of the Sirius is calculated. The Sun's power output is 4×10^{26} W and the absolute magnitude is 4.9. The relative intensity of Sirius with respect to the Sun can be found as

$$I_{Si} = 10^{(M_\odot - M_{Si})/2.5} I_\odot$$
$$= 10^{(4.9-1.38)/2.5} (4 \times 10^{26}) = 1.0 \times 10^{28} \text{ W}$$

As the Sirius is A1 star, its surface temperature must be about 11000 K. So, the power radiated per unit surface area is

$$P = \sigma T^4$$
$$= (5.7 \times 10^{-8}) \times 11000^4$$
$$= 8.3 \times 10^8 \text{ W/m}^2$$

Total surface area of Sirius is

$$\text{surface area} = \frac{1.0 \times 10^{28}}{8.3 \times 10^8}$$
$$= 1.2 \times 10^{19} \text{ m}^2$$

For the radius r of Sirius, we have

$$4\pi r^2 = 1.2 \times 10^{19}$$
$$r = \sqrt{\frac{1.2 \times 10^{19}}{4\pi}}$$
$$= 9.8 \times 10^8 \text{ m}$$

11. Relation between *B - V* Magnitude and Surface Temperature

11.1 UBV Photometric System

The letters U, B, and V stand for the ultraviolet, blue, and visual magnitudes, measured for a star. The photometric system called the UBV is a wide central band of the electromagnetic spectrum, and usually used for the classification of stars according to their colours. Two subtractions, the B-V (B minus V) and U-B (U minus B), are then performed to classify it in the more technical, formulaic values given by this system. For example, the yellowish Sun has a BV index of 0.656, whereas the bluish Rigel has a BV of 0.03 (its B magnitude is 0.09 and its V magnitude is 0.12, so that $BV = 0.03$).

11.2 Colour Index

The colour index in Astronomy is a simple numerical formula to determine the colour of an object. In case of a star, the colour index gives the temperature of the star. The smaller the colour index, the more blue (or hotter) the object is. Conversely, the larger the colour index, the more red (or cooler) the object is. It happens due to the logarithmic magnitude scale, where the brighter objects have smaller (more negative) magnitudes than the dimmer ones.

For measuring the index, one observes the magnitude of an object through two different filters, such as U and B, or B and V, where U is sensitive to ultraviolet rays, B is sensitive to blue light, and V is sensitive to visible (green-yellow) light. The set of passbands or filters is called a photometric system. The difference in magnitudes found with these filters is called the U-B or B-V colour index respectively.

In principle, the temperature of a star can be calculated directly from the *BV* index, and there are several formulae to make this relation. A good approximation can be obtained by considering a star as a black-body, using the following relation:

$$T = 4600\left(\frac{1}{0.92(B-V)+1.17} + \frac{1}{0.92(B-V)+0.62}\right)K$$

In case of distant objects, the colour indices are usually affected by the interstellar extinction, that is, they are redder as compared to those of closer stars.

11.3 Colour-magnitude Diagram

For the stars, the plot of *B-V* versus M_v is known as the colour-magnitude diagram.

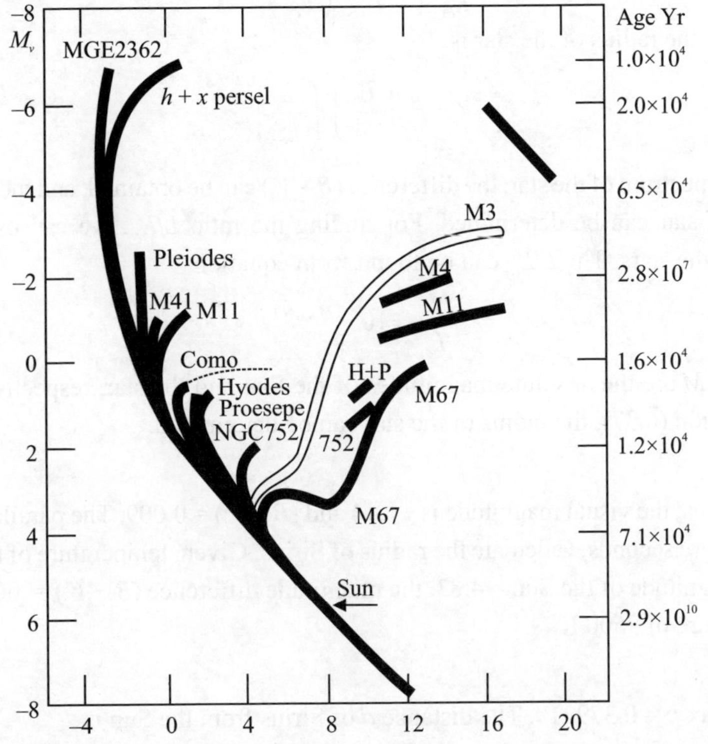

Fig. 6: The composite colour-magnitude diagram for ten galactic and one globular cluster. The age of each cluster whose main sequence termination point is at a given M_v is shown on the right. (*Courtesy:* A. Sandage, ApJ 125 (1956) 435.)

12. Determination of Radius of Star

Radius of a star cannot be measured directly, as even a big star is so far away from us that it appears as a single point, when seen from the Earth. However, there is an indirect method for determination of radius of a star. For a star having surface temperature *T* and radius *R* (assuming spherical in shape), the luminosity *L* is expressed as

$$L = 4\pi R^2 \sigma T^4 \tag{6.35}$$

where σ is the Stefan-Boltzmann constant. For our Sun, if temperature, radius and luminosity are denoted by $T_\odot$, $R_\odot$, $L_\odot$, respectively. Then, we have

$$L_\odot = 4\pi R_\odot^2 \sigma T_\odot^4 \tag{6.36}$$

The parameters for the Sun are known with good accuracy. On dividing equation (6.35) by (6.36), we get

$$\frac{L}{L_\odot} = \left(\frac{R}{R_\odot}\right)^2 \left(\frac{T}{T_\odot}\right)^4$$

or

$$\frac{R}{R_\odot} = \left(\frac{T_\odot}{T}\right)^2 \left(\frac{L}{L_\odot}\right)^{1/2}$$

Therefore, the radius of the star is

$$R = R_\odot \left(\frac{T_\odot}{T}\right)^2 \left(\frac{L}{L_\odot}\right)^{1/2} \tag{6.37}$$

From the spectrum of the star, the difference $(B - V)$ can be obtained, and the corresponding temperature of star can be determined. For finding the ratio $L/L_\odot$, we can use the absolute magnitudes of the stars. The $L/L_\odot$ can be found from equation

$$\frac{L}{L_\odot} = 10^{0.4(M_\odot - M)},$$

where $M_\odot$ and M are the absolute magnitudes of the Sun and the star, respectively. Using the values in equation (6.37), the radius of the star can be determined.

Example 19.

For the star Sirius, the visual magnitude is −1.44 and $(B - V) = 0.009$. The parallax of the Sirius is 379.21 milli arc seconds. Calculate the radius of Sirius. Given, temperature of the Sun = 5800 K, absolute magnitude of the Sun = 4.83, the magnitude difference $(B - V) = .009$ corresponds to the temperature of 9500 K.

Solution:

We have parallax $p = 0.37921''$. The distance d of Sirius from the Sun is

$$d = \frac{1}{p}$$

$$= \frac{1}{0.37921} = 2.637 \text{ pc.}$$

The absolute magnitude of Sirius is

$$M = m - 5 \log d + 5$$

$$= -1.44 - 5 \log(2.637) + 5$$

$$= 1.454$$

We know that the temperature of the Sun is 5800 K. From the chart, the temperature of Sirius is about 9500 K. Our Sun has an absolute magnitude 4.83. The difference in the magnitudes is 3.376. Putting the values, we get

$$\frac{R}{R_\odot} = \left(\frac{T_\odot}{T}\right)^2 \left(\frac{L}{L_\odot}\right)^{1/2}$$

$$= \left(\frac{5800}{9500}\right)^2 (10^{0.4\times3.376})^{1/2}$$

$$= 1.764$$

Sirius has radius approximately 1.764 times that of the Sun.

13. Hertzsprung-Russell Diagram

A diagram obtained by plotting the luminosity L of the stars versus their effective temperature T_e is known as the **Hertzsprung-Russell** (H-R) diagram. Let us plot $\log(L/L_\odot)$ on the ordinate and $\log(T_e)$ on the abscissa, where L and $L_\odot$ denote the luminosities of the star and the Sun, respectively, and T_e is the effective surface temperature of the star (Fig. 7). The luminosity on the ordinate is increasing from down to upward direction, whereas the effective surface temperature is decreasing from left towards right direction. The left side of diagram has hot stars, having extreme surface temperatures up to nearly 10^5 K. The right side has cool stars. At the top of the diagram, the stars are bright whereas at the bottom, the stars are faint.

Such diagram plays a basic role in Astronomy and Astrophysics, as a number of properties of stars can be inferred and discussed in terms of their locations in such diagram. In the diagram, most of the stars are found lying on a diagonal band, extending from the corner corresponding to high L and high T_e (top-left) to the corner corresponding to low L and low T_e (bottom-right). This band is commonly known as the main sequence. The stars on the main sequence are found to have the following common features:

(i) The stars are chemically homogeneous.

(ii) In the core of stars, hydrogen is burning into helium. Fusion of four hydrogen nuclei into helium nucleus is the source of energy generated in such stars.

(iii) These stars having homogeneous distribution of hydrogen, helium, and other heavy elements can steadily shine for the longest period of their luminous lifetime.

In spite of these common features of stars lying on the main sequence, there are some characteristics which make stars to lie at different positions on the main sequence. One possibility could be the initial fractional amount of elements heavier than hydrogen and helium in the stars. The gross properties of the main sequence stars are found to change relatively little on the basis of initial fractional amount of heavier elements. This rules out the possibilities of initial fractional amount of heavier elements to be the reason for various positions of stars in the main sequence.

The second possibility may be that the stars may not be of the same age and therefore the amount of hydrogen which has been burnt into the helium is different for various stars. Estimates have shown that the ratio of the hydrogen to helium in the star can shift its position a little. However, as long as the core hydrogen is not all used up, the star lies inside the band.

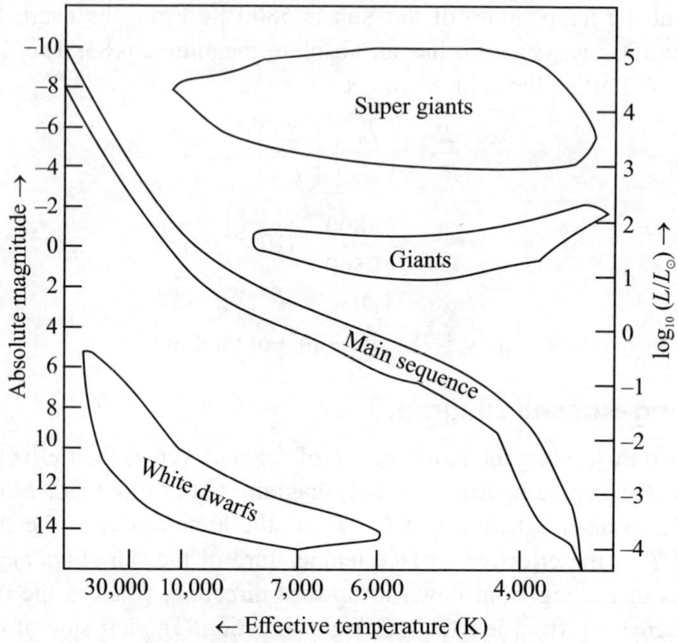

Fig. 7: Hertzsprung-Russell diagram

The major factor which is found responsible for the position of a star in the main sequence is its mass. Thus, it is the mass of a star which determines its position in the main sequence of the H-R diagram. The high mass stars spend energy much faster than the low mass stars.

13.1 Empirical Mass-luminosity Relation

All the stars on main sequence are found to have their luminosity L roughly proportional to the fourth power of the mass of star

$$L \propto M^4$$

or

$$L = kM^4$$

where k is the proportionality constant. However, this relation is not valid for very heavy and very light stars. For those stars, it varies as

$L \propto M^{5.5}/R^{1/2}$ for low to medium mass stars

$L \propto M^3$ for high mass stars

$L \propto M$ for very high mass stars

Let us discuss the consequence of the relation between the luminosity and mass of a star in the main sequence. Suppose, we have two stars, A and B, on the main sequence of masses $1\,M_\odot$ and $10\,M_\odot$, respectively. The ratio of their luminosities, denoted by L_A and L_B, respectively, is

$$\frac{L_B}{L_A} = \left(\frac{10M_\odot}{1M_\odot}\right)^4 = 10^4$$

The time t spent on the main sequence by a star is

$$t = \frac{M}{L} = \frac{M}{kM^4}$$

$$= k'M^{-3}$$

or

$$t \propto M^{-3}$$

It shows that larger the mass of star, lesser the time spent on the main sequence. The ratio of the time spent by the stars A and B on the main sequence is

$$\frac{t_B}{t_A} = \frac{(10M_\odot)^{-3}}{(1M_\odot)^{-3}} = 10^{-3}$$

Thus, the lifetime of the star B on the main sequence is one-thousandth of that of the star A. Assuming that the lifetime of the Sun on the main sequence is 10^{10} years, the star of $10\,M_\odot$ would stay on the main sequence for a period of 10^7 years.

Example 20.

The luminosities of two stars A and B, having masses $1\,M_\odot$ and $10\,M_\odot$, respectively, vary as cube of its mass. Calculate the ratio of their luminosities and the periods they spend on the main sequence of the H-R diagram.

Solution:

We have

$$L \propto M^3$$

and

$$L = kM^3$$

where k is constant. The ratio of luminosities of stars is

$$\frac{L_B}{L_A} = \frac{k(10M_\odot)^{-3}}{k(1M_\odot)^{-3}} = 10^3$$

The time spent on the main sequence is

$$t = \frac{M}{L} = \frac{M}{kM^3} = \frac{M^{-2}}{k}$$

Thus, the ratio of time spent by the stars on the main sequence is

$$\frac{t_B}{t_A} = \frac{(10M_\odot/k)^{-2}}{(1M_\odot/k)^{-2}} = 10^{-2}$$

Example 21.

Obtain the law of variation of luminosity with mass of a star when two stars having different masses spend equal time on the main sequence.

Solution:

Consider two stars A and B having different masses. Suppose, the mass of star B is β times that of the star A and the law of variation of luminosity with the mass of star is

$$L \propto M^\alpha$$

and

$$L = kM^\alpha$$

where k and α are constants. The time spent on the main sequence is

$$t = \frac{M}{L} = \frac{M}{kM^{\alpha}}$$

$$= \frac{M^{1-\alpha}}{k}$$

Thus, the ratio of time spent by the stars on the main sequence is

$$\frac{t_B}{t_A} = \frac{(M_B/k)^{1-\alpha}}{(M_A/k)^{1-\alpha}}$$

$$= \left(\frac{M_B}{M_A}\right)^{1-\alpha}$$

Using $t_B/t_A = 1$ and $M_B/M_A = \beta$, we have

$$1 = \beta^{1-\alpha}$$

It gives $\alpha = 1$. Thus, the law is $L \propto M$.

Example 22.

Luminosity L of a star varies as the fourth power of its mass M. For two stars A and B, the mass of first one is 10 times larger than that of the second. Calculate the ratio of their luminosities.

Solution:

Given, $M_A = 10M_B$. The ratio of luminosities is

$$\frac{L_A}{L_B} = \frac{M_A^4}{M_B^4} = 10^4$$

13.2 H-R Diagram of Clusters

Since all the stars in a cluster formed at the same time, they are all the same age. A very young cluster has a HR diagram with a cluster of T-Tauri stars evolving towards the main sequence. Each star is plotted by its colour and magnitude in the HR diagram.

14. Meghnad Saha

Meghnad Saha (6 October 1893-16 February 1956), was an Indian astrophysicist who has derived the Saha's ionization equation, used to describe the chemical and physical conditions in the stars. His work played important role allowing the astronomers to accurately relate the spectral classes of the stars to their actual temperatures.

Fig. 8: Meghnad Saha (6 October 1893-16 February 1956);
(*Courtesy:* Google)

14.1 Ionized Elements

Ionization is a process by which an atom or a molecule acquires positive charge(s) by losing electron(s). The resulting electrically charged atom or molecule is known as an ion. For removal of an electron from atom or molecule, energy is supplied through some process. The energy required to remove an electron is known as the ionization potential. The electrons can be removed one by one. As an example, the ionized states of iron (Fe) are denoted as Fe^{+}, Fe^{++}, Fe^{+++}, Fe^{4+}, ... In the spectroscopic notations, neutral atom Fe is written as Fe I, and the ionic states are, respectively, written as Fe II, Fe III, Fe IV, Fe V, Thus, Fe^{9+} is Fe X and Fe^{13+} is Fe XIV. 1st-7th ionization potentials in kJ/mol of some atoms are given in the following table. The ionization potential in eV can be obtained by multiplying the value given in the table by a factor 1.03654 × 10–2. It can be noticed that with the increase of the number of ionization of an atom, the value of the ionization potential increases.

Atom	1st	2nd	3rd	4th	5th	6th	7th
H	1312.0						
He	2372.3	5250.5					
Li	520.2	7298.1	11,815.0				
Be	899.5	1757.1	14,848.7	21,006.6			
B	800.6	2427.1	3659.7	25,025.8	32,826.7		
C	1086.5	2352.6	4620.5	6222.7	37,831	47,277.0	
N	1402.3	2856	4578.1	7475.0	9444.9	53,266.6	64,360
O	1313.9	3388.3	5300.5	7469.2	10,989.5	13,326.5	71,330
F	1681.0	3374.2	6050.4	8407.7	11,022.7	15,164.1	17,868
Ne	2080.7	3952.3	6122	9371	12,177	15,23890	19,999.0
Mg	737.7	1450.7	7732.7	10,542.5	13,630	18,020	21,711
Si	786.5	1577.1	3231.6	4355.5	16,091	19,805	23,780
S	999.6	2252	3357	4556	7004.3	8495.8	27,107
Cl	1251.2	2298	3822	5158.6	6542	9362	11,018
Ar	1520.6	2665.8	3931	5771	7238	8781	11,995
Ca	589.8	1145.4	4912.4	6491	8153	10,496	12,270
Fe	762.5	1561.9	2957	5290	7240	9560	12,060
Ni	737.1	1753.0	3395	5300	7339	10,400	12,800

14.2 Shah Equation

In a system (atom or molecule), let us consider two states, A and B, having energies E_A and E_B, and statistical weights g_A and g_B, respectively. In the LTE (Local Thermodynamic Equilibrium), the number of particles in the two states, N_A and N_B, satisfies Boltzmann equation:

$$\frac{N_A}{N_B} = \frac{g_A}{g_B} \exp[-(E_A - E_B)/kT] \tag{6.38}$$

Let us consider an element in two ionic states, "i" and "$i + 1$". The statistical weights of the ground states of the two ions are g_i and g_{i+1}, respectively. Suppose, the number densities, [cm^{-3}], of the two types of ions and the free electrons are n_i, n_{i+1}, and n_e, respectively. Suppose, the ionization potential, i.e., the energy needed to ionize i from the ground state is χ.

We intend to use the Boltzmann equation (6.38) to estimate the number ratio n_{i+1}/n_i. The statistical weight of an ion in the lower ionization state to be used in the equation (6.38) is just g_i. The statistical weight of an ion in the upper ionization state is g_{i+1} multiplied by the number of possible states in which a free electron may be put.

1st-7th ionization energies in kJ/mol of some atoms

For the Planck constant h, in the phase space, each cell has volume h^3, and in each cell, an electron can be found in two possible states, because of its two possible (spin up and spin down) orientations. The energy of a free electron with a momentum p with respect to the ground state of an ion in a lower ionization state is $E = \chi + p^2/2m$. Here, m denotes the mass of electron. The number of cells available for free electrons with a momentum between p and $p + dp$ is $V_e 4\pi p^2 dp/h^3$, where $V_e = 1/n_e$ is the volume in the ordinary space available per electron, and n_e is the free electron number density.

Now, we integrate over all available cells, taking the Boltzmann factor into account

$$\frac{n_{i+1}}{n_i} = \frac{g_{i+1}}{g_i}\frac{2V_e}{h^3}\int_0^\infty e^{-(x+p^2/2m)/kT}\, 4\pi p^2 dp$$

For solving the integral, we put $p^2 = (2mkT)x$, and therefore

$$\frac{n_{i+1}}{n_i} = \frac{g_{i+1}}{g_i}\frac{1}{n_e}\frac{2}{h^3}e^{-\chi/kT}\, 2\pi(2mkT)^{3/2}\int_0^\infty e^{-x}x^{1/2}dx$$

The value of the integral is $\pi^{1/2}/2$, and therefore

$$\frac{n_{i+1}n_e}{n_i} = \frac{(2\pi kT)^{3/2}}{h^3}\frac{2g_{i+1}}{g_i}e^{-\chi/kT} \tag{6.39}$$

This is Saha's ionization equation. For some ions, we have:

For the ionization of hydrogen $\chi = 13.54$ eV and $2g_{i+1}/g_i = 1$

For the first ionization of helium $\chi = 24.48$ eV and $2g_{i+1}/g_i = 4$

For the second ionization of helium $\chi = 54.17$ eV and $2g_{i+1}/g_i = 1$

15. Wilson-Bappu Effect

Two prominent lines H (3968.47 Å) and K (3933.66 Å) produced by singly ionized calcium (Ca II) are generally found in the chromosphere of the Sun-like stars. Wilson and Bappu in 1957 found a remarkable correlation between the width of the emission in the core of the K line and the absolute visual magnitude of late-type stars. This correlation is known as the **Wilson-Bappu effect**. It is found independent of spectral type. The prime application of the Wilson-Bappu effect is for the determination of distance of stars, which are far away for direct measurements.

According to the latest calibration, the relation between the absolute visual magnitude (M_V) expressed in magnitudes and the emission core width W_0, transformed in km/s, is as the following:

$$M_V = 33.2 - 18.0 \log(W_0)$$

The emission core width of the *K* line (W_0) can be measured in distant stars, and using the correlation (Wilson-Bappu effect), we can determine the absolute magnitude of the star. Knowing both the absolute and apparent magnitudes of a star, the distance of the star can be calculated, provided that the interstellar reddening of the star is either negligible or well known.

The Mg II *K* line at 2796.34 Å also found to show the Wilson-Bappu effect. However, as this line lies in the ultraviolet region, and cannot reach the Earth's surface, therefore, it can only be observed using the satellites.

16. Pauli's Exclusion Principle

On the basis of their spin, the particles as well as atoms/ions are classified into two categories:

(i) Fermi particles (Fermions) (ii) Bose particles (Bosons)

The particles having half-integer spin ($I = 1/2, 3/2$, etc.) are known as the Fermions and they follow the Fermi-Dirac statistics. Some examples of the Fermions, are the electron, proton, neutron, μ-meson, $_2He^3$ nucleus. The Pauli exclusion principle is applicable for the Fermions. The particles having integer spin ($I = 0, 1, 2$, etc.) are known as the Bosons and they follow the Bose-Einstein statistics. Some examples of the Bosons, are the photon, π-meson, $_2He^4$ nucleus. These particles do not obey the Pauli exclusion principle.

The Pauli exclusion principle states that no two fermions can occupy the same quantum state. In the quantum theory, a state is described by a set of four quantum numbers, n, l, m_l, m_s. The principal quantum numbers (n) can assume value as non-zero, positive integer, 1, 2, 3, etc. For a given value of n, azimuthal quantum number l can assume integer values from 0 to $(n - 1)$ with the increment of 1. For example, for $n = 2$, l can assume values, 0 and 1. For a given value of n, the number of l-values is n.

For a given value of l, the magnetic orbital quantum number m_l can assume integer values from $-l$ to l with the increment of 1. For example, for $l = 2$, m_l can assume the values, −2, −1, 0, 1, 2. For a given value of l, the number of m_l-values is $(2l + 1)$. The spin s of a fermion can have half-integer value. That is, the spin s of a Fermion can assume the values 1/2, 3/2, 5/2, For a given s, the magnetic spin quantum number m_s can assume the values from $-s$ to s with the increment of 1.

Each of the electron, proton, neutron has spin $s = 1/2$. Hence, the m_s can assume two values, −1/2 and 1/2. Some quantum states for a fermion with $s = 1/2$ are given in Table below. For a given principal quantum number n, the number of total possible states is $2n^2$.

Quantum states

n	l	m_l	m_s
1	0	0	+1/2
1	0	0	−1/2
2	0	0	+1/2

Contd.

2	0	0	−1/2
2	1	−1	+1/2
2	1	−1	−1/2
2	1	0	+1/2
2	1	0	−1/2
2	1	1	+1/2
2	1	1	−1/2

Each quantum state has corresponding energy. Some of the energies may be equal (degenerate state). Since the Pauli exclusion principle is obeyed by the fermions, even at 0 K temperature, all fermions in a system cannot cascade down to the ground state. The highest level which is filled at 0 K temperature is known as the Fermi-level. With the increase of temperature, some particles from the states below the Fermi-level move to the states above the Fermi-level. Since all the fermions cannot be brought to the ground state, there is a pressure in the outward direction, called the **degeneracy pressure**. The electron degeneracy plays an important role for a white dwarf whereas the neutron degeneracy plays similar role for a neutron star.

17. Degenerate Matter

The degenerate matter in Physics is a collection of free, non-interacting fermions with a pressure and other physical characteristics determined by quantum theory effects. The degenerate state of the matter arises at an extraordinarily high density found in the compact stars. It occurs for the matter particles such as electrons, neutrons, protons, and fermions in general, and is referred as the electron-degenerate matter, neutron-degenerate matter, etc. In a mixture of particles, such as ions and electrons in a white dwarf, the electrons may be degenerate, while the ions are not.

In the quantum theory description, in a bound state, energy levels are quantized. That is, the energies of the levels are discrete and are called the quantum states. The Pauli exclusion principle states that no two identical fermions can occupy the same quantum state. Consequently, even at a negligibly small thermal energy of particles, all the lowest energy quantum states are filled. This state is referred as full degeneracy. The pressure (called the degeneracy pressure or the Fermi pressure) remains finite even near the absolute zero temperature. Addition of particles or reduce of volume, forces the particles into higher-energy quantum states. This requires a compression force, and is made manifest as a resisting pressure. The key feature is that this degeneracy pressure does not depend on the temperature and depends only on the density of fermions. It keeps the dense stars in equilibrium irrespective of its thermal structure. The degeneracy pressure acts against the gravitational collapse.

A degenerate matter is also called a Fermi gas or a degenerate gas. A degenerate state with velocities of the fermions, close to the speed of light, is known as the relativistic degenerate matter.

Degenerate gases are composed of fermions at high densities. A fermion gas in which all the energy states below some energy level are filled is called a fully degenerate fermion gas. The electron gas in the interior of the white dwarf stars is an example of a degenerate electron gas. Most of the stars are supported against their own gravitation by normal thermal gas pressure. A white dwarf star is supported by the degeneracy pressure of electron gas in their interior, while for a neutron star the degenerate particles are the neutrons.

17.1 Electron Degeneracy

In an ordinary fermion gas in which the thermal effects dominate, most of the available electron energy levels are unfilled and the electrons are free to move to these states. As particle density is increased, electrons progressively fill the lower energy states and additional electrons are forced to occupy the states of higher energy even at low temperatures. Degenerate gases strongly resist further compression because the electrons cannot move to already filled lower energy levels due to the Pauli exclusion principle. Since the electrons cannot give up energy by moving to the lower energy states, no thermal energy can be extracted. The momentum of the fermions in the fermion gas nevertheless generates pressure, known as the degeneracy pressure.

Under high densities, the matter becomes a degenerate gas when the electrons are all stripped from their parent atoms. In the core of a star, once hydrogen burning in the nuclear fusion reactions stops, it becomes a collection of positively charged ions, largely helium and carbon nuclei, floating in a sea of electrons, which have been stripped from the nuclei. Normal gas exerts higher pressure when it is heated and expands, but the pressure in a degenerate gas does not depend on the temperature. When the gas becomes super-compressed, the particles position right up against each other to produce degenerate gas that behaves more like a solid.

There is an upper limit to the mass of an electron-degenerate object, called the **Chandrasekhar limit**, beyond which electron degeneracy pressure cannot support the object against the gravitational collapse. The limit is approximately 1.44 $M_{\odot}$ for the objects with compositions similar to the Sun. Above this mass limit, a neutron star (supported by the neutron degeneracy pressure) or a black hole may be formed instead.

17.2 Proton Degeneracy

Sufficiently dense matter containing protons experiences proton degeneracy pressure, in a manner similar to the electron degeneracy pressure in the electron-degenerate matter. Because the protons are much more massive than the electrons, the same momentum represents a much smaller velocity for the protons than that for the electrons. As a result, in a matter with approximately equal numbers of protons and electrons, the proton degeneracy pressure is much smaller than the electron degeneracy pressure, and the proton degeneracy is usually modelled as a correction to the equations of state of electron-degenerate matter.

17.3 Neutron Degeneracy

The neutron degeneracy is similar to the electron degeneracy and is demonstrated in the neutron stars, which are primarily supported by the pressure from a degenerate neutron gas. This happens when a stellar core above 1.44 $M_{\odot}$, the Chandrasekhar limit, collapses and is not halted by the degenerate electrons. As the star collapses, the Fermi energy of the electrons increases to the point where it is energetically favourable for them to combine with the protons to produce the neutrons. The result of this collapse is an extremely compact star composed of nuclear matter, which is predominantly a degenerate neutron gas with a small admixture of degenerate proton and electron gases.

There is an upper limit to the mass of a neutron-degenerate object, the Tolman- Oppenheimer-Volkoff limit, which is similar to the Chandrasekhar limit for the electron-degenerate objects. The precise limit is unknown, as it depends on the equations of state of the nuclear matter, for which a highly accurate model is not yet available. Above this limit, a neutron star may collapse into a black-hole, or into other, denser forms of degenerate matter.

18. Hydrostatic Equilibrium

In the universe, most of the bodies are almost spherical in shape. For example, Sun, planets, satellites, stars, etc., are almost spherical. Consider a spherical body of radius R and mass M. Let us consider a spherical thin shell of radius r' and thickness dr'. Volume of the shell is

$$\text{Volume of shell} = \frac{4\pi}{3}(r' + dr') - \frac{4\pi}{3}r'^3$$

$$= 4\pi r'^2 dr' \tag{6.40}$$

where smaller terms are neglected. In general, the density of a body is not uniform. It can, however, be assumed to be radially symmetric, i.e., varying along the radial distance. Suppose, the density of shell is $\rho(r')$, the mass of the shell is

$$dM(r') = \rho(r')\, 4\pi r'^2\, dr' \tag{6.41}$$

Thus, the mass of a spherical volume of radius of r is

$$M(r) = \int_0^r \rho(r')\, 4\pi r'^2 dr' \tag{6.42}$$

Thus, from equation (17.17), we have

$$\frac{dM(r)}{dr} = 4\pi r^2 \rho(r) \tag{6.43}$$

For convenience, we assume that the density of body is uniform and is denoted by $\bar{\rho}$, then from equation (17.18), we get

$$M(r) = \frac{4}{3}\pi r^3 \bar{\rho} \tag{6.44}$$

Similarly, we get

$$M(R) = \frac{4}{3}\pi R^3 \bar{\rho}$$

or

$$M = \frac{4}{3}\pi R^3 \bar{\rho} \tag{6.45}$$

and thus,

$$\frac{M(r)}{M} = \frac{4\pi r^3 \bar{\rho}/3}{4\pi R^3 \bar{\rho}/3}$$

or

$$M(r) = M\left(\frac{r}{R}\right)^3 \tag{6.46}$$

We now consider a volume-element with base area dA and height dr at a distance r from the centre of sphere, as shown in Fig. 9. Mass of the volume-element is

$$dm = \rho(r)\, dA\, dr$$

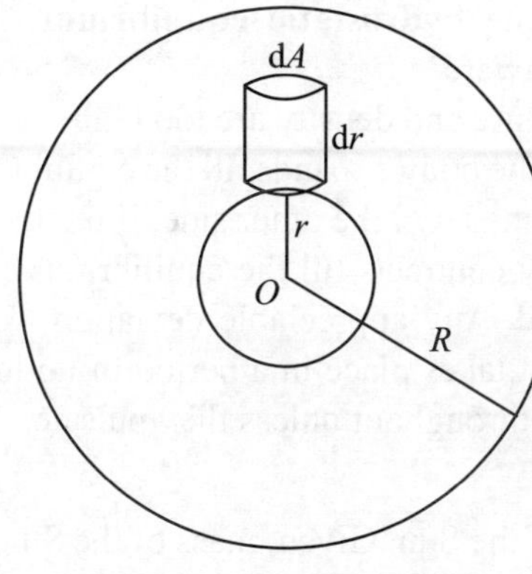

Fig. 9: Spherical body with a volume element.

The mass $M(r)$, of sphere of radius r, produces a gravitational acceleration

$$g(r) = G\frac{M(r)}{r^2}$$

at its surface (i.e., at a distance of r from its centre) in the inward direction. Hence, the inward force on the volume-element is

$$\text{inward force} = \text{mass} \times \text{acceleration}$$
$$= dm \times g(r) = \rho(r)\, dA\, dr\, g(r)$$

This inward force is balanced by some kind of force in the outward direction. This outward force is due to pressure developed inside the body, which decreases radially outward. This pressure depends on the density ρ, temperature T, and mass concentrations X_i of various elements present in the region. Hydrogen and helium mass concentrations are generally denoted by X and Y, respectively. Thus, we have

$$P = P(\rho, T, X, Y, X_i) \qquad i = 1, 2, ..., n$$

Here, X_i denotes the concentration of elements other than hydrogen and helium.

In the absence of such balancing force (pressure), the body continues to shrink and finally would reduce to a point. But, it is not the case because of the pressure gradient inside the body. Let P and $P - dP$ be the pressures at the lower and upper faces of the cylinder. The pressure difference is dP over the distance dr. The force acting on the volume-element in the outward direction is $dP\, dA$. Hence, in the equilibrium state, we have

$$-dP\, dA = \rho(r)\, dA\, dr\, g(r)$$

The negative sign appears as the directions of two forces are opposite to each other. Therefore,

$$\frac{dP}{dr} = -\rho(r)g(r)$$

or

$$\frac{dP}{dr} = -\rho(r)\frac{GMr}{R^2} \tag{6.47}$$

On using equation (6.46) in (6.47), we get

$$\frac{dP}{dr} = -\rho(r)\frac{GMr}{R^3} \tag{6.48}$$

This equation represents the **hydrostatic equilibrium**. The negative sign shows the decrease of pressure radially outward.

In a star, when the temperature and density are too high, the local pressure becomes larger than the gravitational pull, then the body expands till the equilibrium between the local pressure and the gravitational pull is attained. On the other side, if the local pressure is smaller than the gravitational pull, then the body contracts till the equilibrium between the local pressure and the gravitational pull is attained. Any appreciable deviation from the equilibrium leads to a readjustment of equilibrium that takes place in a period of no more than an hour. Most of the stars have pressure equilibrium throughout unless they pulsate.

Example 23.

Calculate the average density of the Sun. Given, mass of the Sun $M_\odot = 1.99 \times 10^{30}$ kg, radius of the Sun $R_\odot = 6.96 \times 10^5$ km.

Solution:

Average density of the Sun is

$$\bar{\rho} = \frac{M_\odot}{4\pi R_\odot^3/3}$$

$$= \frac{1.99 \times 10^{30}}{4\pi(6.96 \times 10^8)^3/3}$$

$$= 1409.8 \text{ kg/m}^3$$

Example 24.

Using the hydrostatic equilibrium, calculate the pressure at half the solar radius. Given, average density $\bar{\rho} = 1410$ kg/m^3.

Solution:

Suppose, the density of the Sun is constant, equal to the average density $\bar{\rho}$. Then the mass within the radius r is

$$M(r) = \frac{4}{3}\pi\bar{\rho}r^3$$

and hydrostatic equilibrium can be written as

$$\frac{dP}{dr} = -\frac{GM(r)\bar{\rho}}{r^2}$$

Using the expression for $M(r)$, we have

$$\frac{dP}{dr} = -\frac{4\pi G\bar{\rho}^2 r}{3}$$

This equation can be integrated from half the solar radius to the surface, where the pressure varies from P to zero.

$$\int_P^0 dp = \frac{4\pi G\bar{\rho}^2}{3}\int_{R_\odot/2}^{R_\odot} r\,dr$$

giving

$$P = \frac{1}{2}\pi G\bar{\rho}^2 R_\odot^2$$

Using the values, the pressure at half the solar radius is

$$P = \frac{1}{2}\pi(6.67\times10^{-11})(1410)^2(6.96\times10^8)^2$$
$$= 1.0 \times 10^{14} \text{ Pa}$$

This estimate is extremely rough, as the density increases strongly inwards.

18.1 Pressure at the Centre of a Spherical Body

In most of the stars, the pressure P is practically equal to the gas pressure P_g. But, in a hot and massive star, it is necessary to include the radiation pressure P_r, so that $P = P_g + P_r$.

If we assume that the density of the object is uniform $\bar{\rho}$, then for a star of radius R and mass M, we have

$$\frac{dP}{dr} = -\bar{\rho}\frac{GMr}{R^3}$$

For the pressure P_c at the centre of the star, we have

$$\int_{P_c}^{0} dP = -\bar{\rho}\frac{GM}{R^3}\int_0^R r\,dr$$

or

$$-P_c = -\bar{\rho}\frac{GM}{R^3}\frac{R^2}{2}$$

Thus, the pressure P_c at the centre of the star is

$$P_c = \frac{\bar{\rho}}{2}\frac{GM}{R}$$

This is a rough estimate for the pressure on the centre of a spherical body of radius R, mass M and uniform density $\bar{\rho}$. Values of the pressure at the centre of some celestial bodies are calculated and are given in Table 3.

Table 3. Pressure at the centre of body

Body	Mass M (kg)	Radius R (m)	Density ρ (kg/m^3)	Pressure P_c (N/m^2)
Sun	1.99×10^{30}	6.96×10^8	1.40×10^3	1.34×10^{14}
Moon	7.35×10^{22}	1.74×10^6	3.34×10^3	4.71×10^9
Mercury	3.28×10^{23}	2.44×10^6	5.43×10^3	2.44×10^{10}
Venus	4.87×10^{24}	6.05×10^6	5.24×10^3	1.41×10^{11}
Earth	5.97×10^{24}	6.37×10^6	5.52×10^3	1.73×10^{11}
Mars	6.39×10^{23}	3.39×10^6	3.94×10^3	2.48×10^{10}
Jupiter	1.90×10^{27}	6.99×10^7	1.33×10^3	1.21×10^{12}
Saturn	5.68×10^{26}	5.82×10^7	7.00×10^2	2.28×10^{11}
Uranus	8.68×10^{25}	2.54×10^7	1.19×10^3	1.36×10^{11}
Neptune	1.02×10^{26}	2.46×10^7	1.66×10^3	2.30×10^{11}

Example 25.

Estimate the pressure at the centre of the Earth. Given, mass of the Earth $M = 5.973 \times 10^{24}$ kg, radius of the Earth $R = 6.378 \times 10^6$ m, average density of Earth $\bar{\rho} = 5.5 \times 10^3$ kg/m³ and universal gravitational constant $G = 6.673\ 10^{-11}$ m³/s² kg.

Solution:

The pressure at the centre of the Earth is

$$P_c = \frac{\bar{\rho}}{2}\frac{GM}{R}$$

$$= \frac{5500}{2}\frac{(6.673\times10^{-11})(5.973\times10^{24})}{6.378\times10^{6}}$$

$$= 1.7 \times 10^{11}\ \text{N/m}^2$$

This value of pressure at the centre of the Earth is about a factor of two smaller than the actual value obtained from the detailed calculations.

Example 26.

Estimate the pressure at the centre of the Sun. Given, mass of the Sun $M_\odot = 1.989 \times 10^{30}$ kg, radius of the Sun $R_\odot = 6.960 \times 10^8$ m, average density of the Sun $\bar{\rho} = 1409$ kg/m³ and universal gravitational constant $G = 6.673 \times 10^{-11}$ m³/s² kg.

Solution:

The pressure at the centre of the Sun is

$$P_c = \frac{\bar{\rho}}{2}\frac{GM}{R}$$

$$= \frac{1409}{2}\frac{(6.673\times10^{-11})(1.989\times10^{30})}{6.960\times10^{8}}$$

$$= 1.3 \times 10^{14}\ \text{N/m}^2$$

Exact model calculations for the Sun yield a central pressure 2.5×10^{16} N/m², which is about a factor of 100 larger than the value obtained from the detailed calculations.

19. Thermal Equilibrium

When the amount of energy emitted by a body is equal to the energy absorbed by the body, the body is said to be in the thermal equilibrium. In case of non-thermal equilibrium, the body warms up (absorbs energy) or cools down (emits energy) until the thermal equilibrium is attained.

19.1 Body Hotter than Atmosphere

Suppose, a hot body of temperature T_1 is placed in an atmosphere of temperature T_0, such that $T_1 > T_0$. The body gives energy to the atmosphere till both of them attain a common temperature. Since the content of atmosphere is quite large, and there is no significant increase in its temperature. That is, the temperature of the atmosphere remains T_0. If T is the temperature of body at time t, the rate of cooling is

$$-\frac{dT}{dt}$$

Now, the rate of cooling of the body is proportional to the temperature difference and therefore, we have

$$-\frac{dT}{dt} \propto (T - T_0)$$

or
$$-\frac{dT}{dt} = k(T - T_0)$$

where k, having dimension of s^{-1}, is the proportionality constant. This equation can be rearranged as

$$\frac{dT}{(T - T_0)} = -k\,dt$$

On integration, we get

$$\ln(T - T_0) = -kt + c \qquad (6.49)$$

The constant of integration c can be obtained by applying the condition that at $t = 0$, we have $T = T_1$ and therefore, we have

$$\ln(T_1 - T_0) = c \qquad (6.50)$$

On substituting the value of c from equation (6.50) in (6.49), we have

$$\ln(T - T_0) = -kt + \ln(T_1 - T_0)$$

or
$$\ln\left(\frac{T - T_0}{T_1 - T_0}\right) = -kt$$

On taking anti-logarithm, we get

$$\frac{T - T_0}{T_1 - T_0} = e^{-k_t}$$

or
$$T = T_0 + (T_1 - T_0)e^{-kt} \qquad (6.51)$$

This relation gives the temperature T of body after a time t.

Note:

Here, it does not matter if the temperatures are expressed in Celsius or Kelvin unit.

Example 27.

A body having temperature of 127°C is placed in an atmosphere having temperature of 27°C. Considering $k = 0.005\ s^{-1}$, calculate the temperature of the body after 5 min.

Solution:

We have $T_1 = 127 + 273 = 400$ K, $T_0 = 27 + 273 = 300$ K, $t = 5 \times 60 = 300$ s, and $k = 0.005$ s–^{1}.
The temperature of body after 300 s is

$$T = T_0 + (T_1 - T_0)e^{-kt}$$
$$= 300 + (400 - 300)e^{-0.005\times 300}$$
$$= 322.31 \text{ K}$$

Temperature of the body is 322.31 − 273 = 49.31°C.

19.2 Body Cooler than Atmosphere

Suppose, a cool body of temperature T_1 is placed in an atmosphere of temperature T_0, such that $T_1 < T_0$. The body takes energy from the atmosphere till both of them attain a common temperature. Since the content of atmosphere is quite large, there is no significant decrease in its temperature. That is, the temperature of the atmosphere remains T_0. If T be the temperature of the body at time t, the rate of heating is

$$\frac{dT}{dt}$$

Now, the rate of heating of the body is proportional to the temperature difference and therefore, we have

$$\frac{dT}{dt} \propto (T_0 - T)$$

or
$$\frac{dT}{dt} = k(T_0 - T)$$

where k, having dimension of s^{-1}, is the proportionality constant. This equation can be rearranged as

$$\frac{dT}{(T_0 - T)} = k\,dt$$

On integration, we get

$$-\ln(T_0 - T) = kt + c \tag{6.52}$$

The constant of integration (c) can be obtained by applying the condition that at $t = 0$, we have $T = T_1$ and therefore, we have

$$-\ln(T_0 - T_1) = c \tag{6.53}$$

On substituting the value of c from equation (6.53) in equation (6.52), we have

$$-\ln(T_0 - T) = kt - \ln(T_0 - T_1)$$

or
$$\ln\left(\frac{T - T_0}{T_1 - T_0}\right) = -kt$$

On taking anti-logarithm, we get

$$\frac{T - T_0}{T_1 - T_0} = e^{-kt}$$

or
$$T = T_0 + (T_1 - T_0)\, e^{-kt} \tag{6.54}$$

This relation gives the temperature T of the body after a time t.

Note:

Here, it does not matter if the temperatures are expressed in Celsius or Kelvin unit.

Example 28.

A body having temperature of –23°C is placed in an atmosphere having temperature of 27°C. Considering $k = 0.005\ s^{-1}$, calculate the temperature of the body after 5 min.

Solution:

We have $T_1 = 23 + 273 = 250$ K, $T_0 = 27 + 273 = 300$ K, $t = 5 \times 60 = 300$ s, and k 0.005 s^{-1}. The temperature of the body after 300 s is

$$\begin{aligned} T &= T_0 + (T_1 - T_0)e^{-kt} \\ &= 300 + (250 - 300)\, e^{-0.005\times 300} \\ &= 288.84 \text{ K} \end{aligned}$$

Temperature of the body is 288.84 − 273 = 15.84 C.

20. Equations of Stellar Structure

20.1 General Equations

Let us consider a spherically symmetric, self-gravitating star, where all the physical quantities depend on two independent variables: the radius and time, (r, t). First, we derive all the equations of stellar structure in a general, non-spherical case, and then we consider the spherically symmetric case. The microscopic properties of matter at any given point may be described with the density ρ, temperature T, and chemical composition, i.e., the abundances of various elements X_i, with $i = 1, 2, 3$ having as many values as there are elements. All the thermodynamic properties and the transport coefficients are functions of (ρ, T, X_i). For example, we have: pressure $P(\rho, T, X_i)$, internal energy per unit volume $U(\rho, T, X_i)$, entropy per unit mass $S(\rho, T, X_i)$, coefficient of thermal conductivity per unit volume $\lambda(\rho, T, X_i)$, and heat source or heat sink per unit mass $\epsilon(\rho, T, X_i)$. All the partial derivatives of P, U, λ, and ϵ are also functions of ρ, T, and X_i. Using these quantities the first law of thermodynamics may be expressed as

$$T\,\mathrm{d}S = \mathrm{d}\left(\frac{U}{\rho}\right) - \frac{P}{\rho^2}\mathrm{d}\rho \tag{6.55}$$

When there are sources of heat, ϵ, and a non-vanishing heat flux $\vec{F}$, then the heat balance equation is written as

$$\rho T\frac{\mathrm{d}S}{\mathrm{d}t} = \rho\epsilon - \operatorname{div}\vec{F} \tag{6.56}$$

The heat flux $\vec{F}$ is directly proportional to the temperature gradient:

$$\vec{F} = -\lambda\,\nabla T. \tag{6.57}$$

The equation of motion (the Navier-Stokes equation of hydrodynamics) may be expressed as

$$\frac{\mathrm{d}^2\vec{r}}{\mathrm{d}t^2} + \frac{1}{\rho}\nabla P + \nabla V = 0 \tag{6.58}$$

where the gravitational potential satisfies the Poisson's equation

$$\nabla^2 V = 4\pi G\rho \tag{6.59}$$

with $V \to 0$ when $r \to \infty$.

In the spherical symmetry these equations may be written as

$$\frac{1}{\rho}\frac{\partial P}{\partial r} + \frac{\partial V}{\partial r} + \frac{\mathrm{d}^2\vec{r}}{\mathrm{d}t^2} = 0 \tag{6.60}$$

$$\frac{1}{r^2}\frac{\partial}{\partial r}\left(r^2\frac{\partial V}{\partial r}\right) = 4\pi G\rho \tag{6.61}$$

$$F = -\lambda\frac{\partial T}{\partial r} \tag{6.62}$$

$$\frac{1}{\rho r^2}\frac{\partial r^2 F}{\partial r} = \epsilon - T\frac{\mathrm{d}S}{\mathrm{d}t} \tag{6.63}$$

It is convenient to introduce a new variable, M_r, defined as

$$M_r = \int_0^r \rho 4\pi r^2 \mathrm{d}r \tag{6.64}$$

which is the total mass within the radius r, and another variable, L_r, expressed as

$$L_r \equiv 4\pi r^2 F \tag{6.65}$$

which is the luminosity, i.e., the total heat flux flowing through a spherical shell with the radius r, and also

$$\kappa = \frac{4acT^3}{3\rho}\frac{1}{\lambda} \tag{6.66}$$

where κ is the coefficient of radiative opacity (per unit mass), c the the speed of light, and a the radiation constant. The last equation is valid if the heat transport is due to the radiation.

Using the definitions and relations (6.64), (6.65), 6.66), we may write the set of equations (6.60), (6.61), (6.62), (6.63) in a more standard form:

$$\frac{1}{\rho}\frac{\partial P}{\partial r} + \frac{GM_r}{r^2} + \frac{\mathrm{d}^2\vec{r}}{\mathrm{d}t^2} = 0 \tag{6.67}$$

$$\frac{\partial M}{\partial r} = 4\pi r^2\rho \tag{6.68}$$

$$\frac{\partial T}{\partial r} = \frac{3\kappa\rho L_r}{16\pi acT^3 r^2} \tag{6.69}$$

$$\frac{\partial L_r}{\partial r} = 4\pi r^2\rho\left(\epsilon - T\frac{\mathrm{d}S}{\mathrm{d}t}\right) \tag{6.70}$$

This set of equations is further written in a somewhat inconvenient way, as all the space derivatives ($\partial/\partial r$) are taken at a fixed value of time, while all the time derivatives ($\mathrm{d}/\mathrm{d}t$) are at the fixed mass zones. For this reason, and also because of the way the boundary conditions are specified (later on), it is convenient to use the mass (M_r) rather than radius (r) as a space-like independent variable. Therefore, we replace all derivatives $\partial/\partial r$ with $4\pi r^2\rho\partial/\partial M_r$, and we obtain the equations

$$\frac{\partial P}{\partial M_r} = -\frac{GM_r}{r^2} - \frac{1}{4\pi r^2}\frac{\partial^2 r}{\partial t^2} \tag{6.71}$$

$$\frac{\partial T}{\partial M_r} = \frac{1}{4\pi r^2\rho} \tag{6.72}$$

$$\frac{\partial T}{\partial M_r} = -\frac{3\kappa L_r}{64\pi^2 acT^3 r^4} \tag{6.73}$$

$$\frac{\partial L_r}{\partial M_r} = \epsilon - T\frac{\partial S}{\partial t} \tag{6.74}$$

The set of fundamental equations (6.71), (6.72), (6.73), and (6.74) describes the time evolution of a spherically symmetric star with a given distribution of chemical composition with mass, $X_i(M_r)$, provided the initial conditions and the boundary conditions are specified.

20.2 Boundary Conditions

We now consider the boundary conditions. At the centre of star, the mass M_r, the radius r, and the luminosity L_r, all vanish. Thus, the inner boundary conditions are:

$$\text{At } r = 0, \quad \text{we have } L_r = 0 \quad \text{and} \quad M_r = 0 \tag{6.75}$$

In most cases, we have interest in the structure and evolution of a star with a fixed total mass, M. At the surface, where $M_r = M$, the density becomes zero, and the temperature has a value that is related to the stellar radius and luminosity. Though the proper outer boundary conditions require rather complicated calculations of a model stellar atmosphere, but we adopt a very simple model atmosphere within the Eddington approximation, which means we use the diffusion approximation to calculate the temperature gradient not only at large optical depth, but also at the small optical depth. The Eddington approximation also means that the surface temperature is $2^{1/4} \approx 1.189$ times lower than the effective temperature. At the surface of the star, the boundary conditions are:

At $M_r = M$,

we have $$\rho = 0$$

and $$T = T_0$$

$$= \left(\frac{L}{8\pi R^2 \sigma}\right)^{1/4} \tag{6.76}$$

where σ is the Stefan-Boltzmann constant. It may be noted that the effective temperature of a star is defined as

$$T_{\text{eff}} = \left(\frac{L}{4\pi R^2 \sigma}\right)^{1/4} = 2^{1/4} T_0 \tag{6.77}$$

20.3 Simplified Equations

Let us consider now an even simpler case, where a star may be described with the equations in which all time derivatives may be neglected, i.e., the star is in the hydrostatic and thermal equilibria. Now, we have no time dependence, and the stellar structure depends on one space-like variable only. Now, we have four ordinary differential equations:

$$\frac{\partial P}{\partial M_r} = -\frac{GM_r}{r^2} \tag{6.78}$$

$$\frac{\partial r}{\partial M_r} = \frac{1}{4\pi r^2 \rho} \tag{6.79}$$

$$\frac{\partial T}{\partial M_r} = -\frac{3\kappa L_r}{64\pi^2 acT^3 r^4} \tag{6.80}$$

$$\frac{\partial L_r}{\partial M_r} = \epsilon \tag{6.81}$$

These have to be supplemented with the boundary conditions (6.75) and (6.76), as well as the total stellar mass M, and the distribution of all elements with the mass, $X_i(M_r)$. We have four ordinary differential equations, four boundary conditions, and four parameters to be found: T_c, ρ_c, R, and L.

21. Integral Theorems on Pressure, Density and Temperature

A gas is described in terms of four physical quantities, the pressure, temperature, density and volume at the macroscopic scale. These quantities of gases in various experiments were measured repeatedly by the scientists. The investigations led to some relations among these physical quantities, known as the gas laws.

21.1 Boyle's Law

Boyle's law, published in 1662 by Robert Boyle, states that for a given amount of a gas at constant temperature, the product of its pressure and volume is constant. As a relation, the Boyle's law can be expressed as

$$P \propto \frac{1}{V}$$

or

$$P = \frac{k_1}{V}$$

where P is the pressure and V the volume of gas, and k_1 is a constant. The statement of Boyle's law is as the following.

For a given amount of gas at a given temperature, its pressure is inversely proportional to its volume.

21.2 Charles' Law

Charles' law, published in 1787 by Jacques Charles, states that for a given amount of a gas at constant pressure, its volume is directly proportional to its absolute temperature. As a relation, the Charles' law can be expressed as

$$V \propto T$$

or

$$V = k_2 T$$

where V is the volume and T the absolute temperature of gas, and k_2 is constant. The statement of Charles' law is as the following.

For a given amount of gas at a given pressure, its volume is proportional to its absolute temperature.

21.3 Gay-Lussac's Law

Gay-Lussac's law, published by Joseph Louis Gay-Lussac in 1809, states that for a given amount of gas at constant volume, its pressure is directly proportional to its absolute temperature. As a relation, Gay-Lussac's law can be expressed as

$$P \propto T$$

or $$P = k_3 T$$

where P is the pressure and T the absolute temperature of gas, and k_3 is constant. The statement of Gay-Lussac's law is as the following.

For a given amount of gas at a given volume, its pressure is proportional to its absolute temperature.

21.4 Avogadro's Law

Avogadro's law, published in 1811 by Amedeo Avogadro, states that at constant pressure and temperature, the volume occupied by a gas is directly proportional to the number of molecules of gas present in the container. The relation can be expressed as

$$V \propto n$$

or $$V = k_4 n$$

where V is the volume of gas and n the number of molecules in gas, and k_4 is constant. The statement of Avogadro's law is as the following.

For a gas at a given pressure and absolute temperature, its volume is proportional to the number of molecules in it.

21.5 Ideal Gas

All the four relations discussed in the preceding sub-sections are valid for an ideal gas. The Boyle's law, Charles' law, and Gay-Lussac's law for 1 mol of gas can be combined in a common relation as

$$\frac{PV}{T} = R$$

where P is the pressure, V the volume, T the absolute temperature of the gas, and R is the gas constant. This relation is known as the ideal gas law for one mole of gas. With the addition of Avogadro's law, the combined gas law develops into the ideal gas law:

$$PV = nRT$$

or $$P = \rho RT$$

where $\rho = n/V$ is the mass density. This equation is valid in the SI units.

22. Homologous Transformations

In the homologous stellar models, the parameters of the stars are scaled in the same way with the fractional mass $x = M(r)/M$. Obviously, at the centre of the star, we have $x = 0$, and at the surface of star, we have $x = 1$.

- For all the homologous models, for some property $X(x)$ (for example, temperature $T(x)$, density $\rho(x)$, etc.), a variation of the $X(x)$ versus x would be the same.
- The homologous models should be a reasonable representation for the zero-age main sequence stars, as they have a uniform chemical composition.
- The stellar structure equations are modified in a manner that they are independent of the absolute mass M of star and depend only on the relative mass x.

- The parameters of interest are therefore expressed as the functions of x with dependencies on M being assumed to be power laws:

$$r = M^{a_1}\, r_s(x) \qquad dr = M^{a_1}\, dr_s(x)$$
$$\rho(r) = M^{a_2}\, \rho_s(x) \qquad d\rho(r) = M^{a_2}\, d\rho_s(x)$$
$$T(r) = M^{a_3}\, T_s(x) \qquad dT(r) = M^{a_3}\, dT_s(x)$$
$$P(r) = M^{a_4}\, P_s(x) \qquad dP(r) = M^{a_4}\, dP_s(x)$$
$$L(r) = M^{a_5}\, L_s(x) \qquad dL(r) = M^{a_5}\, dL_s(x)$$

where the exponents a_i are constants to be determined and the variables $r_s(x)$, $\rho_s(x)$, etc., depend only on the fractional mass x. Also $M(r) = Mx$, and therefore, $dM(r) = Mdx$. We have

$$\kappa(r) = \kappa_0\, \rho^{\alpha}(r)\, T^{\beta}(r)$$
$$= \kappa_0\, \rho_s^{\alpha}(x)\, T_s^{\beta}(x)\, M^{\alpha a_2}\, M^{\beta a_3}$$

and

$$\epsilon(r) = \epsilon_0\, \rho(r)\, T^{\eta}(r)$$
$$= \epsilon_0\, \rho_s(x)\, T_s^{\eta}(x)\, M^{a_2}\, M^{\eta a_3}$$

Substitution of above expressions into the stellar structure equations now allows them to be expressed in terms of the dimensionless mass, x:

- Mass conservation equation:

$$\frac{dr}{dM(r)} = \frac{1}{4\pi r^2 \rho(r)} \quad \text{becomes}$$

$$M^{(a_1-1)}\frac{dr_s(x)}{dx} = \frac{1}{4\pi r_s^2(x)\,\rho_s(x)}M^{-(2a_1+a_2)}$$

or

$$\frac{dr_s(x)}{dx} = \frac{1}{4\pi\, r_s^2(x)\,\rho_s(x)}M^{-(3a_1+a_2-1)}$$

For the homology condition, the expression must be independent of M. That is,

$$3a_1 + a_2 - 1 = 0$$

or

$$3a_1 + a_2 = 1. \tag{6.82}$$

therefore,

$$\frac{dr_s(x)}{dx} = \frac{1}{4\pi r_s^2(x)\,\rho_s(x)}$$

- Hydrostatic equilibrium equation:

$$\frac{dP(r)}{dM(r)} = -\frac{GM(r)}{4\pi r^4} \quad \text{becomes}$$

$$M^{(a_4-1)}\frac{dP_s(x)}{dx} = -\frac{Gx}{4\pi r_s^4}M^{(1-4a_1)}$$

or

$$\frac{dP_s(x)}{dx} = -\frac{Gx}{4\pi r_s^4}M^{(2-4a_1-a_4)}$$

For the homology condition, the expression must be independent of M. That is,

$$2 - 4a_1 - a_4 = 0$$

or

$$4a_1 + a_4 = 2. \tag{6.83}$$

therefore, $$\frac{dP_s(x)}{dx} = -\frac{Gx}{4\pi r_s^4}$$

- Energy production equation:

$$\frac{dL(r)}{dM(r)} = \epsilon(r)$$
$$= \epsilon_0 \rho(r)\, T^{\eta}(r) \text{ becomes}$$
$$M^{(a_5-1)}\frac{dL_s(x)}{dx} = \epsilon_0 \rho_s(x)\, T^{\eta}(x)\, M^{(a_2+\eta a_3)}$$

or $$\frac{dL_s(x)}{dx} = \epsilon_0 \rho_s(x)\, T^{\eta}(x)\, M^{(a_2+\eta a_3+1-a_5)}$$

For the homology condition, the expression must be independent of M. That is,

$$a_2 + \eta a_3 + 1 - a_5 = 0$$

or $$a_2 + \eta a_3 - a_5 = -1. \tag{6.84}$$

therefore, $$\frac{dL_s(x)}{dx} = \epsilon_0 \rho_s(x)\, T^{\eta}(x)$$

- Radiative transport equation:

$$\frac{dT(r)}{dM(r)} = \frac{3\bar{\kappa}_{Ross} L(r)}{16\pi^2 r^4 ac T(r)^3} \text{ becomes}$$
$$M^{(a_3-1)}\frac{dT_s(x)}{dx} = -\frac{3(\kappa_0 \rho_s(x)^{\alpha} T_s(x)^{\beta})L_s(x)}{16\pi^2 r_s(x)^4\, ac\, T_s(x)^3} M^{(a_5+(\beta-3)a_3+\alpha a_2-4a_1)}$$

or $$\frac{dT_s(x)}{dx} = -\frac{3(\kappa_0 \rho_s(x)^{\alpha} T_s(x)^{\beta})L_s(x)}{16\pi^2 r_s(x)^4\, ac\, T_s(x)^3} M^{(a_5+(\beta-4)a_3+\alpha a_2-4a_1+1)}$$

For the homology condition, the expression must be independent of M. That is,

$$a_5 + (\beta - 4)a_3 + \alpha a_2 - 4a_1 + 1 = 0$$

or $$4a_1 + (4 - \beta)a_3 - \alpha a_2 - a_5 = 1. \tag{6.85}$$

- Equation of state:

If gas pressure dominates and we neglect any radial dependence of the mean molecular weight ($\bar{\mu}$):

$$P_{\text{gas}}(r) = \frac{\rho(r)kT(r)}{\bar{\mu}\, m_H} \text{ becomes}$$
$$M^{a_4} P_s(x) = \frac{\rho_s(x)kT_s(x)}{\bar{\mu}\, m_H} M^{(a_2+a_3)}$$

or $$P_s(x) = \frac{\rho_s(x)kT_s(x)}{\bar{\mu}\, m_H} M^{(a_2+a_3-a_4)}$$

For the homology condition, the expression must be independent of M. That is,

$$a_2 + a_3 - a_4 = 0$$

or $$a_2 + a_3 - a_4 = 0. \tag{6.86}$$

Alternatively, if radiation pressure dominates:

$$P_{\text{rad}} = \frac{1}{3}aT^4 \text{ becomes}$$

$$M^{a_4} P_s(x) = \frac{1}{3} a (M^{a_3} T_s(x))^4$$

or $$P_s(x) = \frac{1}{3} a M^{4a_3 - a_4} T_s(x)^4$$

For the homology condition, the expression must be independent of *M*. That is,

$$4a_3 - a_4 = 0$$

or $$a_4 = 4a_3.$$

In a matrix form, the five equations (6.82), (6.83), (6.84), (6.85), and (6.86), (assuming $P_{\text{gas}} >> P_{\text{rad}}$), for the exponents a_i are

$$\begin{bmatrix} 3 & 1 & & & \\ 4 & & & 1 & \\ & 1 & \eta & & -1 \\ 4 & -\alpha & (4-\beta) & & -1 \\ & 1 & 1 & -1 & \end{bmatrix} \begin{bmatrix} a_1 \\ a_2 \\ a_3 \\ a_4 \\ a_5 \end{bmatrix} = \begin{bmatrix} 1 \\ 2 \\ -1 \\ 1 \\ 0 \end{bmatrix}$$

and which can be solved for given values of α, β and η.

Let us consider two cases:

- Low mass stars ($\sim 0.7 \le M \le 2M_\odot$) corresponding to spectral types *F* and later: Adopt Kramers opacity $\kappa \propto \rho T^{-3.5}$ and a nuclear generation rate for the pp-Chain $\epsilon \propto T^4$ so that $\alpha = 1$, $\beta = -3.5$ and $\eta = 4$.
- Higher mass stars ($M \ge 2M_\odot$) corresponding to the spectral types *A* and earlier: Assume opacity to be dominated by the electron scattering and a nuclear generation rate dominated by the CNO cycle with a stronger temperature dependence $\epsilon \propto T^{16}$ so that $\alpha = 0$, $\beta = 0$ and $\eta = 16$.

The system of stellar structure equations in the scaled mass (*x*) representation may be solved numerically, subject to the boundary conditions. But, the beauty of the homology approximation is that the useful conclusions may be derived analytically from the a_i obtained from the above matrix equation and summarized in the table:

Regime	α	β	η	a_1	a_2	a_3	a_4	a_5
Low mass	1	–3.5	4	1/13	10/13	12/13	22/13	71/13
Higher Mass	0	0	16	15/19	–26/19	4/19	–22/19	3

23. Polytropic Gas Sphere - Lane Emden Equation

- We have four stellar structure equations and three auxilary equations, which are highly non-linear, coupled and need to be solved simultaneously with boundary conditions at the centre and at the surface of the star.
- Polytropic models of a star consider that a simple relation between the pressure and density (for example) exists throughout the star. The equations of the hydrostatic equilibrium and the mass conservation may then be solved independently of the other five.

- Obviously, before the advent of computing technology, polytropic models played an important role in the development of stellar structure theory. At present, however, they provide insight.

Take the equation of hydrostatic equilibrium

$$\frac{dP(r)}{dr} = -\frac{GM(r)\rho(r)}{r^2}$$

multiply by $r^2/\rho(r)$ this equation and differentiate with respect to r to get

$$\frac{d}{dr}\left(\frac{r^2}{\rho(r)}\frac{dP(r)}{dr}\right) = -\frac{G\,dM(r)}{dr}$$

Now, using the equation of mass-conservation on the right-hand side, we get

$$\frac{1}{r^2}\frac{d}{dr}\left(\frac{r^2}{\rho(r)}\frac{dP(r)}{dr}\right) = -4\pi G\rho(r)$$

Adopt an equation of state of the form

$$P(r) = K\rho(r)^{\gamma} = K\rho(r)^{(n+1)/n}$$

and

$$\frac{dP(r)}{dr} = K\gamma\rho(r)^{\gamma-1}\frac{d\rho(r)}{dr}$$

where K is a constant and n (not necessarily an integer) is known as the polytropic index.

Using the value of $dP(r)/dr$ in the hydrostatic equilibrium equation combined with the mass conservation equation, and writing ρ for $\rho(r)$ in order to simplify the notation, gives

$$\frac{1}{r^2}\frac{d}{dr}\left(\frac{r^2K}{\rho}\gamma\rho^{\gamma-1}\frac{d\rho}{d\xi}\right) = -4\pi G_{\rho}$$

where we have used $r = \alpha\xi$.

Suppose a radial density dependence

$$\rho = \rho_c\theta(\xi)^n$$

and

$$\frac{d\rho}{d\xi} = \rho_c n\theta(\xi)^{n-1}\frac{d\theta(\xi)}{d\xi}$$

where ρ_c is the central density. Writing θ for $\theta(\xi)$ in order to simplify the notation, the above equation transforms to

$$\frac{K(n+1)}{4\pi G\rho_c^{(1-1/n)}\alpha^2}\frac{1}{\xi^2}\frac{d}{d\xi}\left(\xi^2\frac{d\theta}{d\xi}\right) = -\theta^n$$

As α is arbitrary, we choose

$$\alpha^2 = \frac{K(n+1)}{4\pi G\rho_c^{(1-1/n)}}$$

so that

$$\frac{1}{\xi^2}\frac{d}{d\xi}\left(\xi^2\frac{d\theta}{d\xi}\right) = -\theta^n$$

This equation is known as the **Lane-Emden** equation. It defines the rate of change of density within a stellar interior subject to:

- At the centre of the star where $\xi(0) = 0$, $\theta(0) = 1$ so that $\rho = \rho_c$.
- $dP/dr \to 0$ as $r \to 0$, $d\theta/d\xi = 0$ at $\xi = 0$.
- The outer boundary (surface) is the first location where $\rho = 0$ or $\theta(\xi) = 0$; this location is referred to as ξ_1.

Solutions of the Lane-Emden equation, known as the polytropes, specify $\rho(r)$ although expressed as $\theta(\xi)$. The order of the solution is determined by the index n; in particular it depends only on n and can be scaled by varying the P_c (central pressure) and ρ_c to give the solutions for the stars over a range of total mass and radius. Analytical solutions exist for $n = 0$, 1 and 5; all other solutions need to be obtained numerically.

23.1 Solution of Lane-Emden Equation for *n* = 0

The Lane-Emden equation for $n = 0$ becomes

$$\frac{1}{\xi^2}\frac{d}{d\xi}\left(\xi^2\frac{d\theta}{d\xi}\right)+1=0$$

Rearrangement of this equation and integration once gives

$$\xi^2\frac{d\theta}{d\xi}=C_1-\frac{\xi^3}{3}$$

Dividing both the sides of this equation by ξ^2 and integration again gives

$$\theta(\xi)=C_0-\frac{C_1}{\xi}-\frac{\xi^3}{6}$$

The boundary conditions $\theta(0) = 1$ and $\theta'(0) = 0$ give $C_0 = 1$ and $C_1 = 0$. Thus, we have

$$\theta(\xi)=1-\frac{\xi^3}{6}$$

23.2 Solution of Lane-Emden Equation for *n* = 1

The Lane-Emden equation for $n = 1$ becomes

$$\frac{d^2\theta}{d\xi^2}+\frac{2}{\xi}\frac{d\theta}{d\xi}+\theta=0$$

One assumes a power series solution:

$$\theta(\xi)=\sum_{m=0}^{\infty}a_m\xi^m$$

This leads to a recursive relationship for the expansion coefficients:

$$a_{m+2}=\frac{a_m}{(m+3)(m+2)}$$

This relation can be solved leading to the general solution:

$$\theta(\xi)=a_0\frac{\sin\xi}{\xi}+a_1\frac{\cos\xi}{\xi}$$

The boundary condition for a physical polytrope demands that $\theta(\xi) \to 1$ as $\xi \to 0$. This gives $a_0 = 1$, $a_1 = 0$. Hence, the solution is

$$\theta(\xi) = \frac{\sin\xi}{\xi}$$

23.3 Solution of Lane-Emden Equation for *n* = 5

The Lane-Emden equation for $n = 5$ becomes

$$\frac{1}{\xi^2}\frac{d}{d\xi}\left(\xi^2\frac{d\theta}{d\xi}\right) + \theta^5 = 0$$

The relation

$$\theta(\xi) = \frac{1}{\sqrt{1+\xi^2/3}}$$

is found to satisfy the above relation. Hence, this is the solution of Lane-Emden equation for $n = 5$.

Another solution of Lane-Emden equation for $n = 5$ is

$$\theta(\xi) = \frac{\sin(\ln\sqrt{\xi})}{\sqrt{3}[3 - 2\sin^2(\ln\sqrt{\xi}}$$

24. Estimation of Kinetic Temperature

According to the kinetic theory of gases, the pressure exerted by the atoms hitting and bouncing back from the walls of a gas container is equal to the force exerted per unit area. Let us consider a gas having large number N of molecules, each of mass m, enclosed in a cube of volume $V = L^3$, where L is the side of the container. Consider a gas molecule colliding with the wall of the container perpendicularly, along the x-axis, and bouncing back in the opposite direction with the same speed (an elastic collision), the change in momentum is

$$\Delta p = mv_x - (-mv_x) = 2mv_x$$

where v_x is the velocity along the x-axis (which is the same before and after the collision). The particle strikes one specific side-wall once during the time interval Δt given by

$$\Delta t = \frac{2L}{v_x}$$

The force of this particle's collision with the wall is

$$F = \frac{\Delta p}{\Delta t}$$

$$= \frac{2mv_x}{2L/v_x} = \frac{mv_x^2}{L}$$

The total force on the wall due to collisions by the molecules impacting the walls with a range of possible values of v_x is

$$F = \frac{m\overline{v_x^2}}{L}$$

where the bar denotes an average over the possible velocities of the N particles. The total force on the wall due to collisions by N molecules impacting the walls is

$$F = \frac{Nm\overline{v_x^2}}{L}$$

As the motion of the particles is random and there is no bias applied in any direction, the average squared speed in each direction is equal. That is, $\overline{v_x^2} = \overline{v_y^2} = \overline{v_z^2}$. Thus, the average squared speed v^2 is

$$\begin{aligned} \overline{v^2} &= \overline{v_x^2} + \overline{v_y^2} + \overline{v_z^2} \\ &= 3\overline{v_x^2} \end{aligned}$$

and thus

$$\overline{v_x^2} = \frac{\overline{v^2}}{3}$$

and therefore, the force can be written as

$$\begin{aligned} F &= \frac{Nm\overline{v^2}}{3L} \\ &= \frac{2}{3}\frac{N}{L}E_{av} \end{aligned}$$

with

$$E_{av} = \frac{1}{2}m\overline{v^2}$$

where E_{av} is the average molecular kinetic energy. This force is exerted uniformly on an area L^2. Therefore, the pressure of the gas is

$$P = \frac{F}{L^2} = \frac{2}{3}\frac{N}{V}E_{av}$$

Using this expression for pressure P in the ideal gas equation $P_V = nRT$, we get

$$\begin{aligned} T &= \frac{2}{3}\frac{N}{nR}E_{av} \\ &= \frac{2}{3}\frac{E_{av}}{k} \end{aligned}$$

where k is the Boltzmann constant.

25. Gravitational Energy

When the question about the source of energy being emitted by the Sun continuously was under discussion, scientists thought that the gravitational contraction of the Sun itself could be an effective source of energy. In order to understand it, let us consider a point mass m situated at a distance r from a large mass M. The gravitational force of attraction acting on the mass m is

$$F = -G\frac{Mm}{r^2}$$

Here, the negative sign indicates attractive nature of gravitational force. The potential energy of the system U, which is equal to the amount of work required for moving the mass m from the distance r to an infinite distance, is expressed as

$$U = \int_r^{\infty} F\,\mathrm{d}r$$
$$= \int_r^{\infty} -G\frac{Mm}{r^2}\mathrm{d}r$$
$$= -G\frac{Mm}{r}$$

The work is performed when the motion is along the line joining the two bodies, and no work is performed when the motion is perpendicular to the line. Further, in the conservative field, the amount of work done is independent of the path along which the mass is moved. Obviously, U is the amount of energy lost by the system in moving from an infinite distance to a distance r between the bodies. In other words, $-U$ is the amount of energy which is given to the system in order to separate the bodies at infinite distance from each other.

Let us consider a sphere of radius r, comprising of material of uniform density ρ. Mass of the sphere is

$$M(r) = \frac{4}{3}\pi r^3 \rho$$

Suppose, a thin layer of thickness dr is added to the sphere. Mass of the layer is dm, expressed as

$$\mathrm{d}m = \frac{4}{3}\pi(r + \mathrm{d}r)^3\rho - \frac{4}{3}\pi r^3\rho$$
$$= 4\pi r^2 \rho\,\mathrm{d}r$$

Here, we have neglected smaller terms. Thus, the amount of energy released in adding the layer is

$$\mathrm{d}\phi = -G\frac{M(r)\mathrm{d}m}{r}$$
$$= -G\frac{(4\pi r^3\rho/3)(4\pi r^2 \mathrm{d}r\rho)}{r}$$
$$= -G\frac{16}{3}\pi^2\rho^2 r^4 \mathrm{d}r$$

Thus, the total amount of energy released in the formation of a sphere of radius R is

$$\phi = -\int_0^R G\frac{16}{3}\pi^2\rho^2 r^4 \mathrm{d}r$$
$$= -G\frac{16}{15}\pi^2\rho^2 R^5$$
$$= -\frac{3}{5}\frac{GM^2}{R}$$

where M (= $4\pi R^3\rho/3$) is the mass of the sphere. Thus, the amount of energy released in the formation of sphere of radius R and mass M, known as the gravitational energy E_G, is

$$E_G = \frac{3}{5}\frac{GM^2}{R}$$

The gravitational energy released during the formation of various bodies is given in Table 4.

Table 4. Gravitational energy

Body	Mass (kg)	Radius (m)	Gravitational energy (J)
Sun	1.99×10^{30}	6.96×10^{8}	2.28×10^{41}
Moon	7.35×10^{22}	1.74×10^{6}	1.24×10^{29}
Mercury	3.28×10^{23}	2.44×10^{6}	1.77×10^{30}
Venus	4.87×10^{24}	6.05×10^{6}	1.57×10^{32}
Earth	5.97×10^{24}	6.37×10^{6}	2.24×10^{32}
Mars	6.39×10^{23}	3.39×10^{6}	4.82×10^{30}
Jupiter	1.90×10^{27}	6.99×10^{7}	2.07×10^{36}
Saturn	5.68×10^{26}	5.82×10^{7}	2.22×10^{35}
Uranus	8.68×10^{25}	2.54×10^{7}	1.19×10^{34}
Neptune	1.02×10^{26}	2.46×10^{7}	1.69×10^{34}

Example 29.

Estimate the amount of energy released in the formation of the Sun. Comment if it is sufficient for the existence of Sun as of today. Given, mass of the Sun $M_{\odot} = 1.989 \times 10^{30}$ kg, radius of the Sun $R_{\odot} = 6.960 \times 10^{8}$ m and universal gravitational constant $G = 6.673 \times 10^{-11}$ m^3/s^2 kg.

Solution:

The amount of energy released in the formation of the Sun is

$$E_G = \frac{3}{5}\frac{GM^2}{R}$$

$$= \frac{3}{5}\frac{6.673\times10^{-11}\times(1.989\times10^{30})^2}{6.960\times10^{8}}$$

$$= 2.28 \times 10^{41} \text{ J}$$

Thus, the gravitational energy of 2.28×10^{41} J is released in the formation of Sun.

Example 30.

Estimate the amount of energy released in the formation of the Earth. Given, mass of the Earth $M = 5.973 \times 10^{24}$ kg, radius of the Earth $R = 6.378 \times 10^{6}$ m and universal gravitational constant $G = 6.673 \times 10^{-11}$ m^3/s^2 kg.

Solution:

The amount of energy released in the formation of the Earth is

$$E_G = \frac{3}{5}\frac{GM^2}{R}$$

$$= \frac{3}{5}\frac{6.673\times10^{-11}\times(5.973\times10^{24})^2}{6.378\times10^{6}}$$

$$= 2.24 \times 10^{32} \text{ J}$$

Thus, the gravitational energy of 2.24×10^{32} J is released in the formation of the Earth.

25.1 Estimation of Kinetic Temperature

Let us consider a celestial body consists of a mono-atomic gas with atomic mass μ having radius R and total mass M in hydrostatic equilibrium. According to the Avogadro hypothesis, μ kg of gas has $N = 6.023 \times 10^{26}$ particles. Thus, the number (n) of particles in the body of mass M is

$$n = \frac{N}{\mu} M$$

According to the thermodynamics, at temperature T, the thermal energy per degree of freedom is $kT/2$. Here, k is the Boltzmann constant. A particle of mono-atomic gas has three degrees of freedom. Thus, the energy of each particle is $3kT/2$. Hence, the total energy E_T of n particles is

$$E_T = \left(\frac{N}{\mu} M\right)\left(\frac{3}{2} kT\right)$$

$$= \frac{3NMkT}{2\mu}$$

For a spherical body of mass M and radius R, the gravitational energy is

$$E_G = -\frac{3}{5}\frac{GM^2}{R}$$

Following the Virial theorem for a star in the hydrostatic equilibrium, we have

$$2E_T + E_G = 0$$

thus
$$2\frac{3NMkT}{2\mu} + \left(-\frac{3}{5}\frac{GM^2}{R}\right) = 0$$

The temperature in the core of the body is

$$T = \frac{GM_\mu}{5NkR}$$

Table 5. Temperature in the core of body

Body	Mass (kg)	Radius (*m*)	T/μ (K)
Sun	1.99×10^{30}	6.96×10^{8}	4.59×10^{6}
Moon	7.35×10^{22}	1.74×10^{6}	6.78×10^{1}
Mercury	3.28×10^{23}	2.44×10^{6}	2.16×10^{2}
Venus	4.87×10^{24}	6.05×10^{6}	1.29×10^{3}
Earth	5.97×10^{24}	6.37×10^{6}	1.50×10^{3}
Mars	6.39×10^{23}	3.39×10^{6}	3.03×10^{2}
Jupiter	1.90×10^{27}	6.99×10^{7}	4.36×10^{4}
Saturn	5.68×10^{26}	5.82×10^{7}	1.57×10^{4}
Uranus	8.68×10^{25}	2.54×10^{7}	5.49×10^{3}
Neptune	1.02×10^{26}	2.46×10^{7}	6.66×10^{3}

This is an expression for the temperature in the central core of a body. For calculating its value, we have $G = 6.67 \times 10^{-11}$ m^3/kg s^2, $k = 1.38 \times 10^{-23}$ J/K, $N = 6.023 \times 10^{26}$. Temperatures in the core of some celestial bodies are calculated and are given in Table 5. For the Sun, we have found

$$T = 4.59 \times 10^6 \mu$$

Hence, the temperature of the central core of the Sun is of the order of $10^6 - 10^7$ K. Table 5 also shows that the core of the Jupiter and Saturn are relatively hot and consequently, the material in them is in the fluid form.

Review Questions

1. Discuss the method of trigonometric parallax used for determination of distance of nearby celestial bodies.
2. The angular separation between two apparent positions, at the interval of six months, of a star relative to the distant stars is 30° 45′ 52″. Calculate the distance of the star from the Sun.
3. Show that a system of two moving bodies of masses m_1 and m_2, separated by a distance r is equivalent to a system where one body of mass $M (= m_1 + m_2)$ is stationary and another body of mass $\mu = m_1 m_2/(m_1 + m_2)$ separated by a distance r is in motion.
4. Two components A and B of a binary system have masses m_1 and m_2, respectively. If the position vector and velocity of B relative to A are $\vec{r}$ and $\vec{v}$, respectively, show that total angular momentum of the system is

$$\vec{L} = \mu \vec{r} \times \vec{v}$$

   where μ is the reduced mass of the system.
5. State the Kepler's laws of planetary motion and derive them.
6. For the motion of a planet around the Sun, show that the semi-major axis of the orbit is inversely proportional to the total energy of the planet.
7. Show that the velocity of a planet revolving around the Sun is inversely proportional to the square root of its distance from the Sun.
8. The distance of the Jupiter from the Sun is 5.2 AU. Calculate the time-period of revolution of the Jupiter around the Sun.
9. The distance of the Saturn from the Sun is 9.54 AU. Calculate the time-period of revolution of the Saturn around the Sun.
10. State the Newton's law of gravitation, and derive it from the Kepler's laws of the planetary motion.
11. State and prove the Virial theorem for a system of bodies.
12. Discuss about the Henry-Draper star catalogue. Describe the stellar classification of stars.
13. Describe salient features of black-body spectrum. Write Planck's formula for the black-body spectrum and show that Wien's formula, and Rayleigh and Jeans formula are special cases of it.

14. Using Planck's formula for black-body spectrum

$$I_\lambda d\lambda = \frac{8\pi ch}{\lambda^5}\frac{d\lambda}{\exp(hc/\lambda kT)-1}$$

discuss Wien's displacement law and Stefan-Boltzmann law.
15. Describe the Doppler redshift in radiations coming from the cosmic objects.
16. Show that for non-relativistic motion, the redshift is proportional to the velocity of object.
17. Describe the relation between *B-V* magnitude and surface temperature.
18. Describe the method for determination of radius of a star.
19. Describe salient features of Hertzsprung-Russell diagram.
20. Show that a heavy star stays in the main sequence for a shorter period as compared to that for a lighter star when the luminosity L of star varies as $L \propto M^\alpha$, where M is the mass of star and $\alpha > 1$. Mass of star A is 2.5 times that of B. Calculate the ratio of their periods in the main sequence.
21. Write about Meghnad Saha and his contribution in Astronomy and Astrophysics.
22. Describe the Wilson-Bappu effect and its application for determination of stellar distances.
23. State the Pauli's exclusion principle. Describe its importance in the context of white dwarfs and neutron stars.
24. Describe hydrostatic equilibrium inside a star.
25. Derive an expression for pressure at the centre of a spherical body of mass M, radius R and uniform density ρ.
26. Describe thermal equilibrium. A body having temperature T_1 is placed in an atmosphere having temperature T_0 such that $T_1 > T_0$. Derive expression for temperature of body after a time t.
27. Describe thermal equilibrium. A body having temperature T_1 is placed in an atmosphere having temperature T_0 such that $T_1 < T_0$. Derive expression for temperature of body after a time t.

28. A body having temperature of 177°C is placed in an atmosphere having temperature of 37°C. Considering $k = 0.004\ s^{-1}$, calculate the temperature of body after 6 min.
29. A body having temperature of –17°C is placed in an atmosphere having temperature of 37°C. Considering $k = 0.004\ s^{-1}$, calculate the temperature of body after 4 min.
30. Show that the amount of gravitational energy released in the formation of a celestial body of mass M and radius R is $3GM^2/5R$.
31. Derive an expression for the amount of energy released in the formation of sphere of radius R and mass M.
32. For the Sun, we have mass $M_\odot = 1.989 \times 10^{30}$ kg, radius $R_\odot = 6.960 \times 10^8$ m, luminosity $L_\odot = 3.85 \times 10^{26}$ J/s. Show that the amount of gravitational energy released during the formation of the Sun has not been sufficient for continuous shining of the Sun.
33. Discuss about the equations of stellar structure.
34. Describe the integral theorems on pressure, density and temperature.
35. Discuss about the homologous transformations.

36. Derive an expression for the temperature inside a spherical body of radius R and mass M.
37. For a polytropic star, derive Lane-Emden equation and obtain its solution for $n = 0$.
38. For a polytropic star, derive Lane-Emden equation and obtain its solution for $n = 1$.
39. For a polytropic star, derive Lane-Emden equation and obtain its solution for $n = 5$.
40. Discuss about the gravitational energy of a spherical celestial body.
41. Write notes on the following:
    (i) Parallax
    (ii) Two-body system
    (iii) Kepler's laws of planetary motion
    (iv) Newton's law of gravitation
    (v) Virial theorem
    (vi) Cosmic abundances
    (vii) Spectral classification
    (viii) Black-body radiation
    (ix) Wien's displacement law
    (x) Stefan-Boltzmann law
    (xi) Doppler redshift
    (xii) Relation between B-V magnitude and surface temperature of star
    (xiii) Colour-magnitude diagram
    (xiv) Determination of radius of a star
    (xv) Hertzsprung-Russell diagram
    (xvi) Pauli exclusion principle
    (xvii) Degenerate matter
    (xviii) Electron degeneracy
    (xix) Proton degeneracy
    (xx) Neutron degeneracy
    (xxi) Hydrostatic equilibrium
    (xxii) Thermal equilibrium
    (xxiii) Gravitational energy
    (xxiv) Henry-Draper star catalogue
    (xxv) M-K system of stellar classification
    (xxvi) Harvard scheme of stellar classification
    (xxvii) Meghnad Saha and his contribution
    (xxviii) Wilson-Bappu effect and its application

Multiple-Choice Questions

1. The unit of areal velocity is
   A. cm/sec B. cm^2/sec C. cm/sec^2 D. cm^3/sec **Ans. B**
2. Which does not have the unit of length?
   A. Light year B. Astronomical unit C. Parsec D. Areal velocity
   Ans. D

3. Which is not Kepler's law of planetary motion?
   A. A planet revolves around the Sun in an elliptical orbit.
   B. The force between the Sun and planet is gravitational force.
   C. The line joining a planet and the Sun sweeps equal area in equal time.
   D. The square of time-period of revolution of a planet around the Sun is proportional to the cube of the semi-major axis of the elliptical orbit. **Ans. B**
4. In a binary system of two stars *A* and *B*, having masses $2M_\odot$ and $5M_\odot$, have elliptical orbits of sizes E_A and E_B, respectively. We have
   A. $E_A > E_B$ B. $E_A = E_B$ C. $E_A < E_B$ D. None of them
   Ans. A
5. In a binary of two stars *A* and *B*, having masses $5M_\odot$ and $1.4M_\odot$, have elliptical orbits of sizes E_A and E_B, respectively. We have
   A. $E_A > E_B$ B. $E_A = E_B$ C. $E_A < E_B$ D. None of them
   Ans. C
6. The masses of the Sun and Earth are *M* and *m*, respectively. The reduced mass of the combination is nearly equal to
   A. M B. m C. $M + m$ D. Mm **Ans. B**
7. In a binary system, each of the two stars has mass *m*. The reduced mass of the binary system is
   A. $2m$ B. m C. $m/2$ D. $m/4$ **Ans. C**
8. The sequence of classification of stars is
   A. OAFBGKM B. ABFGKMO C. OMABFGK D. OBAFGKM
   Ans. D
9. Wien's displacement law is
   A. $\lambda m/T$ = const B. λmT = const C. $\lambda m + T$ = const D. $\lambda m - T$ = const
   Ans. B
10. At temperature *T*, total energy emitted by a black-body per unit time, per unit surface area is proportional to
    A. T B. T^2 C. T^3 D. T^4 **Ans. D**
11. For a redshift of a spectral line, the
    A. wavelength of radiation increases.
    B. wavelength of radiation decreases.
    C. source moves towards the observer.
    D. relative velocity between the source and observer is zero. **Ans. A**
12. For a blue-shift of a spectral line, the
    A. wavelength of radiation increases.
    B. wavelength of radiation decreases.
    C. source moves away from the observer.
    D. relative velocity between the source and observer is zero. **Ans. B**

13. A planet moving with velocity v in an elliptical path is at a distance r from the Sun. When this planet is at a distance $3r/2$, its velocity is
A. v B. $3v/2$ C. $2v/3$ D. $v/2$ **Ans. C**
14. A planet moving with velocity v in an elliptical path is at a distance r from the Sun. When this planet is moving with velocity $3v/2$, its distance is
A. r B. $3r/2$ C. $2r/3$ D. $2r$ **Ans. C**
15. The speed of a planet revolving around the Sun in an elliptical orbit increases when it comes closer. This fact is according to which of the Kepler's law of planetary motion.
A. first only B. second only C. third only D. all three **Ans. B**
16. The Earth at 1 AU distance from the Sun takes 1 year to revolve around the Sun. Some planet at a distance of 4 AU from the Sun would require how many years to complete one revolution around the Sun.
A. 1 B. 4 C. 8 D. 16 **Ans. C**
17. The Earth at 1 AU distance from the Sun takes 1 year to revolve around the Sun. Some planet requires 8 years time to complete one revolution around the Sun. The distance of planet in AU from Sun is
A. 1 B. 4 C. 8 D. 16 **Ans. B**
18. A graph between the luminosity and effective temperature of stars is known as
A. phase diagram B. H-R diagram C. Lande diagram D. Planck diagram
Ans. B
19. For a group of stars, the luminosity is found to vary as the fourth power of the mass of star. When the ratio of masses of two stars is $\sqrt{12}$, the ratio of their luminosities is
A. 2 B. 4 C. 8 D. 16 **Ans. B**
20. For a spherical body of radius R, mass M and average density ρ, the pressure at the centre of body is
A. $\rho GM/R$ B. $\rho GM/2R$ C. $3\rho GM/5R$ D. $5\rho GM/3R$ **Ans. B**
21. For a spherical body of radius R and mass M, the gravitational energy is
A. GM^2/R B. $3GM^2/5R$ C. $GM^2/2R$ D. $GM^2/4R$ **Ans. B**

7 Astronomy from the Surface of the Earth

Out of the full electromagnetic spectrum, coming from the outer space towards the Earth, the Earth's atmosphere is transparent for the two bands of electromagnetic spectrum, known as the optical window and the radio window. Thus, from the surface of the Earth, the radiations lying in these two bands (windows) can be studied. An instrument used for gathering the electromagnetic radiations is known as the telescope. Hence, on the surface of the Earth, we have optical and radio telescopes. An optical telescope using the lens cannot be of very large size because of its weight and aberrations. Therefore, optical telescopes of very large size use mirror in place of a lens.

Since the scattering of radiations lying in the optical window is about 10^{20} times larger than of those lying in radio window, the role of radio window becomes very important, in particular, in the context of very distant cosmic objects. However, the resolving power of a single dish radio telescope becomes very poor, which can be improved a lot using the concept of the Very Large Baseline Interferometry (VLBI). In the present chapter, we shall discuss about observations in these two windows.

1. Astronomical Units of Distance

Cosmic objects are, in general, situated at very large distances. Hence, in the Astronomy & Astrophysics (A&A), we need to employ the units for measuring distances which are rather large in magnitude. The units of distances used in the A&A are:

(i) Light year, (ii) Astronomical unit, and (iii) Parsec.

1.1 Light Year (ly)

In the electromagnetic spectrum, all kinds of radiations in vacuum, travel with the same speed c, having the value 3×10^8 m/s. Since the speed of light is finite, it requires finite time to travel from one place to another. For example, it takes 8 minutes and 20 seconds to reach the Earth from the Sun; takes 4.2 years to reach the Earth from the second nearest star, Proxima Centauri; takes 2.5 million years to reach the Earth from the nearest large galaxy, Andromeda (also called M31), situated in the Andromeda constellation. Thus, the distances of astronomical objects may also be expressed in term of the time required for travelling the light from the object to the surface of Earth.

One light year, denoted by ly, is defined as the distance travelled by the light in vacuum in one year. Because of leap years, the average number of days in one year may be taken as 365.25. Thus, we have

$$1 \text{ year} = 365.25 \text{ days}$$
$$= 365.25 \times 24 \times 60 \times 60$$
$$= 3.156 \times 107 \text{ s}$$

The speed of light is 3×10^8 m/s

$$\text{Thus, } 1 \text{ light year} = (3.156 \times 10^7)(3 \times 10^8)$$
$$= 9.47 \times 10^{15} \text{ m}$$
$$1 \text{ ly} = 9.47 \times 10^{15} \text{ m} = 9.47 \times 10^{12} \text{ km}$$

Hence, the distance travelled by the light (photon) in one year in vacuum is 9.47×10^{12} km, which is known as one light year.

Similarly, we can define light day, light hour, light minute, light second. Equivalent of them are given in Table 1.

Table 1. Equivalents

1 light sec	3×10^5 km
1 light min	1.8×10^7 km
1 light hour	1.08×10^9 km
1 light day	2.59×10^{10} km

1.2 Astronomical Unit (AU)

Besides the spin motion about its own axis, the Earth revolves around the Sun in the elliptical orbit in the time-period of one year. The distance between the Earth and the Sun changes with time; its minimum value is 1.47×10^{11} m, around 3 January each year and the maximum value is 1.53×10^{11} m, around 3 July each year[1]. The average distance between the Sun and the Earth is defined as one astronomical unit, denoted by AU. It is equal to 1.50×10^{11} m $= 1.50 \times 10^8$ km. Distances of planets from the Sun in kilometres and astronomical units are given in Table 2.

Table 2. Distances of planets from the Sun

Planet	Distance	
	km	AU
Mercury	5.81×10^7	0.39
Venus	1.08×10^8	0.72
Earth	1.50×10^8	1.00
Mars	2.29×10^8	1.52
Jupiter	7.80×10^8	5.20
Saturn	1.43×10^9	9.54
Uranus	2.88×10^9	19.18
Neptune	4.51×10^9	30.06

[1]One should not get confused with the fact that we have winter season in January and summer season in July. We are in the northern hemisphere. In the southern hemisphere, the seasons are opposite.

Example 1.

Calculate the time taken by light to reach the Earth from the Sun. Given, 1 AU = 1.50×10^{11} m.

Solution:

The average distance between the Earth and the Sun is 1.50×10^{11} m and velocity of light is 3×10^8 m/s. Thus, the time t taken by light for travelling from the Sun to the Earth is

$$t = \frac{1.50 \times 10^{11}}{3 \times 10^8} = 500 \text{ sec}$$

Hence, the light takes 8 min and 20 sec to reach the Earth from the Sun.

1.3 Parsec (pc)

One parsec is equal to the length so that two points, separated by one astronomical unit, subtend an angle of one arc-sec (1″), as shown in Fig. 1.

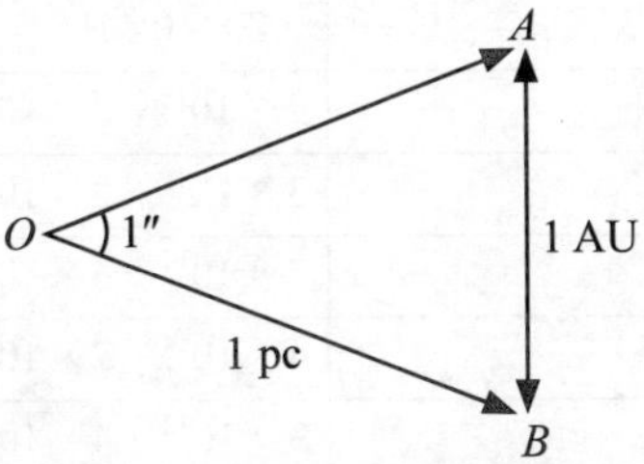

Fig. 1: Two objects A and B, separated by a distance of 1 AU, subtend an angle of 1 arc-sec at O. The distance between O and A, or O and B is defined as 1 parsec.

Now, we have

$$1'' = \frac{\pi}{180} \times \frac{1}{60} \times \frac{1}{60}$$

$$= 4.848 \times 10^{-6} \text{ rad}$$

Thus, we have

$$\text{angle} = \frac{\text{arc}}{\text{radius}}$$

and therefore

$$4.848 \times 10^{-6} = \frac{1 \text{ AU}}{1 \text{ pc}}$$

Therefore, we have

$$1 \text{ pc} = \frac{1.5 \times 10^{11} \text{ m}}{4.848 \times 10^{-6}} = 3.09 \times 10^{16} \text{ m}$$

$$= 3.09 \times 10^{13} \text{ km}$$

Therefore, 1 parsec is equal to 3.09×10^{13} km.

1.4 Relation between pc and ly

We have

$$\frac{1 \text{ pc}}{1 \text{ ly}} = \frac{3.094 \times 10^{18} \text{ cm}}{9.467 \times 10^{17} \text{ cm}} = 3.26$$

Therefore, 1 pc = 3.26 ly.

2. Electromagnetic Spectrum

Radiations coming from the space towards the Earth have full electromagnetic spectrum, consists of γ-rays, x-rays, ultraviolet (UV) radiation, visible light, infrared (IR) radiation, microwaves and radio waves. Though there are no well-defined demarcations for their ranges, but approximate ranges of these regions of radiations are given in Table 3. The frequency of radiations increases from the radio waves to the γ-rays, whereas the wavelength of radiations increases from the γ-rays to the radio waves. The variations of wavelength and frequency of radiations are such that the product of their values in vacuum is constant, equal to the speed of light $c = 3 \times 10^8$ m/s. The frequency ν and wavelength λ are related as $c = \nu\lambda$.

Table 3. Electromagnetic spectrum

Radiation	Frequency
γ-rays	$> 2 \times 10^{19}$ Hz
x-rays	$2 \times 10^{16} - 2 \times 10^{19}$ Hz
EUV	$3 \times 10^{15} - 2 \times 10^{16}$ Hz
UV radiations	$7 \times 10^{14} - 3 \times 10^{15}$ Hz
Visible light	$4 \times 10^{14} - 7 \times 10^{14}$ Hz
IR radiations	$3 \times 10^{12} - 4 \times 10^{14}$ Hz
FIR	$3 \times 10^{11} - 3 \times 10^{12}$ Hz
Microwave	30 – 300 GHz
Radio	0.5 – 30000 MHz

2.1 Optical Window and Radio Window

All of us know that the γ-rays, x-rays, EUV and UV radiations are very energetic radiations, and if any of them could reach the Earth's surface, human life obviously could not survive. It is the Earth's atmosphere which plays a magnificent role by shielding us from merciless, intense radiations coming from the outer space towards the Earth.

It is a matter of great pleasure to notice that the γ-rays and x-rays from the outer space are absorbed in the upper part of the Earth's atmosphere, and if there had been molecules, they dissociate into its constituent atoms; the atoms are stripped of their electrons. As a result of that, the ions (the negatively charged electrons and the positively charged nuclei) are produced there. This region of ions at an altitude of about 500 km is known as the ionosphere. It contains plasma comprising of electrons, protons, and positively charged ions.

Below the ionosphere, there is a layer of ozone (O_3) gas, which absorbs most of the UV radiations, resulting in the depletion of ozone layer. Thus, it is the Earth's upper atmosphere that plays a magnificent role in our life, by shielding us from the γ-rays, x-rays and UV radiations coming from the outer space.[2]

---

[2]Depletion of the ozone layer has been an important issue for discussion at the National as well as the International forums, as it affects the human life on the Earth's surface. The chloroflurocarbon (CFC) produced in the air conditioners, etc., is responsible for the destruction of the ozone layer.

The atmosphere close to the Earth's surface (called troposphere) contains mainly the water (H_2O) and carbon dioxide (CO_2), and thus, it absorbs most of the IR radiations coming from the outer space.

For a plasma, the characteristic frequency (also called the plasma frequency), in Hz, is given by $\nu_p = 9.0\sqrt{Ne}$, where N_e is the density of electrons in m^{-3} in plasma. In the ionosphere, the electron density ranges from 10^{10} to 10^{12} m^{-3}. As a characteristic of plasma, the radiations having frequency less than the plasma frequency are reflected back by the plasma. Hence, the radiations from the outer space whose frequency is less than the characteristic frequency of plasma in the ionosphere are returned back towards the outer space.

Most of the parts of the electromagnetic spectrum coming from the outer space are obviously incapable to transmit through the Earth's atmosphere and to reach the surface of the Earth. There are, however, two bands in which the Earth's atmosphere is transparent. These two bands are known as the optical window and the radio window.

The optical window has a range from 3000 Å to 10,000 Å, whereas the radio window has a range from a few mm to several decameters. The radiations which our eyes can see lie in the optical window. It is the optical window which did not allow human beings to remain ignorant about the beautiful phenomena, occurring in nature, such as the rising and setting of the Sun, the phases of the Moon, the solar and lunar eclipses, the comets, the stars in a night, for example. Thus, on the surface of the Earth, we have only two types of telescopes: the optical telescopes, and the radio telescopes.

2.2 Advantage of Radio Window

The Earth's atmosphere is transparent in the optical window and in the radio window. The radio window has a range from mm to decameter in wavelength. In the context of the distant cosmic objects, the radiations lying in the radio window have remarkable advantages over those in the optical window. The ratio of the average wavelengths in the radio window (λ_R) and in the optical window (λ_O) is[3]

$$\frac{\lambda_R}{\lambda_O} \approx 10^5$$

We know that the scattering (S) of a radiation is inversely proportional to the fourth power of its wavelength (λ).

$$S \propto \frac{1}{\lambda^4}$$

or

$$S = \frac{1}{\lambda^4}$$

where k is constant. Therefore, we have

[3]The mean wavelength in the optical window is about 5000 Å and in the radio window is about 5 cm. Thus, we have

$$\frac{\lambda_R}{\lambda_O} \approx \frac{5 \text{ cm}}{5000 \text{Å}} = 10^5$$

$$S_R = \frac{k}{\lambda_R^4}$$

and
$$S_O = \frac{k}{\lambda_O^4}$$

Thus, the ratio of the scatterings of radiations in the two windows is

$$\frac{S_R}{S_O} = \frac{\lambda_O^4}{\lambda_R^4} \approx 10^{-20}$$

It shows that the probability of reaching the optical radiations to the Earth's surface is about 10^{20} times less than that of the radio radiations. Hence, in the context of the distant cosmic objects, the radio window is more important as compared to the optical window. Hence, it is the radio radiations which can bring valuable information about the cosmic objects, situated at very large distances, e.g., at a distance of several hundred light years. That is, the radiations, which have reached the Earth's surface today, would have started several hundred years ago from that region.

The radio window has importance in the context of communication of information from one place to other. In the past, the communication has been through reflection with ionosphere. In that case, the frequency used in the communication had to be smaller than the plasma frequency.

Nowadays, we communicate with the help of a geostationary satellite. The geostationary satellite is situated at a distance larger than that of the ionosphere. Obviously, the frequency used in the communication should be larger than that of the plasma in the ionosphere, so that the waves could transmit through the ionosphere.

3. Luminosity of Star

The amount of energy emitted per unit time by a cosmic object (star) is defined as its luminosity, and is denoted by L. It is an intrinsic property of the object and it depends on the temperature and composition of the object. For example, the Sun, on an average emits the energy at the rate of 3.826×10^{26} J/s. Thus, the luminosity of the Sun, denoted by $L_\odot$, is 3.826×10^{26} J/s.

Further, the luminosity of an object depends on the frequency band in which the energy is considered/measured. When no frequency band is given, it is the luminosity of full electromagnetic spectrum emitted by the object.

3.1 Radiant Flux of a Star

The amount of energy emitted by an object (star) per unit time is spread over a spherical surface in space whose centre is at the centre of object[4]. If we assume that the atmosphere around the star is isotropic and homogeneous, and no energy is absorbed during its journey from the object to an observer, the amount of energy from the object received by observer per unit time per unit surface area is defined as the radiant flux from the object. The radiant flux F measured at a distance r from the object is

[4]As the size of star is very small in comparison to the distance between the star and an observer, the star may be treated as a point-object.

$$F = \frac{L}{4\pi r^2}$$

where L is the luminosity of object. Thus, the radiant flux depends on both the intrinsic luminosity of the object, and the distance from the object. It shows that the radiant flux received from an object varies as the luminosity and as the inverse square of its distance from the object. Thus, if the distance of an object is large from the observer, the radiant flux received from the object is small. For example, if we increase the distance between the Sun and the observer, the Sun (keeping its luminosity unchanged) may appear even more fainter than a star. The radiant flux from an object is a measure of the brightness of the object.

Example 2.

The luminosity of the Sun is $L = 3.826 \times 10^{26}$ J/s. Calculate the radiant flux received on the surface of the Earth from the Sun, assuming that the absorption of energy in the atmosphere between the Sun and the Earth is negligible. If the Sun had been at a distance of 10 pc from the Earth, what had been radiant flux received on the surface of the Earth. Given, average distance between the Sun and the Earth is 1 AU = 1.5×10^{11} m, and 1 pc = 3.09×10^{16} m.

Solution:

Radiant flux received from the Sun on the surface of the Earth is

$$F = \frac{L}{4\pi(1\ \text{AU})^2}$$

$$= \frac{3.826 \times 10^{26}}{4 \times 3.14 \times (1.5 \times 10^{11})^2}$$

$$= 1.35 \times 10^3\ \text{J/s m}^2$$

If the Sun had been at a distance of 10 pc = 3.09×10^{17} m. The radiant flux received at a distance of 10 pc is

$$F = \frac{3.826 \times 10^{26}}{4 \times 3.14 \times (3.09 \times 10^{17})^2}$$

$$= 3.19 \times 10^{-10}\ \text{J/s m}^2$$

Notice that with the increase of the distance, the flux is decreased.

3.2 Solar Constant

Value of the radiant flux from the Sun received on the surface of the Earth is known as the solar constant. Here, it is assumed that the atmosphere between the Sun and the Earth does not absorb any part of the energy generated by the Sun. The amount of energy received from the Sun on the surface of the Earth per unit surface area per unit time (solar constant) is 1.35×10^3 J/s m^2.

Example 3.

Calculate the solar constant. Given, 1 AU = 1.5×10^{11} m, and solar luminosity 3.826×10^{26} J/s.

Solution:

The solar constant is

$$F = \frac{1}{4\pi(1\ \text{AU})^2}$$

$$= \frac{3.826 \times 10^{26}}{4 \times 3.14 \times (1.5 \times 10^{11})^2}$$

$$= 1.35 \times 10^3 \text{ J/s m}^2$$

3.3 Comparison of two Stars

Let us consider two cosmic objects of luminosities L_1 and L_2, situated at the distances r_1 and r_2, respectively, from an observer. The radiant fluxes received by the observer from the objects are, respectively,

$$F_1 = \frac{L_1}{4\pi r_1^2}$$

and

$$F_2 = \frac{L_2}{4\pi r_2^2}$$

Therefore, the ratio of radiant fluxes is

$$\frac{F_2}{F_1} = \frac{L_2 r_1^2}{L_1 r_2^2}$$

(i) When the luminosities of two objects are equal, $L_1 = L_2$, the ratio of the radiant fluxes is

$$\frac{F_2}{F_1} = \frac{r_1^2}{r_2^2}$$

(ii) When the distances of two objects are equal, $r_1 = r_2$, the ratio of the radiant fluxes is

$$\frac{F_2}{F_1} = \frac{L_2}{L_1}$$

4. Magnitude Scale

When we observe stars (objects) in a clear night, the brightness (radiant flux) received from different stars is not equal. Some of them appear very bright and are visible without any efforts, whereas some are very dim and are barely visible to the naked eyes. (A large number of stars are not visible in the sky to the naked eyes, but they become visible with the help of a telescope.) Other visible stars have brightness in between these two cases.

The stars which are visible to the naked eyes, depending on their apparent brightness, have been classified in terms of a parameter, called the **apparent magnitude**. The most bright stars in the sky were assigned the apparent magnitude $m = +1$, whereas the dimmest ones (which are barely visible to the naked eyes) were assigned the apparent magnitude $m = +6$. Notice that with the increase of m, the brightness of the star decreases. Hence, all the stars, visible to the naked eyes, have the magnitude in the range from +1 to +6.

The correspondence between the magnitude and flux of stars was found to be logarithmic. That is, if m and F are the apparent magnitude and the radiant flux, respectively, of a star, then we have

$$\log_{10}F = -k\,m + C$$

where k and C are constants. The negative sign on the right side indicates that with the increase of the magnitude m, the brightness F decreases. Suppose, the radiant fluxes F_1 and F_2 are received from two stars of the apparent magnitudes m_1 and m_2, respectively. Then, we have

$$\log_{10}F_2 = -k\,m_2 + C$$

and

$$\log_{10}F_1 = -k\,m_1 + C$$

Therefore, we have

$$\log_{10}F_2 - \log_{10}F_1 = -k(m_2 - m_1)$$

or

$$\frac{F_2}{F_1} = 10^{k(m_1-m_2)}$$

According to the present definition (decided by the IAU), a difference of 5 magnitudes corresponds exactly to a ratio of 100 in the brightness of stars. That is, for $m_1 - m_2 = 5$, we have $(F_2/F_1) = 100$. On substituting these values, we get

$$100 = 10^{k5}$$

On taking logarithm base 10, we get

$$2 = k\,5$$

or

$$k = 0.4$$

Thus, the relation between the radiant fluxes received from the stars and their apparent magnitudes is

$$\frac{F_2}{F_1} = 10^{0.4(m_1-m_2)}.$$

The magnitude scale of stars has been extended on both sides. That is, we have stars having negative value of the magnitude, and we have stars having the magnitude more than 6. For example, the apparent magnitude of the star Sirius is −1.4, of the full Moon is −12.6, and of the Sun is −26.73. The Sun is obviously the brightest object, visible to us. On the other side, with the help of the telescopes, we can see stars having magnitude up to + 30. Hubble telescope has located stars with the magnitudes up to +30 at visible wavelengths.

With the help of sensitive instruments (e.g., photometers) it is possible to measure apparent magnitude of an object with an accuracy of ±0.01 magnitude, and difference in magnitudes with an accuracy of ±0.002 magnitude.

Example 4.

Calculate the ratio of fluxes received from two stars whose magnitudes differ by 2.5.

Solution:

The ratio of the fluxes F_1 and F_2 received from two stars of magnitudes m_1 and m_2, respectively, is

$$\frac{F_2}{F_1} = 10^{0.4(m_1-m_2)}$$

Consider that a dim star is of magnitude m_1 and a bright star of m_2. Then, $m_1 - m_2 = 2.5$. Hence, the ratio of their fluxes is

$$\frac{F_2}{F_1} = 10^{0.4\times2.5} = 10$$

Thus, the brighter star is 10 times brighter as compared to the dimmer one.

Example 5.

The ratio of fluxes from two objects is 145. The magnitude of brighter object is 1.2. Calculate the magnitude of other object.

Solution:

For two objects with magnitudes m_1 and m_2, and fluxes F_1 and F_2, respectively, we have

$$\frac{F_2}{F_1} = 10^{0.4(m_1-m_2)}$$

$$145 = 10^{0.4(m_1-m_2)}$$

Thus, $m_1 - m_2 = 5.40$. We have $m_2 = 1.2$. Therefore, $m_1 = 6.6$.

Example 6.

Magnitude of the most bright object in appearance (Sun) is −26.81, and with the help of a good telescope, we can see the stars as faint as having magnitude of 28.19. Calculate the ratio of brightness of two limiting stars.

Solution:

Let $m_1 = 28.19$ and $m_2 = -26.81$. Hence, the ratio of fluxes received form the star (F_1) and from the Sun (F_2) is expressed as

$$\frac{F_2}{F_1} = 10^{0.4(m_1-m_2)}$$

$$= 10^{0.4(28.19+26.81)}$$

$$= 10^{22}$$

Thus, the Sun is 10^{22} times brighter than the star.

4.1 Absolute Magnitude

The radiant flux (and thus the magnitude) of a star (object) depends on its distance form observer. Hence, the comparison of apparent magnitudes of stars cannot provide meaningful information about their relative fluxes. We know that when two stars are at equal distances from an observer, the ratio of their radiant fluxes provides information about their relative fluxes. Thus, when we bring theoretically all the stars at a standard distance from the observer, then the radiant fluxes received from stars provide a real comparison of stars.

This standard distance of an object from an observer is taken as 10 pc, following the decision of the International Astronomical Union (IAU). When an object is brought theoretically at a distance of 10 pc, its apparent magnitude is defined as the **absolute magnitude**. Thus, the apparent magnitude of an object situated at a distance of 10 pc from an observer is known as the absolute magnitude of the star.

Suppose, a star at distance d_1 is moved to a distance d_2, the ratio of the fluxes received F_1 and F_2, respectively, is

$$\frac{F_1}{F_2} = \frac{L/\pi d_1^2}{L/\pi d_2^2} = \left(\frac{d_2}{d_1}\right)^2$$

Since the object is the same, and thus, its luminosity does not change. Let the apparent magnitude of object situated at a distance *d* pc from the observer be *m*. When the object is moved theoretically to a distance of 10 pc, its apparent magnitude, known as absolute magnitude, is denoted by *M*. Hence, the ratio of the radiant fluxes is expressed as

$$\frac{F_{10}}{F_d} = \left(\frac{d}{10}\right)^2$$

and

$$\frac{F_{10}}{F_d} = 10^{0.4(m-M)}$$

From these two relations, we have

$$\left(\frac{d}{10}\right)^2 = 10^{0.4(m-M)}$$

or

$$m - M = 5\log_{10} d - 5$$

Thus, for a star situated at a distance $d < 10$ pc, the absolute magnitude is larger than the apparent magnitude. But, for a star situated at a distance $d > 10$ pc, the absolute magnitude is smaller than the apparent magnitude.

Example 7.

Suppose, two stars of apparent magnitudes m_1 and m_2 are at the distances d_1 and d_2, respectively, from an observer. Calculate the difference between the absolute magnitudes of the stars.

Solution:

Suppose, M_1 and M_2 are absolute magnitudes of two stars having apparent magnitudes m_1 and m_2, respectively. Then we have

$$m_1 - M_1 = 5\log_{10} d_1 - 5$$

and

$$m_2 - M_2 = 5\log_{10} d_2 - 5$$

On taking difference, we get

$$(m_1 - m_2) - (M_1 - M_2) = 5\log_{10} d_1 - 5\log_{10} d_2$$

or

$$(M_1 - M_2) = (m_1 - m_2) - 5\log_{10}\left(\frac{d_1}{d_2}\right)$$

Example 8.

Apparent magnitude of the Sun is $m_\odot = -26.81$. Calculate the absolute magnitude of the Sun. Given, 1 AU = 1.50×10^8 km; 1 pc = 3.09×10^{16} m.

Solution:

The distance of the Sun from the Earth is $d = 1.50 \times 10^8$ km $= 4.848 \times 10^{-6}$ pc, and $m_\odot = -26.81$. Absolute magnitude $M_\odot$ of the Sun is

$$M_\odot = m_\odot - 5\log_{10} d + 5$$

On substituting the values, we get

$$M_\odot = -26.81 - 5\log_{10}(4.848 \times 10^{-6}) + 5$$
$$= 4.76$$

Thus, at a distance of 10 pc, the Sun looks like a faint star with the magnitude +4.76.

Example 9.

The absolute magnitude of a star is +1. Calculate its distance so that the apparent magnitude of the star is −1.

Solution:

We have $m_1 = +1$, $d_1 = 10$ pc and $m_2 = -1$. Thus, we have

$$\left(\frac{d_1}{d_2}\right)^2 = 10^{0.4(m_1-m_2)}$$

or

$$\left(\frac{10}{d_2}\right)^2 = 10^{0.4(1+1)}$$

Therefore, we have

$$d_2 = 10 \times 10^{-0.4} = 3.981 \text{ pc}$$

4.2 Relation between Luminosities and Absolute Magnitudes

The fluxes F_1 and F_2 of two stars of apparent magnitudes m_1 and m_2, respectively, are expressed as

$$\frac{F_2}{F_1} = 10^{0.4(m_1-m_2)}$$

or

$$\frac{L_2/4\pi r_2^2}{L_1/4\pi r_1^2} = 10^{0.4(m_1-m_2)}$$

When each of the r_1 and r_2 are 10 pc, the magnitudes m_1 and m_2 are absolute magnitudes M_1 and M_2. Thus, we have

$$\frac{L_2}{L_1} = 10^{0.4(M_1-M_2)}$$

or

$$M_2 = M_1 - 2.5\log_{10}(L_2/L_1)$$

When $L_2 > L_1$, we have $M_2 < M_1$ and when $L_2 < L_1$, we have $M_2 > M_1$.

Note:

When one of the stars is taken the Sun. Then, $L_1 = L_\odot$, and $M_1 = M_\odot = 4.76$, and thus, luminosity L and absolute magnitude M of star are related as

$$M = 4.76 - 2.5\log_{10}(L/L_\odot)$$

Example 10.

For the Sun, the luminosity is $L_\odot = 3.826 \times 10^{26}$ J/s and absolute magnitude is $M_\odot = 4.76$. Calculate absolute magnitude of a star having luminosity 5.26×10^{26} J/s.

Solution:

The absolute magnitude of star is

$$M_2 = M_1 - 2.5 \log_{10}(L_2/L_1)$$
$$M_2 = 4.76 - 2.5 \log_{10}(5.26 \times 10^{26}/3.826 \times 10^{26})$$
$$= 4.41$$

5. Features of Telescope

There are three main features of a telescope which improve the quality of our vision.

(i) Light-gathering
(ii) Resolution
(iii) Magnification

5.1 Light-gathering

In a telescope, the light falling on the primary (objective) lens is collected. The light-gathering is probably the most important feature of a telescope. This makes the faint images brighter. The amount of light gathered by a lens is proportional to the diameter (or aperture) of the lens. In a telescope, the light is gathered by the primary (objective) lens. With the increase of the size of aperture a, dim objects may be detected, depending on the amount of light gathered. Therefore, the telescopes having larger effective aperture are constructed. The image brightness B is proportional the square of the aperture a. That is,

$$B \propto a^2$$

Example 11.

Two optical telescopes have apertures of 2 m and 8 cm. Compare the brightnesses of images formed by the telescopes under the normal conditions.

Solution:

Under the normal viewing conditions, the ratio of the image brightnesses is expressed as

$$\frac{B_{large}}{B_{small}} = \frac{a^2_{large}}{a^2_{small}}$$
$$= \frac{2^2}{0.08^2} = 625$$

The image produced by the large telescope is thus 625 times brighter than that by the small one.

5.2 Resolution

Suppose, two point sources (objects) are very close together. Then, two diffraction patterns formed by the objects may overlap with each other and hence may be difficult to distinguish them as separate patter. An optical instrument (microscope, telescope, etc.) is said to have resolved the two point sources when two diffraction patterns are sufficiently separated from each other. The ability of an optical instrument just to resolve images of two nearby point sources is known as the **resolving power**.

Rayleigh Criterion

According to Rayleigh, two close spectral lines of equal intensities can be just resolved by an optical instrument when the position of maximum of one spectral line coincides with minimum of the other and vice versa. When they come further closer, the lines can no longer be resolved. The individual intensity distribution curves (line-profiles) of two spectral lines having equal intensities and wavelengths λ and $\lambda + d\lambda$ are shown in Fig. 2 for three different situations.

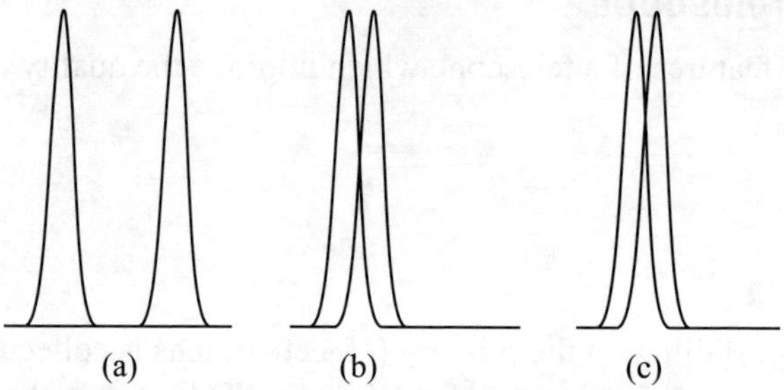

Fig. 2: Individual intensities of two spectral lines having different wavelengths for three different situations.

In the situation (a), separation between the two lines is much larger than their individual line-widths. In the situation (b), the maximum of one line coincides with the minimum of other and vice versa. That is, the separation between the maxima of two lines is equal to half-width of any one of the lines. In the situation (c), the two lines have come further closer. That is, the separation between the maxima of two lines is less than the half-width of any one of the lines.

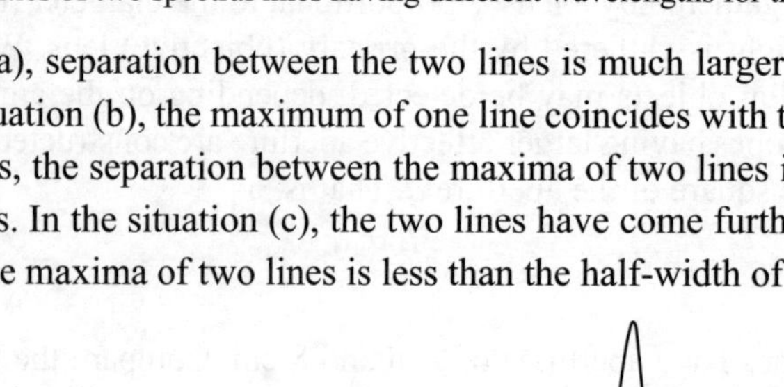

Fig. 3: Individual intensities of two spectral lines having different wavelengths for three different situations are shown by dotted lines. The sum of intensities of two lines are shown by solid line.

In Fig. 3, we have plotted individual intensities of two spectral lines along with the resultant intensity (sum of the intensities). In the situation (a), the resultant is the same as the intensities of individual lines. In the situation (b), the resultant intensity curve shows a dip in the middle of the maxima of the lines. Accordingly to Rayleigh, these spectral lines are distinguished from each other and said to be just resolved. In the situation (c), there is no more dip in the resultant intensity and the two lines cannot be resolved.

Resolving Power of a Telescope

The resolving power of a telescope is measured as the smallest angle that can be resolved. For a telescope having diameter D, the resolving power θ (in radians) is expressed as

$$\theta = 1.22\frac{\lambda}{D}$$

where λ is wavelength of radiation used for detection.

As the wavelength λ of radiation cannot be changed, the only way to improve the resolving power of telescope is to increase its diameter. Another way to increase the resolution is to connect the (radio) telescopes together to make an interferometer. The principle of the Very Large Baseline Interferometer (VLBI) is discussed later on.

Example 12.

Calculate resolving power of a telescope having diameter of 2.34 m when the radiation of wavelength 5500 Å is detected.

Solution:

We have $D = 2.34$ m and $\lambda = 5500 \times 10^{-10}$ m. The resolving power is

$$\theta = 1.22\frac{\lambda}{D}$$

$$= 1.22 \times \frac{5500 \times 10^{-10}}{2.34}$$

$$= 2.87 \times 10^{-7}$$

Example 13.

A 100 m radio-dish is used for the detection of 18 cm radiation of OH molecule. Calculate the resolving power of the radio telescope.

Solution:

We have $D = 100$ m and $\lambda = 18$ cm $= 0.18$ m. The resolving power is

$$\theta = 1.22\frac{\lambda}{D}$$

$$= 1.22 \times \frac{0.18}{100} = 2.2 \times 10^{-3}$$

Example 14.

What size of telescope aperture is required to resolve the bands which run around the Jupiter. Mean radii of the orbits of the Earth and the Jupiter revolving around the Sun are 150×10^6 km and 780×10^6 km, respectively; diameter of Jupiter is 140,000 km.

Solution:

At the closest approach, the separation between the Earth and the Jupiter is 630×10^6 km. For convenience, let us consider that there are 14 strips, each of thickness 10,000 (=140,000/14) km. Hence, the telescope must be capable to resolve 10,000 km from a distance of 630×10^6 km. As the angle θ is small and given in radian, we have

$$\theta = \frac{10{,}000}{630 \times 10^6} = 1.6 \times 10^{-5}$$

The average wavelength of visible spectrum is 5000 Å. Thus, the aperture of the telescope is

$$D = \frac{1.22\lambda}{\theta}$$

$$= \frac{1.22(5000\times10^{-10})}{1.6\times10^{-5}} = 0.038 \text{ m}$$

Hence, a telescope with an aperture of 4 cm is sufficient to make out the *i* surface features on Jupiter.

Example 15.

What size of telescope is required to resolve the size of a sun-like star situated at a distance of 4×10^{13} km. Diameter of star is 1.4×10^{6} km.

Solution:

The separation between the observer and object is 4×10^{13} km. The angular size of star in radian is

$$\theta = \frac{1.4\times10^{6}}{4\times10^{13}} = 3.5 \times 10^{-8} \text{ rad}$$

The average wavelength of visible spectrum is 5000 Å. Thus, the aperture of the telescope is

$$D = \frac{1.22\lambda}{\theta}$$

$$= \frac{1.22(5000\times10^{-10})}{3.5\times10^{-8}} = 17.4 \text{ m}$$

This value of aperture for an optical telescope is very large.

5.3 Magnification

An optical instrument (microscope, telescope, etc.) produces image of an object. The ratio of the size of image formed to the apparent size of object is known as the magnification of apparatus. In case of the optical instrument, the magnification is larger than 1. The magnifying power of a telescope is described as

$$\text{magnifying power} = \frac{\text{focal length of objective}}{\text{focal length of eyepiece}}$$

Example 16.

In an optical telescope, the focal lengths of objective and eyepiece are 24 cm and 12 mm, respectively. Calculate the magnifying power of the telescope.

Solution:

Magnifying power of the telescope is

$$\text{magnifying power} = \frac{\text{focal length of objective}}{\text{focal length of eyepiece}}$$

$$= \frac{24}{1.2} = 20$$

6. Atmospheric Distortion

Many people may believe that the astronomers want to build a telescope on high mountains or to put it in space, above the Earth's atmosphere, so that it is closer to the objects to be observed and studied. This opinion is not correct, as the distance of a telescope from the surface of the Earth is much smaller than the distance of object from the surface of the Earth.

The reason why the telescopes are built on high mountains or put in space is to get away from the distortion of starlight due to the atmosphere around the Earth. The atmospheric distortion is in the from of poor seeing, reddening, extinction and the addition of absorption lines to the stellar spectra.

There is no way out to control the atmosphere. However, one can minimize the effect of the atmosphere by selecting a proper site and by putting the telescopes on a high mountain, so that the radiation has to travel through the minimum amount of the atmosphere. This is why the optical telescopes are installed at the selected places on the Earth.

6.1 Seeing

The air is constantly in fast turbulent motion and light from a celestial object is bent randomly in many ways, because of the variation of the refractive index of the medium. Images dance about (twinkle) and are blurred. The atmospheric blurring distorts the view of the astronomical object. This atmospheric effect is known as the **seeing**. The light reaches different parts of the telescope's objective from slightly different directions and at different times. Good seeing is when the air is stable (little turbulence) and the twinkling is small. Poor seeing happens when the air is turbulent so the images dance about and the details smaller than 2 to 3 arc seconds in size cannot be seen. The more atmosphere there is above a telescope, the greater is the turbulent motion and the poorer is the seeing. This is one of the reasons why telescopes for research are located on very high mountains.

6.2 Reddening and Extinction

The air also absorbs and scatters electromagnetic radiations by an amount that varies with wavelength. The redder (longer wavelength) light is scattered less by the atmosphere (molecules and dust) than the bluer (shorter wavelength) light. This effect is known as the **reddening**[5]. All wavelengths of light are scattered or absorbed by some amounts. This effect is called the extinction. Some wavelength bands suffer more extinction than the others.

6.3 Plate Scale

The plate scale of a telescope provides a relation between the angular separation of an object with the linear separation of its image at the focal plane. If the focal length f is measured in mm, the plate scale in radians per mm is expressed by the angular separation θ and the linear separation of image at the focal plane s, or by simply the focal length f:

[5]This effect explains why the Sun appears orange or red when it is close to the horizon. The other colours of sunlight are scattered out of your line of sight so that only the orange and red colours make it through the atmosphere to your eyes. This effect also explains why the sky is blue. Since blue light is scattered more, you see more blue light scattered back to your eyes when you look in a direction away from the Sun.

$$p = \frac{\theta}{s} = \frac{1}{f}$$

because $$s = f\theta$$

The plate scale is usually expressed in arc-seconds per mm as:

$$p = \frac{206265}{f}$$

where the focal length f is in mm.

6.4 Diffraction Limit

The minimum angular separation of two sources that can be distinguished by a telescope depends on the wavelength of light being observed and the diameter of the telescope. This angle is called the diffraction limit.

Example 17.

Find the largest telescope aperture that is limited by diffraction rather than by scintillation in good viewing conditions.

Solution:

Take scintillation to limit the resolution of a telescope to 1″ (4.8×10^{-6} radians). The average wavelength of visible spectrum is 5000 Å. The telescope aperture that provides the same resolution can be found as

$$D = \frac{1.22\lambda}{\theta}$$

$$= \frac{1.22(5000 \times 10^{-10})}{4.8 \times 10^{-6}}$$

$$= 0.13 \text{ m}$$

The resolution of a telescope with an aperture greater than 13 cm is thus likely to be limited by the atmosphere rather than by the diffraction.

7. Optical Telescopes

The main purpose of an optical telescope is to collect the electromagnetic radiations lying in the optical window (i.e., visible light). The light collecting surface in an optical telescope may be either a lens or a mirror. Hence, the optical telescopes may be classified into two categories:

(i) Lens telescopes or refractors (ii) Mirror telescopes or reflectors

The amount of light, from a source, collected by a telescope is proportional to the area of aperture $\pi D^2/4$, where D is the diameter of the aperture of the telescope. Now, the linear size of image of object is proportional to the focal length f of the lens/mirror. Therefore, the image area must be proportional to f^2. Hence, the amount of light energy focused per second onto a unit area of the resolved image is proportional to D^2/f^2. The focal-ratio n of the telescope is defined as

$$\text{focal-ratio} = \frac{\text{focal length}}{\text{diameter}}$$

or $$n = \frac{f}{D}$$

For a telescope having 10 m diameter primary mirror with focal length of 17.5 m, the focal ratio is 1.75. This telescope is referred to as 10 m $f/1.75$ telescope.

Example 18.

Suppose, the eyepiece of a refracting telescope has focal length of 50 mm. What is the minimum focal length of objective, so that for the eye to resolve the bands which run around the Jupiter? Mean radii of the orbits of the Earth and the Jupiter revolving around the Sun are 150×10^6 km and 780×10^6 km, respectively. Diameter of the Jupiter is 140,000 km.

Solution:

At the closest approach, the separation between the Earth and the Jupiter is 630×10^6 km. For convenience, let us consider that there are 14 strips, each of thickness 10,000 (=140,000/14) km. Hence, the telescope must be capable to resolve 10,000 km from a distance of 630×10^6 km. As the angle θ is small and given in radian, we have

$$\theta = \frac{10{,}000}{630 \times 10^6}$$
$$= 1.6 \times 10^{-5} \text{ rad}$$

The average wavelength of the visible spectrum is 5000 Å. The resolving power of telescope is

$$R = \frac{1.22\lambda}{f}$$
$$= \frac{1.22(5000 \times 10^{-10})}{5 \times 10^{-3}}$$
$$= 1.2 \times 10^{-4} \text{ rad}$$

For the eye to resolve the bands, the image should be magnified by a factor of $(1.2 \times 10^{-4})/(1.6 \times 10^{-5}) = 7.5$. The focal length of objective is

$$f_o = M f_e$$
$$= 7.5 \times 0.05 = 0.38 \text{ m}$$

To resolve the bands, the focal length of the objective is 0.38 m. The total length of telescope needs to be half a metre with an aperture of 4 cm and such telescope is easy to build.

Example 19.

Suppose, the eyepiece of a refracting telescope has focal length of 50 mm. What is the minimum focal length of the objective, so that for the eye to resolve the size of a sun-like star situated at a distance of 4×10^{13} km. Diameter of the star is 1.4×10^6 km.

Solution:

The separation between the observer and the object is 4×10^{13} km. The angular size of the star in radian is

$$\theta = \frac{1.4 \times 10^6}{4 \times 10^{13}}$$
$$= 3.5 \times 10^{-8} \text{ rad}$$

The average wavelength of the visible spectrum is 5000 Å. The resolving power of the telescope is

$$R = \frac{1.22\lambda}{D}$$

$$= \frac{1.22(5000\times10^{-10})}{5\times10^{-3}}$$

$$= 1.2 \times 10^{-4}\ \text{rad}$$

The angular magnification required is $(1.2 \times 10^{-4})/(3.5 \times 10^{-8}) = 3400$. The focal length of objective is

$$f_e = Mf_o$$

$$= 3400 \times 0.05$$

$$= 170\ \text{m}$$

The focal length of the objective must therefore be about 170 m for the eye to see the star.

8. Lens Telescope or Refractor

A lens telescope, shown in Fig. 4, consists of two lenses, called the objective (*O*) and the eyepiece (*E*). The objective having large diameter is situated towards the object, and the eyepiece is situated towards the observer. The size of the objective cannot be increased to very large size, as its weight increases tremendously and therefore, sags down. Thus, the diameter of a lens telescope is limited. It may be some centimetres (say, 10 cm).

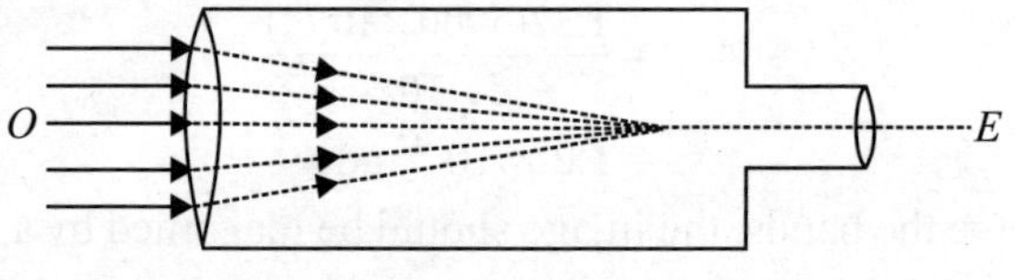

Fig. 4: Sketch of lens telescope.

9. Aberrations

One of the basic problems of a lens is the imperfect quality of image formed by it. The deviations in size, shape, position of a colour (wavelength) in the actual image produced by a lens are known as aberrations (defects) produced by the lens. The aberrations may be classified into two categories:

(i) chromatic aberrations (ii) monochromatic aberrations

The chromatic aberrations are the distortions of image due to dispersion of light in a lens used in a optical telescope, when the white light (various colours; various wavelengths) is used. Thus, the chromatic aberration may be defined as a defect of colour image formed by the lens while using the white light. When a monochromatic light is used then the chromatic aberrations are automatically absent. However, there are some defects which are still present. Such defects are known as the monochromatic aberrations. The monochromatic aberrations are due to the large

(i) aperture of the lens,
(ii) angle subtended by the rays with the principal axis, and
(iii) size of the object.

Examples of the monochromatic aberrations are (a) spherical aberration, (b) astigmatism, (c) coma, (d) curvature, and (e) distortion. In the present text, we have discussed about the chromatic aberration and the spherical aberration.

9.1 Chromatic Aberration

The rays of light of different wavelengths (colours) have different velocities in a refracting medium. The refractive index μ of medium for a colour is defined as $\mu = c/v$. Here, v and c are velocities of radiation in the medium and in the vacuum, respectively. Thus, the medium has different refractive index for each colour (wavelength). In the visible region of electromagnetic spectrum, the refractive index is the lowest for the red colour and the highest for the violet colour[6]. It shows that the velocity of the red colour is the largest and that of the violet colour is the smallest[7].

As the focal length of a lens depends on the refractive index of the medium through the relation

$$\frac{1}{f} = (\mu - 1)\left(\frac{1}{R_1} - \frac{1}{R_2}\right) \tag{7.1}$$

the focal length for the red colour is the largest and that of the violet colour is the smallest[8]. Thus, when the white light is used, a single lens will produce not only one image of an object but one image for each colour (wavelength) at different position and of different size. The variation of distance of image from the lens with refractive index measures the axial or longitudinal chromatic aberration.

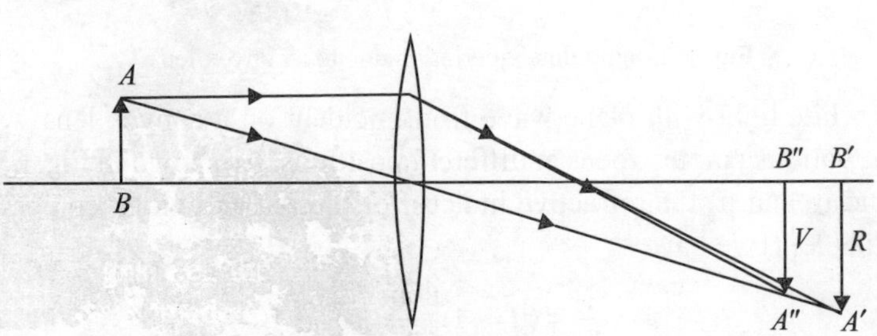

Fig. 5: Chromatic aberration.

In Fig. 5, *AB* is an object illuminated by white light. Its images corresponding to red and violet colours are *A'B'* and *A''B''*, respectively. The images corresponding to other colours are

---

[6]The refractive index varies with the wavelength (colour) as

$$\mu_v > \mu_i > \mu_b > \mu_g > \mu_y > \mu_o > \mu_r$$

[7]The velocity of radiation varies with the wavelength (colour) as

$$v_v < v_i < v_b < v_g < v_y < v_o < v_r$$

[8]The focal length varies with the wavelength (colour) as

$$f_v < f_i < f_b < f_g < f_y < f_o < f_r$$

formed in between these two limiting images. The size of an image depends on the wavelength (colour). Hence, the magnification depends on wavelength (colour). The difference $B''B'$ is the measure of longitudinal chromatic aberration and the difference $(A'B' - A''B'')$ measures the lateral chromatic aberration.

Example 20.

For a refracting medium, decide if the following statement is correct or not.

(i) In the electromagnetic spectrum, the velocity of red colour is larger than that of violet colour.

(ii) In the electromagnetic spectrum, the refractive index of red colour is larger than that of the violet colour.

(iii) Focal length of a lens for red colour is larger than that for the violet colour.

Answer:

(i) Yes, (ii) No, (iii) Yes

9.2 Chromatic Aberration of a Lens for an Object Placed at Infinity

The chromatic aberration of a lens for an object placed at infinity can be calculated in the following manner.

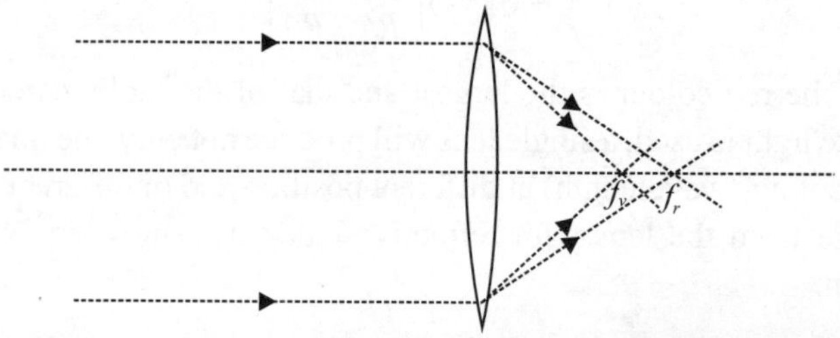

Fig. 6: Longitudinal aberration due to a convex lens.

Consider a white light with plane wave front incident on a convex lens. After refraction through the lens, various colours focus at different positions. As shown in Fig. 6, let f_r and f_v be focal lengths, and μ_r and μ_v the refractive indices for the red and violet colours, respectively. Now, for the red colour, we have

$$\frac{1}{f_r} = (\mu_r - 1)\left(\frac{1}{R_1} - \frac{1}{R_2}\right) \tag{7.2}$$

and for the violet colour, we have

$$\frac{1}{f_v} = (\mu_v - 1)\left(\frac{1}{R_1} - \frac{1}{R_2}\right) \tag{7.3}$$

where R_1 and R_2 are radii of curvatures of the two faces of lens. For the average focal length f and refractive index μ of lens, we have

$$\frac{1}{f} = (\mu - 1)\left(\frac{1}{R_1} - \frac{1}{R_2}\right) \tag{7.4}$$

From equations (7.2) and (7.4), we have

$$\frac{f}{f_r} = \frac{\mu_r - 1}{\mu - 1}$$

or
$$\frac{1}{f_r} = \frac{\mu_r - 1}{f(\mu - 1)} \tag{7.5}$$

and from equations (7.3) and (7.4), we have

$$\frac{f}{f_v} = \frac{\mu_v - 1}{\mu - 1}$$

or
$$\frac{f}{f_v} = \frac{\mu_v - 1}{f(\mu - 1)} \tag{7.6}$$

Subtracting equation (7.5) from (7.6), we get

$$\frac{1}{f_v} - \frac{1}{f_r} = \frac{(\mu_v - \mu_r)}{f(\mu - 1)} = \frac{\omega}{f} \tag{7.7}$$

where $\omega = (\mu_v - \mu_r)/(\mu - 1)$ is known as the dispersion power of the material of lens. Equation (7.7) can be rearranged as

$$\frac{f_r - f_v}{f_r f_v} = \frac{\omega}{f}$$

or
$$f_r - f_v = \frac{\omega}{f} f^2 = \omega f \tag{7.8}$$

where $f = \sqrt{f_r f_v}$ is the geometrical average focal length of lens. Thus, the longitudinal chromatic aberration ($f_r - f_v$) is equal to the product of the dispersive power and the mean focal length of the lens, which may be taken as for the yellow colour. It is obvious that a single lens with white light cannot produce an image which is free from the chromatic aberration.

9.3 Condition for Achromatism of two Lenses in Contact

Chromatic aberration may be eliminated by keeping two lenses, made of different materials, in contact in such a way that all the rays of different colours are brought to focus by the combination at a common point. In practice, the chromatic aberration cannot be removed completely. Usually, the achromatism is achieved for two prominent colours.

A achromatic combination is made by placing in contact two lenses of different materials and of suitable focal lengths, so that the focal length of the combination is equal for both the extreme colours, red and violet. The refraction formula for a lens is

$$\frac{1}{f} = (\mu - 1)\left(\frac{1}{R_1} - \frac{1}{R_2}\right) \tag{7.9}$$

where μ is the refractive index of the material of the lens, and R_1 and R_2 are the radii of curvatures of two surfaces of the lens. On differentiating equation (7.9), we get

$$-\frac{df}{f^2} = d\mu\left(\frac{1}{R_1} - \frac{1}{R_2}\right) \tag{7.10}$$

Diving equation (7.10) by (7.9), we get

$$-\frac{df}{f^2} = \frac{d\mu}{f(\mu - 1)} \tag{7.11}$$

If μ_v and μ_r are the refractive indices for the violet and red colours, respectively, then $d\mu$ is the change in the refractive index such that $d\mu = \mu_v - \mu_r$. Therefore, the dispersive power of lens is

$$\omega = \frac{\mu_v - \mu_r}{\mu - 1} = \frac{d\mu}{\mu - 1} \tag{7.12}$$

Using equation (7.12) in (7.11), we have

$$-\frac{df}{f^2} = \frac{\omega}{f} \tag{7.13}$$

When two lenses of focal lengths f_1 and f_2 are placed in contact, the focal length f of the combination is

$$\frac{1}{f} = \frac{1}{f_1} + \frac{1}{f_2}$$

On differentiating this equation, we get

$$-\frac{df}{f^2} = \frac{df_1}{f_1^2} - \frac{df_2}{f_2^2}$$

In order that the rays of different colours are brought to focus at a common point, the change in the focal length of the combination should be zero, i.e., $df = 0$. Thus, we have

$$-\frac{df_1}{f_1^2} - \frac{df_2}{f_2^2} = 0$$

Using equation (7.13) here, we get

$$\frac{\omega_1}{f_1} + \frac{\omega_2}{f_2} = 0 \tag{7.14}$$

where ω_1 and ω_2 are the dispersive powers for the materials of the lenses, respectively, and are always positive. Equation (7.14) thud shows that f_1 and f_2 must have opposite signs, i.e., the combination should consist of one concave lens and the other convex lens, generally made of different materials. Such combination is often known as the achromatic doublet.

9.4 Achromatic Doublet

An achromatic doublet consists of one convex lens and one concave lens, placed in contact with each other, as shown in Fig. 7. One of them is made of, say, crown glass and the other of, say, flint glass. The dispersive powers of the materials of these lenses are ω_1 and ω_2, respectively.

From the condition of the achromatism, we have

$$-\frac{f_2}{f_1} = \frac{\omega_2}{\omega_1}$$

A negative sign here appears as the focal lengths of lenses have opposite signs. We know that the focal length of a convex lens is taken as positive whereas that of a concave lens is taken as negative.

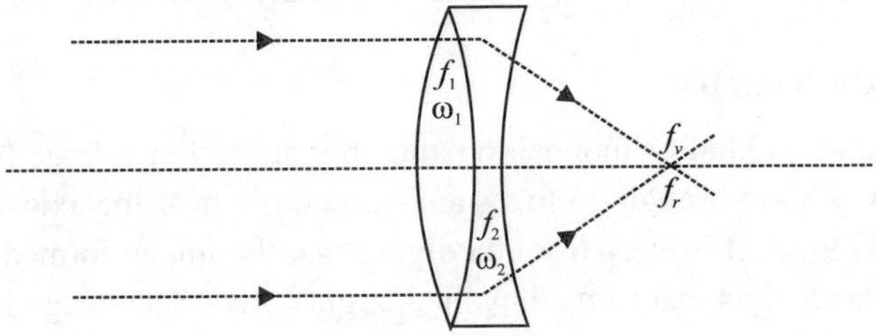

Fig. 7: Achromatic doublet.

Example 21.

Calculate focal length of a lens having dispersive power 0.03 which should be kept in contact with a convex lens of focal length 80 cm and dispersive power 0.023 so that the combination is achromatic.

Solution:

Given, $f_1 = 80$ cm, $\omega_1 = 0.023$, and $\omega_2 = 0.03$. For the achromatism, we have

$$-\frac{f_2}{f_1} = \frac{\omega_2}{\omega_1}$$

Using the values, we get

$$-\frac{f_2}{80} = \frac{0.03}{0.023}$$

so that $$f_2 = -104.34 \text{ cm}$$

Thus, the required focal length of a concave lens is -104.34 cm.

Example 22.

A convex lens of focal length 80 cm and dispersive power 0.041 is kept in contact with a concave lens of focal length 60 cm so that the combination is achromatic. Calculate the dispersive power of the concave lens.

Solution:

We have $f_1 = -80$ cm, $\omega_1 = 0.041$ and $f_2 = 60$ cm. The dispersive power ω_2 of the concave lens is

$$\omega_2 = -\frac{\omega_1 f_2}{f_1}$$

$$= -\frac{0.041(-60)}{80} = 0.031$$

Example 23.

Dispersive powers of the convex and concave lenses in a achromatic combination are 0.03 and 0.04, respectively. The focal length of the convex lens is 75 cm. Calculate the focal length of the concave lens.

Solution:

We have $\omega_1 = 0.03$, $\omega_2 = 0.04$ and $f_1 = 75$ cm. The focal length of the concave lens is

$$f_2 = -\frac{\omega_2 f_1}{\omega_1} = -\frac{0.04(75)}{0.03} = -100 \text{ cm}$$

9.5 Spherical Aberration

Spherical aberration is a kind of monochromatic aberration. For a large lens, the rays which are away from the axis are brought to focus at F_p and those near the axis are brought to focus at F_c, as shown in Fig. 8. It is clear from the figure that the image formed by the central rays lies at a longer distance than that formed by the marginal rays. The image formed is, therefore, not sharp at any point on the axis. If a screen is placed perpendicular to the axis at F_p, the outer portions of the object are in focus, whereas at F_c, the inner portions of the object are in focus. The spherical aberration for a convex lens is positive. When the screen is placed in between F_p and F_c, the image appears a circular patch. This patch is known as the circle of least confusion and corresponds to the position of the best image.

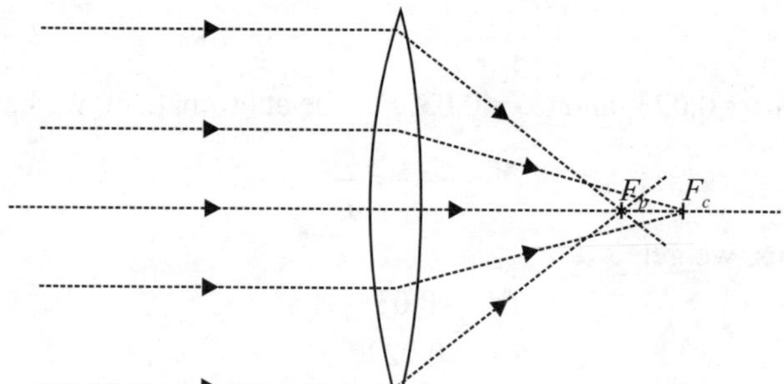

Fig. 8: Spherical aberration due to convex lens.

The distance F_cF_p measures the axial or longitudinal spherical aberration. The radius of the circle of least confusion, lying between F_c and F_p, is a measure of the lateral spherical aberration.

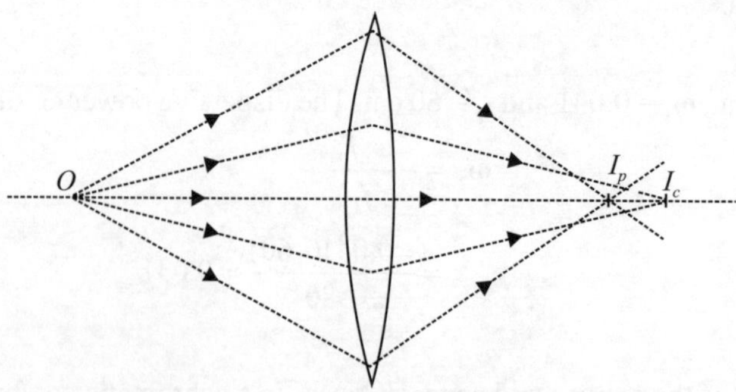

Fig. 9: Formation of the image of a point object O with the help of a convex lens using the monochromatic light.

It is obvious that for an point-object O placed on the axis, the image formed by the central rays is I_c whereas that formed by the marginal rays is I_p (Fig. 9). Thus, the image extends from I_p to I_c. That is, the image is not sharp. This defect is known as the spherical aberration and is due to the fact that different zones (parts) of the lens have different focal lengths.

9.6 Removal of Spherical Aberration

For a single lens, spherical aberration can be minimized under the following conditions.

(i) The spherical aberration depends on the aperture of the lens. It can therefore be reduced by reducing the aperture of the lens with the help of a co-axial aperture stop and therefore using only the central portion of the lens.

(ii) By using a specially designed lens, called the crossed lens, the radii of curvatures R_1 and R_2 of the two surfaces of the lens should satisfy the relation

$$\frac{R_2}{R_1} = \frac{\mu(2\mu+1)}{\mu(2\mu-1)-4}$$

where μ is the refractive index of the material of the lens relative to the outer medium (air).

10. Limitations of Lens Telescope

Since the main purpose of a telescope is to collect a large amount of radiations from an object, we would like to increase the size of the objective lens of the telescope. With the increase of the size not only the spherical aberration appears in picture, but the weight of the lens increases tremendously. For large weight, it becomes difficult to hold the lens with the help of a frame around it, and the lens breaks down.

Moreover, it is not possible to make the telescope free from the chromatic aberration for full range of the visible light. In a lens telescope, the radiations are collected after the refraction through the lens. We are aware of the fact that the radiations can also be collected through the reflection by a spherical mirror.

11. Mirror Telescope or Reflector

A mirror telescope can be constructed of very large size, as the mirror of the telescope can be supported on the back side (Fig. 10). Under such circumstances, we make use of the reflection property of light with the help of a mirror. In case of the reflection of light, we do not encounter the problem of aberrations. Further, as we can put an external support on the back side of the mirror, it is possible to increase the size of the telescope.

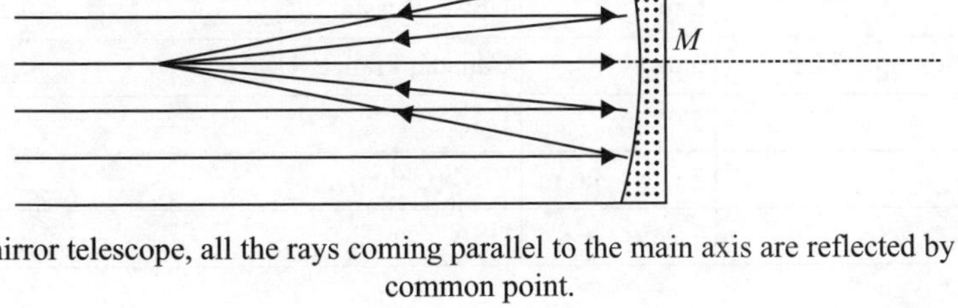

Fig. 10: In a mirror telescope, all the rays coming parallel to the main axis are reflected by the mirror to a common point.

The most common telescope type used in the astronomical research is mirror telescope or reflector. As a light-collection surface, it employs a mirror coated with a thin layer of aluminium. The shape of the mirror is generally parabolic, which reflects all the light rays entering the telescope parallel to the main axis, into a common focal point. The image formed at this point is observed with the help of an eyepiece or is registered on a detector.

One of the advantages of the reflectors is the absence of chromatic aberration, and all the wavelengths are reflected to a common point. Examples of large size reflectors are the Soviet 6-m BTA telescope in the Caucasus (1975) and the American 5-m Hale telescope on Mount Palomar (1947).

In a very large telescope, the observer can sit with his instruments in a special cage at the primary focus (Fig. 11) without eclipsing too much of the incoming light. In a smaller telescope, this is not possible, and the image must be inspected from outside the telescope. Information about the sizes of some big optical telescopes is given in Table 4.

Table 4. Some big optical telescopes

Name	Size	Sponsor
Gran Telescopio Canarias	10.4 m	Mexico, USA
Keck-1	10 m	USA
Keck-2	10 m	USA
Southern African Large Telescope	9.2 m	South Africa, USA, UK, Germany, Poland, New Zealand
Hobby-Eberly Telescope	9.2 m	USA, Germany
Large Binocular Telescope	8.4 m × 2	USA, Italy, Germany
Subaru (INLT)	8.2 m	Japan
VLT UT1 (Antu)	8.2 m	ESO countries, Chile
VLT UT2 (Kueyen)	8.2 m	ESO countries, Chile
VLT UT3 (Melipal)	8.2 m	ESO countries, Chile
VLT UT4 (Tepun)	8.2 m	ESO countries, Chile
Gemini North (Gillett)	8.1 m	USA, UK, Canada, Chile, Australia, Argentina, Brazil
Gemini South	8.1 m	USA, UK, Canada, Chile, Australia, Argentina, Brazil
MMT	6.5 m	USA
Magellan 1 (Walter Baade)	6.5 m	USA
Magellan 2 (London Clay)	6.5 m	USA
BTA-6	6 m	USSR/Russia
Large Zenith Telescope (ZLT)	6 m	Canada, France, USA
Hale Telescope	5.08 m	USA
LAMOST	4.9 m - 3.6 m	PRC (China)
MMT	4.7 m	USA

Discovery Channel Telescope	4.3 m	USA
William Herschel Telescope	4.2 m	UK, Netherlands, Spain
SOAR	4.1 m	USA, Brazil

11.1 Prime Focus

All the rays coming parallel to prime axis, after reflection from the mirror, reach a common point, called prime focus, as shown in Fig. 11.

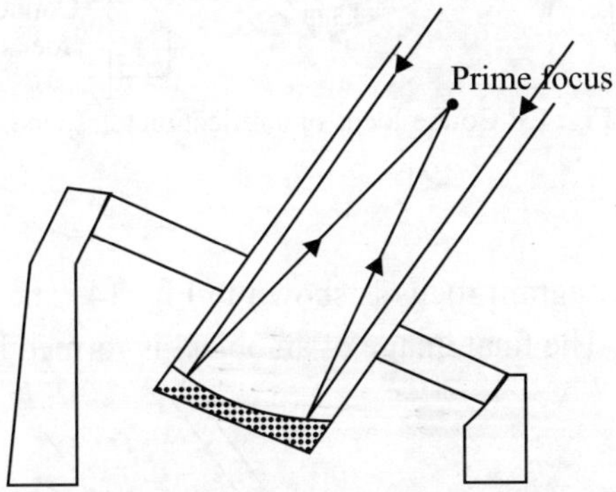

Fig. 11: Prime focus of reflection telescope.

11.2 Newton Focus

Isaac Newton, who invented the reflector, guided the light perpendicular out from the telescope with the help of a small flat mirror M_1, which is inclined at an angle of 45° with the prime axis of telescope. Therefore, the focus of the image in such a system is outside the tube of the telescope. The arrangement of the Newton focus is shown in Fig. 12.

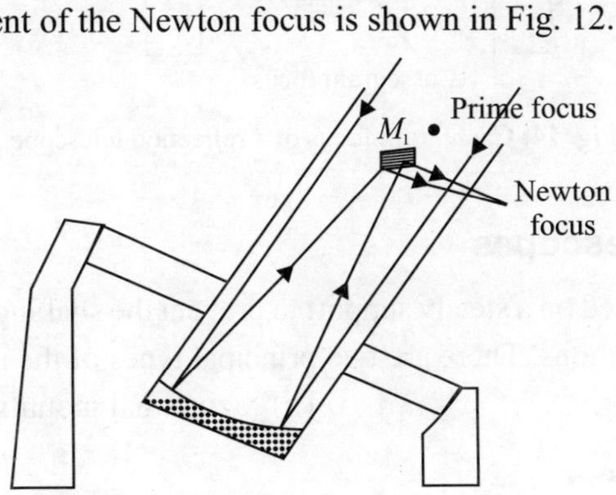

Fig. 12: Newton focus of reflection telescope.

11.3 Coude Focus

The arrangement of the Coude focus is shown in Fig. 13. There are two flat mirrors M_1 and M_2. The mirror M_1 is normal to the prime axis of the telescope and the mirror M_2 is inclined at an angle of 45° with the prime axis of the telescope.

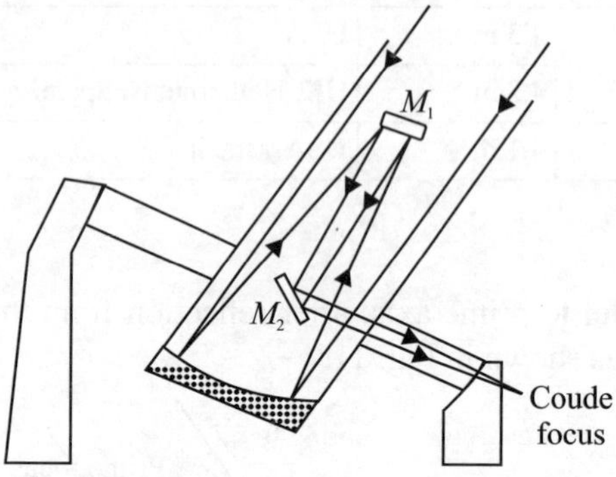

Fig. 13: Coude focus of a reflection telescope.

11.4 Cassegrain Focus

The arrangement of the Cassegrain focus is shown in Fig. 14. The mirror M_1 is normal to the prime axis of the telescope. The final image of an object is formed behind the telescope.

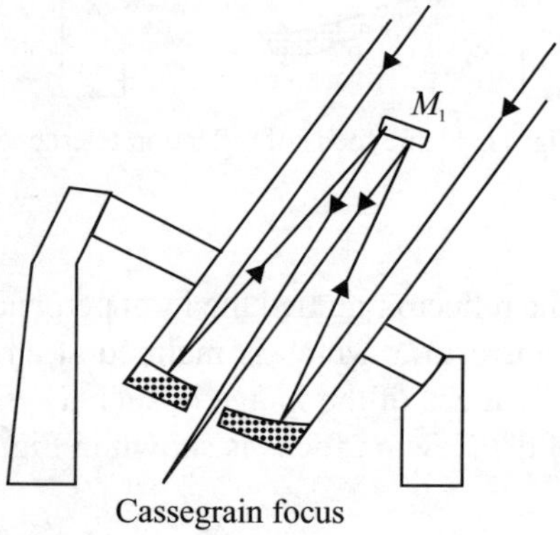

Fig. 14: Cassegrain focus of a reflection telescope.

12. Mounting of Telescopes

A telescope is to be mounted on a steady support to prevent the shaking, and it must be smoothly rotated during the observations. There are two principal types of the mounting:

(i) equatorial mounting (ii) azimuthal mounting

12.1 Equatorial Mounting

In the equatorial mounting (Fig. 15), one of the axis is directed towards the celestial pole. It is known as the polar axis or hour axis. The other one, known as the declination axis, is perpendicular to the polar axis. As the hour axis is parallel to the axis of the Earth, the apparent rotation of the sky is compensated by turning the telescope around that axis at a constant rate.

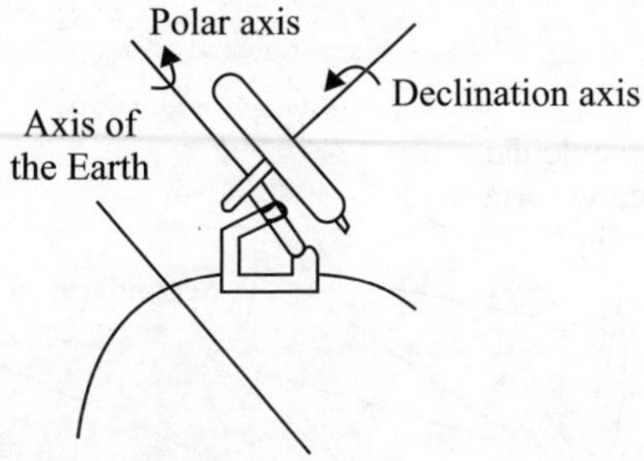

Fig. 15: Equatorial mounting of an optical telescope.

12.2 Azimuthal Mounting

In the azimuthal mounting (Fig. 16), one of the axis is vertical and the other is horizontal. This mounting is easier to construct than the equatorial mounting, and is more stable for the very large telescopes. In order to follow the rotation of sky, the telescope must be turned around both the axes with the changing velocities. The field of view also rotates, and is compensated when the telescope is used for the recording.

Most of the telescopes in the world were equatorially mounted until the development and use of computers which made possible the more complicated guidance needed for the azimuthal mounting.

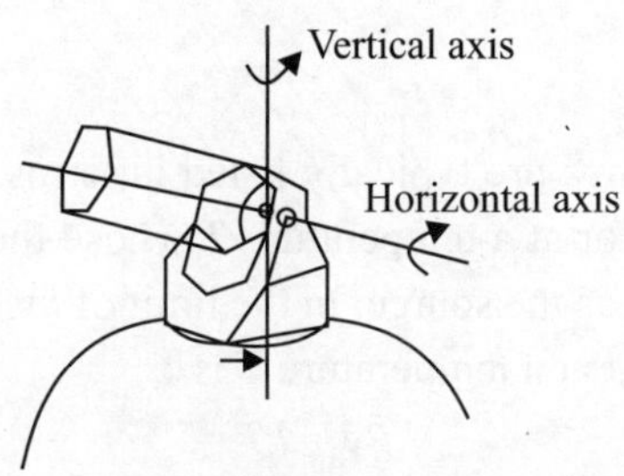

Fig. 16: Azimuthal mounting of an optical telescope.

13. Radio Telescopes

Schematic diagram of a radio telescope is shown in Fig. 17. The paraboloidal dish (Fig. 17) serves several important functions. First, it is a collector which reflects radiation from a beam equal to its own diameter to the receiver. As we always deal with very weak signals, a reflector with a large diameter focuses large amounts of energy.

13.1 Resolution

The resolution is expressed as how close things can be together and still be distinguished as separate objects. The angular resolution of a telescope depends on the diameter (D) of the dish and the wavelength (λ) of radiations observed with the telescope. The angular resolution of the telescope is expressed as $\sim \lambda/D$. The radio waves having large wavelengths have poor resolution. So, to improve the resolution of the radio telescopes, they are made as large as possible or two radio telescopes situated at two places, separated by a large distance are used as an interferometer.

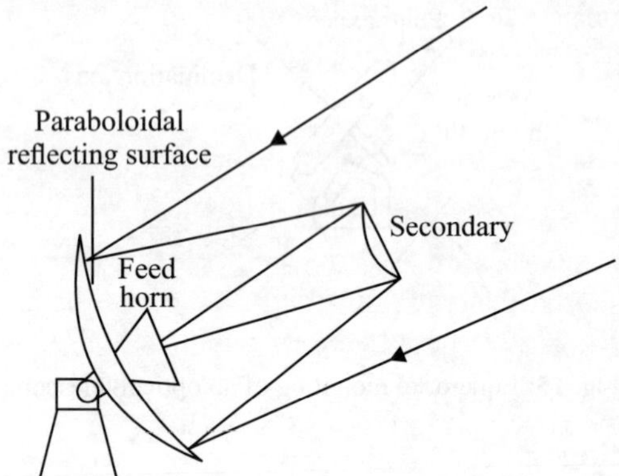

Fig. 17: Schematic diagram of a radio telescope.

13.2 Sensitivity

The sensitivity of a radio telescope is described as the ability to measure weak sources of radio emission. It depends on the area and efficiency of the antenna, the sensitivity of the radio receiver used to amplify and detect the signals, and the duration of the observation.

13.3 Noise Temperature

Noise sources, in radio astronomy, are typically listed in terms of the temperature. The power of noise is compared to a resistor at a temperature T whose thermal noise would produce the same power per unit bandwidth as the source. In the limit of $h\nu << kT$, the noise power per unit bandwidth of a resistive element at a temperature T is

$$P\nu = kT$$

This expression is known as the Nyquist formula. Thus, the noise temperature is

$$T_N = P_\nu / k$$

13.4 Limitation of Single Dish Radio Telescope

When a radio telescope has a single dish, the resolving power of the radio telescope is very poor. For example, for an optical radiation of 5000 Å observed by an optical telescope of diameter 2 m, the resolving power of optical telescope is

$$\text{R.P.} \approx \frac{\lambda}{D}$$

$$= \frac{5000 \times 10^{-8}\,\text{cm}}{2 \times 10^{2}\,\text{cm}}$$

$$= 2.5 \times 10^{-7}\,\text{rad}$$

Now, for a radio wave of 18 cm wavelength observed by a radio telescope of diameter 100 m, the resolving power of radio telescope is

$$\text{R.P.} \approx \frac{\lambda}{D}$$
$$= \frac{18 \text{ cm}}{100 \times 10^2 \text{ cm}}$$
$$= 1.8 \times 10^{-3} \text{ rad}$$

Obviously, the resolving power of a single dish radio telescope is very poor in comparison to that of an optical telescope. Thus, the resolving power of an optical telescope is 10^4 times more than that of a radio telescope. However, resolving power of a radio telescope may be enhanced tremendously by making the use of the Very Long Baseline Interferometry (VLBI) technique in which two radio telescopes situated at two different places, separated by a large distance, are used to behave like two outer segments of a large telescope. The two telescopes behave like two outer most segments of a single large telescope. The distance between the two telescopes is known as the baseline. The largest baseline on the surface of the Earth can be equal to the diameter of the Earth, which is about 12,700 km. Now, the resolving power of telescope using the VLBI technique becomes

$$\text{R.P.} \approx \frac{\lambda}{D}$$
$$= \frac{18 \text{ cm}}{12{,}700 \text{ km}}$$
$$= 1.42 \times 10^{-8} \text{ rad}$$

which is better than that of an optical telescope. For further improvement in the resolving power of a radio telescope, one of the radio telescopes used in the VLBI technique may be installed in the sky on an geostationary satellite, orbiting around the Earth.

14. Radio Sources and their Spectrum

In the outer space, a radio source is an object which emits strong radio waves, which could be produced by a wide variety of sources. In the universe, such objects are among the most extreme and energetic physical processes.

The electromagnetic spectrum having frequencies from 30 Hz to 300 GHz is known as the radio spectrum or radio waves. The radio waves are widely used in the modern technology, particularly in the telecommunication. Karl Jansky in 1932 noticed radio waves coming from an unknown source in the centre of our galaxy. In order to prevent interference between different users, the generation and transmission of radio waves is strictly regulated by the national laws, coordinated by an international body, called the International Telecommunication Union (ITU). Different segments of the radio spectrum are allocated by the ITU for different radio transmission technologies and applications.

14.1 Thermal and Non-thermal Mechanisms for Radio Emissions

The processes responsible for the generation of the continuum radio emissions are:

(i) Thermal processes

(ii) Non-thermal processes

1. The thermal processes are those which depend on the temperature of the source (e.g., black-body radiation).
2. The non-thermal processes are those which do not depend on the temperature of the source (e.g., synchrotron radiation).

15. Interferometer

The resolving power of a single dish radio telescope is very poor as compared to that of optical telescopes. For improvement of the resolving power in radio astronomy, two radio telescopes separated by a large distance are employed in an arrangement known as the Very Long Baseline Interferometry (VLBI).

15.1 Principle of VLBI

The Very Long Baseline Interferometry (VLBI) helps for improving the resolving power in radio astronomy. The idea of the VLBI can be understood in the following manner. When the angular size of an object is very small (few milli arc seconds), in order to study it in detail, a radio telescope with high spatial resolution is required. The resolving power of a telescope varies as λ/D, where λ is the wavelength of the radiation observed and D the size of the disc (mirror) of the telescope. The only possibility to improve the resolving power of a telescope is to increase the size D of the disc. The VLBI experiments involving radio telescopes across the United States were performed for the first time in 1968. The principle of VLBI is shown in Fig. 18.

A large single mirror (shown in A) can be considered as a set of a number of smaller mirrors (shown in B). Since the resolving power of a telescope, for a given wavelength, depends only on its aperture, therefore, retaining only the outermost segments (shown in C) would give the same resolving power as of the full mirror (shown in A). However, the intensity of radiations decreases as the less number of photons are received. The separation between the outermost segments is known as the baseline. In practice, the two telescopes T_1 and T_2 situated at two distant stations (shown in D) behave as two outermost segments.

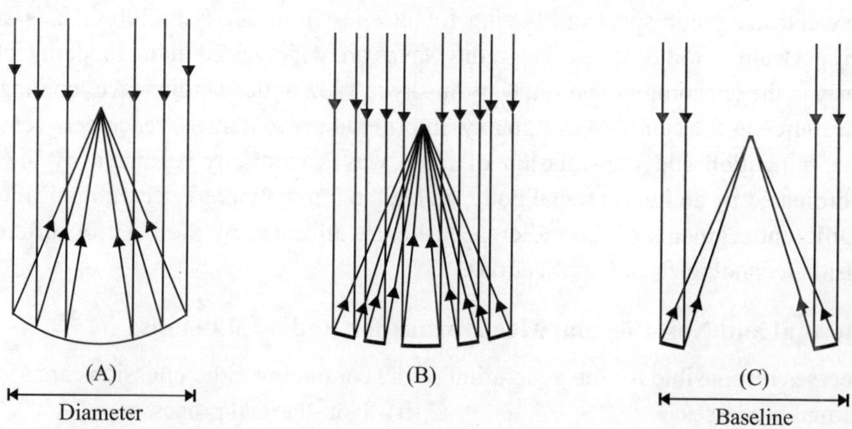

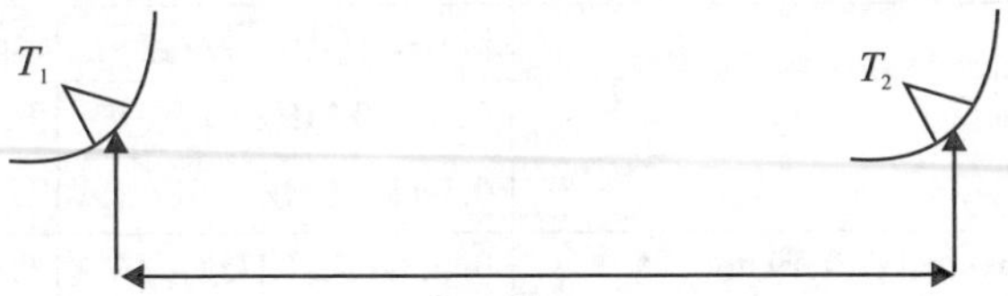

(D)

Fig. 18: Principle of very long baseline interferometry.

In the VLBI experiments, the signals from a common source are recorded simultaneously at two stations separated by a continental or intercontinental baselines and later on correlated. The maximum baseline which can be used on the Earth's surface is equal to the Earth's diameter (12,700 km). It corresponds to a resolution 10^{-8} arc seconds for a wavelength of few centimetres, which is better than the resolution that can be achieved with the optical telescopes.

Information about some radio telescopes is given in Table 5.

Table 5. Some radio telescopes

Name	Frequency range	Location
HartRAO 26 m	1.66-23 GHz	South Africa
HartRAO XDM	2.3-8.65 GHz	South Africa
Indlebe	1420 MHz	South Africa
KAT-7	1200-1950 MHz	South Africa
MeerKAT	0.58-14.5 GHz	South Africa
C-BASS South	4.5-5.5 GHz	South Africa
Delingha 13.7 m	85-115 GHz	China
Sheshan	1660 MHz	China
Nanshan 25 m	1.4-18 GHz	China
Giant Meterwave Radio Telescope	50-1420 MHz	India
Ooty Radio Telescope	326.5 MHz	India
Gauribidnaur Radio Observatory	40-150 MHz	India
Nobeyama Radio Observatory	17-34 GHz	Japan
Siberian Solar Radio Telescope	5.7 GHz	Russia
Badary Radio Astronomical Observatory	1.4-22 GHz	Russia
Qitai Radio Telescope	0.3-117 GHz	China
Australian Square Kilometre Array Pathfinder (ASKAP)	0.7-1.8 GHz	Australia
Molonglo Observatory Synthesis Telescope (MOST)	0.6-1.2 GHz	Australia

Contd.

RATAN-600	0.61-30 GHz	Russia
Nancy Radio Heliographe	150-450 MHz	France
Allen Telescope Array	0.5-11.5 GHz	USA
Dominion Radio Astrophysical Observatory	0.408-1.42 GHz	Canada
OVRO 40 metre Telescope	15 GHz	USA
Very Large Array	0.1-0.2 GHz	USA

16. Govind Swarup

Govind Swarup (March 23, 1929 - September 7, 2020) was the first person to start and establish radio astronomy in India. He was the key scientist behind the concept, design and installation of both the Ooty Radio Telescope (Ootacamund, India) and the Giant Metrewave Radio Telescope (GMRT) near Pune. He was the founding director of the National Centre for Radio Astrophysics (NCRA) at the Tata Institute of Fundamental Research (TIFR), where he established a strong group in radio astrophysics, which is comparable to the best in the world. He contributed to the fields of solar radio emission, radio galaxies, quasars, pulsars, interplanetary scintillation, dark matter and cosmology.

Fig. 19: Govind Swarup,
(*Courtesy*: Google website)

16.1 Radio Telescopes in India

India has two big radio telescopes: (i) Ooty Radio Telescope (Ootacamund, India) and (ii) Giant Metrewave Radio Telescope (GMRT) near Pune. The key scientist behind the concept, design and installation of both the radio telescopes has been Prof. Govind Swarup.

Ooty Radio Telescope

The Ooty Radio Telescope (ORT) is located in Muthorai near Ooty, in the south part of India. It is first big radio telescope established in India during 1963 - 1970, and it belongs to the National Centre for Radio Astrophysics (NCRA), Pune of the Tata Institute of Fundamental Research (TIFR). The radio telescope is a 530 metres long and 30 metres tall cylindrical parabolic antenna. It operates at a frequency of 326.5 MHz with a maximum bandwidth of 15 MHz at the front end.

Fig. 20: Ooty Radio Telescope.
(*Courtesy:* Google website)

Fig. 21: Giant Metrewave Radio Telescope (GMRT).
(*Courtesy:* Google website)

Giant Metrewave Radio Telescope (GMRT)

The Giant Metrewave Radio Telescope (GMRT), located near Pune, in India, is an array of thirty fully steerable parabolic radio telescopes of 45 metre diameter, observing at metre wavelengths. It is operated by the National Centre for Radio Astrophysics (NCRA), Pune, a part of the Tata Institute of Fundamental Research (TIFR), Mumbai. It was conceived and built under the direction of Govind Swarup during 1984 and 1996. It is an interferometric array with baselines of up to 25 km, and works at 50,1500 MHz frequencies. It was recently upgraded with new receivers, and after that it is also known as the Upgraded Giant Metrewave Radio Telescope (uGMRT).

17. Thirty Metre Telescope

Fig. 22: Thirty metre telescope, (*Courtesy:* Google website)

The Thirty Metre Telescope (TMT) is an under-construction extremely large telescope (ELT) located on Mauna Kea, on the island of Hawaii. The TMT would become the largest optical telescope on Mauna Kea. Scientists initiated plans about the ELTs since the mid-1980s. In 2000, astronomers considered the possibility of a telescope with a light-gathering mirror having diameter larger than 20 metres. Though the technology to construct a mirror larger than 8.4 m does not exist; instead scientists considered using either small segments that create one large mirror, or a grouping of larger 8-metre mirrors working as one unit. The US National Academy of Sciences recommended a 30-metre telescope be the focus of U.S. interests, seeking to have within the decade. Scientists at the University of California and Caltech began development of a design that would eventually become the TMT, consisting of a 492-segment primary mirror. It is considered to be nine times the power of the Keck Observatory. Due to its light-gathering power and the optimal observing conditions which exist atop Mauna Kea, the TMT would enable astronomers to conduct research which is not feasible with the current instruments. The TMT is designed for near-ultraviolet to mid-infrared (0.31 to 28 μm wavelengths) observations, featuring adaptive optics to assist in correcting the image blur. The TMT will be at the highest altitude of all the proposed ELTs. This project is being supported by several nations.

18. Square-kilometre Array (SKA)

Fig. 23: Square-kilometre array (SKA). (*Courtesy:* Google website)

The Square Kilometre Array (SKA) is an intergovernmental radio telescope project being planned to be built in Australia and South Africa. Conceived in the 1990s, and further developed and designed by the late-2010s, when completed it is planned to have a total collecting area of approximately one square kilometre sometime in the 2020s. It will operate over a wide range of frequencies and its size is such that it would be 50 times more sensitive than any other radio instrument. For its proper functioning, its requirement would be to have very high performance central computing engines and long-haul links with a capacity greater than the global Internet traffic as of 2013. According to its planning, it should be able to survey the sky more than ten thousand times faster than before.

The receiving stations of the SKA would be extended out to a distance of at least 3,000 km from a concentrated central core, and the SKA will exploit the ability of radio astronomy so that the resolution of the resolution images would be the highest. The SKA is proposed to construct in the southern hemisphere, with cores in South Africa and Australia, where the view of the Milky Way Galaxy is the best and radio interference is least. The headquarters of the project are located at the Jodrell Bank Observatory in the UK.

Initial construction contracts of the SKA started in 2018. Scientific observations with the fully completed array are not expected any earlier than 2027.

19. Detectors

In the Astronomy, a detector is an instrument used for capturing radiations coming from the outer space. Our eyes can be considered the first and one of the detectors. However, our eyes can see the radiations having wavelengths in the visible part of electromagnetic spectrum. Radiations of other wavelengths cannot be seen by our eyes. Further, the image of an object seen by our eyes is recorded in the memory of an observer, and cannot be shown to the others. Detection with the help of eyes has been the only way to make the observations till the end of nineteenth century.

In order to see the dim objects, one takes the help of a telescope. In a telescope, image of an object is formed. Now, the prime object is to record the information received in the form of an image of the object. By looking through a telescope, only a limited amount of information about the object, from which the radiations are coming, can be obtained.

The invention of photography brought a revolution in Astronomy. The next important step in the forward direction in optical astronomy has been the development of photoelectric photometry in the 1940s and 1950s. A new revolution, comparable to that caused by the invention of photography, took place in the middle of 1970s with the introduction of different semiconductor detectors. The sensitivity of such detectors has grown tremendously large. The detectors used in the modern Astronomy are the charge-coupled devices.

19.1 Charge-coupled Device (CCD)

A charge-coupled device (CCD) is a device for the movement of electrical charge, usually, within the device, from a place to an area where the charge can be manipulated, e.g., conversion into a digital value (Fig. 24). This is achieved by shifting the signals between the stages within the device one at a time. A CCD moves the charge between capacitive bins in the device, with the shift allowing for the transfer of charge between the bins.

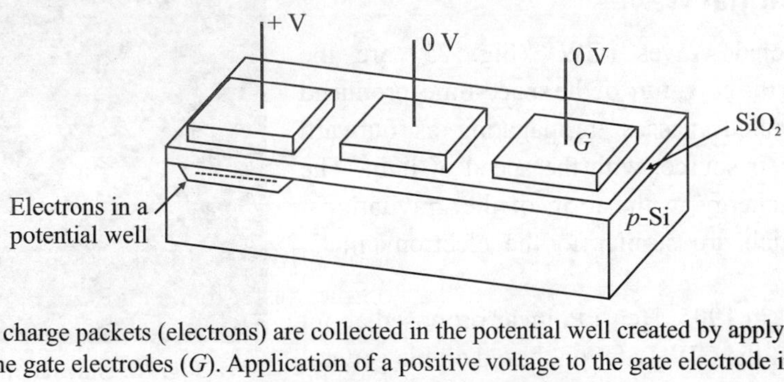

Fig. 24: The charge packets (electrons) are collected in the potential well created by applying a positive voltage at the gate electrodes (*G*). Application of a positive voltage to the gate electrode in the correct sequence transfers the charge packets.

The CCD is a major piece of technology in digital imaging. In a CCD image sensor, pixels are represented by the p-doped MOS capacitors. These capacitors are biased above the threshold for inversion when the image acquisition begins, allowing the conversion of incoming photons into electron charges at the semiconductor-oxide interface; the CCD is then used to read out these charges. Although the CCDs are not the only technology to allow for the light detection, CCD image sensors are widely used in the medical and scientific applications, where high-quality image data are required.

Basics of operation

The charge packets (electrons) are collected in the potential wells created by applying a positive voltage at the gate electrodes (G). Application of positive voltage to the gate electrode in the correct sequence transfers the charge packets. Visualization of the process may be found at the website:

https://en.wikipedia.org/wiki/Charge-coupled_device

In a CCD for capturing the images, there is a photoactive region (an epitaxial layer of silicon), and a transmission region made out of a shift register (the CCD, properly speaking).

An image is projected through a lens onto the capacitor array (the photoactive region), causing each capacitor to accumulate an electric charge proportional to the light intensity at that location. A one-dimensional array, used in line-scan cameras, captures a single slice of the image, whereas a two-dimensional array, used in the video and still cameras, captures a two-dimensional picture corresponding to the scene projected onto the focal plane of the sensor. Once the array has been exposed to the image, a control circuit causes each capacitor to transfer its contents to its neighbour (operating as a shift register). The last capacitor in the array dumps its charge into a charge amplifier, which converts the charge into a voltage. By repeating this process, the controlling circuit converts the entire contents of the array in the semiconductor to a sequence of voltages. In a digital device, these voltages are then sampled, digitized, and usually stored in the memory; in an analog device (such as an analog video camera), they are processed into a continuous analog signal (e.g., by feeding the output of the charge amplifier into a low-pass filter), which is then processed and fed out to other circuits for transmission, recording, or other processing.

20. Gravitational Waves

The Gravitational Waves (GW) (Fig. 25) are the disturbances in the curvature of the space-time, produced by the accelerated masses, propagating as outward waves from their source, with the speed of light. The GW transport energy in the form of the gravitational radiations, which are similar to the electromagnetic radiation.

Fig. 25: Gravitational waves.
(*Courtesy:* Google website)

First time, in 1905, Henri Poincar proposed about the existence of the GW. Later on, based on his general theory of relativity, Albert Einstein in 1916 predicted the existence of the GW. Universal Newton's law of gravitation cannot support the GW as it considers instantaneous (at infinite speed) propagation of the interactions between the masses.

The observed orbital decay of the Hulse Taylor binary pulsar is the first indirect evidence for the existence of the GW. The decay matched with the decay predicted by general relativity, where the energy is lost to the gravitational radiation. For this discovery of GW, Russell A. Hulse and Joseph H. Taylor Jr. received the Nobel Prize in Physics in 1993. The first direct observation of the GWs was made in 2015, when a signal generated by the merger of two black holes was received by the LIGO gravitational wave detectors. The 2017 Nobel Prize in Physics was awarded to Rainer Weiss, Kip Thorne and Barry Barish for their role in the direct detection of the GWs.

The purpose of the observations of the gravitational waves, in the GW Astronomy, is to decide about the sources of the GWs. The sources that can be studied in this manner include the binary star systems, consists of the white dwarfs, neutron stars, and black holes; and the events such as the supernovae, and the formation of the early universe shortly after the Big Bang.

The GWs carry energy away from their sources and, in the case of the orbiting bodies, this is associated with an in-spiral or decrease in orbit. To understand the situation, let us consider the Earth-Sun system, which is although revolving slowly compared to the speed of light, in the circular orbits about their centre-of-mass. Further, the Earth-Sun system may not produce GW; it is just an example of binary system. Theoretically, the loss of energy through the gravitational radiation could eventually drop the Earth into the Sun.

The total energy of the Earth orbiting the Sun (kinetic energy + gravitational potential energy) is about 1.14×10^{36} joules. According to the model, out of that only 200 watts (joules per second) is lost through the gravitational radiation, leading to a decay in the orbit by about 1×10^{15} m/day. At this rate, it would take the Earth to spiral onto the Sun, approximately 1×10^{13} times more than the current age of the universe. This estimate overlooks the decrease in r over time, but the radius varies slowly for most of the time and plunges at later stages, as $r(t) = r_0[1 - (t/t_c)]^{1/4}$, where r_0 is the initial radius, and the t_c is the total time required for fully coalesce.

More generally, the rate of orbital decay can be approximated by

$$\frac{dr}{dt} = -\frac{64}{5}\frac{G^3}{c^5}\frac{m_1 m_2 (m_1 + m_2)}{r^3}$$

where r is the separation between the bodies, t the time, G the gravitational constant, c the speed of light, and m_1 and m_2 the masses of the bodies. This leads to an expected time, to the merger, of

$$t = \frac{5}{256}\frac{c^5}{G^3}\frac{r^4}{m_1 m_2 (m_1 + m_2)}$$

A device designed for the detection of the GWs can be described a GW detector. Since the 1960s, various kinds of GW detectors have been built and improved constantly. New generation laser interferometers have reached the necessary sensitivity, required to detect the GWs from the astronomical sources.

21. Neutrino Astronomy

The study of the astronomical objects producing the neutrinos comes under the neutrino Astronomy. Neutrinos are produced, for example, in nuclear reactions in the cosmic objects, such as the Sun, during the hit of the cosmic rays with the atoms in the atmosphere. Due to the weak interaction of the neutrinos with the matter, there is a unique opportunity to observe the processes, which are not accessible in other branches of the Astronomy.

As neutrinos interact only very rarely with the matter, the enormous flux of the solar neutrinos moving through the Earth is sufficient to produce only 1 interaction for 1036 target atoms, and each interaction generates only a few photons or one transmuted atom. The observation of the neutrino interactions requires a large detector mass, along with a sensitive amplification system. For very weak signal, sources of background noise must be reduced as much as possible. The detectors must be shielded by a large shield mass, and so are constructed deep underground, or underwater.

Review Questions

1. Describe light year, astronomical unit and parsec. Out of them, which is the biggest.
2. Define light year and calculate its value.
3. Define astronomical unit and give distances of the planets from the Sun in terms of astronomical unit.
4. Define parsec and calculate its value.
5. Describe electromagnetic spectrum of radiation, and its absorption in the Earth's atmosphere.
6. Describe the optical window and radio window in the Earth's atmosphere. Which one is advantageous for investigation of distant cosmic objects?
7. Describe luminosity and radiant flux of a star.
8. Describe radiant flux of a star. How the radiant fluxes of two stars are compared?
9. Define luminosity and radiant flux of a star. Luminosity of a star is 4.26×10^{28} J/s. Calculate radiant flux received at a distance of 5.4×10^{15} m, assuming that the absorption of energy in the atmosphere between the star and the observer is negligible.
10. What are the apparent and absolute magnitudes of a shining object? Derive a relation between them.
11. What is the range of apparent magnitude for the stars visible to the naked eyes?
12. Derive relation between luminosities and absolute magnitudes of two stars.
13. Discuss the magnitude scale for the stars and explain the absolute magnitude.
14. The absolute magnitude of a star is +1.5. Calculate its distance so that the apparent magnitude of star is −1.5.
15. Describe Rayleigh criterion for resolution of two objects.
16. Discuss the atmospheric distortion of starlight.
17. Discuss in brief about salient features of a telescope.
18. What is f-ratio of a telescope. The diameter of primary mirror of a telescope is 234 cm and its focal length is 4.2 m. Calculate the f-ratio of the telescope.

19. Describe a lens telescope and its limitations.
20. Derive an expression for the chromatic aberration of a lens for an object placed at infinite distance.
21. Describe a mirror telescope and discuss about its resolving power.
22. Describe a radio telescope.
23. Discuss limitation of a single dish radio telescope. Describe the principle of the Very Large Baseline Interferometry.
24. Write about Govind Swarup and his contribution in the form of two biggest radio telescopes in India.
25. Discuss about the gravitational waves.
26. Discuss about the neutrino astronomy.
27. Write notes on the following:
    (i) Light year
    (ii) Astronomical unit
    (iii) Parsec
    (iv) Electromagnetic spectrum
    (v) Transparency of Earth's atmosphere
    (vi) Optical window and radio window
    (vii) Luminosity of a star
    (viii) Radiant flux of a star
    (ix) Solar constant
    (x) Magnitude scale
    (xi) Apparent magnitude and absolute magnitude
    (xii) Salient features of a telescope.
    (xiii) Rayleigh criterion for resolution
    (xiv) Atmospheric distortion
    (xv) Lens telescope
    (xvi) Mirror telescope
    (xvii) Prime focus
    (xviii) Newton focus
    (xix) Coude focus
    (xx) Cassegrain focus
    (xxi) Equatorial mounting
    (xxii) Azimuthal mounting
    (xxiii) Radio telescope
    (xxiv) VLBI
    (xxv) Radio telescopes in India
    (xxvi) Thirty metre telescope
    (xxvii) Square-kilometre array
    (xxviii) CCD
    (xxix) Gravitational waves
    (xxx) Neutrino astronomy

Multiple-Choice Questions

1. Light year is the unit of
   A. time B. reflection C. distance D. refraction **Ans. C**
2. 1 light year is equivalent to (in km)
   A. 1.50×10^8 B. 9.47×10^{12} C. 3.09×10^{13} D. 6.02×10^{16} **Ans. B**
3. 1 astronomical unit is equivalent to (in km)
   A. 1.50×10^8 B. 9.47×10^{12} C. 3.09×10^{13} D. 6.02×10^{16} **Ans. A**
4. 1 parsec is equivalent to (in km)
   A. 1.50×10^8 B. 9.47×10^{12} C. 3.09×10^{13} D. 6.02×10^{16} **Ans. C**
5. Which radiations have the largest frequency.
   A. IR radiation B. UV radiation C. γ-rays D. x-rays **Ans. C**
6. Which radiations have the largest wavelength.
   A. IR radiation B. UV radiation C. γ-rays D. x-rays **Ans. A**
7. The light reaches the Earth from the Sun in
   A. 100 s B. 200 s C. 500 s D. 800 s **Ans. C**
8. Our eyes can see radiations having wavelengths in the range
   A. 4000 – 7500 cm B. 4000 – 7500 mm C. 4000 – 7500 nm D. 4000 – 7500 Å **Ans. D**
9. Which radiations from interstellar medium can be studied on the surface of the Earth.
   A. γ-rays B. x-rays C. UV D. visible light **Ans. D**
10. Which radiations from interstellar medium can be studied on the surface of the Earth.
    A. γ-rays B. x-rays C. IR D. radio waves **Ans. D**
11. In which layer of the Earth's atmosphere, the γ-rays coming from interstellar medium are absorbed
    A. ionosphere B. troposphere C. exosphere D. mesosphere **Ans. A**
12. In which layer of the Earth's atmosphere, the x-rays coming from interstellar medium are absorbed
    A. ionosphere B. troposphere C. exosphere D. mesosphere **Ans. A**
13. In which layer of the Earth's atmosphere, the infra-red radiations from interstellar medium are absorbed
    A. ionosphere B. troposphere C. exosphere D. mesosphere **Ans. B**
14. Two objects *A* and *B* of luminosities *L* and 2*L* are at the distances *r* and 2*r*, respectively, from observer. The ratio of their fluxes F_A/F_B is
    A. 1/8 B. 1/4 C. 1 D. 2 **Ans. D**
15. The range of magnitudes of stars which can be seen without the help of a telescope is from
    A. from 1 to 6 B. from 0 to 6 C. from 1 to 10 D. from 0 to 9 **Ans. A**

16. The distance of a star is increased to 10 times of its initial distance. Then, its apparent magnitude
    A. increases by 10 B. increases by 5 C. decreases by 10 D. decreases by 5
    Ans. B
17. The distance of a star is increased by a factor of 2. Then, the flux of radiations received
    A. increases by a factor 2 B. increases by a factor 4
    C. decreases by a factor 2 D. decreases by a factor 4 **Ans. D**
18. For absolute magnitude, the distance of objects from observer is
    A. 1 AU B. 10 AU C. 1 pc D. 10 pc **Ans. D**
19. Which of the following features does not pertain to a telescope?
    A. Light-gathering B. Resolution C. Magnification D. Dispersion **Ans. D**
20. A radiation of wavelength λ is detected by a telescope having diameter a. The resolving power depends on
    A. λ only B. λ and a both C. a only D. none of them
    Ans. B
21. A telescope has focal length 20 m and diameter 10 m. The focal-ratio of telescope is
    A. 0.5 B. 30 m C. 2 D. 10 m **Ans. C**
22. The type of telescope having chromatic aberration is a
    A. lens B. lens and mirror both
    C. mirror D. neither lens nor mirror **Ans. A**
23. The type of telescope having 2 m diameter is a
    A. lens B. lens and mirror both
    C. mirror D. neither lens nor mirror **Ans. C**

8 Astronomy from Space

Earth's atmosphere is transparent in two windows, known as the optical window (OW) and the radio window (RW). The electromagnetic radiations in other regions (gamma-rays, x-rays, UV radiations, IR radiations) are in general not allowed to transmit through the Earth's atmosphere and to reach the Earth's surface. Thus, for the astronomical observations in the regions, other than OW and RW, scientists have in general to plan to send instruments/apparatus in the space.

1. Gamma-ray Astronomy

The most energetic radiations of electromagnetic spectrum lie in the gamma-ray band. Though it is difficult to make a demarcation, but the radiations having photon energies above 100 keV are considered the gamma radiations and is the subject of gamma-ray astronomy. The x-rays are classified as radiation below 100 keV.

The gamma-rays produced in the solar flares have been found to have energy in the range from MeV to GeV, whereas those produced in the Earth's atmosphere have energy in the MeV range. In the earlier models developed and used it had been believed that the GeV gamma-rays do not originate in the Solar System. The models are based on the assumption that GeV gamma-rays are produced by the extra-solar, and especially the extra-galactic sources. But, after finding the GeV photons in the solar flares, the models and the conclusions drawn would be affected. The mechanisms responsible for the generation of gamma-rays are diverse, mostly identical with those generating x-rays but at higher energies. The prime mechanisms for gamma-rays production are:

- electronpositron annihilation,
- inverse Compton effect

In some cases also the decay of radioactive material (gamma decay) in space reflecting extreme events such as supernovae and hypernovae, and the behaviour of matter under the extreme conditions, as in pulsars and blazars.

The highest photon-energies measured are in the TeV range, the record being held by the Crab Nebula in 2004, yielding photons with as much as 80 TeV.

1.1 Detector Technology

First time, the observation of gamma-rays could became possible in the 1960s because these rays are comparatively rare, their observation is therefore much more problematic as compared to that of the x-rays or of the visible light. Even a bright source takes an observation-time of several minutes before it is even detected. As the gamma-rays are difficult to focus, resulting in a very poor resolution. The production of the gamma-ray telescopes (2000s) have a resolution of the order of 6 arc minutes in the GeV range (seeing the Crab Nebula as a single pixel). This resolution is considered very poor if we compare it with that of the x-rays; resolution of 0.5 arc seconds is found in the low energy x-ray (1 keV) range by the Chandra x-ray Observatory (1999), and about 1.5 arc minutes in the high energy x-ray (100 keV) range seen by the High-Energy Focusing Telescope (2005).

Very energetic gamma-rays, with photon energies over ~30 GeV, can also be detected by the ground-based experiments. The extremely low photon fluxes at such high energies require detector effective areas which are impractically large for current space-based instruments. Such high-energy photons produce extensive showers of the secondary particles in the atmosphere that can be observed on the Earth's surface, both directly by the radiation counters and optically via the Cherenkov light, emitted by the ultra-relativistic shower particles. The Imaging Atmospheric Cherenkov Telescope technique currently achieves the highest sensitivity.

In 1989, Gamma radiation generated in the TeV range from the Crab Nebula was first detected by the observatory in Arizona in the USA. The most energetic photons (up to 16 TeV) observed from an extragalactic object originate from the blazar[1], Markarian 501 (Mrk 501). These measurements were made by the High-Energy- Gamma-Ray Astronomy (HEGRA). The Gamma-ray Astronomy observations are so far limited. Larger area detectors and better background suppression are essential for the progress in the field.

1.2 Cherenkov Radiation

Soviet scientist Pavel Cherenkov detected experimentally the Cherenkov radiations, and they now named after him. He shared the Nobel Prize in Physics in 1958. The process of Cherenkov light can be understood in the following manner. In the vacuum, the light travels with the speed $c = 299{,}792{,}458$ m/s, whereas in a material, the presence of medium reduces the speed of light and its value v may be significantly less. In water, for example, we have $v = 0.75c$. During the nuclear reactions and in particle accelerators, however, the matter can accelerate beyond this speed v (although less than c). The Cherenkov radiation results when a charged particle, most commonly an electron, travels through a dielectric medium with a speed greater than v.

In the beginning, a clear concept was that theoretical explanation of the Cherenkov radiation cannot be given based on any common mechanism such as the interaction of the fast electron with the individual atom or as the radiative scattering of electrons on the atomic nuclei. But, later on, an idea emerged that the phenomenon can be explained both qualitatively and quantitatively by considering the fact that an electron moving in a medium does radiate light even if it is moving uniformly provided that its velocity is greater than the velocity of light in the medium. The emission angle for the Cherenkov radiations can be explained as the following.

Emission angle

In the medium having refractive index n, a radiation n travels with the velocity $v = c/n$. The geometry of production of Cherenkov radiation is shown in Fig. 1. A particle (on the left) travels in a medium with speed v_p such that

$$v < v_p < c$$

[1]A blazar is an active galactic nucleus (AGN) with a relativistic jet. The name "blazar" was coined in 1978 by the Astronomer Edward Spiegel.

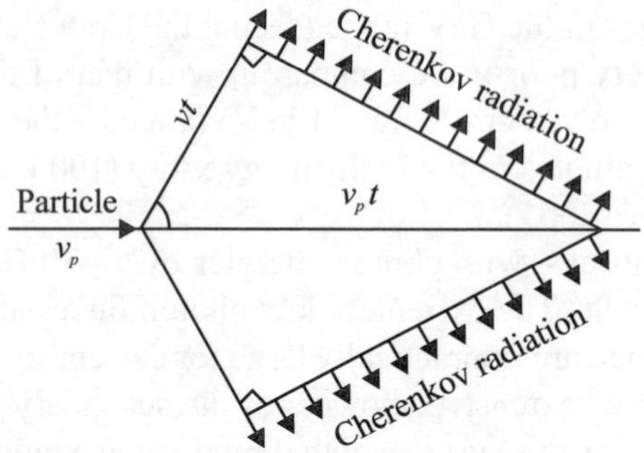

Fig. 1. The geometry of the Cherenkov radiation shown for the ideal case of no dispersion.

Notice that velocity of particle v_p is larger than v. The ratio of the speed of particle and the speed of light is $\beta = v_p/c$. The emitted light waves make an angle θ with the direction of the particle and travel at speed

$$v = c/n$$

Notice that the radiation is moving with the velocity v. The left corner of the triangle represents the location of the superluminal particle at some initial moment ($t = 0$). The right corner of the triangle is the location of the particle at some later time t. In the given time t, the particle travels the distance

$$x_p = v_p t = \beta c t$$

whereas the emitted electromagnetic waves travel the distance

$$x_{em} = vt = \frac{c}{n}t$$

So the emission angle θ is expressed as

$$\cos\theta = \frac{vt}{v_p t} = \frac{1}{n\beta}$$

It may be noted that this relation is independent of time. Thus, for the arbitrary times, one can achieve similar triangles. Since the angle θ stays the same, the subsequent waves generated between the initial time $t = 0$ and final time t are from the similar triangles where the right endpoint moves.

Cherenkov technique is used in Astrophysics observatories to measure air showers to determine the properties of astronomical objects that emit Very High Energy gamma rays, such as supernova remnants and blazars.

2. X-ray Astronomy

X-rays are electromagnetic waves with frequencies in the range from 3×10^{16} to 3×10^{19} Hz, wavelengths in the range from 0.1 to 100 Å, and energies in the range 100 eV to 100 keV. The astronomical study dealing with the x-rays is known as the x-ray astronomy.

The Earth's atmosphere absorbs the x-rays coming from the outer space, and therefore, the x-rays cannot transmit through the Earth's atmosphere and reach the Earth's surface. Therefore, for the detection of x-rays, the instruments must be taken to a high altitude by the balloons, sounding rockets, and satellites. For the study of x-rays, the form of the space telescopes is very different from the optical, radio or other kind of telescopes.

X-rays are expected from astronomical objects having extremely hot gases at the temperatures from about a 10^6 kelvin to 10^8 kelvin. It has been difficult to verify the theoretical prediction that the Sun and the stars are the prominent x-ray sources, because the x-rays from the outer space cannot transmit through the Earth's atmosphere for arriving at the surface of the Earth. Thus, until the methods for sending the instrument packages to the high altitude were developed, it was not possible to study the x-ray sources.

The existence of the solar x-rays was confirmed after making the use of the rockets. Since 1958, the mission of launching the satellites has been for the detection of extraterrestrial x-ray sources. A sounding rocket[2] discovered the first cosmic (beyond the solar system) x-ray source in 1962. Since then, several thousands of x-ray sources have been discovered. Further, the space between the galaxies in the galaxy clusters is filled with a very hot, but very dilute gas at a temperature from 10^8 kelvin to 10^9 kelvin. The total amount of hot gas is five to ten times the total mass in the visible galaxies. The Scorpius X-1 (Sco X-1) has 10,000 times larger x-ray emission than its visual emission, the Sun has a million times less x-ray emission than its visual emission. The energy output in the x-rays is 100,000 times greater than the total emission of the Sun in all the wavelengths.

2.1 Sounding Rocket Flights

The first sounding rocket flights for the x-ray research were performed in New Mexico with a V-2 rocket in January 1949. A detector was installed in the nose cone section and the rocket was launched in a suborbital flight to an altitude just above the atmosphere.

U.S. Naval Research Laboratory detected solar x-rays in an experiment on board. An Aerobee 150 rocket was launched in June 1962 and the first x-rays from other celestial sources (Scorpius X-1) were detected. It is now known that such x-ray sources as the Sco X-1 are the compact stars, such as the neutron stars or the black holes. The material falling into a black hole may emit x-rays, but the black hole itself does not. The energy source for the x-ray emission is the gravity. The strong gravitational fields of these objects and other celestial objects heat the infalling gas and dust.

The largest drawback to the rocket flights is their very short duration (just a few minutes above the atmosphere before the rocket falls back to the Earth) and their limited field of view.

2.2 Balloons

A solution to the limitation of flight-time of sounding rocket, was found by the scientists in the form of the balloon flights. The instruments boarded on the balloon flights can be carried up to 40 km above the sea level, where the almost Earth's atmosphere is covered. Contrary to few minutes duration of data collection by the sounding rockets, a baloon can stay aloft for much longer time. However, even at such altitudes, much of the x-ray spectrum is still absorbed by

[2]A rocket designed for carrying the instruments, taking measurements and perform scientific experiments during its suborbital flight.

the atmosphere. x-rays with energies less than 35 keV cannot reach the balloons. Using the scintillation counter flown on a balloon, the Crab Nebula supernova remnant was discovered in July 1964 to be a hard x-ray (15-60 keV) source. This was likely the first balloon-based detection of x-rays from a discrete cosmic x-ray source.

2.3 X-ray Telescope

An x-ray telescope (XRT) is specially designed to observe remote objects in the x-ray spectrum. In order to put the telescope above the Earth's atmosphere, it is mounted on high altitude rockets, balloons or artificial satellites.

The optics (instruments) and the detector are two basic components of the x-ray telescope. The optics (focusing or collimating) collects the radiations entering the telescope and focuses them on the detector, where the radiations are recorded and measured. These components are used in a variety of designs and technologies. In most of the x-ray telescopes, the optics are the grazing incidence mirrors and collimated apertures.

The utilization of x-ray mirrors allows to focus the incident radiation on the detector. Different geometries have been suggested or employed. The mirrors are generally made of ceramic or metal foil coated with a thin layer of a reflective material (typically gold or iridium). The mirrors having this construction work according to the total reflection of radiation at grazing incidence. (When the beam of radiation is nearly parallel to a surface, this small angle is called the grazing angle.) Grazing incidence diffraction is a fundamental method used in the x-ray spectroscopy, so that significant reflection can be achieved only at small values of the grazing angle. The grazing angle is usually measured in milliradians.

For the detectors in the x-ray telescopes, several technologies have been used, ranging from the counters like the ionization chambers, Geiger counters or scintillators to the imaging detectors like CCDs or CMOS sensors.

2.4 High-energy Focusing Telescope

The high-energy focusing telescope (HEFT) is a balloon-borne experiment to image astrophysical sources of hard x-rays (20-100 keV). The angular resolution of the HEFT is about 1.5′. Rather than using a grazing-angle x-ray telescope, the HEFT makes the use of a novel tungsten-silicon multilayer coatings to extend the reflectivity of nested grazing-incidence mirrors beyond 10 keV. The HEFT was launched for a 25-hour balloon flight in May 2005. The instrument performed within the specification and observed the Tau X-1, the Crab Nebula.

3. UV Astronomy

The ultraviolet (UV) radiations of electromagnetic spectrum have wavelengths between approximately 100 and 3200 Å. The UV radiations coming from the outer space are studied under the UV astronomy. Similar to the gamma-rays and x-rays, the UV radiations are not visible to the human eyes and most of the UV radiations are absorbed by the Earth's atmosphere, so the observations at these wavelengths also must be performed from the upper atmosphere of the Earth or from the space.

The UV universe looks quite different from the familiar stars and galaxies seen in the visible light. Most of the stars are actually relatively cool objects emitting much of their electromagnetic radiation in the visible or the near-infrared part of the electromagnetic spectrum. The UV radiation is the signature of the hotter objects, typically in the early and the late stages of their evolution. The UV telescopes have used in space-based solar observatories such as SDO and SOHO to view the activities on the Sun and its corona. Most of the countries are making observations. India has launched the AstroSat in 2015 working in the range from 1300 to 5300 Å.

4. IR Astronomy

The infrared (IR) radiations of electromagnetic spectrum have wavelengths approximately from 0.75 to 300 micrometers. The IR radiations coming from the outer space are studied under the IR astronomy. The IR falls in between visible radiation and the submillimetre waves.

Although the IR radiation was discovered in 1800, the IR astronomy began in the 1830s. In the beginning, much attention was not paid until the early 20th century when the IR detections of the astronomical objects other than the Sun and the Moon were made.

A positive and encouraging aspect is that the IR and optical astronomy can be performed using the same telescopes, as the same mirrors or the lenses are usually effective over a wavelength range that includes both the visible and the IR light. Both of them also use solid state detectors, though the specific type of solid state photodetectors used are different. At many wavelengths, the IR radiations are absorbed by water vapour in the Earth's atmosphere, so most of the IR telescopes are situated at high elevations in the dry places, above as much of the atmosphere as possible.

In the space also, there are infrared observatories, such as Spitzer Space Telescope and the Herschel Space Observatory.

5. Chandra Telescope

As the Earth's atmosphere absorbs most of the x-rays coming from the outer space, they cannot be detected by the Earth-based telescopes. So, an x-rays telescope is to be installed in the space.

Chandra x-ray Observatory (CXO) (Fig. 2), is a space telescope launched aboard the Space Shuttle Columbia by NASA on July 23, 1999. Chandra is sensitive to the x-ray sources which are 100 times fainter than those detected by any previous X-ray telescope, enabled by the high angular resolution of its mirrors. Chandra is an Earth satellite in a 64-hour orbit, and its mission is ongoing as of 2021.

Fig. 2: Chandra telescope.
(*Courtesy:* Google website)

This telescope is named after the Nobel Prize-winning Indian-American astrophysicist Subrahmanyan Chandrasekhar. It has very good angular resolution. Chandra is one of the Great Observatories, along with the Hubble Space Telescope, Compton Gamma Ray Observatory (1991-2000), and the Spitzer Space Telescope (2003-2020).

6. AstroSat

The Astrosat (Fig. 3) is the India's first dedicated multi-wavelength space telescope. It was launched on a PSLV-XL on 28 September 2015. Inspired with the success of this satellite, Indian Space Research Organization (ISRO) has proposed the launching of AstroSat-2 as a successor for the Astrosat.

The Astrosat is a general purpose proposal-driven observatory, with the main scientific focus on:

- Simultaneous multi-wavelength monitoring of intensity variations in a broad range of cosmic sources.
- Monitoring the x-ray sky for the new transients.
- Sky surveys in the hard x-ray and the UV bands.
- Broadband spectroscopic studies of x-ray binaries, AGN, SNRs, clusters of galaxies, and the stellar coronae.
- Studies of periodic and non-periodic variability of x-ray sources.

Fig. 3: AstroSat.
(*Courtesy:* Google website)

Astrosat performs the multi-wavelength observations, covering the spectral bands from the radio, optical, IR, UV, and x-ray wavelengths. Both the individual studies of the specific sources of interest and the surveys are undertaken. While the radio, optical, and IR observations would be coordinated through the ground-based telescopes, the high energy regions, i.e., the UV, x-ray and visible wavelength, would be covered by the dedicated satellite-borne instrumentation of the Astrosat.

7. Cosmic Rays

The Cosmic Rays (CR) consists of high-energy protons and atomic nuclei moving through the space at nearly the speed of light. They were discovered by Victor Hess in 1912 in balloon experiments. Cosmic rays can be divided into two types:

(i) galactic cosmic rays, and
(ii) extragalactic cosmic rays.

The primary cosmic rays mostly originate from outside the Solar System and sometimes even the Milky Way. When they interact with the Earth's atmosphere, they are converted to secondary particles (rays). The CR mainly originate from the sun, from outside of the solar system, and from distant galaxies. Since the launch of the first satellites in the late 1950s, it could be feasible to have the direct measurement of the cosmic rays, especially at lower energies. Particle detectors similar to those used in the nuclear and the high-energy physics are used on the satellites and the space probes for the research into the cosmic rays. On the incident with the Earth's atmosphere, cosmic rays can produce the showers of the secondary particles that sometimes reach the Earth's surface.

Observations from the Fermi Space Telescope (2013) have been interpreted as an evidence that a significant fraction of primary cosmic rays originate from the supernova explosions of stars. Based on the observations of the neutrinos and the gamma-rays from blazar TXS 0506 + 056 in 2018, it has been inferred that the active galactic nuclei also appear to produce the cosmic rays.

8. Hubble Space Telescope (HST)

The most famous space telescope is the Hubble space telescope (HST) (Fig. 4) which was launched aboard the space shuttle Discovery on 24 April 1990 and its orbit lies at 569 km above the Earth's surface. It completes one rotation around the Earth in 95.47 minutes (15.08 rev/day), travelling at a speed of approximately 7.59 km/s. Some orbital parameters of HST are given in Table 1. During its journey, its mirrors capture light and other radiations and direct them onto its several instruments.

Fig. 4: Hubble space telescope.
(*Courtesy:* Google website)

Table 1. Orbital parameters

Semi-major axis	6919 km
Inclination	28.47°
Period	95.47 min
Mean motion	15.08 rev/day
Velocity	7.59 km/s

The HST has a Cassegrain reflector with a primary mirror of diameter 2.4 m. Although the mirror diameter is smaller than those of the earth-based telescopes (which are up to 10 metres in diameter) its location above the Earth's atmosphere gives it excellent clarity. The resolution of the images produced is at least 10 times better than those from any telescope on the Earth's surface.

Besides a reflector, the HST carries other instrumentation also. One of them is the Wide Field Camera 3 (WFC3). This allows it to see the radiations from three different wavelengths; near the ultraviolet, the visible light and near the infrared, however not at the same time. In order to avoid the contamination of the Earth's atmosphere, the radiations from the interstellar medium can be studied by the boarding an instrument on a geostationary satellite orbiting around the Earth.

Review Questions

1. Discuss about the gamma-ray astronomy.
2. Discuss about the x-ray astronomy.
3. Discuss about the UV astronomy.
4. Discuss about the IR astronomy.
5. Discuss about the Chandra telescope.
6. Discuss about the AstroSat.
7. Discuss about the cosmic rays.
8. Discuss about the Hubble Space Telescope
9. Write notes on the following:
   (i) Cherenkov light
   (ii) Gamma-ray astronomy
   (iii) X-ray astronomy
   (iv) UV astronomy
   (v) IR astronomy
   (vi) Chandra telescope
   (vii) AstroSat
   (viii) Cosmic rays
   (ix) Hubble Space Telescope

Multiple-Choice Questions

1. What is the range of highest energy of gamma-ray photons?
   A. keV B. MeV C. GeV D. TeV **Ans. D**
2. What is the highest energy of gamma-rays produced in the solar flares?
   A. keV B. MeV C. GeV D. TeV **Ans. C**
3. Which radiations are recorded in the Chandra telescope?
   A. gamma-rays B. x-rays C. UV radiations D. IR radiations **Ans. B**
4. Which is not an electromagnetic radiation?
   A. gamma-rays B. x-rays C. UV radiations D. gravitational waves **Ans. D**
5. Which is not an electromagnetic radiation?
   A. gamma-rays B. x-rays C. UV radiations D. cosmic rays **Ans. D**
6. Which are particles?
   A. gamma-rays B. x-rays C. UV radiations D. cosmic rays **Ans. D**

9 Stars

Though the boundaries of the universe are not known, scientists believe that there are several hundred billion galaxies. Our galaxy, the Milky Way, is one of them. In a galaxy, more than 50% stars are in the form of groups of two or more stars. A majority of them is in the form of binary system, i.e., a group of two stars. Groups of three or more stars are very small in number. A well known group of three stars is the alpha-Centauri system whose third component is the Proxima Centauri, which is the next nearest star to us, after the Sun.

It is well established that the stars are made of hot gases – mostly hydrogen and helium. The hydrogen is being converted continuously into the helium in the shining stars. The stars are away from the Earth and the light from them passes through the atmosphere before reaching the Earth's surface. During this journey, they undergo the bending or refraction, due to the variation of density (refractive index) in the atmosphere. When there is wind, the magnitude of the bending of light-rays varies with time. This makes the stars to twinkle.

The stars which are relatively rich in the heavy elements, such as Sun, are known as the population I stars. The stars which are relatively poor in the heavy elements, such as globular clusters, are known as the population II stars. It is commonly believed that the population I stars represent a later generation of the stars whereas the population II stars represent an earlier generation. That is, the population II stars are older than the population I stars. No stars with the zero abundance of heavy elements are found. However, some astronomers theoretically consider their existence in the universe and they are known as the population III stars. In this chapter, we have discussed about the various types of the stars.

1. Formation of Stars

The existing stars did not exist forever, is a common opinion. Therefore, about the birth of the stars, speculations may be made. Further, the stars which we consider to be young are found associated with the clouds of the interstellar dust and gas. It may therefore be argued that such clouds of the dust and gas are contracting slowly, giving rise to increasingly compact condensations, and some of them finally collapse down to a stellar size. This indicates the formation of the stars out of the dust and gas. Such scenario may help for understanding the formation of the stars. Some aspects about the formation of the stars are as the following.

1.1 Cooling and Collapsing of Cloud

In a cloud out of which a star is formed, the initial gas is considered to consist of hydrogen atoms. In the cloud, the dust grains can radiate heat in a very efficient manner. When a hydrogen atom collides with a dust grain, the kinetic energy of the hydrogen atom is transferred to the dust grain and the temperature of the grain increases slightly. This energy is radiated away fast by the dust grain in the form of infrared radiations, which can escape from the cloud easily. In this manner, the kinetic energy of the gas is reduced. When the gas loses the kinetic energy, it falls towards the centre of the cloud under the gravitational attraction. During the fall, the gas acquires some kinetic energy which is partly transferred to a dust grain, and the said cooling cycle repeats.

The atom transfers also some of its centrally directed momentum to the grain. This causes the grains to drift towards the centre of the contracting cloud. As a result of many such interactions, the cloud as a whole contracts.

Radiation from the grain is not the only process responsible for the loss of the energy of the protostar. During the collapse, the temperature of the protostar cloud increases and due to the collisions, the molecules and atoms are excited. These excited molecules and atoms return back to their ground state, producing photons. This way the energy of the collapse is converted into the radiations which can escape from the cloud into the outer space, giving rise to the cooling. Though this explanation of the cooling and the collapse of a cloud looks attractive, but there are difficulties with this picture, as besides the loss of energy, there must be

(i) loss of angular momentum, and

(ii) conservation of magnetic field.

1.2 Problem of Angular Momentum

There are observational evidences about the rotation of the interstellar molecular clouds where the star formation is in progress. Thus, the contracting cloud has a large angular momentum from the early stage. In the process, the matter of the interstellar cloud collapses into a star. For the formation of a star of density 10^9 atoms/m^3, there should be collapse of gas from a volume whose initial radius r should be $\sim 10^{16}$ m. Over such distances, for a cold galactic molecular cloud, the velocity v must be $\sim 3 \times 10^2$ m/s. Thus, the initial angular momentum of cloud per unit mass is $l = mvr \sim 3 \times 10^{18}$ m^2/s.

But, the observations for typical stars show that the angular momentum per unit mass varies from 10^{12} to 2×10^{14} m^2/s. It is many orders of magnitude lower than the initial value. Further, in case of the Sun, the angular momentum per unit mass is only $\sim 10^{11}$ m^2/s. Even if we consider the whole solar system, it is 10^{13} m^2/s. In the solar system, most of the angular momentum belongs to the motions of the planets, and in particular, in the motion of the Jupiter orbiting around the Sun.

It shows that for becoming a star or a planetary system, the initially contracting cloud of interstellar matter must lose almost all of its angular momentum and only less than 0.1% can remain. How the angular momentum is lost is not yet understood.

1.3 Problem of Magnetic Field

Another problem is about the conservation of magnetic field present in the initial cloud. The magnetic lines of force act as if they are frozen to the gaseous material, and the number of these lines of force threading through the cross-sectional area of the contracting cloud remain constant. Thus, the magnetic field B is proportional to r^{-2} in such a contraction[1]. Suppose, some weak magnetic field present in the initial cloud is predominantly oriented along some direction. Then, after the formation of a star or a planetary system, the magnetic field must have the same direction and the strong intensity. Suppose, initially there is weak magnetic field B of the order of 10^{-7} gauss (which is smaller than the observed value). In the contraction of a cloud from a radius of 10^{16} m to 10^{9} m, the magnetic field becomes 10^{14} times stronger, i.e., it becomes of the order of 10^{7} gauss.

The actual magnetic field found on the surface of the stars are of the order of 1 gauss; some peculiar stars have the magnetic fields of the order 10^{4} gauss. It is very small as compared to the estimated value[2]. How this loss of the magnetic field occurs is a major unsolved part of the problem of the star formation.

2. Evolution of Stars

A star evolves with time after its formation. The lifetime of a star depends on its mass. A rough estimate shows that the lifetime may vary from a few million years for a most massive star to a few trillion years for a least massive star. Note that the age of least-massive star is considerably longer than even the age of the universe (~ 13.8×10^{9} years), showing no death of the least-massive stars so far.

Stars are considered to be born out of the collapsing clouds having gas and dust, often called the nebulae or the molecular clouds. Over a period of millions of the years, these protostars settle down into a state of equilibrium, becoming what is known as a main sequence star. Before reaching the main sequence, typical variation of the paths, depending on the initial mass on the main sequence is shown in Fig. 1, where the luminosity versus effective temperature of the star are plotted. In the Hertzsprung-Russell diagram, a star appears on the right side and after moving to left reaches the main sequence. For most of its life on the main sequence, a star is powered through nuclear fusion occurring in its core. In the initial phase, the energy is generated through the fusion of hydrogen atoms at the core of the main sequence star. The fusion of the hydrogen atoms generates helium atoms, and the stars, like Sun, begin to fuse hydrogen along a spherical shell surrounding the core. During this process, the stars gradually grow in size, pass through the sub-giant stage until it reaches the red giant phase.

Stars having at least 0.5 $M_{\odot}$ can also begin to generate energy through the fusion of helium at their core, whereas a more massive star can fuse heavier elements along a series of concentric shells, producing a picture of onion shells, where the hydrogen is burning in the outermost shell, and then helium, carbon, oxygen, neon, magnesium, silicon, and iron, in the inner shells in sequence. For a very heavy star (say, with mass 25 $M_{\odot}$), onion-like structure of its core, where the elements up to iron are formed, is shown in Fig. 2. In the outermost shell, the hydrogen is burning and around that is the region containing hydrogen. Starting from the hydrogen burning shell, up to which shell in the inward direction goes, depends on the mass of the star.

[1]The magnetic field is inversely proportional to the surface area of the spherical surface.

[2]Very compact white dwarf stars and even the denser neutron stars, respectively, have the magnetic fields believed to be of the order of 10^{5} gauss and 10^{12} gauss. These are the field strengths we expect when the Sun could shrink to the size of such compact stars.

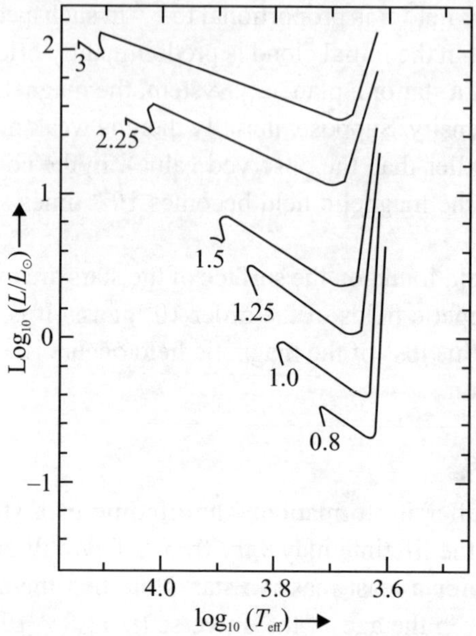

Fig. 1. A diagram showing the evolution of stars of various masses, written in the diagram in unit of solar mass $M_{\odot}$, till they arrive the main sequence. The evolution is from the right to the left.

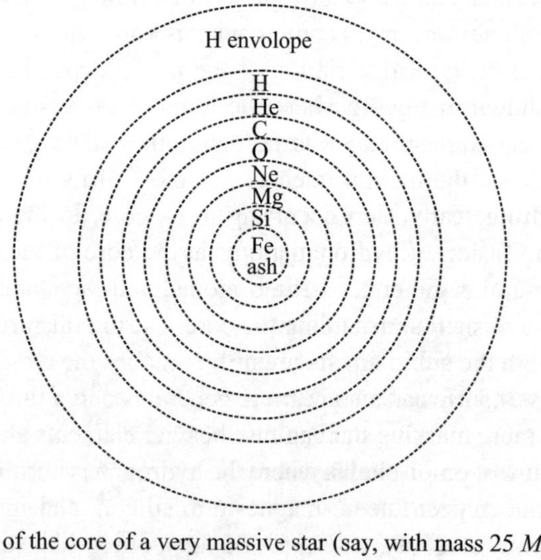

Fig. 2. Onion like structure of the core of a very massive star (say, with mass 25 $M_{\odot}$) where the elements up to the iron are formed.

After the exhaust of the nuclear fuel, in a sun-like star, the core collapses into a dense white dwarf and the outer layers are expelled as a planetary nebula. A star having more than 10 $M_{\odot}$ can explode in a supernova as their inert iron cores collapse into an extremely dense neutron star or a black hole. Although the universe is not so old that any of the smallest red dwarfs to have reached the end of their lives, but the stellar models tell that they would slowly become brighter and hotter, before running out of the hydrogen-fuel and becoming the low-mass white dwarfs.

It is obvious that the stellar evolution cannot be studied by observing the life of a single star, as the most stellar changes occur too slowly to be detected, even over many centuries. The astrophysicists therefore come to understand how the stars evolve by observing numerous stars at various stages in their lifetime, and by simulating the stellar structure using the computer models.

2.1 Protostars

As stated above, with the gravitational collapse of a giant molecular cloud, the formation and the evolution of the stars starts. A typical giant molecular cloud is roughly 100 ly (9.5×10^{14} km) across and contains up to $6 \times 10^6\ M_\odot$ (1.2×10^{37} kg). During the process of the collapse, a giant molecular cloud breaks into a large number of segments. In each of these segments, the collapsing gas releases the gravitational energy in the form of heat. On the increase of its temperature and pressure, a fragment condenses into a rotating sphere of superhot gas, known as a protostar. A protostar continues to grow by the accretion of gas and dust from the molecular cloud, becoming a pre-main sequence star as it reaches its final mass. After reaching the main sequence, further evolution of the star depends on its mass.

The protostars have dust as component, and therefore are easily visible at the infrared wavelengths.

2.2 Hayashi Lines

The Hayashi track is a luminosity-temperature relationship obeyed by the infant stars of less than $3\ M_\odot$ in the pre-main sequence phase of the stellar evolution. In the Hertzsprung-Russell (H-R) diagram, the track is a nearly vertical curve (Fig. 1). The Hayashi line is defined as the locus in the H-R diagram of fully convective stars of given parameters (mass and chemical composition). Note that for each set of the parameters, there is a separate Hayashi line. After a protostar ends its phase of rapid contraction begins and it becomes a T Tauri star, which is extremely luminous. The star continues to contract, but very slowly. While slowly contracting, the star follows the Hayashi track downwards, becoming several times less luminous but staying at roughly the same surface-temperature, until a radiative zone develops, at which point the star starts nuclear fusion, marking the beginning of main sequence.

The shape and the position of the Hayashi track in H-R diagram depends on the mass of the star and the chemical composition.

2.3 Brown Dwarfs

A protostar having a mass less than $0.08\ M_\odot$ (1.6×10^{29} kg) never reaches a temperature high enough for the initiation of the nuclear fusion of hydrogen into helium. These protostars are known as the brown dwarfs. As per definition given by the International Astronomical Union, a brown dwarf is a star having mass enough to fuse deuterium at some point in its life. Typical mass of a brown dwarf is 2.5×10^{28} kg or $0.0125\ M_\odot$. Objects having smaller than this mass are classified as sub-brown dwarfs (but when they orbit around another stellar object they are classified as the planets). Both types, deuterium-burning and not, shine dimly and die away slowly, cooling gradually over hundreds of millions of years. These brown dwarfs never reach the main sequence in the H-R diagram.

2.4 Main Sequence

For a more-massive protostar, the core temperature eventually reaches 10 million kelvin, sufficient for the initiation of the proton-proton chain reaction which fuses hydrogen first into the deuterium and then into the helium. In a star having mass slightly larger than 1 $M_\odot$, the carbon-nitrogen-oxygen fusion reaction (CNO-cycle) contributes a large portion for the energy generation. In the nuclear fusion, the energy released maintains a high gas pressure which balances the weight of the star's matter and prevents further gravitational collapse. The star thus evolves rapidly to a stable state, beginning the main sequence phase of its evolution.

Depending on its mass, the newly born star arrives at a specific position on the main sequence in the H-R diagram. Small, relatively cold, low-mass red dwarfs fuse hydrogen slowly and remains on the main sequence for hundreds of billions of years or longer, whereas the massive, hot *O*-type stars leave the main sequence after just a few million years. A mid-sized yellow dwarf star, like the Sun, remains on the main sequence for about 10 billion years. The Sun is thought to be in the middle of its main sequence lifespan.

2.5 Mature Stars

At the end of life on the main sequence, the core exhausts its supply of hydrogen and the star begins to evolve off from the main sequence, where the outward pressure generated by the fusion of hydrogen is absent. The core of the star contracts under the force of gravity until either the electron degeneracy pressure becomes sufficient to oppose the gravity or the core becomes hot enough (about 100 million kelvin) for the helium fusion to begin. Which of these happens first depends upon the mass of star as the following.

When there is no more production of energy through the fusion, what happens to a low-mass star. The age of the universe (about 13.8 billion years) is less (by several orders of magnitude, in some cases) than it takes for stopping the fusion in such stars. Therefore, no further information about such stars is available. Astrophysical models suggest that the red dwarfs of 0.1 $M_\odot$ may stay on the main sequence for six to twelve trillion years, gradually increasing in both the temperature and the luminosity, and take several hundred billion more to collapse, slowly, into a white dwarf. These stars never become red giants as they are fully convective and do not develop a degenerate helium core with a shell burning hydrogen. Instead, the hydrogen fusion proceeds until almost the whole star is the helium.

Slightly more massive stars do expand into the red giants, but their helium cores are not massive enough showing that their temperatures attain the value required for the helium fusion. Thus, they never reach the tip of the red giant branch. At the end of the hydrogen burning, these stars move directly off the red giant branch like a post-asymptotic-giant-branch (AGB) star, but at the lower luminosity, to become a white dwarf. A star with an initial mass above ~ 0.8 $M_\odot$ is able to attain the temperature high enough to fuse the helium, and these stars go on to further stages of evolution beyond the red giant branch.

Mid-sized stars

Stars having mass in the range of roughly 0.8 – 10 $M_{\odot}$ become the red giants, which are large non-main sequence stars of the stellar classification *K* or *M*. Red giants due to their red colour and large luminosity lie along the right edge of the H-R diagram.

The mid-sized stars are red giants during two different phases of their post-main sequence evolution: red-giant-branch stars, with inert cores made of the helium and the hydrogen-burning shells, and the asymptotic-giant-branch stars, with the inert cores made of carbon and helium-burning shells inside the hydrogen-burning shells. In between these two phases, the stars spend a period on the horizontal branch with a helium-fusing core. Many of these helium-fusing stars cluster towards the cool end of the horizontal branch as *K*-type giants and are known as the red clump giants.

Subgiant phase

After the exhaustion of the hydrogen in its core, the star leaves the main sequence and begins to fuse the hydrogen in a shell outside the core. As the shell produces more helium, the mass of the core increases. Depending on the mass of the helium core, this continues for several million to one or two billion years, with the star expanding and cooling at a similar or slightly lower luminosity to its main sequence state. Finally, either the core becomes degenerate, in the stars having around the mass of the Sun, or the outer layers cool sufficiently to become opaque, in the more massive stars. Either of these changes causes the hydrogen shell to increase in the temperature and the luminosity of the star, at which point the star expands onto the red giant branch.

2.6 Red-giant Branch Phase

The helium core continues to grow on the red-giant branch. It is no longer in the thermal equilibrium, so it increases in the temperature which causes the rate of fusion in the hydrogen shell to increase. The luminosity of star increases towards the tip of the red-giant branch. Red giant branch stars with a degenerate helium core all reach the tip with very similar core masses and very similar luminosities, although the more massive of the red-giants become hot enough to ignite the helium fusion before that point.

2.7 Horizontal Branch

In the helium cores of stars having mass from 0.8 to 2.0 $M_{\odot}$, which are largely supported by the electron degeneracy pressure, the helium fusion ignites on a timescale of days in a helium flash. In the nondegenerate cores of more massive stars, the ignition of the helium fusion occurs relatively slow with no flash. The nuclear power released during the helium flash is very large, of the order of 10^8 $L_{\odot}$ for a few days and 10^{11} $L_{\odot}$ for a few seconds. However, the energy is consumed by the thermal expansion of the initially degenerate core and thus cannot be seen from outside the star. Due to the expansion of the core, the hydrogen fusion in the overlying layers slows and the total energy generation decreases. The star contracts, although not all the way to the main sequence, and it migrates to the horizontal branch in the H-R diagram, gradually shrinking in radius and increasing its surface temperature.

2.8 Asymptotic-giant-branch Phase

After the consumption of helium at the core of the star, hydrogen and helium fusion continues in shells around a hot core of carbon and oxygen. The star follows the asymptotic giant branch on the H-R diagram, paralleling the original red giant evolution, but with the even faster energy generation (which lasts for a shorter time). Although the helium is burning in a shell, the majority of the energy is produced by the hydrogen burning in a shell farther from the core of star. From these hydrogen burning shells, the helium drops towards the centre of star and periodically the energy output from the helium shell increases dramatically. This is known as a thermal pulse and they occur towards the end of the asymptotic-giant-branch phase, sometimes even into the post-asymptotic-giant-branch phase.

There may be several to hundreds of thermal pulses, depending on the mass and composition of the star. On the ascent of the asymptotic-giant-branch, there is a phase where a deep convective zone forms and can bring carbon from the core to the surface of star. This is known as the second dredge up. In some stars, there may be even a third dredge up. In this way a carbon star is formed. It is very cool and strongly reddened having strong carbon lines in their spectra. A process known as the hot bottom burning may convert the carbon into the oxygen and nitrogen before it is dredged to the surface.

Another well known class of asymptotic-giant-branch stars are the Mira variables, which pulsate with well-defined periods of tens to hundreds of days and large amplitudes up to about 10 magnitudes (in the visual, total luminosity changes by a much smaller amount). For a more massive star, the star becomes more luminous and the pulsation period is longer, leading to an enhanced mass loss, and the star becomes heavily obscured at visual wavelengths.

2.9 Post-AGB

The mid-range stars finally reach the tip of the asymptotic-giant-branch and run out of fuel for the shell burning. They do not have sufficient mass so that the full-scale carbon fusion could be started, and therefore, they contract again, going through a period of post-asymptotic-giant-branch superwind to produce a planetary nebula with an extremely hot central star. Then, the central star cools to a white dwarf. The expelled gas is relatively rich in heavy elements generated within the star and depending on the type of star, may be particularly then oxygen or carbon enriched. The gas forms an expanding shell, called the circumstellar envelope, which cools as it moves away from the star, allowing the dust particles and molecules to form. With the high infrared energy input from the central star, ideal conditions are formed in these circumstellar envelopes for the maser excitation.

In the massive stars, the core is already large enough at the onset of the hydrogen burning that the helium ignition occurs before the electron degeneracy pressure has a chance to become prevalent. Therefore, when these stars expand and cool, they do not brighten as much as the lower-mass stars. They are however much brighter than the low-mass stars to begin with, and are thus still brighter than the red giants formed from less-massive stars. These stars are unlikely to survive as the red super giants; instead they would destroy themselves as the type II supernovae.

Extremely massive stars (more than about 40 $M_{\odot}$), which are very luminous and thus have very rapid stellar winds, lose mass so rapidly due to the radiation pressure that they tend to strip off their own envelopes before they can expand to become red supergiants, and thus retain extremely high surface temperatures from their main sequence time onwards. The largest stars of the current generation are about 100-150 $M_{\odot}$ because the outer layers would be expelled by the extreme radiation.

The core grows hotter and denser as it gains the material from the fusion of hydrogen at the base of the envelope. In all the massive stars, the electron degeneracy pressure is insufficient to halt collapse by itself, so as each major element is consumed in the centre, progressively heavier elements ignite, temporarily halting the collapse. If the core of the star is not too massive (less than approximately 1.4 $M_{\odot}$, taking into account mass loss that has occurred by this time), it may then form a white dwarf (possibly surrounded by a planetary nebula) as described above for the less massive stars, with the difference that the white dwarf is composed mainly of the oxygen, neon, and magnesium.

Note:

As an example, Fig. 3 shows the post-main sequence evolutionary track of a sun-like star.

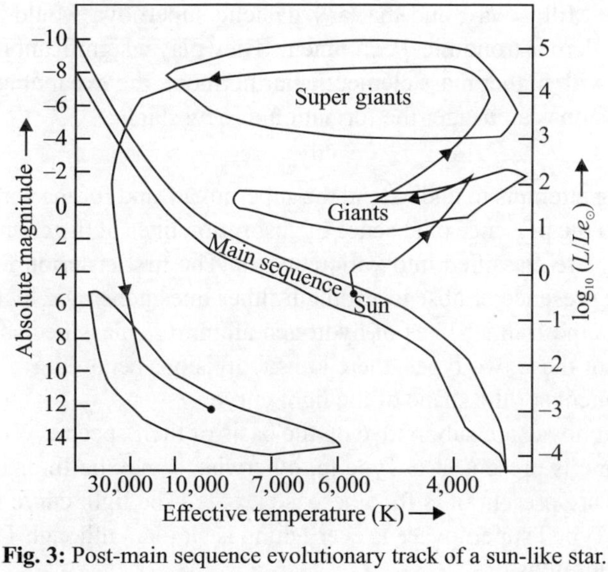

Fig. 3: Post-main sequence evolutionary track of a sun-like star.

2.10 Supernova

Once the nucleosynthesis process arrives at the iron-56, the continuation of this process consumes the energy. If the mass of the core exceeds the Chandrasekhar limit, 1.4 $M_{\odot}$, the electron degeneracy pressure is insufficient to support its weight against the force of gravity, and the core undergoes a sudden, catastrophic collapse to form a neutron star or a black hole. Through a process, that is not completely understood so far, some of the gravitational energy is released by this core collapse and is converted into a supernova.

A supernova (SN) is a stellar explosion that in a very short period outshines an entire galaxy, radiating energy equivalent to the total energy of the Sun or of an ordinary star, which later on fades out in a period of several weeks or months. The extremely luminous burst of radiation pushes outward much or all the material of the star at a velocity of up to 3×10^4 km/s (10% of the speed of light). It consequently drives a shock wave into the surrounding interstellar medium, which sweeps up an expanding shell of gas and dust called a supernova remnant.

There can be one of the following two possible ways for the triggering of a supernova.

1. By the sudden reignition of nuclear fusion in a degenerate star. Here, a degenerate white dwarf may accumulate sufficient material from a companion, either through the accretion or via a merger, to raise its core temperature, responsible for the ignition of carbon fusion, and trigger runaway nuclear fusion, completely disrupting the star.
2. By the gravitational collapse of the core of a massive star, the core of a massive star may undergo a sudden gravitational collapse, releasing gravitational potential energy, responsible for the creation of a supernova explosion.

The last directly observed supernova in the Milky Way was the Kepler's Star of 1604 (SN 1604); remnants of two more recent supernovae have been found retrospectively. Observations in other galaxies indicate that the supernovae should occur on average about three times every century in the Milky Way, and that any galactic supernova would almost certainly be observable in the modern astronomical equipment. They play a significant role in enriching the interstellar medium with higher mass elements. Furthermore, the expanding shock waves from the supernova explosions can trigger the formation of new stars.

Classification

Scientists are making attempts to understand the supernovae, and for the convenience, based on their light curves and the presence or absence of absorption lines of different chemical elements in their spectra, they are classified into various types. The first criterion for the classification of supernovae is the presence or absence of the Balmer lines generated by the hydrogen atom. When supernova has the Balmer lines of hydrogen atom, it is classified as Type II; otherwise it is Type I. In each of these two types, there are subdivisions according to the presence of the lines from other elements or the shape of the light curve.

The Type I supernovae are subdivided on the basis of their spectra. When a strong ionized silicon absorption line is present, it is Type Ia, otherwise it may be Ib or Ic. When the strong neutral helium lines are present, it is Ib, otherwise it is Ic. The light curve is not important for the classification of Type I supernovae, as every thing is similar, although Type Ia are generally brighter at the peak luminosity.

The Type II supernovae can also be subdivided on the basis of their spectra. Most of the Type II supernovae have very broad emission lines, showing the expansion velocities of many thousands of km/s. However, some have relatively narrow lines in their spectra and they are classified as Type IIn supernovae. Here, the 'n' indicates the 'narrow'.

A few supernovae, such as the SN 1987K and the SN 1993J, appear to change from Type Ib to Type II, as, at the early times, they show the lines of hydrogen, but after a period of some weeks to months, they have dominated helium lines. Such supernovae are classified as Type IIb, indicating a combination of Ib and II.

Type II supernovae with the normal spectra, dominated by the broad hydrogen lines that remain for the life of the decline, are classified on the basis of their light curves. The most common type shows a distinctive plateau in the light curve shortly after peak brightness where the visual luminosity stays relatively constant for a period of several months before the decline starts. These are called the Type II-P referring to the plateau. Some supernovae do not have distinct plateau, and they are classified as Type II-L, where the 'L' signifies the 'linear' although the light curve is not actually a straight line.

Supernovae that do not fit into the normal classification are designated as peculiar, or 'pec'. However, some scientists attempted to classify the supernovae into the Types III, IV and V, taking a few examples.

3. Variable Stars & Asteroseismology

Asteroseismology is a branch of science which deals with the oscillation in stars. As different oscillation modes of a star are sensitive to different parts of the star, they provide information about the internal structure of star, which otherwise cannot be derived from overall properties like brightness and surface temperature. When the same concepts of science are applied to the Sun, the asteroseismology is called the helioseismology. In case of the Sun more information is available as its surface can be resolved, which is not possible in case of distant stars.

3.1 Photometry of Variable Stars

Photometry is a technique used in Astronomy to measure the flux or intensity of light radiated by astronomical objects. This light is measured with the help of a telescope using a photometer. An electronic photometer, such as CCD photometer, is constructed using electronic devices. On other side, in a photoelectric photometer the light is converted into an electric current by the photoelectric effect. On calibration against standard stars (or other light sources) of known intensity and colour, the photometers can be used for measuring the brightness or apparent magnitude of a celestial object.

According to the wavelength region, a method of photometry is selected. In principle, the photometry is conducted by gathering light and passing it through specialized photometric optical bandpass filters, and then collecting and recording the light energy using a photosensitive instrument.

For variable stars also, photometry is used, where differential photometry or relative photometry are adopted. In the differential photometry, one measures simultaneously the brightness of a target object and nearby stars in the starfield. In the relative photometry, one compares the brightness of the target object with the stars having known fixed magnitudes. A plot of magnitude against time produces a light curve, providing considerable information about the physical process causing the brightness changes.

3.2 Extinction Coefficient

The absorption of light peopagating through a medium is measured in terms of extinction coefficient. Suppose, a radiation having intensity I_0 propagates through a medium, and after propagation through a diatance z, the intensity of radiation becomes I, then we have

$$I = I_0 \, e^{-kz}$$

where k having dimension of $(\text{length})^{-1}$ is called the attenuation coefficient. As in general k is positive and a real number, this expression shows that the intensity of radiation decreases exponentially.

3.3 Classes of Variable Stars

Variable stars may be classified into two different categories:

(i) Intrinsic variables (ii) Extrinsic variables

In case of intrinsic variables, the luminosity changes physically due to pulsations, eruptions or through swelling and shrinking. In case of extrinsic variables, the luminosity changes because of being eclipsed by stellar rotation or by another star or planet.

3.4 Period-luminosity Relationship

In astronomy, a period-luminosity relation is a relationship between the luminosity of pulsating variable stars and their pulsation period. The best-known relation is the direct proportionality law holding for Classical Cepheid variables, sometimes called the Leavitt law. For Population I Cepheid, a relation between the period P and its mean absolute magnitude M_v

$$M_v = (-2.43 \pm 0.12)(\log_{10} P - 1) - (4.05 \pm 0.02)$$

where P is measured in days. This relation was established from Hubble Space Telescope trigonometric parallaxes for 10 nearby Cepheids.

3.5 Pulsation Mechanisms

The driving mechanism behind the changes in luminosity of many types of pulsating variable stars is considered the kappa opacity mechanism. The parameter kappa (κ) denotes the radiative opacity at a point in the stellar atmosphere.

In case of a normal star, an increase in compression of atmosphere causes an increase in temperature and density, which produces a decrease in the opacity of atmosphere, which allows the energy to escape more rapidly. In this case an equilibrium condition is maintained. But, in case where the opacity of atmosphere increases with temperature, the atmosphere becomes unstable against pulsations. When a layer of a stellar atmosphere moves inward, it becomes denser and more opaque, causing heat flow to be checked. As a result, this heat increase causes a build-up of pressure that pushes the layer back out again. This is a cyclic process as the layer repeatedly moves inward and then is forced back out again.

Stellar non-adiabatic pulsation because of κ-mechanism occurs in the regions where hydrogen and helium are partly ionized, or where there are negative hydrogen ions. An example of such a zone is in RR Lyrae variables where the partial second ionization of helium occurs. In Mira variables, hydrogen ionization is most likely the cause of pulsation activity, rapidly oscillating Ap stars (roAp) and ZZ Ceti variables. In Beta Cephei variables, stellar pulsations occur at a depth where the temperature reaches approximately 200,000 K and there is an abundance of iron.

4. Compact Stars

In view of the war between the gravity and the thermodynamics in a star, regarding the end state (death) of a star, astronomers believe that there are four possibilities.

(a) Nothing is left. (b) A degenerate dwarf is left.
(c) A neutron star is left. (d) A black hole is left.

The term compact star (sometimes compact object) is used to refer collectively to the white dwarfs, neutron stars, and black holes. They are all small in volume for their mass, giving them a very high density. The white dwarfs and neutron stars may also be called the degenerate stars.

4.1 White Dwarf

A white dwarf is a very dense stellar remnant which is composed mostly of the electron-degenerate matter. A typical white dwarf has a mass comparable to that of the Sun, whereas its volume is comparable to that of the Earth. The nearest known white dwarf is the Sirius B, situated at a distance of 8.6 ly. A white dwarf is thought to be the final evolutionary state of a star whose mass is not high enough to become a neutron star. In our Milky Way, the masses of over 97% stars is such that they finally evolve into the white dwarfs.

During the process of fusion of hydrogen into helium, a low or medium mass star spends life on the main sequence of the HR diagram. When the process of fusion of hydrogen into the helium is almost over, the star expands to form a red-giant where the fusion of helium to carbon and oxygen takes place in its core through the processes.

$$3\ {}_2He^4 \rightarrow {}_6C^{12}$$

and

$$4\ {}_2He^4 \rightarrow {}_8O^{16}$$

After blowing out its outer layer to form a planetary nebula, the star leaves behind the core, called a white dwarf. Consequently, the white dwarfs generally consist of carbon and oxygen.

As the material in a white dwarf no longer undergoes fusion and therefore the star has no source of energy and has no support of heat, generated by the fusion, against the gravitational collapse. It is supported only by the electron degeneracy pressure, causing it to be extremely dense.

In a white dwarf, the electron degeneracy creates a barrier to the gravitational collapse. When the electrons are squeezed too close together, the Pauli's exclusion principle requires them to have different energy levels. If we want to add another electron to a given volume, it raises an electron's energy level to make room and this requirement for the energy to compress the material manifests as a pressure. The electron degeneracy pressure in a material is

$$P = \frac{3^{2/3} h^2 \rho_e^{5/3}}{20\pi^{2/3} m_e} = \frac{0.048 h^2 \rho_e^{5/3}}{m_e}$$

where h stands for the Planck's constant, m_e is electron mass, ρ_e is the electron density. This expression is for non-relativistic case. For the relativistic case, a modified formula is required. This electron degeneracy pressure is always present and is in addition to the normal pressure $P = NkT/V$.

The electron degeneracy pressure works like a spring against the gravitational collapse. When the collapsing mass is large, the size of the spring is small. Consequently, with the increase of the mass, the size of the white dwarf decreases. Chandrasekhar investigated the problem of electron degeneracy pressure and obtained a limiting mass of a white dwarf to be about 1.4 $M_\odot$. Beyond this, the white dwarf cannot be supported by the electron degeneracy pressure.

The relation between the mass *M* and radius *R* of a white dwarf is obtained as

$$R = 0.114 \frac{h^2}{Gm_e m_p^{5/3}} \left(\frac{Z}{A}\right)^{5/3} M^{-1/3}$$

where m_e and m_p are the masses of electron and proton, respectively, *Z* and *A* are the average atomic number and atomic weight of ions in a white dwarf. The point to be noted here is that the volume of a white dwarf is inversely proportional to its mass. If we compare two white dwarfs of different masses, the more massive white dwarf has smaller radius. This result is very different from that in case of an ordinary body, where the mass is proportional to its volume.

At the time of its formation, a white dwarf is very hot. As there is no more source of energy inside it, it gradually radiates away its energy and cools down with time. Consequently, its radiation which initially has a high colour temperature, decreases and reddens with time. After a long time, a white dwarf cools to a temperature at which it no longer emits significant heat or light, and it becomes a cold black dwarf. However, the estimated length of time it takes to convert from a white dwarf to a black dwarf is longer than the current age of the universe. As no white dwarf can be older than the age of the universe, it is thought that no black dwarfs yet exist. The older white dwarfs still radiate at temperature of few thousand kelvin.

Fowler

William A. Fowler (August 9, 1911-March 14, 1995) (Fig. 4) was an American nuclear physicist, later astrophysicist, who won the 1983 Nobel Prize in Physics with Subrahmanyan Chandrasekhar. He is famous for his theoretical and experimental research in the field of nuclear reactions within stars and the energy elements produced in the process.

Fig. 4. W.A. Fowler (August 9, 1911 - March 14, 1995).

(*Coutesy:* Google website)

Edington

Arthur S. Eddington (28 December 1882-22 November 1944) (Fig. 5) was an English astronomer, physicist, and mathematician. He was also a philosopher of science and a populariser of science. The Eddington limit, the natural limit to the luminosity of stars, or the radiation generated by accretion onto a compact object, is named in his honour. When the source of stellar energy was a complete mystery, around 1920, Eddington foreshadowed the discovery and mechanism of nuclear fusion processes in stars. Thus, Eddington was the first to correctly speculate that the source was fusion of hydrogen into helium.

Fig. 5. A.S. Edington (28 December 1882-22 November 1944).
(*Courtesy:* Google website)

Chandrasekhar

Subrahmanyan Chandrasekhar (19 October 1910-21 August 1995) (Fig. 6) was an Indian-American astrophysicist who spent his professional life in the United States. He shared the 1983 Nobel Prize for Physics with William A. Fowler. His mathematical treatment of stellar evolution yielded many of the current theoretical models of the later evolutionary stages of massive stars and black holes. The **Chandrasekhar limit** is named after him.

Chandrasekhar limit

Chandrasekhar investigated theoretically that when the mass of star was big enough, the radius of white dwarf formally waits to shrink to zero. The limiting radius is reached to zero at a mass now known as the Chandrasekhar limit, expressed as

$$M_{ch} = 0.20\left(\frac{Z}{Am_p}\right)^2\left(\frac{hc}{G}\right)^{3/2}$$

where Z and A are the average atomic number and atomic weight of ions in the white dwarf, and other symbols have their usual meanings.

Fig. 6. S. Chandrasekhar (19 October 1910 - 21 August 1995).
(*Courtesy:* Google website)

White dwarfs are an end state of the stellar evolution where we can expect all the hydrogen to have been converted to some heavier chemical elements. Since the ratio Z/A for the most heavier elements (until we approach and pass the iron group) is 0.5, the numerical value of Chandrasekhar limit can be calculated to be

$$M_{ch} = 1.4M_{\odot}$$

In other words, the maximum possible mass for a white dwarf is only 40% more than the mass of the Sun. All the white dwarfs whose mass have been observationally measured have masses less than this maximum value.

Example 1.

We are given mass of proton $m_p = 1.67 \times 10^{-27}$ kg, universal gravitational constant $G = 6.673 \times 10^{-11}$ m^3/s^2 kg, Planck's constant $h = 6.62 \times 10^{-34}$ Js, speed of light $c = 3 \times 10^8$ m/s. Calculate the Chandrasekhar limit.

Solution:

The Chandrasekhar limit is

$$M_{ch} = 0.20\left(\frac{Z}{Am_p}\right)^2 \left(\frac{hc}{G}\right)^{3/2}$$

$$= 0.20\left(\frac{0.5}{1.67\times10^{-27}}\right)^2 \left(\frac{6.62\times10^{-34}\times3\times10^8}{6.673\times10^{-11}}\right)^{3/2}$$

$$= 2.91 \times 10^{30} \text{ kg}$$

$$= \frac{2.91\times10^{30}}{1.99\times10^{30}} = 1.4M_{\odot}$$

Schonberg-Chandrasekhar limit

In stellar astrophysics, the Schönberg-Chandrasekhar limit is the maximum mass of a non-fusing, isothermal core that can support an enclosing envelope. It is named after the astrophysicists Subrahmanyan Chandrasekhar and Mario Schönberg, and is expressed as the ratio of the core mass to the total mass of the core plus envelope. The value of limit depends on the model used and the considered chemical compositions of the core as well as envelope; typical values are from 0.10 to 0.15 (10% to 15% of the total stellar mass). This is the maximum to which a helium-filled core can grow, and if this limit is exceeded, as can only happen in massive stars, the core collapses, releasing energy that causes the outer layers of the star to expand to become a red giant.

4.2 Neutron Stars

A neutron star is a type of stellar remnant that can result from a gravitational collapse of a massive star (with an initial mass of around 10 $M_{\odot}$ or above) and its almost entire composition is the neutrons. Neutron stars are the densest (from 3.7×10^{17} to 5.9×10^{17} kg/m^3) and tiniest (with a radius of only about 12-13 km) stars known to exist in the universe. A typical neutron star has a mass between 1.4 and 3.2 solar mass with a surface temperature of about 6×10^5

K. That is, the mass of a neutron star is more than the Chandrasekhar limit ($1.4M_\odot$) for the white dwarfs. A neutron star having diameter of about 10 km may have a mass of several times the solar mass. Obviously, the neutron stars are the end points of stars whose inert core's mass after the nuclear burning is larger than $1.4M_\odot$, but the mass is not enough to overcome the neutron degeneracy pressure and to become the black holes. The neutron stars have mass between $1.4M_\odot$ and $3M_\odot$. The maximum observed mass of the neutron stars is about $2M_\odot$. The gravitational field at the star's surface is about 2×10^{11} times stronger than that on the Earth.

In case of a neutron star, the mass is larger than that of a white dwarf and the electron degeneracy pressure is insufficient to control the gravitational collapse. In the neutron stars, we have neutrons, which also are fermions. There neutron degeneracy pressure comes in the picture. The neutron degeneracy pressure in a material is

$$P = \frac{3h^2 \rho_n^{5/3}}{4\pi^{2/3} m_n} = \frac{0.35 h^2 \rho_n^{5/3}}{m_n}$$

where m_n is the neutron mass and ρ_n is the neutron density. The neutron degeneracy pressure also works like a spring with a large spring constant against the gravitational collapse. When the collapsing mass is large, the size of spring is small. Consequently, with the increase of mass, the size of the neutron star decreases. (Similar behaviour has been obtained in case white dwarfs.) The relation between the number of neutrons N and radius R of a neutron star is obtained as

$$R = \left(\frac{81\pi^{14}}{16}\right)^{1/3} \frac{h^2}{4Gm_n^3} N^{-1/3}$$

The point to be noted here is that the volume of a neutron star is inversely proportional to its mass (number of neutrons in the star). Neutron stars are very hot and are supported against further collapse by the quantum degeneracy pressure due to the neutrons, following the Pauli's exclusion principle.

The discovery of pulsars in 1967 suggested the existence of neutron stars. They are born in supernova explosion and have radius of about 12-13 km. They spin about 642 times per second. Neutron stars have overall densities of 3.7×10^{17} to 5.9×10^{17} kg/m^3 which is comparable to the density of an atomic nucleus of $\approx 3 \times 10^{17}$ kg/m^3. In case of the neutron stars also it has been found that the mass of a star is inversely proportional to its volume. That is, a neutron star having large mass is small in size.

Some neutron stars rotate very rapidly and emit beams of electromagnetic radiation as the pulses. Non-rotating and non-accreting neutron stars are virtually undetectable.

Pulsars

Radio astronomers have discovered first the celestial objects, called the pulsars, which correspond to the neutron stars. In 1967, scientists discovered pulsing radio sources which are named as the pulsars. Further investigations revealed that the pulsars are magnetized. These pulsars are the spinning neutron stars. A new born neutron star can rotate several times a second.

Baade and Zwicky hypothesis

Baade and Zwicky, in 1934, coined the term supernova, hypothesized the existence of neutron stars, and knit them together with the origin of cosmic rays. Supernovae are the source of many of the elements of nature, and their blasts produce the interstellar medium in ways that plays important role in the star formation. They are the origin of most cosmic-rays, and these energetic rays have pronounced effects in the galaxy. Bright supernovae can be seen across the universe, and a majority of them give birth to impressively dense neutron stars and black holes. Indeed, the radio and x-ray pulsars are rapidly spinning neutron stars injected into the galaxy upon the eruption of a supernova.

4.3 Black Holes

A black hole is defined as a region of space-time having so high gravitational field which does not allow the escaping of any thing including light. The theory of general relativity says that a sufficiently compact mass deforms the space-time to form a black hole. Around a black hole, there is a mathematically defined surface, called the **event horizon** that marks the **point of no return**. That is, any thing which has gone inside the event horizon never comes back outside. The hole is given the name *black* as it absorbs every thing, including the light, that incident the event horizon. It is similar to a black body in the thermodynamics which absorbs all kinds of electromagnetic radiations falling on it.

At the end of the life cycle of a very massive star, a black hole of stellar mass is expected to form. After the formation of a black hole, it continues to grow by absorbing the mass from its surroundings. By absorbing other stars and merging with other black holes, suppermassive black holes of millions of solar masses may form. There is general consensus that the supermassive black holes exist in the centres of most galaxies.

Despite its invisible interior, the presence of a black hole can be inferred through its interaction with other matter and with the electromagnetic radiation, such as the light. Matter falling into a black hole can form an accretion disk heated by friction, forming some of the bright objects in the universe. If there are other stars orbiting a black hole, their orbit can be used to determine its mass and location. In this way, the astronomers have identified numerous stellar black hole candidates in the binary systems, and established that the core of the Milky Way contains a supermassive black hole of about 4.3 million $M_\odot$.

In case of a neutron star, the escape speed from its surface is typically a good fraction of the speed of light. Now, in case of a black hole, the radius R of an object of mass M to have the escape speed from the surface equal to the speed of light has been obtained. This radius, known as the **Schwarzschild radius**, is

$$R_{\text{sch}} = \frac{2GM}{c^2}$$

Numerical value of Schwarzschild radius for a 1 $M_\odot$ is $R_{sch} = 3$ km.

Example 2.

For a star having 1 solar mass, calculate the Schwarzschild radius. Given, solar mass $M_\odot = 1.99 \times 10^{30}$ kg, gravitational constant $G = 6.673 \times 10^{-11}$ m^3/s^2 kg and speed of light $c = 3 \times 10^8$ m/s.

Solution:

The Schwarzschild radius is

$$R_{\text{sch}} = \frac{2GM}{c^2}$$

$$= \frac{2(6.673 \times 10^{-11})(1.99 \times 10^{30})}{(3 \times 10^8)^2}$$

$$= 2.9 \times 10^3 \text{ m}$$

Example 3.

Estimate the smallest possible Schwarzschild radius for a black hole of $3M_{\odot}$ that forms as the radius of a supernovae. The mass of the Sun is 2×10^{30} kg.

Solution:

The Schwarzschild radius of a black hole is proportional to the mass of collapsing core. The smallest possible Schwarzschild radius is therefore expressed by the smallest possible mass that is capable of overcoming the neutron degeneracy pressure, thought to be about 3 solar mass. Thus, we have

$$R_{\text{sch}} = \frac{2GM}{c^2}$$

$$= \frac{2(6.67 \times 10^{-11})(6 \times 10^{30})}{(3 \times 10^8)^2}$$

$$\approx 9000 \text{ m}$$

A Schwarzschild radius of 9 km is the smallest possible.

Example 4.

In the model for the core of an active galaxy an accretion disk forms around the super giant black hole. This is composed of former stellar material ripped apart by the tidal forces of the black hole and has a radius of about 10^{14} m. Calculate the upper limit for the power limited by the system when about one solar mass moves from the accretion disk the black hole (of 10^8 solar masses) per year.

Solution:

The Schwarzschild radius of the black hole is

$$R_{sch} = \frac{2GM}{c^2}$$

$$= \frac{2(6.67 \times 10^{-11})(10^8 \times 2 \times 10^{30})}{(3 \times 10^8)^2}$$

$$= 3 \times 10^{11} \text{ m}$$

The energy released after the matter has entered within Schwarzschild radius is not observable and so, in the calculations the release of the gravitational potential energy, this represents the 'end point'. The rate at which the matter enters inside the Schwarzschild radius is 2×10^{30} kg/year or $(2 \times 10^{30}/(365 \times 24 \times 60 \times 60) =)$ 6×10^{22} kg/s. The rate of gravitational energy released is given by the gravitational energy of matter while in the accretion disk (radius r_a) minus that at the Schwarzschild radius is

rate of energy released, $$E = \frac{GMm}{R_{\text{sch}}} - \frac{GMm}{r_a}$$

$$= GMm\left(\frac{1}{R_{\text{sch}}} - \frac{1}{r_a}\right)$$

$$= (6.67 \times 10^{-11})(10^8 \times 2 \times 10^{30})(6 \times 10^{22})\left(\frac{1}{3\times10^{11}} - \frac{1}{10^{14}}\right)$$

$$= 3 \times 10^{39} \text{ J/s}$$

This is equal to the amount of energy released per second, in other words, the active galaxy has a potential output of more than 10^{39} W.

Example 5.

Compare the densities of a white dwarf, a neutron star and a main sequence star where all three bodies are of about 1.4 solar mass.

Solution:

A white dwarf of 1.4 solar mass corresponds to the most massive white dwarf that can withstand the electron degeneracy pressure and therefore has the smallest possible radius for a white dwarf. This is similar to the Earth's size (radius ~ 10 m). The density is

$$\text{density} = \frac{\text{mass}}{\text{volume}}$$

$$= \frac{3\times10^{30}}{4\pi(10^7)^3/3} \approx 10^9 \text{ kg/m}^3$$

The largest neutron star (those of minimum mass 1.4 solar mass) have a radius of 20 km. The density is

$$\text{density} = \frac{\text{mass}}{\text{volume}}$$

$$= \frac{3\times10^{30}}{4\pi(2\times10^4)^3/3} \approx 10^{17} \text{ kg/m}^3$$

A main sequence star of 1.4 solar mass has a similar density to the Sun (which shows that slightly more massive main sequence stars than the Sun have slightly larger radii). The Sun density is

$$\text{density} = \frac{\text{mass}}{\text{volume}}$$

$$= \frac{2\times10^{30}}{4\pi(7\times10^8)^3/3} \approx 10^3 \text{ kg/m}^3$$

The densities of three bodies of the same mass are thus radically different.

4.4 Accretion Process

In astrophysics, accretion is a process which accumulates particles, typically gaseous matter in an accretion disk, into a massive object, where gravitational attraction between particles is responsible. Accretion process is considered responsible for formation of most astronomical objects, such as galaxies, stars, and planets.

4.5 Suppermasive Blackholes

When the mass of a black hole is of the order of millions to billions of times the solar mass, it is known as a supermassive black hole. There are observational evidences that almost each large galaxy has a supermassive black hole at its centre. For example, our galaxy Milky Way has a supermassive black hole in its Galactic Centre. Accretion of interstellar gas onto supermassive black holes is considered responsible for the formation of a suppermasive blackhole.

5. Star Clusters

A group having large number of stars is called a star cluster. Two distinct categories of star clusters are well known:

(i) Open clusters (ii) Globular clusters

5.1 Open Clusters

A group having up to a few thousand stars belongs to the category of open cluster. The stars in an open cluster are considered formed from the same giant molecular cloud and have roughly the same age. In the Milky Way, more than 1,100 open clusters have been discovered, and there may be many more. Because of large separations between the stars, the open clusters are loosely bound by mutual gravitational attraction. As per estimates, the open clusters are young (usually a few tens of millions or hundreds of millions of years old). Open clusters are generally found in the disks of spiral galaxies, like the Milky Way.

5.2 Globular Clusters

A group having nearly 10 million stars bound tightly together belongs to the category of globular clusters. A globular cluster is generally spherical in shape with large diameter up to 300 light years. The stars in a globular cluster are very tightly bound by gravity, with a high concentration of stars towards their centres. As per estimates, the globular clusters are very old (typically about 12-13 billion years old). Globular clusters are found in nearly all galaxies. In spiral galaxies like the Milky Way, they are mostly found in the galactic halo. The Milky Way has over 150 known globulars, and there may be many more.

6. Clusters of Galaxies

6.1 Galactic Clusters

A galactic cluster is a structure of galaxies from hundreds to thousands that are bound together by gravity with typical masses ranging from 10^{14}-10^{15} solar masses. They are large known gravitationally bound structures in the universe and were believed to be the largest known structures in the universe until the discovery of superclusters in 1980s.

6.2 Supercluster

Superclusters are among the largest known structures of the universe. A supercluster is a large group of smaller galaxy clusters or galaxy groups. The Milky Way is a part of the galaxy group, called Local Group having more than 54 galaxies. This Local Group is a part of the Virgo Supercluster, and this Supercluster is a part of the Laniakea Supercluster. Because of large size and low density of superclusters, they, unlike clusters, expand with the Hubble expansion. As per estimation, the number of superclusters in the observable universe is 10 million.

7. Jeans Criterion

In physics, the Jeans instability causes the collapse of an interstellar gas cloud and subsequent formation of star. It occurs when the internal gas pressure is not strong enough to prevent the gravitational collapse of a region filled with the matter. For the stability, the cloud must be in the hydrostatic equilibrium, which in case of a spherical cloud is described through the relation

$$\frac{dP}{dr} = \frac{G\rho(r)M(r)}{r^2}$$

where $M(r)$ is the mass enclosed in the sphere of radius r, P the pressure, $\rho(r)$ the spherically symmetric density of gas at a distance r from the centre of star, G is the gravitational constant and r the radius. The equilibrium is stable if the small perturbations are damped and is unstable if the perturbations are amplified. In general, the cloud is unstable if it is either very massive at a given temperature or very cool at a given mass for the gravity to overcome the gas pressure.

7.1 Jeans Mass

Jeans considered the process of gravitational collapse within a gaseous cloud, and he was able to show that, under the appropriate conditions, a cloud, or a part of cloud, would become unstable and begin to collapse when it does not have sufficient gaseous pressure support to balance the force of gravity. The cloud is stable for sufficiently small mass (at a given temperature and radius), but once this critical mass is exceeded, it begins a process of runaway contraction until some other force can oppose the collapse. He derived a formula for the calculation of this critical mass as a function of its density and temperature. The greater the mass of cloud, the smaller its size, and the colder its temperature, the less stable it is against the gravitational collapse.

The approximate value of Jeans mass may be derived through a simple physical argument. One begins with a spherical gaseous region of radius R, mass M, and with a gaseous sound speed c_s. Imagine that we compress the region slightly. It takes a time,

$$t_{\text{sound}} = \frac{R}{c_s}$$

$$\simeq (5\times10^5\ \text{yr})\left(\frac{R}{0.1\ \text{pc}}\right)\left(\frac{c_s}{0.2\ \text{km/s}}\right)^{-1}$$

for the sound waves to cross the region, and attempt to push back and re-establish the system in the pressure balance. At the same time, the gravity attempts to contract the system even further, and does so on a free-fall time,

$$t_{ff} = \frac{1}{\sqrt{G\rho}}$$

$$\simeq (2\text{Myr})\left(\frac{n}{10^9\,\text{m}^{-3}}\right)^{-1/2}$$

where G is the universal gravitational constant, ρ is gas density within the region, and $n = \rho/\mu$ is the gas-number density for the mean mass per particle $\mu = 3.9 \times 10^{-27}$ kg, appropriate for the molecular hydrogen with 20% helium by number. Now, two situations may arise:

(i) When the sound-crossing time is less than the free-fall time, pressure forces temporarily overcome the gravity, and the system returns to a stable equilibrium.

(ii) When the free-fall time is less than the sound-crossing time, the gravity overcomes the pressure forces, and the region undergoes the gravitational collapse. The condition for the gravitational collapse is therefore:

$$t_{ff} < t_{\text{sound}}.$$

The resultant Jeans length λ_J is approximately:

$$\lambda_J = \frac{c_s}{\sqrt{G\rho}} \simeq (0.4\text{pc})\left(\frac{c_s}{0.2\text{ km/s}}\right)\left(\frac{n}{10^9\,\text{m}^{-3}}\right)^{-1/2}$$

This length scale is known as the Jeans length. All the scales larger than the Jeans length are unstable to the gravitational collapse, whereas the smaller scales are stable. The Jeans mass M_J is just the mass contained in a sphere of radius R_J (R_J is half the Jeans length):

$$M_J = \left(\frac{4\pi}{3}\right)\rho R_J^3$$

$$= \left(\frac{\pi}{6}\right)\frac{c_s^3}{G^{3/2}\rho^{1/2}}$$

$$\simeq (2M_\odot)\left(\frac{c_s}{0.2\text{ km/s}}\right)^3\left(\frac{n}{10^9\,\text{m}^{-3}}\right)^{-1/2}$$

Later on, it was pointed out by other astrophysicists that in fact, the original analysis used by the Jeans was flawed, for the following reason. In his formal analysis, Jeans assumed that the collapsing region of the cloud was surrounded by an infinite, static medium. In fact, because all the scales greater than the Jeans length are also unstable to collapse, any initially static medium surrounding a collapsing region in fact also is collapsing. Therefore, the growth rate of the gravitational instability relative to the density of the collapsing background is slower than that predicted by the Jeans' original analysis. This flaw has come to be known as the Jeans swindle.

7.2 Jeans Length

The Jeans length is the critical radius of a cloud (typically a cloud of the interstellar dust) where thermal energy, which causes the cloud to expand, is counteracted by the gravity, which causes the cloud to collapse. The formula for the Jeans length is:

$$\lambda_J = \sqrt{\frac{15kT}{4\pi G\mu\rho}},$$

where k is Boltzmann constant, T is the temperature of cloud, r is the radius of the cloud, μ is mass per particle in the cloud, G the gravitational constant and ρ the cloud's mass density (i.e., the cloud's mass divided by the cloud's volume).

Perhaps the easiest way to conceptualize the Jeans length is in terms of a close approximation, where the factors 15 and 4π are discarded and ρ is rephrased as M/r^3. Then, the Jeans length is expressed as

$$\lambda_J \approx \sqrt{\frac{kTr^3}{GM\mu}}.$$

It follows immediately that when $kT = GM\mu/r$, we have $\lambda_J = r$, i.e., the cloud's radius is Jeans length when thermal energy per particle equals the gravitational work per particle. At this critical length the cloud neither expands nor contracts. It is only when the thermal energy is not equal to the gravitational work that the cloud either expands and cools or contracts and warms, a process that continues until equilibrium is attained.

8. Transport of Energy Inside a Star

Prime source of energy in the stars is thermonuclear reactions which produce the energy in their deep interior as there are the pressures and the temperatures high enough to sustain thermonuclear reactions. However, most of the luminous energy of a star is radiated from a thin region at its surface, called the photosphere. Hence, an important issue in this context is the mode of the transport of energy from the deep interior to the photosphere. There are 4 important processes discussed by the scientists for the transport of such energy:

(i) Radiation of photons
(ii) Conduction
(iii) Convection
(iv) Radiation of neutrinos

8.1 Radiation of Photons

The most common method of energy transport in the normal stars is through the photons of various wavelengths, ranging from the γ-rays to the radio waves, depending on the kinetic temperature in the region. For the propagation of radiations, no medium is essentially required. However, a medium is a kind of obstacle and it reduces the speed of radiation. The stellar material in the deep interior is very opaque, and therefore a radiation travels only a small distance before it is absorbed. It is then reemitted in a random direction, absorbed after a small distance, reemitted, and so on until it reaches the photosphere. For example, in the case of the Sun, the average distance travelled by a photon between two consecutive absorptions is about a centimetre, and it takes perhaps hundreds of thousands of years for the energy released in the deep interior to make its way to the surface.

8.2 Conduction

Conduction is the way in which metals transport heat from one place to another. For the conduction process, a medium is required which should be a good conductor of heat. Conduction is not important in most normal stars because the atmosphere inside a star is approximately an ideal gas, which is a thermal insulator.

8.3 Convection

If the rate of the production of energy in a star is much larger than the rate of transport through the radiation, the stellar matter may boil. In the boiled material, the transport of energy is through convection. The convection is a familiar process from the boiling water. The convection is a very efficient method of energy transport because it involves the vertical motion of a large packets of gas.

Remark

In most normal stars the energy transport is by radiation unless the rate at which energy is being produced in the interior exceeds a critical value, in which case the transport becomes convective. In many stars the both may operate: some regions of the interior may transport heat by the convection and some by the radiative transport.

8.4 Radiation of Neutrinos

In a massive star, late in its life the amount of energy that must be transported is sometimes larger than either radiation of photons or convection. In these cases, significant amounts of energy may be transported from the centre to the space by the radiation of neutrinos. This is the dominant method of cooling of stars in advanced burning stages, and also plays a central role in the events like supernovae associated with the death of massive stars.

9. Simple Stellar Model

Here, we have presented a simple stellar model where the radial variation of density is expressed as

$$\rho(r) = \rho_c\left(1 - \frac{r}{R}\right)$$

where ρ_c is the density at the centre of star and R is the radius of star. For this spherically symmetric density profile, let us calculate various parameters of star.

9.1 Mass

For spherically symmetric density $\rho(r)$, the mass of a star within a radius r is expressed by the equation of mass continuity.

$$M(r) = \int_0^r \rho(r) 4\pi r^2 \mathrm{d}r$$

Using the expression for $\rho(r)$, we have

$$\begin{aligned} M(r) &= \int_0^r \rho_c\left(1 - \frac{r}{R}\right) 4\pi r^2 \mathrm{d}r \\ &= \frac{4\pi r^3}{3}\rho_c\left(1 - \frac{3r}{4R}\right) \end{aligned}$$

The total mass of star is

$$\begin{aligned} M &= M(R) \\ &= \frac{\pi R^3}{3}\rho_c \end{aligned}$$

Thus, the central density in terms of total mass is

$$\rho_c = \frac{3M}{\pi R^3}$$

9.2 Pressure

For the hydrostatic equilibrium, the pressure gradient is expressed as

$$\frac{\mathrm{d}P}{\mathrm{d}r} = -\frac{GM(r)\rho(r)}{r^2}$$

Using the expressions for $M(r)$ and $\rho(r)$, we get

$$\frac{\mathrm{d}P}{\mathrm{d}r} = \pi G\rho_c^2\left(-\frac{4r}{3}+\frac{r^2}{R}+\frac{4r^2}{3R}-\frac{r^3}{R^2}\right)$$

On integration, we get

$$P(r) - P_c = \int_0^r \pi G\rho_c^2\left(-\frac{4r}{3}+\frac{r^2}{R}+\frac{4r^2}{3R}-\frac{r^3}{R^2}\right)\mathrm{d}r$$

$$P(r) = P_c - \pi G\rho_c^2\left(\frac{2r^2}{3}-\frac{7r^2}{9R}+\frac{r^4}{4R^2}\right)$$

The central pressure, can be obtained by considering the fact that the pressure at the surface must be zero. That is,

$$P(R) = P_c - \pi G\rho_c^2\left(\frac{2R^2}{3}-\frac{7R^3}{9R}+\frac{R^4}{4R^2}\right) = 0$$

This relation gives

$$P_c = \pi G\rho_c^2\left(\frac{5R^2}{36}\right)$$

9.3 Temperature

The temperature $T(r)$ at distance r from the centre of star, can be calculated in terms of the central density ρ_c and mean molecular weight by assuming that the star obeys the ideal gas relation. The ideal gas law says in terms of the mean molecular weight μ as

$$P(r) = \frac{\rho(r)kT(r)}{\mu m_H}$$

or

$$T(r) = \frac{P(r)\mu m_H}{\rho(r)k}$$

where $P(r)$ is the pressure at a distance r from the centre of the star, μ the mean molecular weight, m_H the mass of hydrogen atom and k the Boltzmann constant. On using the expressions for $\rho(r)$ and $P(r)$, we get

$$T = \frac{\mu m_H\left[P_c - \pi G\rho_c^2\left(\frac{2r^2}{3}-\frac{7r^3}{9R}+\frac{r^4}{4R^2}\right)\right]}{k\rho_c(1-r/R)}$$

So the temperature at the centre ($r = 0$) is

$$T_c = T(0) = \frac{\mu m_H P_c}{k\rho_c}$$

Example 6.

A gas contains, by number, 80% molecular hydrogen and 20% helium. Calculate the mean mass per particle.

Solution:

In amu, the mass of hydrogen molecule is 2 and that of the helium is 4. Thus,

$$\text{Average atomic mass} = \frac{80 \times 2 + 20 \times 4}{100} = 2.4$$

$$\text{Mean mass per particle} = 2.4 \times (1.66 \times 10^{-27})$$

$$= 3.984 \times 10^{-27} \text{ kg}$$

9.4 Polytropic Stellar Model

In astrophysics, a polytrope refers to a solution of Lane-Emden equation in which the pressure P depends upon the density ρ in the form

$$P = K\rho^{(n+1)/n}$$

where K is a constant and the constant n is known as the polytropic index. This relation need not be interpreted as an equation of state. This is simply a relation that expresses an assumption about the change of pressure with radius in terms of the change of density with radius. For this pressure, the Lane-Emden equation is solved.

10. Binary Stars

A binary system is a pair of stars which are bound to each other under the mutual gravitational attraction. In some binary systems, both the stars may be of comparable masses whereas in others, the mass of one star may be very small in comparison to that of the other. The heavier component is known as the primary, whereas the lighter one as the secondary. Both the components of binary system rotate around their centre-of-mass in the Keplerian orbits. The primary makes a small ellipse with focus on the centre-of-mass whereas the secondary makes a large ellipse with focus on the same point (centre-of-mass).

Both the ellipses lie in a common plane known as the orbital-plane. This orbital-plane is generally inclined to the sky-plane[3].When the orbital-plane is projected on the sky-plane, the elliptic orbit makes another ellipse, but the focus of the original ellipse (i.e., centre-of-mass) may not be at the focus of the projected ellipse.

For a binary system, it is possible to get information about the masses and sizes of component stars. This information about the stars is required as input for the theories of origin and evolution of stars. Some binaries having a compact object, like a white dwarf, a neutron star or a black hole as one component may help for the verification of General Theory of

[3]A plane whose normal is the line-of-sight is known as the sky-plane.

Relativity. Depending on the method used for its study, the binary may be called a visual binary, a spectroscopic binary or an eclipsing binary. In some cases, a binary may be visual and spectroscopic or eclipsing; it may be spectroscopic and eclipsing.

The centre-of-mass of a binary may not be stationary. Since there is no net external force acting on the system, the centre-of-mass may move along a straight line and the components of the system trace out the serpentine paths. Period of revolution of a binary range from 5-10 years to thousands of years. Hence, it may not be possible for an astronomer to observe a complete orbit of a binary system.

10.1 Visual Binary

When a binary system can be seen (resolved) unaided or with the help of a telescope, the system is known as visual binary. Mizar (ς UMa) has been the first visual binary discovered in 1650.

10.2 Astrometric Binary

In a visual binary, it is possible that one of its components is not visible and only the brighter one is visible. The brighter component may be the primary or the secondary of system. The presence of the faint component can however be inferred from the wiggly motion of visible component with respect to the background stars. Such a system is known as the astrometric binary.

Sirius (Fig. 7) is a famous visual binary in which Sirius A is a main sequence star of the spectral type $A1$ and Sirius B is a white dwarf of the spectral type $A5$ (which can be detected with the efforts), orbiting about one another with a period of 49.9 years.

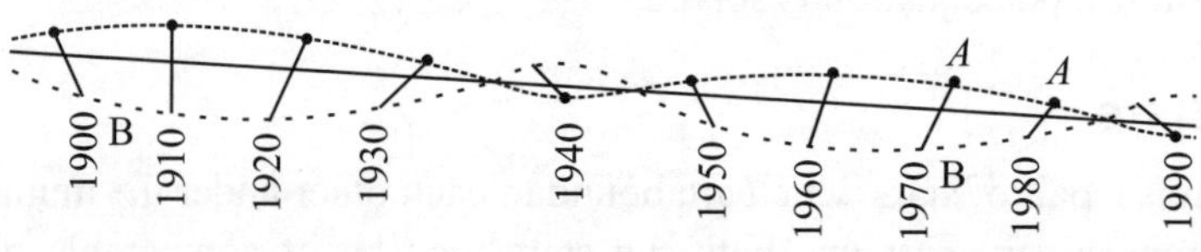

Fig. 7. Schematic diagram for the centre-of-mass of Sirius (a binary system).

Even with the help of a good telescope, it is not easy to see the white dwarf in the glare of much brighter companion. The faint companion of Sirius was discovered later on.

10.3 Spectroscopic Inary

When the two components of a binary system cannot be resolved or seen even with the help of a telescope, the system may be studied through the oscillations of their spectral lines due to the Doppler effect. Such a study is only possible when the angle between the orbital-plane and the sky-plane is reasonably large to have an appreciable component of the velocity along the line-of-sight.

Spectroscopic binaries are usually too distant and their components cannot be resolved even with the help of a telescope. The first star of this type was ς UMa A, discovered in 1889, the brighter component of the first observed visual binary.

In order to understand the situation, let us consider a simple view (Fig. 8) where more the bright star A and the less bright star B of a binary system are moving in the circular orbits around the centre-of-mass, *C*. Let the angle between the plane of the orbits and the sky-plane is 90°, i.e., the line-of-sight is in the orbital-plane. (This is the best situation from the point of view of observation of a spectroscopic binary.) Let us consider four positions of the stars *A* and *B* denoted by 1, 2, 3 and 4. The directions of motions of stars are shown by the arrows. Here, the primary star *A* is assumed to be heavier than the secondary star *B*.

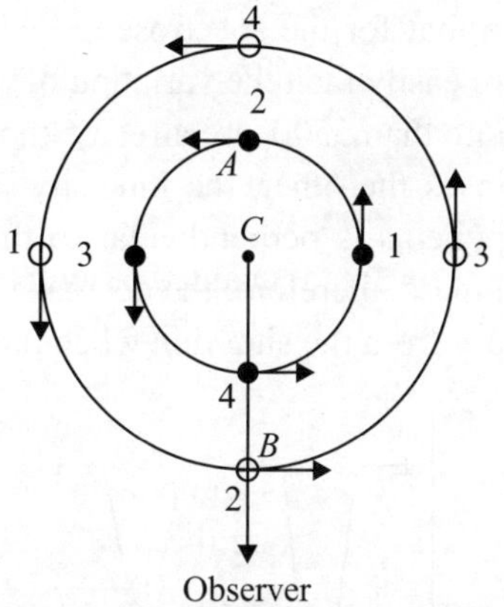

Fig. 8. A simple view of a binary system where the primary star *A* and the secondary star *B* revolve in the circular orbits around the centre-of-mass, *C*. The line-of-sight of the observer is in the orbital-plane.

When the stars are in the position 2 as well as in the position 4, they are moving across the line-of-sight and the spectral lines from the two stars overlap, as there is no Doppler shift, and we see only one set of the spectral lines. When the stars are in the position 1, the star *A* moves away from the observer whereas the star *B* moves towards the observer.

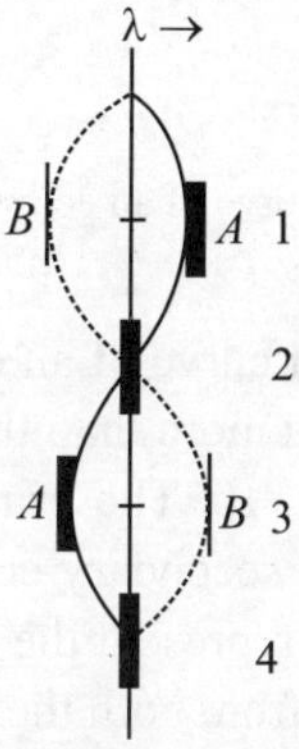

Fig. 9. Shift of a spectral line from the rest wavelength λ_0 is shown. Here, 1, 2, 3 and 4 correspond to the positions of the stars *A* and *B*, given in Fig. 8.

Owing to the Doppler effect, the star *A* gives the redshifted set of the spectral lines whereas the *B* gives the blueshifted set of the spectral lines. Thus, we see the two sets of the spectral lines of different intensities, corresponding to the brightness of the stars. When the stars are in the position 3, the star *A* moves towards the observer whereas the star *B* moves away from the observer. Now, the star *A* gives the blueshifted set of the spectral lines whereas the *B* gives the redshifted set of the spectral lines. Thus, we see the two sets of the spectral lines of reverse intensities. Therefore, when the stars revolve around the centre-of-mass, each spectral line oscillates, as shown in Fig. 9.

When one of the components of the binary is too faint for its spectral lines, we see only one set of the oscillating spectral lines. Such binaries are called the single lined spectroscopic binaries, in contrast to the double lined binaries.

10.4 Eclipsing Binary

It is the case where the orbital-plane and the sky-plane are inclined in such a manner that the components of the binary system eclipse with each other. Algol was the first eclipsing binary discovered in 1782. The first x-ray eclipsing binary, Hercules-X1, was discovered in 1971. This situation arises when the inclination angle is very large (for the ideal situation, it is 90°). This condition is more stringent than that for the spectroscopic binaries. But, the variation of the brightness can be measured more easily than the variation of radial velocity. Therefore, a large number of eclipsing binaries, more than 5500, are already known.

When one component eclipses the other, the intensity of a line becomes less than that when the line joining the components is perpendicular to the line-of-sight. Further, the two components have different brightness, therefore, the decrease in the intensity is different when the primary eclipses the secondary from the situation when the secondary eclipses the primary.

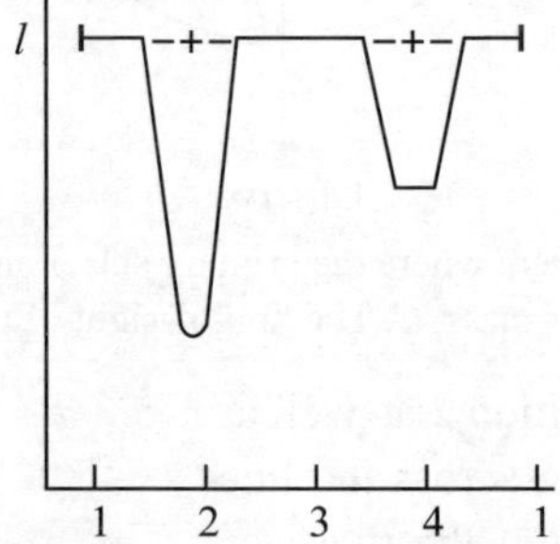

Fig. 10. Light curve of an eclipsing binary. The primary is eclipsed by the secondary at the position 2 whereas the secondary is eclipsed by the primary at position 4.

The light curve of an eclipsing binary is shown in Fig. 10. Assuming that the intensity of the primary is more than that of the secondary, schematic light-curve of an eclipsing binary is shown in Fig. 10. The primary eclipse occurs at position 2 (we get light from the secondary only) and the secondary eclipse occurs at the position 4 (we get light from the primary only). These points represent the conjunction. The positions 1 and 3 are called the quadratures when the full light from both the primary and secondary is obtained.

10.5 Determination of Masses of Components

Consider a binary system having two stars *A* and *B* of masses m_1 and m_2, respectively, as shown in Fig. 11. The two stars revolve about the centre-of-mass, *C*. Let the distances of the stars *A* and *B* from the point *C* be d_1 and d_2, respectively.

If *T* is the time-period of revolution of each of the stars, the Kepler's third law of planetary motion gives

$$M = \frac{4\pi^2 r^3}{GT^2} \tag{9.1}$$

where $M = (m_1 + m_2)$ is the total mass of the system and $r = (d_1 + d_2)$ the separation between the components *A* and *B*.

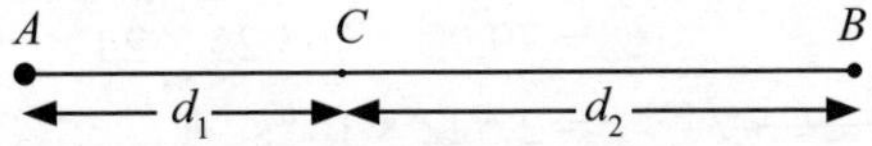

Fig. 11. Binary system having two stars *A* and *B*.

We have

$$m_1 d_1 = m_2 d_2$$

or
$$m_1 = \frac{m_2 d_2}{d_1} \tag{9.2}$$

Using equation (9.2) in (9.1), we get

$$\frac{m_2 d_2}{d_1} + m_2 = \frac{4\pi^2 (d_1 + d_2)^2}{GT^2}$$

or
$$m_2 = \frac{4\pi^2 d_1 (d_1 + d_2)^2}{GT^2} \tag{9.3}$$

Using equation (9.3) in (9.2), we get

$$m_1 = \frac{4\pi^2 d_2 (d_1 + d_2)^2}{GT^2} \tag{9.4}$$

Thus, for the known values of d_1, d_2 and *T*, the masses m_1 and m_2 of individual stars in a binary system can be determined with the help of equations (9.4) and (9.3), respectively.

Note:

Here, we have assumed that the orbits of the stars are circular. Generally, they are elliptical. For the elliptical orbits, the analysis is similar, but becomes complicated.

Binary system with limited information

For the determination of masses of the stars in a binary system, we often find that the information available is insufficient to apply the method discussed above. Such a situation arises because of the following.

- The orbits of the stars of the binary system may not be mapped exactly on the celestial sphere. It is true when the binary is astrometric where we can see only one star of the binary system and the second star is realized due to the wavy motion of the first star.
- Even when the orbits of both stars in a binary system can be mapped, we have two-dimensional projections on the celestial sphere of true three-dimensional orbits. Hence, more information is required for construction of true orbits.

Under such cases, we can determine only the sum of the masses of the stars or we may put limits on the masses rather than to determine the individual masses.

Example 7.

The α-Centauri system is 1.338 pc distant with a period of 79.92 years. The *A* and *B* components have a mean separation of 23.7 AU. Assuming that the orbits are circular, calculate the total mass of the system.

Solution:

We have

$$T = 79.92 \text{ yr}$$

$$= 79.92 \times 365.25 \times 24 \times 60 \times 60$$
$$= 2.522 \times 10^9 \text{ s}$$
$$r = (d_1 + d_2)$$
$$= 23.7 \text{ AU} = 23.7 \times (1.5 \times 10^{11})$$
$$= 3.555 \times 10^{12} \text{ m}$$

Total mass

$$M = \frac{4\pi^2 r^3}{GT^2}$$
$$= \frac{4 \times 3.14^2 \times (3.555 \times 10^{12})^3}{6.67 \times 10^{-11} \times (2.522 \times 10^9)^2}$$
$$= 4.18 \times 10^{30} \text{ kg}$$

Example 8.

The α-Centauri system is 1.338 pc distant with a period of 79.92 years. The *A* and *B* components have a mean separation of 23.7 AU. If the α-Cent A has mean distance of 11.2 AU from the system's barycenter, assuming that the orbits are circular, calculate the masses of the components.

Solution:

We have

$$d_1 + d_2 = 23.7 \text{ AU}$$
$$= 23.7 \times (1.5 \times 10^{11})$$
$$= 3.555 \times 10^{12} \text{ m}$$

and

$$d_1 = 11.2 \text{ AU}$$
$$= 11.2 \times (1.5 \times 10^{11})$$
$$= 1.68 \times 10^{12} \text{ m}$$

Therefore,

$$d_2 = 3.555 \times 10^{12} - 1.68 \times 10^{12}$$
$$= 1.875 \times 10^{12} \text{ m}$$

The time-period is

$$T = 79.92 \text{ yr}$$
$$= 79.92 \times 365.25 \times 24 \times 60 \times 60$$
$$= 2.522 \times 10^9 \text{ s}$$

The masses of the components are

$$m_1 = \frac{4\pi^2 d_2 (d_1 + d_2)^2}{GT^2}$$
$$= \frac{4 \times 3.14^2 \times (1.875 \times 10^{12}) \times (3.555 \times 10^{12})^2}{6.67 \times 10^{-11} \times (2.522 \times 10^9)^2}$$
$$= 2.197 \times 10^{30} \text{ kg}$$

and $$m_2 = \frac{4\pi^2 d_1(d_1+d_2)^2}{GT^2}$$

$$= \frac{4\times 3.14^2\times(1.68\times10^{12})\times(3.555\times10^{12})^2}{6.67\times10^{-11}\times(2.522\times10^9)^2}$$

$$= 1.969\times 10^{30}\text{ kg}$$

Example 9.

The approximate masses of *A* and *B* components of Sirius are about 2.1 and 1.0 solar masses, respectively. If the separation between Sirius *A* and Sirius *B* at a particular time is 20 AU, calculate the distances of the stars from the centre-of-mass of the system.

Solution:

Suppose, m_A and m_B are the masses, and d_A and d_B the distances from the centre-of-mass of *A* and *B* components of Sirius, respectively. We have $m_A = 2.1\ M_\odot$, $m_B = 1.0\ M_\odot$ and $d_A + d_B = 20$ AU. Therefore,

$$\frac{d_A}{d_B} = \frac{m_A}{m_B} = 2.1$$

or $$d_A = 2.1 d_B$$

Therefore, we have

$$2.1 d_B + d_B = 20$$

or $$d_B = \frac{20}{3.1} = 6.45\text{ AU}$$

Hence, we have

$$d_A = 20 - 6.45$$

$$= 13.55\text{ AU}$$

Example 10.

If the total mass of Sirius binary is 3.2 solar masses and the period for the rotation is 50.1 years, calculate the mean separation between the two stars. Given, mass of Sun = 1.99×10^{30} kg.

Solution:

We have total mass

$$M = 3.2\times(1.99\times10^{30})$$

$$= 6.368\times10^{30}\text{ kg}$$

Time-period of revolution

$$T = 50.1\text{ yr}$$

$$= 50.1\times365.25\times24\times60\times60$$

$$= 1.58\times10^9\text{ s}$$

The mean distance *d* between the components of the binary is

$$d = \left(\frac{MGT^2}{4\pi^2}\right)^{1/3}$$

$$= \left[\frac{6.368\times10^{30}(6.67\times10^{-11})(1.58\times10^{9})^2}{4\times3.14^2}\right]^{1/3}$$

$$= 2.996 \times 10^{12} \text{ m}$$

Example 11.

Observations of Sirius show it to be a binary star. The Sirius is visible with the naked eye and the motion of both stars can be tracked by a telescope. The separation of stars is (on average) 20 AU and the orbital period is 50 years.

Observations of Sirius A show it to have an orbital radius of 7 AU. Calculate the masses of Sirius *A* and *B*. Given, $G = 6.67 \times 10^{-11}$ m^3/s^2 kg.

Solution:

We have time-period

$$T = 50 \text{ yrs}$$
$$= 50 \times 365 \times 24 \times 60 \times 60$$
$$= 1.5768 \times 10^9 \text{ s}$$
$$r = 20 \text{ AU}$$
$$= 20 \times 1.5 \times 10^{11}$$
$$= 3 \times 10^{12} \text{ m}$$
$$r_1 = 7 \text{ AU}$$
$$= 7 \times 1.5 \times 10^{11}$$
$$= 1.05 \times 10^{12} \text{ m}$$
$$r_2 = 13 \text{ AU}$$
$$= 13 \times 1.5 \times 10^{11}$$
$$= 1.95 \times 10^{12} \text{ m}$$

The mass m_1 is

$$m_1 = \frac{4\pi^2 r^2 r_2}{GT^2}$$
$$= \frac{4(3.14)^2 \times (3\times10^{12})^2 \times (1.95\times10^{12})}{6.67\times10^{-11} \times (1.5768\times10^{9})^2}$$
$$= 4.2 \times 10^{30} \text{ kg}$$

The mass m_2 is

$$m_2 = \frac{4\pi^2 r^2 r_2}{GT^2}$$
$$= \frac{4(3.14)^2 \times (3\times10^{12})^2 \times (1.05\times10^{12})}{6.67\times10^{-11} \times (1.5768\times10^{9})^2}$$
$$= 2.2 \times 10^{30} \text{ kg}$$

11. Interstellar Medium

The matter between stars is known as the interstellar medium (ISM) and it has two components: (i) gas, and (ii) dust. It fills the interstellar space and blends smoothly into the surrounding intergalactic space. The energy that occupies the same volume, in the form of the electromagnetic radiation, is the interstellar radiation field. By mass, 99% of the ISM is gas in any form, and 1% is dust.

11.1 Interstellar Gas

The interstellar gas is in the form of atoms, ions and molecules. The interstellar medium is composed of multiple phases, distinguished by whether the matter is ionic, atomic, or molecular, and the temperature and density of the matter. The interstellar medium is composed primarily by hydrogen followed by helium with trace amounts of carbon, oxygen, and nitrogen comparatively to hydrogen.

In the cool, dense regions of the ISM, matter is primarily in molecular form, and reaches the number densities of 10^6 molecules/cm^3. In the hot, diffuse regions of the ISM, the matter is primarily ionized, and the density may be as low as 10^{-4} ions/cm^3. Compare this with a number density of roughly 10^{19} molecules/cm^3 for the air, and 10^{10} molecules/cm^3 for a laboratory high-vacuum chamber. Of the gas in the ISM, by number, 91% of atoms are hydrogen and 9% are helium, with 0.1% being atoms of elements heavier than hydrogen or helium, known as metals in the astronomical language. By mass this amounts to 70% hydrogen, 28% helium, and 1.5% heavier elements. The hydrogen and helium are primarily a result of the primordial nucleosynthesis, while the heavier elements in the ISM are mostly a result of the enrichment in the process of stellar evolution.

11.2 Interstellar Dust

The dust found in the space is known as the cosmic dust or the interstellar dust. Most part of it consists of small dust particles which are of the size from few molecules to 0.1 μm. A smaller fraction of dust consists of larger refractory minerals that condensed as the matter left the stars. It is called the star-dust. The dust density in the local ISM of the Local Bubble is approximately 10^{-6} dust grain/m^3 with each grain having a mass of approximately 10^{-17} kg.

Cosmic dust was once solely an annoyance to astronomers, as it obscures objects they wish to observe. When infrared astronomy began, those dust particles were observed to be significant and vital components of the astrophysical processes. Their analysis can reveal the information about phenomena like the formation of our solar system.

Review Questions

1. Discuss the process for the formation of a star, and problems encountered in it.
2. Describe the evolution of stars.
3. What is a white dwarf? Discuss about the mass-radius relationship in a white dwarf.
4. What is a neutron star? Discuss about the mass-radius relationship in a neutron star.
5. What is a black hole? Discuss about the Schwarzschild radius of a black hole.

6. Describe the Jeans criterion for the formation of stars.
7. Discuss about the transport of energy inside a star.
8. A star has spherically symmetric density

$$\rho(r) = \rho_c\left(1 - \frac{r}{R}\right)$$

where ρ_c is the density at the centre of star and R is the radius of star. For the star, calculate total mass, and pressure and temperature at the centre.

9. Discuss various types of binary systems. How are masses of components of a visual binary system calculated?
10. What are supernovae? Explain their classification.
11. Two components A and B of a binary system have masses m_1 and m_2, respectively. If the position vector of B relative to A is $\vec{r}$, show that the position- vectors of A and B, relative to the centre-of-mass, respectively, are

$$\vec{r}_1 = -\frac{m_2}{m_1 + m_2}\vec{r}$$

and

$$\vec{r}_2 = \frac{m_1}{m_1 + m_2}\vec{r}$$

12. Two components A and B of a binary system have masses m_1 and m_2, respectively. If the velocity of B relative to A is $\vec{v}$, show that the position-vectors of A and B, relative to that of the centre-of-mass, respectively, are

$$\vec{v}_1 = -\frac{m_2}{m_1 + m_2}\vec{v}$$

and

$$\vec{v}_2 = -\frac{m_1}{m_1 + m_2}\vec{v}$$

13. Comment on the following:
    (i) When the angle between the orbital-plane of a binary system and the sky-plane is zero, the system cannot be studied if it is not a visual binary.
    (ii) For spectroscopic and eclipsing binary systems, the angle between the plane of orbit of a binary system and the sky-plane should be as large as possible.
    (iii) When a visual binary cannot be a spectroscopic and an eclipsing binary.
14. Write notes on the following:
    (i) Protostars
    (ii) Hayashi lines
    (iii) Brown dwarfs
    (iv) Main sequence stars
    (v) Red-giant-branch phase
    (vi) Asymptotic-giant-branch phase
    (vii) Post-AGB stars
    (viii) Supernova
    (ix) White dwarf
    (x) Chandrasekhar limit

(xi) Neutron stars
(xii) Pulsars
(xiii) Black holes
(xiv) Jeans criterion
(xv) Conduction
(xvi) Convection
(xvii) Polytropic stellar model
(xviii) Binary systems
(xix) Visual binary
(xx) Astrometric binary
(xxi) Spectroscopic binary
(xxii) Eclipsing binary
(xxiii) Interstellar medium
(xxiv) Interstellar dust
(xxv) Jeans criterion
(xxvi) Jeans mass
(xxvii) Jeans length
(xxviii) Interstellar medium

Multiple-Choice Questions

1. Which is not a category of binary system?
   A. visual B. gravitational C. spectroscopic D. eclipsing **Ans. B**
2. In Fig 9.6 the limit is
   A. 0.8 $M_\odot$ B. 1.4 $M_\odot$ C. 2.2 $M_\odot$ D. 3.1 $M_\odot$ **Ans. B**
3. In the spectrum of a supernova, Balmer lines of hydrogen atom are present. The type of supernova is
   A. I B. II C. I and II both D. neither I nor II
   Ans. B
4. In the spectrum of a supernova, Balmer lines of hydrogen atom are absent. The type of supernova is
   A. I B. II C. I and II both D. neither I nor II
   Ans. A
5. In the spectrum of a supernova, Balmer lines of hydrogen atom are absent, strong ionized silicon absorption line is present. The type of supernova is
   A. Ia B. Ib or Ic C. all the Ia, Ib, Ic D. none of Ia, Ib, Ic
   Ans. A
6. In the spectrum of a supernova, Balmer lines of hydrogen atom as well as the absorption line of ionized silicon both are absent. The type of supernova is
   A. Ia B. Ib or Ic C. all the Ia, Ib, Ic D. none of Ia, Ib, Ic
   Ans. B

7. In the spectrum of a supernova, Balmer lines of hydrogen atom and absorption line of ionized silicon both are absent, the strong lines of neutral helium are present. The type of supernova is
   A. Ib B. Ic C. both Ib and Ic D. neither Ib nor Ic
   Ans. A
8. In the spectrum of a supernova, Balmer lines of hydrogen atom, absorption line of ionized silicon, strong lines of neutral helium all are absent. The type of supernova is
   A. Ib B. Ic C. both Ib and Ic D. neither Ib nor Ic
   Ans. B
9. Mass of a white dwarf is
   A. proportional to its volume B. inversely proportional to its volume
   C. proportional to its radius D. inversely proportional to its radius **Ans. B**
10. Mass of a neutron star is
   A. proportional to its volume B. inversely proportional to its volume
   C. proportional to its radius D. inversely proportional to its radius **Ans. B**
11. Mass of a neutron star can be
   A. 0.8 $M_\odot$ B. 1.2 $M_\odot$ C. 2.2 $M_\odot$ D. 10.4 $M_\odot$ **Ans. C**
12. For the study of which type of binary, the Doppler effect is used.
   A. Visual B. Astrometric C. Spectroscopic D. Eclipsing **Ans. C**
13. White dwarfs are generally composed of
   A. Hydrogen and Helium B. Carbon and Oxygen
   C. Magnesium and Calcium D. Iron and Nickel **Ans. B**
14. In which type of binary, both the components can be seen?
   A. Eclipsing B. Spectroscopic C. Astrometric D. Visual **Ans. D**
15. In which type of binary, only one component is seen.
   A. Eclipsing B. Spectroscopic C. Astrometric D. Visual **Ans. C**
16. The best situation for the study of a binary system with the help of the Doppler effect is when the angle between the orbital plane and the sky-plane is
   A. 0° B. 45° C. 90° D. 180° **Ans. C**
17. The best situation for the study of a binary system with the help of the eclipsing method is when the angle between the orbital plane and the sky-plane is
   A. 0° B. 45° C. 90° D. 180° **Ans. C**
18. In a degenerate gas, the pressure depends on
   A. Volume B. Volume and temperature both
   C. Temperature D. Neither volume nor temperature **Ans. B**
19. The term 'electron degeneracy pressure' is used in the context of
   A. white dwarf B. neutron star C. pulsar D. black hole **Ans. A**
20. The term 'neutron degeneracy pressure' is used in the context of
   A. white dwarf B. neutron star C. Sun D. black hole **Ans. B**
21. Population I stars are rich in
   A. Hydrogen B. Helium C. Heavy elements D. Carbon **Ans. C**

22. Stars having zero population of heavy elements are known as population
    A. I stars B. II stars C. III stars D. Zero stars **Ans. C**
23. The stars which are rich in the heavy elements are known as the population
    A. I stars B. II stars i C. III stars D. zero stars **Ans. A**
24. The stars which are poor in the heavy elements are known as the population
    A. I stars B. II stars i C. III stars D. zero stars **Ans. B**
25. In a black hole, the relation between the temperature T of the event-horizon and the mass M of the black hole is
    A. $T \propto M^{1/2}$ B. $T \propto M$ C. $T \propto M^{-1}$ D. $T \propto M^{-1/2}$ **Ans. C**
26. Our universe does not have
    A. white dwarf B. neutron star C. black dwarf D. brown dwarf **Ans. C**
27. Mass of a neutron star cannot be
    A. 1.6 $M_\odot$ B. 2 $M_\odot$ C. 2.4 $M_\odot$ D. 1 $M_\odot$ **Ans. D**
28. Which one of the processes for the transport of energy in a star from the deep interior to its photosphere is less probable.
    A. Radiation B. Conduction C. Convection D. All of them **Ans. B**
29. When the energy produced through the thermonuclear reaction in the deep interior of a star cannot be transported through the radiations at the same rate, the additional mode of the transport of energy becomes
    A. Conduction B. Convection C. Both of them D. None of them **Ans. B**
30. According to the Jeans criterion, under the gravitational collapse, the size of cosmic structure to be unstable is (λ_J denotes Jeans length)
    A. equal to λ_J B. larger than λ_J C. smaller than λ_J D. half of λ_J **Ans. B**
31. According to the Jeans criterion under the gravitational collapse, the size of cosmic structure to be stable is (λ_J denotes Jeans length)
    A. equal to λ_J B. larger than λ_J C. smaller than λ_J D. twice of λ_J **Ans. C**

10 Sun

It is unanimously accepted that the universe was created about 13.8 billion years ago through the Big Bang and it is expanding continuously. Though the boundaries of the universe are not known, scientists believe that it has one hundred billion (10^{11}) galaxies, and our galaxy, Milky Way, is one of them. Estimations are that the Milky Way has one hundred billion (10^{11}) stars and our Sun is one of them. It provides an idea, how small is our solar system in the universe, and our Earth is a part of the solar system. Obviously, the Sun is the nearest star to us and it has been studied in detail. The next nearest star is the **Proxima Centauri** at a distance of 4.2 ly. The Sun is situated at a distance of 27,000 ly away from the centre of our galaxy, Milky Way.

Scientists have obtained good amount of knowledge about the Sun. The Sun is considered to be situated at the centre of our solar system. The Sun is a ball of hot gas held together through its gravity. It consists mainly of hydrogen and helium which are in the ionized state, because of very high temperature in the core of the Sun. In addition to the hydrogen and helium, there are other elements, like carbon, nitrogen, oxygen, neon, etc., in very small proportions. The estimated mass of the Sun is 1.99×10^{30} kg and its average radius is 6.96×10^5 km. The visual magnitude of the Sun is –26.74 (extremely bright) and the absolute magnitude is + 4.8. That is, the Sun can be visible to the naked eyes from a distance of 10 pc. Some physical parameters of the Sun are summarized in Table 1.

Table 1. Physical parameters of the Sun

Age	4.5×10^9 years
Mass ($M_{\odot}$)	1.99×10^{30} kg
Radius ($R_{\odot}$)	6.96×10^5 km
Mean density	1.4×10^3 kg/m^3
Mean distance from the Earth	1.50×10^8 km
Escape velocity	618 km/s
Luminosity ($L_{\odot}$)	3.86×10^{26} W
Equatorial rotation period	26 days
Polar rotation period	31 days
Mass loss rate	10^9 kg/s

The Sun rotates about its own axis. However, it does not rotate like a rigid body, but its different layers rotate with different velocities. The layer along its equator rotates with the rotation period of about 26 days. The period of rotation goes on increasing away from the equator and it is about 31 days at the poles. The Sun is a magnetically active star. It supports the strong, changing magnetic field that varies from year to year and reverses its polarity every time in a period of about eleven years. The Sun's magnetic field gives rise to many effects/ phenomena which are collectively known as the solar activities, including the sunspots on the solar surface, solar flares, and variations in the solar wind which carries material moving through the solar system. Since the electron density in the ionosphere around the Earth varies with the solar activity, the radio communication (where the reflection through the ionosphere is utilized) depends on the solar activity.

Solar radiations are the primary source of energy for our Earth. The Sun radiates its energy in all directions. The amount of energy received at 1 AU from the Sun (i.e., at the distance of the Earth) is approximately equal to 1.37 kW per square metre. This amount of energy reaching per square metre per second on the surface of Earth is known as the solar constant. The sunlight arriving the Earth's surface is attenuated by the Earth's atmosphere, and very little part of the solar energy reaches the Earth's surface.

Although the Sun, being the nearest star to the Earth, has been studied intensively for quite a long time, many questions about it are still waiting for the answers. For example, the existence of the Sun; the reversibility of the polarity of the magnetic field of the Sun. For convenience, the study of the Sun can be divided into two parts: (i) solar interior, and (ii) solar atmosphere.

1. Solar Interior

The interior part of the Sun is completely opaque and we cannot see into it. There are no direct observations about the solar interior. However, theoretical models suggest that there are three major regions in the solar interior, known as the core, radiative zone, and convective zone. A schematic view of general structure of the Sun is shown in Fig. 1.

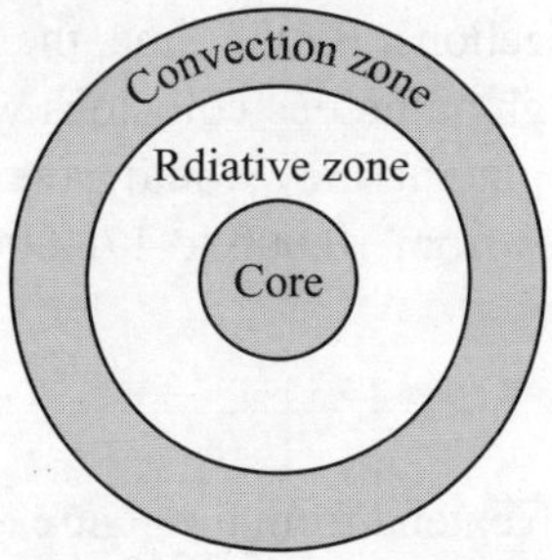

Fig. 1: Schematic structure of the Sun.

The Sun does not have a definite boundary. The density of its gases drops approximately exponentially with increasing the distance from its centre. The solar radius, denoted by R⊙, is measured from its centre to the edge of the photosphere. At any point in the interior, the pressure associated with the thermal energy of the gas exactly balances the weight of the overlying material, a balance known as the hydrostatic equilibrium. The equations of radiative and hydrostatic equilibrium, together with the perfect gas law, govern the structure of solar interior from the edge of the core out a distance equal to about one-third of solar radius.

The escape of radiation from the core is not completely blocked. Photons proceed from one collision to the next, diffusing outwards in a manner that looks like a random walk. A temperature gradient is established between the hot core and the cool outer layers which promotes the outward flow of heat. The mean energy of photons decreases with the decrease of temperature. The variation of temperature in the solar interior is shown in Fig. 2.

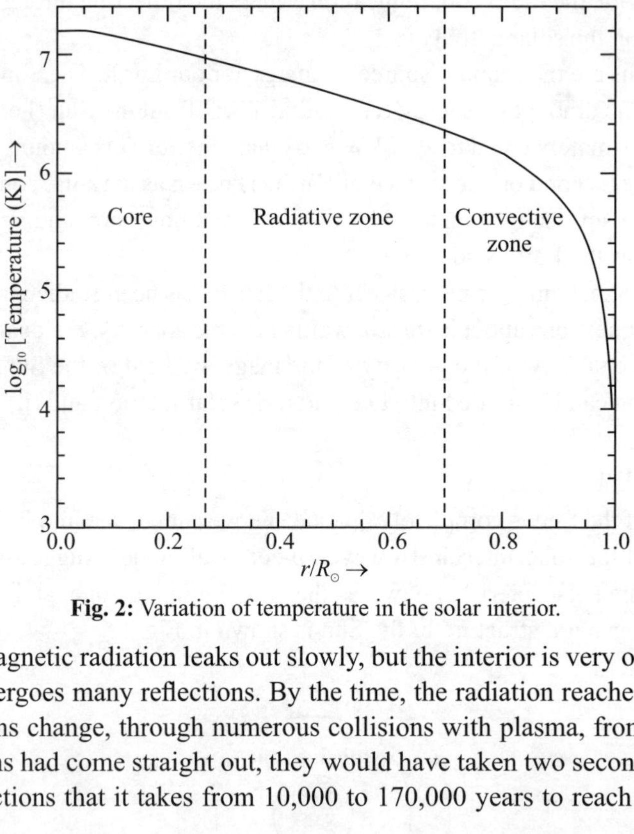

Fig. 2: Variation of temperature in the solar interior.

The electromagnetic radiation leaks out slowly, but the interior is very opaque, like a thick fog, so that it undergoes many reflections. By the time, the radiation reaches the photosphere, most of the photons change, through numerous collisions with plasma, from x-rays to visible light. If the photons had come straight out, they would have taken two seconds only. But, there are so many reflections that it takes from 10,000 to 170,000 years to reach the surface of the Sun from the core.

1.1 Core

The core of the Sun is considered to extend from the centre to about 0.2-0.25 $R_\odot$ and it contains about 34% of the total solar mass. It has a density of up to 1.5×10 kg/m^3 and a temperature of close to 15.7×10^6 K (by contrast, the surface temperature of the Sun is about 5800 K). During most of the solar life, the energy is produced by nuclear fusion through a series of steps called the p-p (proton-proton) chain where the hydrogen is converted into helium. The core is the only location in the Sun where the temperature is sufficient to initiate the thermonuclear reactions for the production of an appreciable amount of heat. The rest of the star is heated by the energy that is transferred outward from the core. All of the energy produced by the fusion in the core must travel through many successive layers to the solar photosphere before it escapes into the space in the form of the sunlight or the kinetic energy of the particles.

The rate of nuclear fusion depends strongly on the density and temperature, and therefore, the fusion rate in the core is in a self-correcting equilibrium: a slightly higher rate of fusion would cause the core to heat up more and to expand slightly against the weight of outer layers.

The high-energy photons (gamma rays) released in the fusion reactions are absorbed in only a few millimetres of the solar plasma and then re-emitted. After the final trip through the convective outer layer to the transparent surface of the photosphere, the photons escape in the form of visible light. Each gamma radiation in the Sun's core is converted into several million visible light photons before escaping into the space.

Energy generation in the Sun

The Sun shines and emits tremendous amount of energy per second. Scientists thought that the self-energy of Sun could be the source of energy being emitted from the Sun. The self energy of the Sun is

$$E_s = \frac{3}{5}\frac{GM_\odot^2}{R_\odot}$$

Using the values, we have

$$E_s = \frac{3}{5}\frac{6.67\times10^{-11}(1.99\times10^{30})^2}{6.96\times10^8}$$

$$= 2.28 \times 10^{41}\text{ J}$$

Luminosity of the Sun $L_\odot = 3.85 \times 10^{26}$ J/s. The period for which the self-energy of the Sun is sufficient is

$$T = \frac{2.28\times10^{41}}{3.85\times10^{26}}$$

$$= 5.92 \times 10^{14}\text{ s}$$

$$= 1.88 \times 10^7\text{ yr}$$

Thus, the self-energy of the Sun is sufficient for its illumination for a period of 1.88×10^7 years. On the other side, the life of the Sun is 4.5×10^9 years. It shows that the self-energy of the Sun is not sufficient source for its illumination.

Then, it was thought that the source of such energy could be the conversion of its own material (nuclear reaction) into the illumination energy. Nuclear reactions can be initiated at a temperature of the order of 10^6-10^7 K. For a celestial body consisting of mono-atomic gas with atomic mass μ having radius R and total mass M in the hydrostatic equilibrium, the temperature in its core is

$$T = \frac{GM\mu}{5NkR}$$

where $N = 6.023 \times 10^{26}$ is the Avogadro number. Using the parameters for the Sun, we have

$$T = 4.59 \times 10^6\mu$$

Hence, the temperature of the central core of the Sun is of the order of 10^6-10^7 K, which is sufficient for the initiation of nuclear reactions.

When two light nuclei combine to form a single larger nucleus, the process is known as the nuclear fusion. The central temperature and density are so high that the nuclear fusion reaction takes place and the hydrogen is burnt into the helium through the reaction known as the p-p chain reaction. It is represented by the following set of reactions. Two protons combine to from a deuteron.

$$_1H^1 + {}_1H^1 \rightarrow {}_1H^2 + e^+ + \nu \tag{10.1}$$

Here, ν denotes a neutrino particle and e^+ is positron. A positron combines with an electron to produce a pair of photons (**pair-production** process).

$$e^+ + e^- \rightarrow \gamma + \gamma \tag{10.2}$$

The deuteron combines with a proton to form the helium isotope.

$$_1H^2 + {}_1H^1 \rightarrow {}_2He^3 + \gamma \tag{10.3}$$

Two helium isotopes obtained from two sets of reactions combine to form a helium nucleus

$$_2He^3 + {}_2He^3 \rightarrow {}_2He^4 + {}_1H^1 + {}_1H^1 \tag{10.4}$$

When we consider the combination of reactions: 2 (10.1) + 2 (10.2) + 2 (10.3) + (10.4), the net effect is

$$4\,{}_1H^1 + 2e^- \rightarrow {}_2He^4 + 2\nu + 6\gamma$$

In each reaction, the energy is released. Thus, four protons combine to form one $_2He^4$ nucleus with the release of energy.

In some part of the core of the Sun, the temperature may be sufficient to produce the fusion of three helium nuclei (**triple-alpha reaction**) to form a carbon nucleus.

$$3\,{}_2He^4 \rightarrow {}_6C^{12}$$

With the formation of carbon, another set of reactions, called the carbon-nitrogen-oxygen (CNO) cycle, has also been suggested for the energy generation in the Sun. This set of reactions is as the following. The carbon nucleus combines with a proton to form an isotope of nitrogen.

$$_6C^{12} + {}_1H^1 \rightarrow {}_7N^{13} + \gamma \tag{10.5}$$

This isotope of nitrogen converts into an isotope of carbon.

$$_7N^{13} \rightarrow {}_6C^{13} + e^+ + \nu \tag{10.6}$$

This isotope of carbon combines with a proton to form the nitrogen.

$$_6C^{13} + {}_1H^1 \rightarrow {}_7N^{14} + \gamma \tag{10.7}$$

The nitrogen atom combines with a proton to form an isotope of oxygen.

$$_7N^{14} + {}_1H^1 \rightarrow {}_8O^{15} + \gamma \tag{10.8}$$

This isotope of oxygen converts into an isotope of nitrogen.

$$_8O^{15} \rightarrow {}_7N^{15} + e^+ + \nu \tag{10.9}$$

This isotope of nitrogen combines with a proton to form the carbon nucleus.

$$_7N^{15} + {}_1H^1 \rightarrow {}_6C^{12} + {}_2He^4 \tag{10.10}$$

Combining the reactions (10.5) through (10.10), we get

$$4\,{}_1H^1 \rightarrow {}_2He^4 + 2e^+ + 2\nu + 3\gamma$$

In each reaction here energy is released. This is again a net fusion of four protons to form a helium nucleus. It liberates the gamma-rays, neutrinos and thermal energy. In this set of reactions, carbon is not destroyed in the process but it acts only as a catalyst.

For initiation of this reaction, million kelvin temperature is required. Hence, at the centre or in the core, the temperature is about 15 million kelvin. The estimates show that 600 million tones of hydrogen is transformed into helium per second and 5 million tonnes of material is lost in the form of energy per second. Though this weight appears very large, but in comparison to the solar mass it is insignificant. So far the Sun has consumed only a few hundredth of 1% of the original mass of the Sun. The Sun will run out of hydrogen in the next 5 billion years.

Example 1.

In the core of the Sun, suppose a helium nucleus is formed through the fusion of four protons. Calculate the amount of energy released per reaction. Given, mass of proton $m_p = 1.0079$ amu, mass of helium nucleus $m_{\text{He}} = 4.0026$ amu. 1 amu $= 1.66 \times 10^{-27}$ kg.

Solution:

$$\text{Mass of four protons} = 4.0316 \text{ amu}$$

$$\text{Mass defect} = \text{mass of four protons} - \text{mass of helium nucleus}$$
$$= 4.0316 - 4.0026 = 0.0290 \text{ amu}$$

$$\text{Mass defect } \Delta m = 0.0290 \times 1.66 \times 10^{-27}$$
$$= 4.814 \times 10^{-29} \text{ kg}$$

$$\text{Energy released} = \Delta mc^2$$
$$= 4.814 \times 10^{-29} (3 \times 10^8)^2$$
$$= 4.3326 \times 10^{-12} \text{ J}$$

Example 2.

In the core of the Sun, suppose a helium nucleus is formed through the fusion of four protons. Calculate the amount of hydrogen converted into helium per year. Given, mass of proton $m_p = 1.673 \times 10^{-27}$ kg, mass of helium nucleus $m_{\text{He}} = 6.664 \times 10^{-27}$ kg, luminosity of the Sun $L_\odot = 3.86 \times 10^{26}$ J/s.

Solution:

$$\text{Mass of four protons} = 4\,(1.673 \times 10^{-27})$$
$$= 6.692 \times 10^{-27} \text{ kg.}$$

$$\text{Mass defect} = \text{mass of four protons} - \text{mass of helium nucleus}$$
$$= 6.692 \times 10^{-27} - 6.664 \times 10^{-27}$$
$$= 2.8 \times 10^{-29} \text{ kg}$$

$$\text{Energy released} = 2.8 \times 10^{-29} \times (3 \times 10^8)^2$$
$$= 2.52 \times 10^{-12} \text{ J}$$

$$\text{Luminosity of the Sun } L_\odot = 3.86 \times 10^{26} \text{ J/s.}$$

$$\text{Number of reactions per second} = 3.86 \times 10^{26}/2.52 \times 10^{-12}$$
$$= 1.5317 \times 10^{38}$$

$$\text{Mass converted into energy per second} = (2.8 \times 10^{-29})(1.5317\times 10^{38})$$
$$= 4.289 \times 10^9 \text{ kg}$$

Further,

$$1 \text{ year} = 365 \times 24 \times 60 \times 60 = 3.1536 \times 10^7 \text{ s.}$$

$$\text{Mass converted into energy per year} = (4.289 \times 10^9)(3.1536 \times 10^7) = 1.3526 \times 10^{17} \text{ kg}$$

Composition of the Sun

The relative abundance (RA) of element (logarithmic base ten) and its first ionization potential (FIP) in eV in the Sun are given in Table 2. Here, the abundance of hydrogen is taken as 10^{12}. Table 2 shows that there is variety of elements, though the abundance is small, have been found in the Sun. The mass concentrations of hydrogen and helium are generally denoted by X and Y, respectively. The mass concentration of rest of the elements is sometimes denoted by Z or by X_i, where the i indicates various elements other than hydrogen and helium.

Example 3.

Calculate the mean molecular weight of the Sun.

Solution:

In the outer layers of the Sun, the initial chemical composition has not changed by the nuclear reactions. In this case, one uses the values $X = 0.71$, $Y = 0.27$, $Z = 0.02$. The mean molecular weight is

$$\mu = \frac{1}{2 \times 0.71 + 0.75 \times 0.27 + 0.5 \times 0.02} = 0.61$$

When the hydrogen is completely exhausted, we have $X = 0$ and $Y = 0.98$, and hence

$$\mu = \frac{1}{0.75 \times 0.98 + 0.5 \times 0.02} = 1.34$$

Table 2. Relative abundance and first ionization potential

Element	Z	RA	FIP
H	1	12.0	13.6
He	2	10.9	24.6
C	6	8.5	11.3
N	7	8.0	14.5
O	8	8.8	13.6
Ne	10	7.9	21.6
Mg	12	7.4	7.6
Si	14	7.5	8.2
S	16	7.2	10.4
A	18	6.8	15.8
Ca	20	6.3	6.1
Fe	26	7.6	7.9
Ni	28	6.3	7.6

Example 4.

Calculate the temperature of the Sun at half of the solar radius. The surface value of the mean molecular density $\mu = 0.61$.

Solution:

The temperature of the Sun can be estimated from the perfect gas law as

$$T = \frac{\mu m_H P}{k\rho}$$

$$= \frac{0.61(1.67 \times 10^{-27})(1.0 \times 10^{14})}{(1.38 \times 10^{-23})(1410)}$$

$$= 5 \times 10^6 \text{ K}$$

1.2 Radiative Zone

The radiative zone inside the Sun is considered to exist from 0.25 $R_\odot$ to 0.7 $R_\odot$. The plasma inside the Sun is sufficiently hot and dense to transfer the intense heat from the core to outward. In this zone, there is no thermal convection. With the increase of the distance, the plasma cools down. The heat is transferred through the radiations, and the ions of hydrogen and helium emit photons, which travel a brief distance before being reabsorbed by the other ions. In this manner, the energy makes its way very slowly in the outward direction.

The opacity, is the measure of collisional resistance of gas to the flow of the radiant energy. The flow of the energy is measured in terms of the flux, the rate at which the energy crosses a unit area perpendicular to the given direction. When the opacity is large, the radial heat flux due to the flow of radiation τ_r at a distance r from the centre of the Sun is described by the radiative conduction equation

$$\tau_r = \frac{16\sigma T^3}{3\bar{k}} \frac{dT}{dr}$$

where T is the temperature, σ the Stefan-Boltzmann constant and $\bar{k}$ the mean opacity (averaged over all wavelengths). Outside the core, where no further heat is produced by the nuclear reactions, the total heat flux carried out by the photons moving outwards must be constant, so that

$$\frac{T^3}{\bar{k}} \frac{dT}{dr} = \text{constant.}$$

This condition is known as the Radiative Equilibrium. The opacity depends on the density or equivalently on the number of particles per unit volume. The density decreases steadily as we move outwards from the centre of the Sun. The variation of density with distance in the solar interior is shown in Fig. 3.

1.3 Convective Zone

The convective zone is the outer layer of the interior of the Sun. In the convective zone, the plasma is neither dense enough nor hot enough to transfer the heat energy from the interior to outward through the radiation. The temperature of plasma decreases with the increase of the

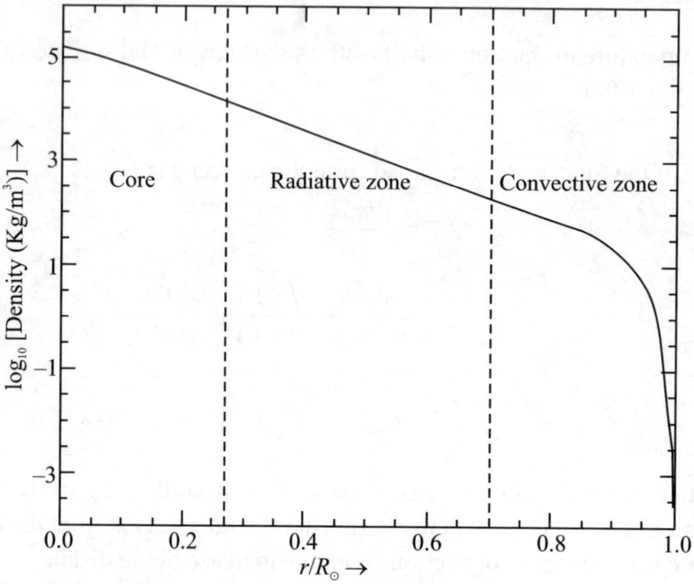

Fig. 3: Variation of density in the solar interior.

distance so rapidly that the situation becomes unstable and produces the turbulent convective motions. As a result, the thermal convection takes place, as the thermal columns carry hot material to the solar surface (photosphere). The thermal columns in the convective zone generate the imprints on the solar surface, in the form of the solar granulation and super-granulation.

The turbulent convection of this outer part of the solar interior gives rise to a small-scale dynamo that produces the magnetic north and south poles all over the solar surface. The thickness of the layer, about 1.5×10^5 km, corresponds to $\simeq 0.2R_\odot$. Above this zone, the hydrogen is practically neutral; within the zone, it is partially ionized; and below the zone, it is completely ionized.

2. Solar Atmosphere

The region above the photosphere of the Sun is referred to the solar atmosphere. It can be viewed with the help of various telescopes, operating in various parts of the electromagnetic spectrum. It can be broadly divided into five principal zones, called the photosphere, chromosphere, transition region, corona and heliosphere. The heliosphere, which may be considered the tenuous outer atmosphere of the Sun, extends outward past the orbit of Pluto. The transition region is not easily visible from the Earth's surface, but is readily observable from the space with the help of the instruments which are sensitive to the far ultraviolet portion of the electromagnetic spectrum.

The atmosphere of the Sun is composed of the layers of hot gases. These layers, however, are not sharply defined. In fact, each one of them is separated by a transition zone of several kilometres thickness and they greatly differ from each other in the physical characteristics. Starting from the visible solar surface upwards these layers are the chromosphere and corona, respectively. It transmits the energy, mainly in the form of radiation, coming from below and delivers it out into the space.

The particles of the solar wind move quickly enough to escape from the solar system. Moreover, their speed is such that the disturbances are unable to propagate back to the Sun. The surface layers are unaware of the what is happening out in the wind. For this reason, we do not consider the solar wind to be a part of the solar atmosphere. However, the atmosphere transforms only gradually into the solar wind and it is thus hard to draw an exact boundary between the two regions. The atmosphere may consist of three main regions:

1. Photosphere
2. Chromosphere
3. Corona

The variation of temperature and density of the solar atmosphere is shown in Fig. 4.

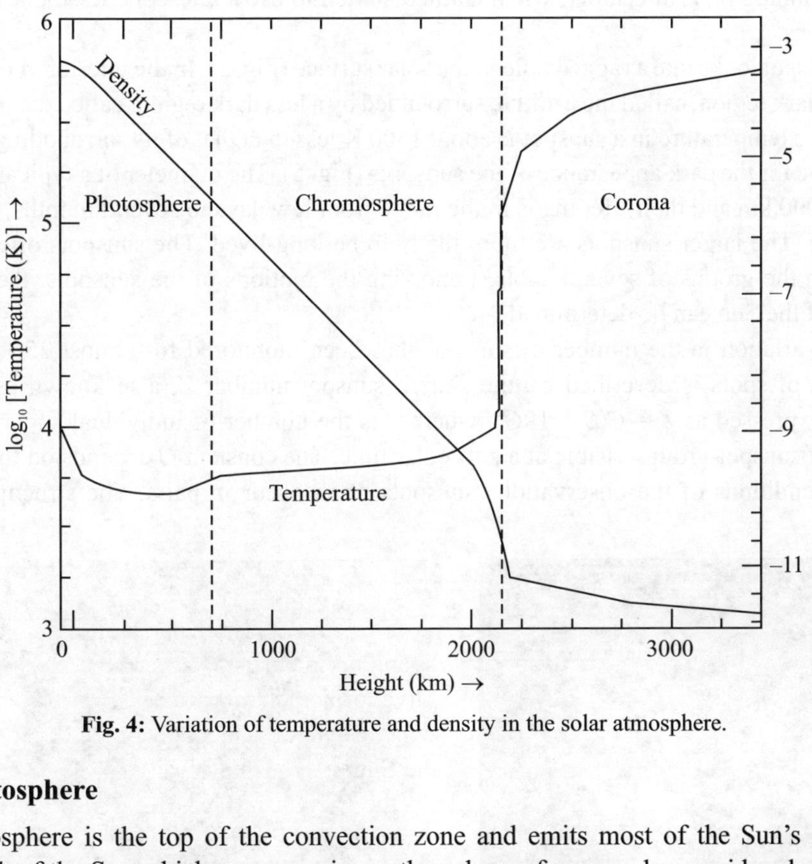

Fig. 4: Variation of temperature and density in the solar atmosphere.

2.1 Photosphere

The photosphere is the top of the convection zone and emits most of the Sun's light. The visible disk of the Sun which we recognize as the solar surface may be considered as the base of photosphere. This is the innermost layer of solar atmosphere. The thickness of photosphere is 0.5 Mm with a density of 10^{23} particles/m^3, and thus, the mass density is ~ 10^{-4} kg/m^3. It is the layer below which the Sun becomes opaque to the visible light. Above the photosphere, the radiation in all the wavelengths is free to propagate into the space, and the entire energy of the

Sun escapes outward. The change in the opacity is due to the decreasing amount of H^- ions, which can absorb the visible light easily.

The photosphere is actually 10-100 km thick, being slightly less opaque than the air on the Earth. Because the upper part of the photosphere is cooler than the lower one, an image of Sun appears brighter in the centre than on the edge or limb of the solar disk, and this phenomenon is known as the limb darkening. Sunlight has approximately a black-body spectrum that indicates its kinetic temperature of about 5800 K.

2.2 Sunspots

The clearest visible sign of the solar activity is the sunspot (Fig. 5). The existence of the sunspots is known for a long time, as the largest ones can be seen even with the naked eye by looking at the Sun through a suitably dense layer of fog. More precise observations became available in the beginning of 17th century, when Galileo started to use a telescope for the astronomical observations.

A sunspot looks like a ragged hole in the solar surface (Fig. 5). In the interior of the sunspot, there is a dark region, called the **umbra**, surrounded by a less dark region, called the **penumbra**. The surface temperature in a sunspot is about 1500 K less than that of its surroundings, and it is responsible for the dark appearance of the sunspots (Fig. 5). The diameter of a typical sunspot is about 10,000 km and their lifetime is in the range from few days to several months, depending on its size. The larger sunspots are more likely to be long-lived. The sunspots often occur in pairs or in the groups of several spots. Following the motions of the sunspots, the period of rotation of the Sun can be determined.

The variation in the number of sunspots has been monitored for almost 250 years. The frequency of spots is described by the Zurich sunspot number Z, also known as the Wolf number, expressed as $Z = C(S + 10G)$, where S is the number of individual spots and G the number of sunspot-groups visible at a particular time. The constant C depends on the observer and the conditions of the observation. Sunspots often occur in pairs. The structure of such

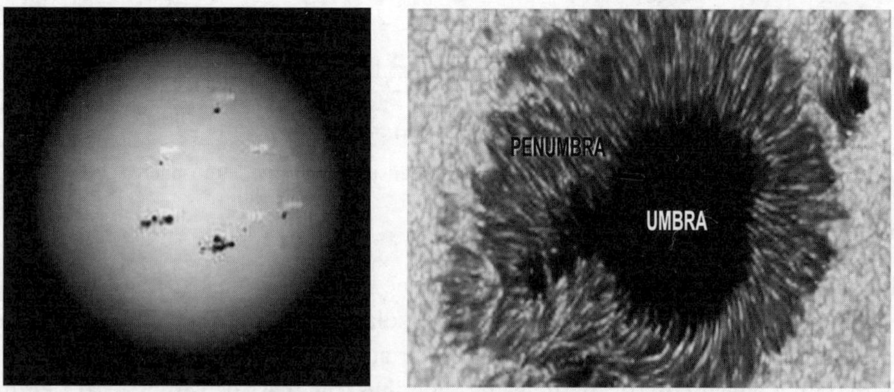

Fig. 5: Sunspots.

(*Courtesy:* Google website)

bipolar groups can be understood if the field rises in the form of a loop above the solar surface, connecting the components of the pair. If the gas is streaming along such a loop, it becomes visible as a loop prominence. The periodic variation in the number of sunspots gives a variation in the general solar magnetic field.

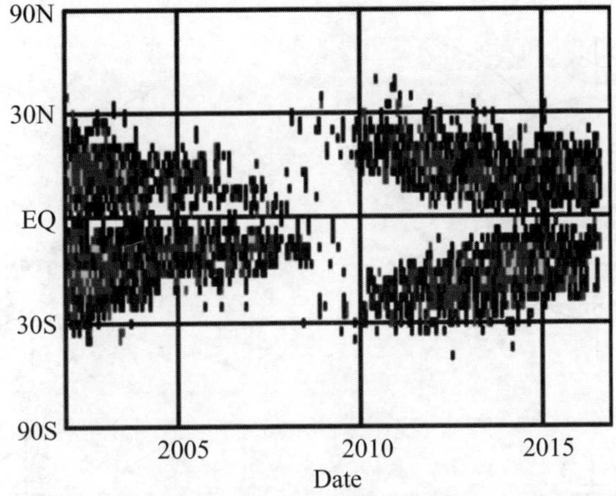

Fig. 6: Butterfly diagram of sunspots.
(*Courtesy:* Google website)

At the beginning of a new activity cycle, spots first begin to appear at the latitudes of about 40°. The characteristic pattern in which the spots appear, is known as the Maunder butterfly diagram or butterfly diagram (Fig. 6). Spots of the next cycle begin to appear while those of the previous one are still present near the equator. Spots belonging to the new cycle have a polarity opposite to that of the old ones. (Spots in the opposite hemispheres also have opposite polarity.) Since the field is reversed between consecutive 11 years cycle, the complete period of solar magnetic activity is 22 years.

Babcock model of sunspot formation

George Hale was the first to find the magnetic fields in the sunspots in 1908, and he linked the magnetic fields with the sunspots. Hale suggested 22 years for the sunspot cycle, covering two polar reversals of the solar magnetic dipole field. Horace Babcock proposed a qualitative model for the solar dynamics in 1961. Though the model could describe major features of the solar cycle, but could not yet provide adequate explanation for many important details of solar activity. The Sun has an oscillatory magnetic field, with a quasi-steady periodicity of 22 years. This oscillation is known as the Babcock-Leighton dynamo cycle, showing an oscillatory exchange of energy between the poloidal and the toroidal solar magnetic field components.

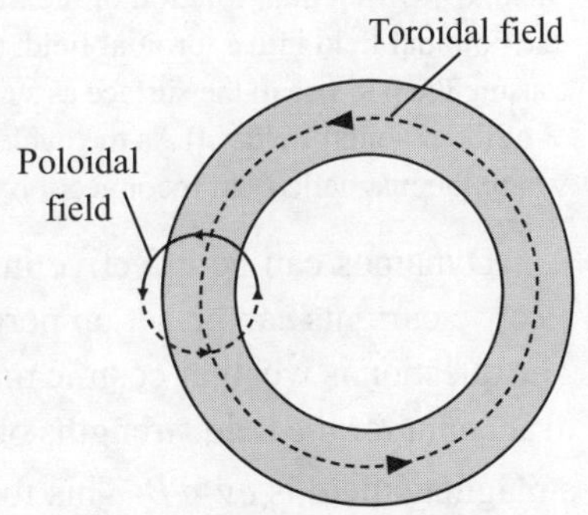

Fig. 7: A diagram depicting the poloidal (θ) and toroidal (ϕ) directions of the magnetic field.

The topology of the poloidal and the toroidal magnetic field can be viewed with the help of Fig. 7. For a toroid, the field along the axis of the toroid (along ϕ direction) is the toroidal field and the field perpendicular to the toroidal field (along θ direction) is the poloidal field. In the magnetic dynamo model of the Sun, the initial poloidal magnetic field (Fig. 8 (a)) converts to the toroidal magnetic field (Fig. 8 (b)).

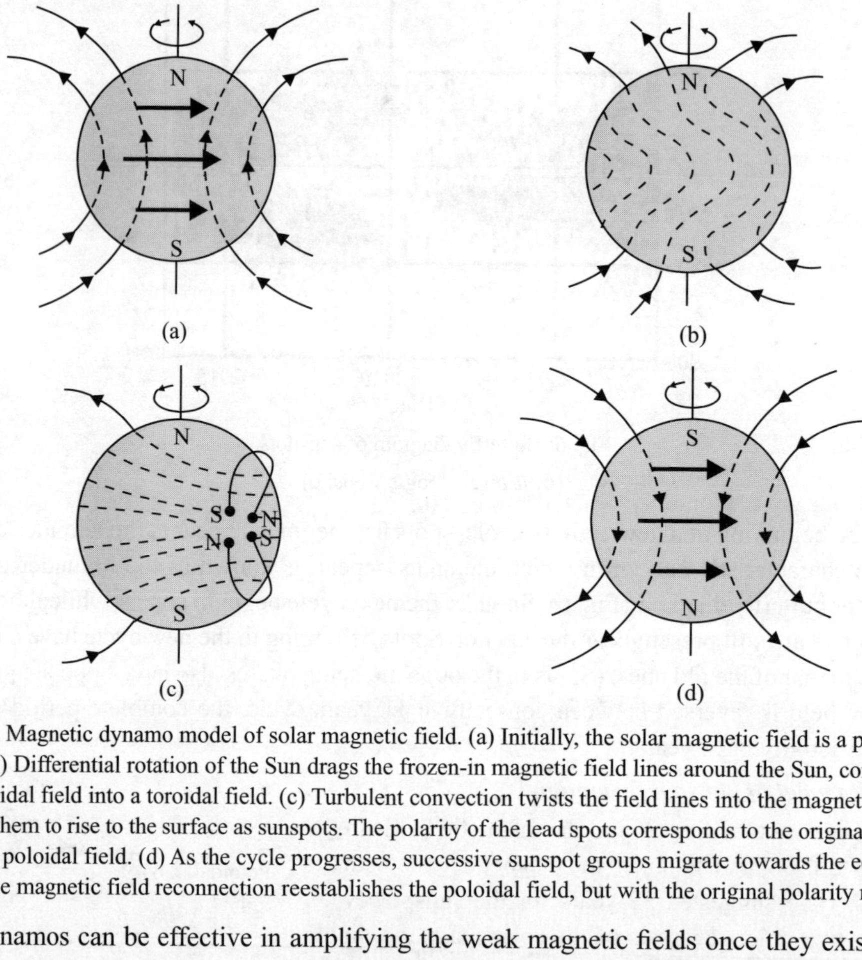

Fig. 8: Magnetic dynamo model of solar magnetic field. (a) Initially, the solar magnetic field is a poloidal field. (b) Differential rotation of the Sun drags the frozen-in magnetic field lines around the Sun, converting the poloidal field into a toroidal field. (c) Turbulent convection twists the field lines into the magnetic ropes, causing them to rise to the surface as sunspots. The polarity of the lead spots corresponds to the original polarity of the poloidal field. (d) As the cycle progresses, successive sunspot groups migrate towards the equator where the magnetic field reconnection reestablishes the poloidal field, but with the original polarity reversed.

Dynamos can be effective in amplifying the weak magnetic fields once they exist. If the electric currents can be set up perpendicular to the direction of a seed field, the field can grow. The question is whether cosmic magnetic fields can grow sufficiently fast by the dynamo action to account for the field strengths observed today. We saw that the force on an electron moving in a magnetic field is $e\vec{v} \times \vec{B}$. This force gives rise to a current just as an electric field would. The current density in the absence of varying displacement fields, therefore, becomes

$$\vec{J} = \sigma[\vec{E} + \vec{v} \times \vec{B}]$$

$$\vec{E} = \left[\frac{\vec{J}}{\sigma} - \vec{v} \times \vec{B}\right]$$

We know

$$\nabla \times \vec{E} = -\frac{\partial \vec{B}}{\partial t}$$

$$\nabla \times \vec{B} = \mu_0 \vec{J}$$

Inserting $\vec{E}$, we get

$$\frac{\partial \vec{B}}{\partial t} = -\nabla \times \vec{E}$$

$$= -\nabla \times \left[\frac{\vec{J}}{\sigma} - \vec{v} \times \vec{B} \right] = \nabla \times (\vec{v} \times \vec{B}) - \frac{\nabla \times \vec{J}}{\sigma}$$

$$= \nabla \times (\vec{v} \times \vec{B}) - \frac{\nabla \times (\nabla \times \vec{B})}{\sigma \mu_0} \tag{10.11}$$

Using the identity $\nabla \times (\nabla \times \vec{B}) = \nabla(\nabla \cdot \vec{B}) - \nabla^2 \vec{B}$, and $\nabla \cdot \vec{B} = 0$, we have

$$\frac{\partial \vec{B}}{\partial t} = \nabla \times (\vec{v} \times \vec{B}) + \frac{\nabla^2 \vec{B}}{\sigma \mu_0}$$

From equation (10.11), we see that the magnetic field grows only if the velocity transverse to the field is appreciable

$$v \geq c^2 \epsilon_0 / \sigma \mu L$$

where L is the dimension over which the magnetic field changes significantly. To note what this means, we need to evaluate the conductivity σ.

The Babcock model shown in Fig. 5 explains the Maunder butterfly diagram. The field lines wind up and tangle first at the higher latitudes, so initially the sunspots are formed there. In the further progression of cycle, the field winds up, the tangling progresses to the latitudes nearer the equator and the sunspots from closer to the equator.

Any complete picture of the solar cycle requires a full treatment of the MHD equations in the solar environment, along with differing rotation rates with the latitude and the depth in the Sun, convection, solar oscillations, heating of the upper atmosphere, and the mass loss. Though all these processes are not likely to play equally important roles for the study of the solar cycle, but it is important to understand the degree to which each of them contributes to the particular phenomenon under investigation.

As shown in Fig. 8, as the magnetic field lines are frozen into the gas, the differential rotation of the Sun drags the lines along, converting a poloidal field (essentially a simple magnetic dipole field) to one that has a significant toroidal component (field lines that are wrapped around the Sun). The turbulent convection zone then has the effect of twisting the lines, creating the regions of intense magnetic fields, called the magnetic ropes. The buoyancy produced by the magnetic pressure causes the ropes to rise to the surface, appearing as the sunspot groups. The polarity of the sunspots is due to the direction of the magnetic field along the ropes; consequently, every lead spot in one hemisphere has the same polarity whereas the lead spots in the other hemisphere have the opposite polarity.

Initially, the little twisting that does develop occurs at the higher latitudes; during the sunspot minimum. As the differential rotation continues to drag the field lines along and the convective turbulence ties them in the form of knots, developing more sunspots at the intermediate latitudes, producing a sunspot maximum. It would seem that ultimately the greatest amount of twisting and the largest number of sunspots should develop near the equator. However, the sunspots from the two hemispheres tend to cancel out near the equator because the polarities of their leading spots are opposed. Therefore, the number of the sunspots appearing near the equator is small.

The cancellation of magnetic fields near the equator causes the poloidal field to be reestablished, but with its original polarity reversed. This process takes about 11 years. The entire procedure repeats continuously, with the polarity of the magnetic field returning to its original orientation every other cycle. Hence, the entire solar cycle is actually 22 years long when the magnetic field polarities are considered.

2.3 Faculae

A bright area around the sunspots and covering a fraction of solar disk is known as the faculae (Fig. 9). These are phenomena belonging to the transition region between the photosphere and chromosphere. Photospheric faculae are definitely associated with the sunspots, and they are found to appear before the spots and live for several months.

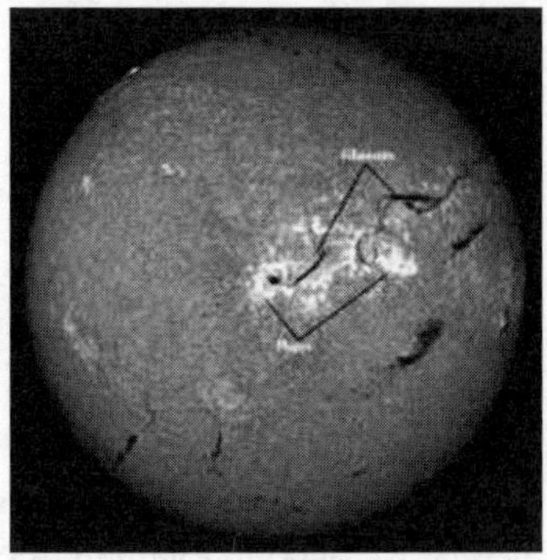

Fig. 9: Faculae.
(*Courtesy:* Google website)

The average lifetime of a photospheric faculae is 90 days. They are brighter than their surroundings, and are believed to be hotter (temperature in the faculae is higher than that of the photosphere) and thus they radiate more energy. For this reason the faculae are, important for the energy balance between the sunspots and the photosphere. There are also faculae which are not associated with the sunspots, but they are invariably smaller, fainter and relatively short-lived. The chromospheric faculae have been mostly observed in Hα and Ca II lines.

2.4 Granulation

Typical grain-like fine structures on the visible disc of the Sun are known as the photospheric granulation (Fig. 10). The granulation of the Sun was known to the observers for many decades, but its detailed visual studies had been difficult, because the fine structures were not revealed due to bad seeing, because of the Earth's atmospheric turbulence. In 1959, a part of the effect of the bad seeing was overcome by taking photographs of the Sun from a balloon sent to a height of 5486 m above the Earth's surface.

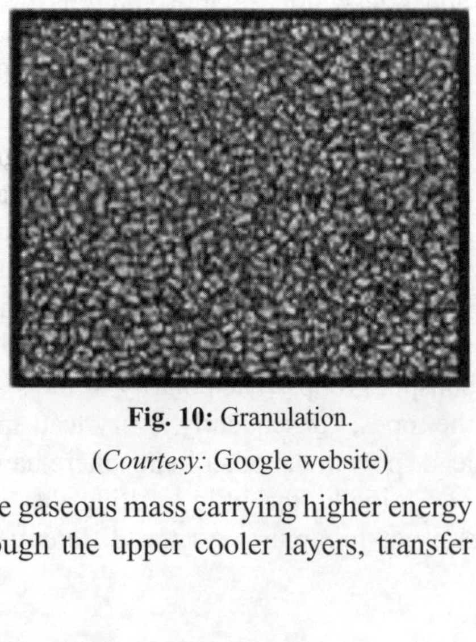

Fig. 10: Granulation.
(*Courtesy:* Google website)

In the envelope of the stars, like the Sun, the radiative energy transport has to be supplemented by the convective transport of energy. Convection cells of the gaseous mass carrying higher energy from the hotter deeper layers of the envelope rise through the upper cooler layers, transfer there the excess energy and subsequently sink down again to the deeper layers. The process is repeated incessantly.

The diameters of the granules vary from 300 km to 18,000 km, with a mean value of about 700 km. Their shapes resemble with the irregular polygons or cells which are separated from one another by a rather narrow and dark filament-like regions. Granules are found to vary in both the size and the number with the 11 years solar cycle of the magnet.

2.5 Chromosphere

Above the photosphere lies a second major layer of solar atmosphere, known as the chromosphere. It is named chromosphere because of its appearance as a pink flash during the total solar eclipse. The chromosphere word is derived from the Greek word 'chroma' meaning colour, because the chromosphere is visible as a coloured flash.

The chromosphere is the region between the temperature minimum and the corona. The layer extends to nearly 20,000 km above the photosphere with gas-density gradually decreasing with the distance, but the temperature rises rapidly upwards. The atmosphere becomes transparent in the continuum, so the chromospheric continuum can be observed at the limb during the total solar eclipse.

The chromosphere is rarer and more transparent with a particle density of 1017 m^{-3}, and thus the mass density is about 10^{-10} kg/m^3. Minimum temperature of ~ 4000 K exists in the chromosphere. This part of the solar atmosphere is cool enough to support the existence of simple molecules, such as CO and H_2O, which can be identified through their absorption spectra.

The temperature in the chromosphere increases gradually with the altitude, ranging up to around 10^5 K near the top. When we study the chromosphere in H_α, we often find small transient features. These include surges and eruptions, small brightening called the micro-flares, transient small filaments separating small bipolar elements.

2.6 Spicules

During the total solar eclipse or with the use of a coronagraph, spiky structures of hot, tenuous gas are seen to rise from the lower chromosphere and is extended up to 10,000 km, or more, sometimes even up to 20,000 km. The diameters of these spikes are, on an average between 500 to 600 km with the larger ones having diameter about 2000 to 5000 km. These projected chromospheric streamers are known as the spicules (Fig. 11). The velocity of their upward movement is about 30 km/s, and their average lifetime is about 10 minutes. Spicules are observed as the dark mottles in H_α spectroheliograms to cover a good fraction of solar disc.

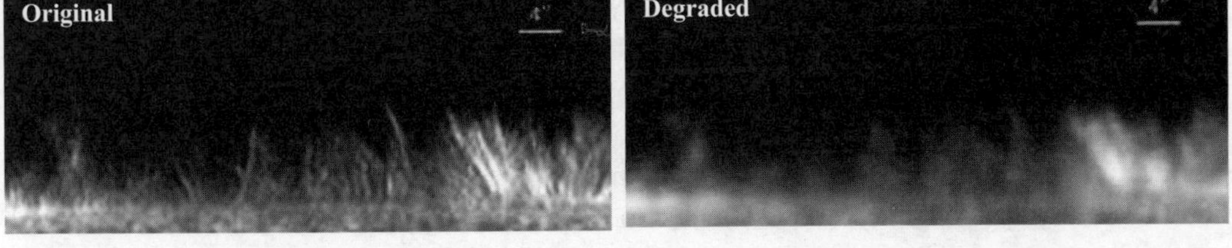

Fig. 11: Spicules.

(*Courtesy:* Google website)

When viewed near the solar limb, a large number of such structures are seen together in the projection as to give the appearance of a 'forest of spikes'. They appear to be a moving disturbance through the solar chromosphere driven by the enhanced magnetic field and probably, they play an important role in transferring the mass from the lower chromosphere to the corona.

2.7 Plages and Filaments

Large bright and dark clouds were revealed in the chromosphere when the spectroheliograms in the monochromatic light of H_α and Ca II K were examined. These are, respectively, called the plages (Fig. 12) and the filaments (Fig. 13). In the spectroheliograms, they appear as the clouds of hydrogen and calcium, but in fact they are just the regions of chromospheric gas which radiate more intensely at the wavelengths of H_α and Ca II K line.

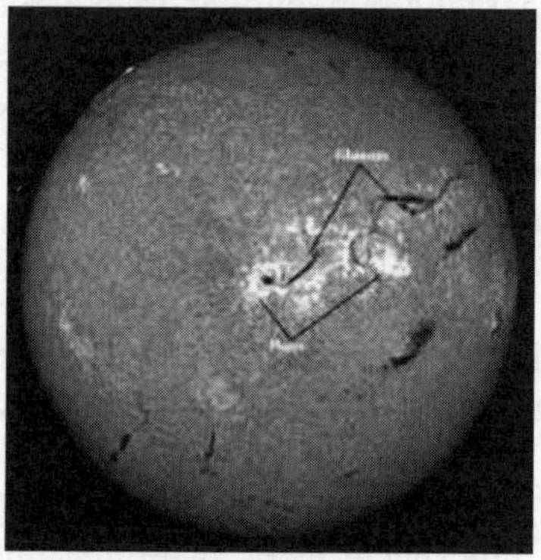

Fig. 12: Plages.
(*Courtesy:* Google website)

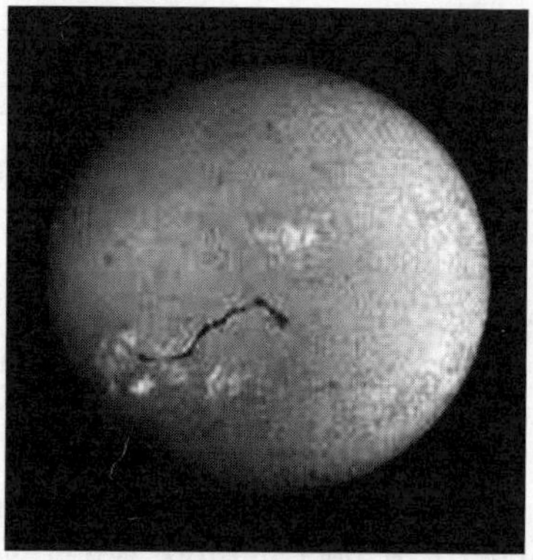

Fig. 13: Filaments.
(*Courtesy:* Google website)

Like the faculae and spicules, the plages are also associated with the enhanced magnetic field of the Sun. They are mostly seen in the regions of sunspots but sometimes are also seen where there are not visible sunspots.

2.8 Corona

The corona is a Latin word, meaning 'crown'. It is an atmosphere around the Sun, containing plasma and extending millions of kilometres into the space. It can be easily seen during the total solar eclipse as shown in Fig. 14, and can also be observed with the help of an instrument, called the coronagraph, which creates an artificial solar eclipse.

Two spectral lines at wavelengths 5303 Å and 6374 Å having very high intensity were found in the spectrum of corona in the year 1869. Scientist attempted to assign them to some elements or their ions. Since the temperature of the photosphere is 5800 K, the scientists had opinion that the temperature in the atmosphere of the Sun should not be more than 5800 K. They produced spectrum of the elements and their singly ionized ions in laboratories, but could not find these two lines in their spectra. Then, some scientists suggested that these two lines were emitted by some unknown element, named, the 'coronium'. Now, the point was to describe the details of the coronium.

Fig. 14: Structure of corona (white region) during the total solar eclipse. The dark sphere is the Sun which is covered by the Moon.
(*Courtesy:* Google website)

Some scientists were not satisfied with this concept, and they went on producing spectra of highly ionized ions in their laboratories. B. Edlen in the year 1940 found that the lines of the wavelength 5303 Å and 6374 Å were produced by Fe XIV and Fe X, respectively.[1] That is, it took 70 years to find the source for these two lines. Some other lines which have been found in the corona are assigned to various elements as given in Table 3.

Table 3. Wavelengths and elements

Wavelength (Å)	Element	Wavelength (Å)	Element
3328	Ca XII	5303	Fe XIV
3388	Fe XIII	6374	Fe X
3601	Ni XVI	6702	Ni XV
3643	Ni XIII	7060	Fe XV
3987	Fe XI	7892	Fe XI
4086	Ca XIII	8024	Ni XV
4231	Ni XII	10747	Fe XIII
5116	Ni XIII	10798	Fe XIII

These highly ionized elements are produced at a temperature of several million kelvin. These spectral features have been traced to highly ionized iron (Fe XIV) which indicates a plasma temperature more than 10^6 K. The corona is observed at the time of eclipse as a faint halo of very low density of 10^{15} particles/m^3, and thus the mass density is about 10^{-12} kg/m^3, it extends to the Earth's orbit where the density is 10^7 m^{-3} (i.e., the mass density is about 10^{-20} kg/m^3) and beyond.

[1]As per convention, for example, neutral iron element is written as Fe I; singly ionized calcium is written as Ca II. Thus, Fe X is nine times ionized iron, i.e., Fe^{+9} and Fe XIV is thirteen times ionized iron, i.e., Fe^{+13}.

The temperature of the corona is several million kelvin. In spite of large number of attempts, no complete theory yet proposed to explain the temperature of corona; at least some of its heat is known to be from the magnetic reconnection.

Temperature of corona

For quite a long time, some lines in the spectra of the Sun could not be assigned to any known atom or ion. Among these lines, the two at $\lambda = 5303$ Å and $\lambda = 6374$ Å were quite prominent. In absence of any species to which these lines could be assigned, it was thought that there could be an element, named the coronium, in the corona, which is emitting these lines. However, no one was in a position to tell about the physical characteristics of the coronium. Later on, *B*. Edlen found that these lines could be generated by the highly ionized ions. For example, the lines of 5303 Å and 6374 Å could be generated by the Fe XIV and Fe X ions, respectively.

These highly ionized ions can be produced at a temperature of million kelvin. In the beginning it had been unbelievable that the temperature in the solar corona could be of million kelvin. Later on, it was found that there was no other option, but to accept it. Though the temperature of solar atmosphere is 10^6 K, still the spectrum for the Sun corresponds to a temperature of about 6000 K, which is the temperature of photosphere of the Sun. It is because the density of corona is very low.

The temperature of the solar corona generally lies in the range of 1-5 million kelvin. Such a hot plasma is subject to lose energy through the radiation in all directions and the thermal conduction to the transition region below. As the coronal temperature is maintained at several million kelvin, these energy losses must obviously be compensated by an energy source. The explanation of mechanism through which the energy is supplied to the corona is a great challenge to the scientists.

Many solutions have been proposed, but none of them has yet achieved a consensus. The corona can be divided into three segments as the following.

1. K-corona (K stands for continuum)
2. E-corona (E stands for emission)
3. F-corona (F stands for Fraunhofer)

1. **K-corona:** The name K-corona is derived from the German word 'Kontinum', means continuum. The inner corona which is sometimes known as the real corona extends from 1.03 $R_{\odot}$ to 2.5 $R_{\odot}$. This part of the corona, known as the K-corona, imitates the continuous spectrum of the photosphere, but the Fraunhofer lines are absent. This may be explained by considering that the free electrons in highly ionized coronal gas scatter photospheric light to produce the observed continuum. It has been suggested that the Thomson scattering is operative in the process.
2. **E-corona:** The E-corona is due to the spectral emission lines produced by the ions, present in the coronal plasma. These emission lines include highly ionized atoms of iron, calcium, and some other elements. It may be observed in the broad or hot spectral emission lines and is the main source of the information about the corona's composition. The E-corona is much fainter than the K and F coronae and therefore its observation is very difficult.
3. **F-corona:** The F-corona which lies at $R > 2.5\ R_{\odot}$ displays the solar spectrum with the Fraunhofer lines superimposed on the continuum. This is sometimes known as the 'false'

part of the corona and the prefix *F* stands for Fraunhofer. The spectrum of the F-corona is produced by the Sun's light scattered by the tiny dust particles in the interplanetary space. The lines are assumed to be formed by the forward scattering produced by the particles with the sizes somewhat larger than the wavelengths scattered.

The F-corona together with the K-corona constitute what is known as the white corona, which merges into the interplanetary space with decreasing brightness.

2.9 Prominences

A quiescent prominence is an immense, dense, cool vertical sheet of plasma in the corona with a density $n \sim (0.5 - 1.0) \times 10^{17}$ particles/m^3, about 500 times more than the ambient coronal value and a temperature (5000 – 10,000 K) about 500 times less. It changes slowly in shape and is long-lived. The main observational features of prominences are as the following:

1. The geometry of a prominence is a thin vertical sheet. The length of the sheet lies between 60 and 600 Mm (typically 200 Mm), while its height ranges between 10 and 100 Mm (typically 50 Mm) and its width between 4 and 15 Mm (typically 6 Mm).
2. The particle density is between 10^{16} and 10^{17} particles/m^3, and thus the mass density is about $10^{-11} - 10^{-10}$ kg/m^3. The temperature is usually between 5000 K and 8000 K. Both the density and temperature tends to decrease with height.
3. The magnetic field has a magnitude between 3 and 30 gauss (typically 5 G) and usually increases with the increase of height.
4. Prominences are long-lived, lasting for 1-300 days, with high-latitude ones enduring typically 140 days.
5. Prominences tend to form where there are converging flows and shearing flows.

2.10 Coronal Mass Ejection

Coronal mass ejection (CME) is a phenomenon in which a significant amount of plasma and accompanying magnetic field is released from the Sun's corona into the solar wind. The CMEs are often associated with the solar flares and other forms of solar activity, but no acceptable theoretical explanation of these relationships has been given, so far. The CME entering the interplanetary space, is called the interplanetary coronal mass ejection (ICME). The ICMEs are often capable of reaching and colliding with the Earth's magnetosphere, where they can create geomagnetic storms, aurorae, and in rare cases damage to electrical power grids.

2.11 Heating of Solar Chromosphere and Corona

The solar corona has temperature of millions of degrees, which is several orders of magnitude larger than the temperature of photosphere. Because of high temperature in the corona, there should be losses due to conduction and radiation, but the coronal temperature is maintained at million degrees. The explanation for replenishing the energy loss has been challenging task for scientists. Probably, the answer to the question lies in the crucial magnetic connection through the atmosphere of the Sun. The magnetic field that threads the corona extends below the solar pho- tosphere, where the convective motions drag the magnetic field footpoints, tangling and twisting them.

2.12 Force-free Magnetic Fields

In a plasma, when the plasma pressure is negligibly small as compared to the magnetic pressure, then only the magnetic pressure is considered, and the magnetic field is the force-free magnetic field. In absence of gravity, the Navier-Stokes equation for a plasma, in steady state, becomes

$$0 = -\nabla p + \vec{j} \times \vec{B}$$

where p is the plasma pressure, $\vec{B}$ the magnetic field and $\vec{j}$ the electric current. Considering that plasma pressure is negligibly small compared to the magnetic pressure, i.e., $p << B^2/2\mu$, then the pressure term can be neglected. Here, μ is the magnetic permeability of plasma. Thus, we have

$$\vec{j} \times \vec{B} = 0$$

showing that the current density is either zero or parallel to the magnetic field. It may be expressed as $\mu_0 \vec{j} = \alpha \vec{B}$, where α is a spatial-varying function to be determined. Combining this equation with Maxwell's equations:

$$\nabla \times \vec{B} = \mu \vec{j}$$

and

$$\nabla \cdot \vec{B} = 0$$

and the vector identity: $\nabla . (\nabla \times \vec{B}) = 0$, leads to a pair of equations:

$$\vec{B} \cdot \nabla\alpha = 0$$

and

$$\nabla \times \vec{B} = \alpha \vec{B}$$

3. Solar Neutrinos

As discussed, in the thermonuclear reactions occurring in the core of the Sun, neutrinos are produced. Raymond Davis began measuring the neutrino flux from the Sun using a detector located about 1.5 km below the Earth's surface, in the Homestake Gold Mine in Lead, South Dakota. He found that the number of neutrinos detected on the Earth was almost one-third of the expected value. Because of very low cross section of the neutrino interactions with other matter, the neutrinos can easily travel through all the matters, and even through the Earth, while the other particles being originated from the space cannot. Thus, the underground detector assured of measuring the neutrinos created 500 seconds earlier in the core of the Sun.

The Davis neutrino detector had 100,000 gallons of C_2Cl_4 (perchloroethylene). One isotope of chlorine, the $_{17}Cl^{37}$ is capable of interacting with the neutrinos of sufficient energy to produce a radioactive isotope of argon, having a half-life of 35 days. The reaction is

$$_{17}\mathrm{Cl}^{37} + \nu_e \rightarrow {}_{18}\mathrm{A}^{37} + e^-$$

The threshold energy for this reaction, 0.814 MeV, is less than the energies of neutrinos produced in each step of the p-p chain except for the reaction

$$_1\mathrm{H}^1 + {}_1\mathrm{H}^1 \rightarrow {}_1\mathrm{H}^2 + e^+ + \nu_e$$

However, the reactions that account for 77% of the neutrinos detected in the Davis experiment is due to the decay of $_5B^8$

$$_5B^8 \rightarrow {}_4\mathrm{Be}^8 + e^+ + \nu_e$$

Unfortunately this reaction is very rare. Once every few months, the atoms of the argon produced were counted. The counting rate is measured in terms of the solar neutrino unit (SNU). 1 SNU is defined as 10^{-38} reactions per target atom per second. With approximately 2.2 $\times 10^{30}$ atoms of $_{17}Cl^{37}$ atoms in the tank, when only one argon atom is produced each day, this rate would correspond to 5.35 SNU.

Standard solar models predict that the outcome of the experiment should get a capture rate of 7.9 SNU whereas the actual data give 2.23 ± 0.26 SNU. That is, only one argon atom is produced every two to three days. This discrepancy in the production and detection rates of the solar neutrinos is known as the **solar neutrino puzzle**.

Other neutrino experiments, fundamentally different from the $_{17}Cl^{37}$ experiment, confirmed the existence of the solar neutrino puzzle.

Scientists continued attempts to resolve the discrepancy between the experiment and theory ever since the problem first appeared. Given the small number of neutrinos that are detected, one obvious solution is that there is some thing wrong with the experiments or with the analysis of the resulting data. However, all the attempts to find the errors that could produce symmetrically low counts failed.

Can the standard solar model be fundamentally flawed in some way? Do we understand solar interior well enough to predict the neutrino fluxes accurately?

The search for a theoretical resolution to the problems of solar neutrinos and oscillations has considered two general approaches: Either some fundamental physical process operating in the standard models is incorrect, or something happens to neutrinos on their way from the Sun's core to the Earth. The first of these possibilities resulted in the re-examination of a host of features of the solar model. One point was to enhance the helium abundance in the core. It improved the agreement between the theory and observations with regard to the solar oscillations, but it made the solar neutrino problem even worse.

An elegant solution to the solar neutrino problem suggests that the standard solar models are essentially correct but that the neutrinos produced in the Sun's core actually change before they reach the Earth. The Mikheyev-Smirnov-Wolfenstein (or MSW) effect involves the transformation of neutrinos from one type to another.

The neutrinos produced in various branches of the p-p chain are all electron neutrinos (ν_e). Two other type of neutrinos also exist: the muon neutrinos (ν_μ) and the tau neutrinos (ν_τ). The MSW effect suggests that neutrinos oscillate between the electron neutrinos, muon neutrinos, or tau neutrinos during their passage through the Sun. The neutrino oscillations are caused by the interactions with the electrons as the neutrinos travel towards the solar surface.

One testable consequence of the MSW effect is that the neutrinos would necessarily have mass, with $m(\nu_e) << m(\nu_\mu) << m(\nu_\tau)$. The required mass of the electron neutrino is well within the upper mass limit of approximately 7.2 eV established by the experiment.

4. End State of the Sun

The age of the Sun is about 5 billion years. In the next period of about 5 billion years, the Sun will cool down and will expand outward to a size having diameter several times larger than the current value (becoming a red giant), before blowing out its outer layers in the form of the planetary nebula, and leaving behind a solid body, known as the white dwarf. This white dwarf will cool down and its colour will become brown; called the brown dwarf. Finally, it will exhaust its all energy and will become a part of the dark matter in the universe.

5. Magnetohydrodynamics

The magnetohydrodynamics (MHD) is a physical-mathematical framework about the dynamics of magnetic fields in the electrically conducting fluids, e.g., in the plasmas and liquid metals. The word magnetohydrodynamics is comprised of the words magneto – meaning magnetic, hydro – meaning water (or liquid) and dynamics referring to the movement of an object under forces.

The central point of the MHD theory is that the conductive fluids can support the magnetic fields. The presence of the magnetic fields leads to the forces which in turn act on the fluid (typically a plasma), and therefore changing the geometry (or topology) and strength of the magnetic fields themselves. A key issue for a particular conducting fluid is the relative strength of advecting motions in the fluid, compared to the diffusive effects caused by the electrical resistivity.

The standard non-relativistic form of the MHD equations are the basic conservation laws, which in the SI units are as the following.

Continuity equation

$$\frac{\partial \rho}{\partial t} + \nabla \cdot (\rho \vec{u}) = 0$$

where ρ is the mass density and $\vec{u}$ the fluid bulk velocity.

Momentum equation

$$\rho\left(\frac{\partial}{\partial t} + \vec{u} \cdot \nabla\right)\vec{u}) = \vec{j} \times \vec{B} - \nabla p$$

where p is the gas pressure, $\vec{B}$ the magnetic field (magnetic flux density), $\vec{j} = \nabla \times \vec{B}/\mu_0$ the current density, and μ_0 the vacuum permeability.

Ideal Ohm's law

$$\vec{E} + \vec{u} \times \vec{B} = 0$$

Faraday's law

$$\frac{\partial \vec{B}}{\partial t} = -\nabla \times \vec{E}$$

Conservation of magnetic field

$$\nabla \cdot \vec{B} = 0$$

Energy equation

$$\frac{\mathrm{d}}{\mathrm{d}t}\left(\frac{p}{\rho^{\gamma}}\right) = 0$$

where $\gamma = 5/3$ is the ratio of the specific heats for an adiabatic equation of state.

These MHD equations are to be solved simultaneously, either analytically or numerically. An important dimensionless parameter in the MHD theory is the plasma beta β, defined as the ratio of the gas pressure p to the magnetic pressure

$$\beta = \frac{p}{B^2/2\mu_0}$$

When $\beta \ll 1$, the magnetic field dominates in the fluid, and the fluid is forced to move along with the field. When $\beta \gg 1$, the magnetic field is weak and the field is swirled along by the fluid.

In the ideal MHD, the topology of the magnetic field (connectivity of the field lines) does not change. If a set of magnetic field lines is twisted into a knot, they remain unchanged. This means that it is very difficult to change the magnetic topology of some field configuration by the MHD reconnection, if the resistivity is very low. It is therefore possible to store a large amount of energy in the magnetic fields by twisting them into a complex topology by the fluid motions. This energy can subsequently be released into the heat and the kinetic energy, if the scales of the magnetic field become very small leading to a large magnetic gradients or if the resistivity increases. In this case, the ideal MHD breaks down allowing the magnetic reconnection to occur.

5.1 Applications of MHD

Magnetism is found throughout the universe. Magnetic fields are known to exist in the planets, stars, accretion disks, interstellar medium, galaxies and in the active galactic nuclei. Often these magnetic fields are generated and maintained by the magnetohydrodynamics dynamo action (e.g., the solar dynamo or the planetary dynamos, which are described by the hydromagnetic dynamo theory), and in many cases, the magnetic field is dynamically dominant, determining the evolution of the object. On the Sun and other stars, these magnetic structures may take the form of the cool spots (sunspots) and the magnetic bright points.

Magnetic structures can store enormous amounts of energy that may be released through the reconnection of the magnetic field lines (MHD reconnection) leading to the solar flares and the stellar or even the galactic eruptions. It is in this way that the Sun causes the storms in the heliospherical magnetic field and ultimately aurorae and space weather.

5.2 Hall effect

The Hall effect was discovered by Edwin Hall in 1879, and it is the production of a voltage difference (called the Hall voltage) across an electrical conductor which is perpendicular to both the electric current in the conductor and the applied external magnetic field, which is also perpendicular to the current.

In a conductor, the flowing current consists of electrons, holes, and ions, which move approximately along a straight line in the direction of applied potential difference. When an external magnetic field, having component perpendicular to the direction of current, is applied, these charges experience the Lorentz force. The direction of Lorentz force is perpendicular to both the direction of current and the direction of applied magnetic field. Consequently, the paths of the charges are curved, thus moving charges accumulate on one face of the material. This leaves equal and opposite charges exposed on the opposite face. The separation of charges on these two opposite faces establishes an electric field that opposes the arrival of further charges on the faces. Thus, a steady electric potential between the faces is established as long as the current is flowing and magnetic field is applied.

Suppose, the current is flowing parallel to *x*-axis and an external magnetic field is applied along z-axis. The hall voltage is developed along the negative *y*-axis. Suppose, the electric field $\vec{E}$ is developed due to Hall voltage. The force $\vec{F}$ on a charge *q* moving with velocity $\vec{v}$ under the combined influence of electric field $\vec{E}$ and magnetic field $\vec{E}$ is expressed as

$$\vec{F} = q(\vec{E} + \vec{v} \times \vec{B})$$

In the steady state, the net force on the charge is zero, $\vec{F} = 0$. Thus, we get $0 = E_y - v_x B_z$. If *d* is the separation between the two faces where Hall voltage V_H is developed, $E_y = V_H/d$. Thus, we get

$$V_H = v_x B_z d$$

The conventional current due to holes is in the negative direction of the current due to electrons.

6. Helioseismology

In the helioseismology, we study of the propagation of wave oscillations, in particular the acoustic pressure waves (*p*-waves), in the Sun. Contrary to the seismic waves on the Earth, the solar waves have practically no shear component (*s*-waves). The solar pressure waves are considered to be produced through the turbulence in the convection zone near the solar surface. Certain frequencies are amplified by the constructive interference. The acoustic waves are transmitted to the outer photosphere of the Sun. These oscillations are detectable on almost any time series of solar images, but are best observed by measuring the Doppler shift of the photospheric absorption lines. Changes in the propagation of the oscillation waves through the Sun reveal inner structures and help the astrophysicists to develop an extremely detailed profiles for the interior conditions in the Sun.

Helioseismology can be used to image the far side of Sun from the Earth, including the sunspots. In the simple terms, the sunspots absorb the helioseismic waves. This sunspot absorption causes a seismic deficit.

The analysis of the data from the SOHO spacecraft shows the seismic images of the central portion of the solar far side being produced nearly continuously since late 2000.

6.1 Solar Oscillations

Individual oscillations in the Sun are damped and they die out within a few periods. However, the interference between these localized waves produces global standing waves, also known as the normal modes. Analysis of these overlapping modes are studied under the discipline of global helioseismology. These modes are essentially divided up into three categories, according to the restoring force that drives them: the acoustic, gravity, and surface-gravity wave modes.

- For the *p*-mode or the acoustic waves, the pressure is their restoring force, and they are called the *p*-mode. Their dynamics are determined by the variation of the speed of sound, inside the Sun. The *p*-mode oscillations have frequencies > 1 mHz and are very strong in the 2-4 mHz range, where they are often referred to as the 5-minute oscillations[2]. The *p*-modes at the solar surface have amplitudes of hundreds of kilometres and can be detected easily with Doppler imaging or with the sensitive spectral line intensity imaging.
- For the *g*-mode or gravity waves, the gravity is their restoring force, and they are called the *g*-mode. The *g*-mode oscillations are low frequency waves (0-0.4 mHz) and they are confined to the interior of Sun below the convection zone (which extends from 0.7-1.0 solar radius), and practically cannot reach the surface. Since the 1980s, there have been several claims of the *g*-mode detection, none of which have been confirmed.
- *f*-mode or surface gravity waves are also gravity waves, but occur at or near the photosphere, where the temperature gradient drops below the adiabatic lapse rate. Some f-modes of moderate and high degree, between $l = 117$ and $l = 300$, have been observed by Michelson Doppler Imager (MDI).

6.2 Analysis of Oscillation Data

The data from the time-series of solar spectra show all the oscillations overlapping. Thousands of modes have been detected (with true number perhaps being in millions). The mathematical technique of Fourier analysis is used to recover the information about individual modes from this mass of data.

The simplest modes to analyze are the radial ones; however most solar modes are non-radial. A nonradial mode is characterized by three wave numbers: the spherical-harmonic degree l and azimuthal order m which determine the behaviour of the mode over the surface of star and the radial order n which reflects the properties in the radial direction.

In general, the frequencies ω_{nlm} of stellar oscillations depend on all the three wave numbers. It is convenient, however, to separate the frequency into the multiplet frequency ω_{nl}, obtained as a suitable average over the azimuthal order m and corresponding to the spherically symmetric structure of the star, and the frequency splitting $\delta\omega_{nlm} = \omega_{nlm} - \omega_{nl}$.

Analysis of the oscillation data must attempt to separate these different frequency components. In case of the Sun, the oscillations can be observed directly as functions of position on the solar disk as well as time. Thus, here it is possible to analyze their spatial properties. This is done by means of a generalized 2-dimensional Fourier transform in position on the solar surface, to isolate particular values of l and m. This is followed by a Fourier transform in time which isolates the frequencies of modes of that type. In fact, the average over the stellar surface implicit in the observations of the stellar oscillations can be thought of as one example of such a spatial Fourier transform.

7. Solar Wind

Besides the electromagnetic radiations, the Sun continuously emits a stream of charged particles, called the solar wind. Main part of solar wind consists of electrons, protons and alpha particles

[2]Note: The 5 minutes per cycle is 1/300 cycles per second = 3.33 mHz.

with kinetic energy between 0.5 and 10 keV. The solar wind also includes a mixture of material found in the solar plasma, such as trace amounts of heavy ions and atomic nuclei such as C, N, O, Ne, Mg, Si, S, and Fe. Rarer traces of some other nuclei and isotopes such as P, Ti, Cr, ^{54}Fe and ^{56}Fe, and ^{58}Ni, ^{60}Ni, and ^{62}Ni also have been found in the solar wind. Superposed with the solar-wind plasma is the interplanetary magnetic field. The release of solar wind depends on the solar latitude and longitude (i.e., position on the surface of Sun) and shows a variation in density, temperature and speed with time. Its particles can escape the Sun's gravity; the solar wind remains supersonic up to a distance of few solar radii, and then its speed decreases and no longer is supersonic.

7.1 Parker's Solar Wind Model

In 1958, Parker proposed a model for solar wind. which can be understood in the following manner. The basic steady equations are

$$\nabla \cdot (\rho \vec{v}) = 0 \tag{10.12}$$

$$\rho(\vec{v} \cdot \nabla)\vec{v} = -\nabla p + \rho \vec{g} \tag{10.13}$$

$$p = \rho R T \tag{10.14}$$

$$T = T_0 \tag{10.15}$$

Let us consider that the velocity is purely radial so that $\vec{v} = v\hat{r}$ and the gravitational acceleration obeys the inverse square law, $g = -GM_\odot/r^2$, where $M_\odot$ is mass of the Sun. In spherical polar coordinates and assuming a steady flow, equation (10.12) gives

$$\frac{\mathrm{d}}{\mathrm{d}r}(r^2 \rho v) = 0$$

$$\Rightarrow \qquad r^2 \rho v = a \text{ (const)} \tag{10.16}$$

The raidial component of equation (10.13) becomes

$$\rho v \frac{\mathrm{d}v}{\mathrm{d}r} = -\frac{\mathrm{d}p}{\mathrm{d}r} - \frac{GM_\odot \rho}{r^2} \tag{10.17}$$

For isothermal sound speed defined as $c_s = (p/\rho)^{1/2}$, we have

$$p = c_s^2 \rho \tag{10.18}$$

and substituting equations (10.16) and (10.18) into (10.17) we get

$$\rho v \frac{\mathrm{d}v}{\mathrm{d}r} = -c_s^2 \frac{\mathrm{d}p}{\mathrm{d}r} - \frac{GM_\odot \rho}{r^2}$$

Then dividing by ρ we obtain

$$v \frac{\mathrm{d}v}{\mathrm{d}r} = -\frac{c_s^2}{\rho}\frac{\mathrm{d}p}{\mathrm{d}r} - \frac{GM_\odot}{r^2}$$

Now, using $\rho = a/r^2 v$ from equation (10.16), we have

$$v \frac{\mathrm{d}v}{\mathrm{d}r} = -c_s^2 r^2 v \frac{\mathrm{d}}{\mathrm{d}r}\left(\frac{1}{r^2 v}\right) - \frac{GM_\odot}{r^2}$$

or

$$v \frac{\mathrm{d}v}{\mathrm{d}r} = -c_s^2 r^2 v \left[-\frac{2}{r^3 v} - \frac{1}{r^2 v^2}\frac{\mathrm{d}v}{\mathrm{d}r}\right] - \frac{GM_\odot}{r^2}$$

or
$$\left(v - \frac{c_s^2}{v}\right)\frac{dv}{dr} = 2\frac{c_s^2}{r^2}(r - r_c)$$

where $r_c = GM_\odot / 2c_s^2$. On integration, we get

$$\frac{v^2}{2} - c_s^2 \ln v = 2c_s^2\left[\ln r + \frac{r_c}{r}\right] + A \qquad (10.19)$$

where A is constant of integration. Considering that at $r = r_c$, we have $v = k\,c_s$. The parameter k decides that at $r = r_c$, the speed of solar wind is subsonic ($k < 1$), sonic ($k = 1$) or supersonic ($k > 1$). Calculating the value of A at this point from equation (10.19), and using the value of A in equation (10.19), we get

$$\left(\frac{v}{c_s}\right)^2 - \log\left(\frac{v}{c_s}\right)^2 = 4\left[\log\left(\frac{r}{r_c}\right) + \frac{r_c}{r}\right] + C \qquad (10.20)$$

where $C = k^2 - 2\ln k - 4$ is a constant. The variation of C versus k is shown in Fig. 15. It shows that the lowest allowed value of C is –3, which corresponds to the sonic speed of solar wind at $r = r_c$. For other values of C, the speed of solar wind can be sub-sonic or super-sonic.

Solutions of equation (10.20) for three different values of C are given in Fig. 16.

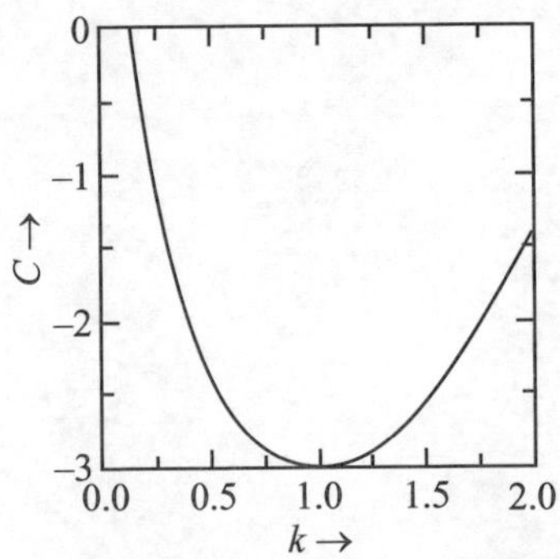

Fig. 15: Variation of C versus k.

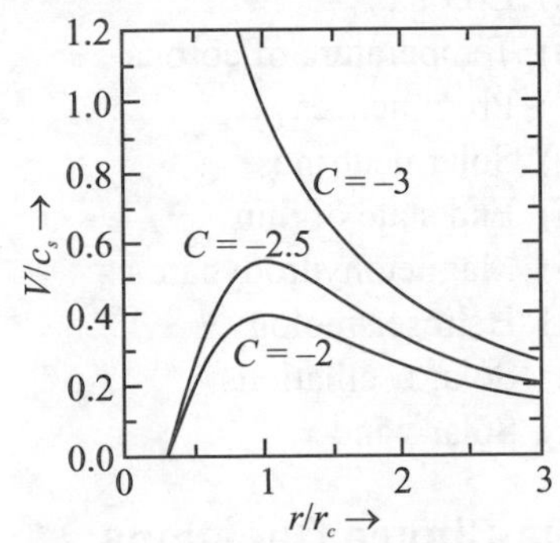

Fig. 16: Variation of v/c_s versus r/r_c.

Review Questions

1. Provide a brief description about the Sun.
2. Describe the interior of the Sun.
3. Describe the atmosphere of the Sun.
4. Explain how the temperature of corona was decided.
5. What is the source of energy in the core of the Sun?
6. Discuss about the solar neutrinos.
7. In the core of the Sun, suppose one helium nucleus is formed through the fusion of four protons. Calculate the amount of energy released per reaction. Given, mass of proton is 1.0079 amu, mass of helium nucleus is 4.0026 amu. 1 amu = 1.66×10^{-27} kg.
8. In the core of the Sun, suppose one helium nucleus is formed through the fusion of four protons. Calculate the amount of hydrogen converted into helium per year. Given, mass of

proton = 1.673×10^{-27} kg, mass of helium nucleus = 6.664×10^{-27} kg, luminosity of Sun $L_\odot = 3.86 \times 10^{26}$ J/s.

9. Describe various modes of helioseismology.
10. Describe the Babcock model of sunspot formation.
11. What is solar wind? Discuss Parker's explanation for the solar wind.
12. Write notes on the following:
    (i) Solar interior
    (ii) Energy generation in the Sun
    (iii) p-p chain
    (iv) CNO cycle
    (v) Photosphere
    (vi) Sunspots
    (vii) Faculae
    (viii) Granulation
    (ix) Chromosphere
    (x) Spicules
    (xi) Plages and filaments
    (xii) Corona
    (xiii) Temperature of corona
    (xiv) Prominences
    (xv) Solar neutrinos
    (xvi) End state of Sun
    (xvii) Magnetohydrodynamics
    (xviii) Helioseismology
    (xix) Solar oscillations
    (xx) Solar wind

Multiple-Choice Questions

1. A photon produced in the core of the Sun reaches the photosphere in about
   A. 2 sec B. 8 m 20 sec C. 1 year D. 10 million years
   Ans. D
2. Which is not the interior part of the Sun?
   A. core B. convective zone C. corona D. chromosphere
   Ans. D
3. Which is not the part of the solar atmosphere?
   A. chromosphere B. transition region C. corona D. convective zone
   Ans. D
4. Which is not the part of the solar atmosphere?
   A. chromosphere B. transition region C. corona D. radiation zone
   Ans. D
5. Umbra and penumbra are the parts of
   A. corona B. chromosphere C. solar wind D. sunspot **Ans. D**

6. Suppose, the Sun is taken to a distance of 10 pc from the Earth. Can the Sun be seen by unaided eyes?
   A. No B. only in the morning
   C. Yes D. only at the noon **Ans. C**
7. The sunspots lie up to the latitudes of about
   A. 10° B. 40° C. 60° D. 90° **Ans. B**
8. The time-period of the solar magnetic activity for getting the same polarity is
   A. 5 yr B. 11 yr C. 22 yr D. 44 yr **Ans. C**
9. The time-interval for having the consecutive maximum number of sunspots is
   A. 5 yr B. 11 yr C. 22 yr D. 44 yr **Ans. B**

11 Solar System

According to the heliocentric system, the Sun is the main object around which all the planets, asteroids (small planets) and other smaller bodies are revolving. Hence, each cosmic body revolving around the Sun is not a planet. A planet must have at least specific mass (size). Because of this reason, the Pluto has been removed from the list of the planets. A body revolving around the Sun may have its satellites revolving around it. A simple exercise of numbers gave the idea of the missing planet, which led to the discovery of asteroids. All these bodies are directly or indirectly controlled by the gravitational field of the Sun. Besides them, there are some bodies which reach to the region between the planets, such as the comets and the meteors. All these bodies together constitute a group, called the solar system. The solar system may consists of:

1. planets
2. asteroids (minor planets)
3. other minor planets
4. comets
5. meteors

In this chapter, we have discussed about the solar system. Students in the schools are given information about the solar system.

1. Planets

In the language of a common person, a planet is a natural body which revolves around the Sun. This is incomplete definition, as a specific mass (or size) of the revolving body is required in order to call it a planet. A body having more than the specific mass (or size) and revolving around the Sun is known as a planet. We know that there are eight planets:[1] Mercury, Venus, Earth, Mars, Jupiter, Saturn, Uranus, and Neptune. These planets revolve around the Sun in distinct elliptical orbits, with the Sun at one (common) focus of all the ellipses. Besides the revolution around the Sun, a planet spins about its own axis. Let us first look into some physical parameters of the planets.

1.1 Mass

Planets are physical bodies, and thus each of them has mass. Estimated values of the masses of the planets are given in Table 1. The Jupiter is 318 times heavier than the Earth. That is, out of the material of the Jupiter, 318 Earth-like bodies can be constructed. Table 1 shows that the heaviest planet is the Jupiter and the lightest planet is the Mercury.

---

[1]We have later on discussed why the Pluto no more belongs to the list of the planets. Before 24 August 2006, there have been 9 planets: Mercury, Venus, Earth, Mars, Jupiter, Saturn, Uranus, Neptune and Pluto.

1.2 Density

A planet has its finite volume and finite mass. Though the distribution of mass in a planet, in general, is not uniform, we can however calculate the average density of a planet. Estimated values of the mean densities of planets are given in Table 1. The Earth has the largest density and the lowest density is of the Saturn. Interesting to note that the density of Saturn is smaller than that of the water. That is, if all the planets could be put in water, only the Saturn would float.

Table 1. Some parameters for the planets

Planet	Mass (kg)	Density (10^3 kg/m^3)	Distance (AU)	Orbital period
Mercury	3.28×10^{23}	5.43	0.39	87.97 days
Venus	4.87×10^{24}	5.24	0.72	224.71 days
Earth	5.97×10^{24}	5.52	1.00	365.25 days
Mars	6.39×10^{23}	3.94	1.52	1.88 yr
Jupiter	1.90×10^{27}	1.33	5.20	11.86 yr
Saturn	5.68×10^{26}	0.70	9.54	29.46 yr
Uranus	8.68×10^{25}	1.19	19.18	84.02 yr
Neptune	1.02×10^{26}	1.66	30.06	164.77 yr

1 AU = 1.496×10^8 km

1.3 Mean Distance from Sun

Each of the planets revolves around the Sun in an elliptical path. The Sun is situated at one of the foci of ellipse of the path. Each planet has its own elliptical path. Thus, there is a common point, where one focus of each elliptical orbit coincides. This common point is the position of the Sun. Values of the mean distances of planets from the Sun is reported in Table 1 in AU (Astronomical Unit). The distance of a planet is according to its velocity of revolution and vice versa. Suppose, a planet of mass *m* revolves around the Sun of mass *M* with the velocity *v*, the equilibrium is governed by the gravitational force and the centrifugal force.

$$\frac{GMm}{r^2} = \frac{mv^2}{r}$$

or

$$v = \sqrt{\frac{GM}{r}}$$

where *r* is the distance between the Sun and the planet. Notice that the velocity does not depend on the mass of the planet. It shows that a planet closer to the Sun revolves with higher speed. For convenience, the orbit of planet is generally taken as circular.

1.4 Orbital Period

The time taken by a planet to make one revolution around the Sun is known as the orbital period. The orbital period *T* can be obtained as

$$T = \frac{2\pi r}{v} = \frac{2\pi r^{3/2}}{\sqrt{GM}}$$

or
$$T^2 = \frac{4\pi^2 r^3}{GM}$$

This relation $T^2 \propto r^3$ shows the Kepler's third law of planetary motion. It shows that a planet at a larger distance from the Sun has a larger period of revolution. The periods of revolution of the planets are given in Table 1. It is interesting to note that the Neptune takes 164.77 years to complete one revolution around the Sun.

1.5 Orbital Eccentricity

Suppose, semi-major axis and semi-minor axis of the elliptical path of a planet revolving around the Sun are a and b, respectively. Then, the eccentricity of the orbit is expressed as

$$e = \sqrt{1 - b^2/a^2}$$

For a circular orbit $a = b$ and the eccentricity is zero. Values of the orbital eccentricities of the paths of the planets are given in Table 2. It shows that the path of any planet is not circular in shape. Venus has the lowest eccentricity.

Table 2. Some parameters for the planets

Planet	Orbital eccentricity	Rotation period	Equatorial radius (km)	Polar oblateness	Escape velocity (km/s)
Mercury	0.2056	58.6462 days	2439.99	0.0	4.2
Venus	0.0068	243.026† days	6051.813	0.0	10.4
Earth	0.0167	$23^h56^m4^s.1$	6378.137	0.003352	11.2
Mars	0.0934	$24^h37^m22^s.1$	3397.515	0.005186	5.0
Jupiter	0.0485	$9^h55^m29^s.1$	71398	0.064809	59.6
Saturn	0.0557	$10^h39^m24^s$	60330	0.107621	35.4
Uranus	0.0472	$17^h14^m24^s$	25559	0.024	21.3
Neptune	0.0086	16^h3^m	24750	0.0259	23.4

† denotes the retrograde motion

1.6 Rotation Period

Besides the revolution around the Sun, a planet spins about an axis passing through its centre of gravity. This spin axis intersects the surface of planet at two points, called the poles. Values of the rotation period of planets are given in Table 2.

Members of the solar system, in general, move from west to east. According to the convention, the motion from west to east is taken as positive. When the direction of motion is from east to west, it is known as the retrograde motion.

1.7 Equatorial Radius

A plane perpendicular to the spin axis and passing through the centre of gravity of the planet intersects its surface in a circle, called the equator. Owing to the spin motion, the planet is oblate spheroid and the radius of equator (called the equatorial radius) is in general larger than the distance of a pole from its centre of gravity. Values of the equatorial radius of the planets are given in Table 2.

1.8 Polar Oblateness

Suppose, equatorial and polar radii of a planet are, respectively, R_e and R_p. The polar oblateness (or flattening ratio) of the planet is expressed as

$$\text{Polar oblateness} = \frac{R_e - R_p}{R_p}$$

Values of the polar oblateness of the planets are given in Table 2. Zero value of polar oblateness for a planet (Mercury and Venus) shows that the planet is spherical in shape, and the equatorial and polar radii of planet are equal.

1.9 Escape Velocity

Escape velocity is defined as the minimum velocity that an object need to acquire to escape from the surface of a heavenly body, such as a planet, satellite. Let us consider a spherical body of mass M and radius R. Suppose, an object of mass m is thrown with velocity v from the surface of body. When the object is at a distance r from the centre of the body, the force of attraction between the body and the object is

$$F = -G\frac{Mm}{r^2}$$

The potential energy V is expressed as

$$F = -\frac{\mathrm{d}V}{\mathrm{d}r}$$

or

$$V = -\int F\,\mathrm{d}r + C$$

where C is a constant of integration. Using the expression for F, we get

$$V = -G\frac{Mm}{r} + C$$

Since $V = 0$ when $r = \infty$, we have $C = 0$ and therefore

$$V = -G\frac{Mm}{r}$$

When the object is on the surface of body, the potential energy is

$$V = -G\frac{Mm}{R}$$

Thus, in order to send the object out of the gravitational field of the body at least this much amount of kinetic energy be given to the object. The minimum velocity v_e given to the object so that it goes away from the body's gravitational field is known as the escape velocity. Thus, we have

$$\frac{1}{2}mv_e^2 = G\frac{Mm}{R}$$

or
$$v_e = \sqrt{\frac{2GM}{R}} \tag{11.1}$$

It is interesting to note that the escape velocity does not depend on the mass of the object. If g denotes the acceleration due to gravity of the body, we have

$$g = \frac{GM}{R^2} \tag{11.2}$$

From equations (11.1) and (11.2), we have

$$v_e = \sqrt{2gR}$$

Values of escape velocity for the planets are given in Table 2.

Example 1.

Mass and radius of the Moon are 7.34×10^{22} kg and 1.74×10^6 m, respectively. Calculate the escape velocity for the Moon. (Given, $G = 6.67 \times 10^{-11}$ N m^2/kg.)

Solution:

The escape velocity is

$$v_e = \sqrt{\frac{2GM}{R}}$$

$$= \sqrt{\frac{2\times 6.67\times 10^{-11}\times 7.34\times 10^{22}}{1.74\times 10^6}}$$

$$= 2372 \text{ m/s}$$

The escape velocity of the Moon is 2.4 km/s. It is much smaller than that for the Earth. Because of this low value of the escape velocity, the Moon does not have any atmosphere, as all the material from its surface escaped away.

Existence of atmosphere around a celestial body

For a heavenly body of mass M radius R and acceleration due to gravity g, the escape velocity is

$$v_e = \sqrt{\frac{2GM}{R}} = \sqrt{2gR}$$

That is, an object having velocity more than v_e escapes from the surface of the heavenly body. Suppose, T is the temperature of the atmosphere around the body. The gas of molecular mass m in the atmosphere has velocity expressed as

$$\frac{1}{2}mv^2 = \frac{3}{2}kT$$

or
$$v = \sqrt{\frac{3kT}{m}}$$

Thus, if $v > v_e$, the molecules escape from the atmosphere of the body. Consequently, the body does not have atmosphere. For example, the satellite Moon has no atmosphere.

However, at a given temperature T, all the molecules do not have the same velocity v. Some molecules may have large velocity and some may have small velocity. The molecules follow the Maxwellian distribution of velocity and the average velocity is

$$\bar{v} = \sqrt{\frac{8kT}{\pi m}}$$

Following the theory of J.H. Jeans, if a gas is to remain on a body for a period of 1 billion years, the mean speed $\vec{v}$ should be at least $v_e/5$. Thus, for the existence of atmosphere around a celestial body, we have a condition

$$\sqrt{\frac{8kT}{\pi m}} \leq \frac{v_e}{5}$$

or

$$\frac{8kT}{\pi m} \leq \frac{2GM}{25R}$$

Thus, the minimum molecular mass m that can remain in the atmosphere of a celestial body is

$$m = \frac{100kTR}{\pi GM}$$

For $T = 300$ K on the Earth, we have

$$m = \frac{100 \times 1.38 \times 10^{-23} \times 300 \times 6.4 \times 10^{6}}{3.14 \times 6.67 \times 10^{-11} \times 6 \times 10^{24}} = 2.10 \times 10^{-27} \text{ kg}$$

Since 1 amu = 1.66×10^{-27} kg, we have $m = 1.27$ amu. Thus, there is no possibility of hydrogen ($m = 1$) to exist in the Earth's atmosphere. However, in the upper regions of the atmosphere, the temperature is very low and therefore, the atomic hydrogen can exist in the Earth's atmosphere.

Let us look for the Moon with the atmospheric temperature $T = 400$ K. We have

$$m = \frac{100 \times 1.38 \times 10^{-23} \times 400 \times 1.7 \times 10^{6}}{3.14 \times 6.67 \times 10^{-11} \times 7.3 \times 10^{22}}$$

$$= 6.14 \times 10^{-26} \text{ kg} = 37 \text{ amu}$$

It shows that the gases nitrogen (28), oxygen (32), which are the predominant constituents of the Earth's atmosphere cannot be present on the Moon.

1.10 Equatorial Gravity

As the equatorial radius R_e is larger than the polar radius R_p, the gravity at the equator is smaller than that at a pole. Table 3 has the values of equatorial gravity of a planet.

1.11 Surface Temperature

With the spin motion of a planet, surface temperature at a point on the planet changes with time. This change is quite appreciable in case of inner planets (Mercury, Venus, Earth and Mars)[2]. The surface temperature is given in Table 3. Some planets have thick atmosphere, and

[2]Planets may be classified into two categories: (i) the inner planets, and (ii) the outer planets. The orbit of missing planet (asteroids) is the boundary between the planets of two categories.

therefore, the temperature of the outer layer of cloud around the planet is reported, instead of the surface temperature.

Table 3. Some parameters for the planets

Planet	Equatorial gravity (m/s²)	Surface temp (K)	Albedo	Obliquity of orbit	Obliquity of rotation
Mercury	3.704	90-700	0.13	7° 0′ 15″	0° .0
Venus	8.870	740-250*	0.72	3° 23′ 40″	−2° .7
Earth	9.780	223-328	0.45	0°	23° 27′
Mars	3.694	136-243	0.16	1° 51′	25° 11′
Jupiter	22.65	130*	0.43	1° 18′ 17″	3° 7′
Saturn	8.792	94*	0.61	2° 29′ 22″	26° 43′
Uranus	8.699	52*	0.35	0° 46′ 23″	−82° 8′
Neptune	11.33	53*	0.40	1° 46′ 22″	29° 34′

* denotes the temperature at the top of the atmosphere

Example 2.

The power output of the Sun is 4×10^{26} W. Calculate the temperature of Venus situated at a distance of 110 million km from the Sun. Radius of Venus = 6×10^6 m.

Solution:

Power received by the Venus per unit area is

$$\text{Power per unit area} = \frac{\text{Total power}}{\text{Total area}}$$

$$= \frac{4\times10^{26}}{4\pi(1.1\times10^{11})^2} = 2600 \text{ W/m}$$

The surface area that Venus presents to the Sun is that of a disk with a radius equal to that of Venus. This area is

$$\pi \times (6 \times 10^6)^2 = 1.1 \times 10^{14} \text{ m}^2$$

so that the total power received by the Venus is the product of power per unit area and the presented area:

$$2600 \times 1.1 \times 10^{14} = 2.9 \times 10^{17} \text{ W}.$$

The albedo of the Venus is 0.72. That is only 28% of the incident radiation reaches the planet. The energy reaching the planet is

$$0.28 \times 2.9 \times 10^{17} = 8 \times 10^{16} \text{ W}.$$

For the Venus to have a steady state temperature, it must radiate this same amount of energy from its surface. For the unknown temperature, we have

$$P = \sigma T^4 \times 4\pi r^2$$

$$= (5.67 \times 10^{-8})T^4 \times 4\pi(6 \times 10^6)^2$$

$$= 2.6 \times 10^7 T^4$$

The power received is equal to the power radiated. Thus, we have

$$8 \times 10^{16} = 2.6 \times 10^{7} T^{4}$$

and

$$T = 240 \text{ K}$$

This value of 240 K is a bit lower than a more sophisticated calculation. It may be noted that the internal heating also plays a role in increasing the ambient temperature.

1.12 Albedo

Solar energy (sunlight) continuously falls on the surface of a planet. A part of that energy is reflected back, whereas the rest is absorbed by the planet. The Albedo of a planet is defined as the ratio of the amount of energy reflected by the planet to the amount of energy incident on the planet.

$$\text{Albedo} = \frac{\text{Energy reflected}}{\text{Energy incident}}$$

The albedo of the planets are reported in Table 3. The albedo of the Venus is the largest and therefore it appears as the most beautiful planet.

1.13 Obliquity of Orbit

The elliptical path of the Earth around the Sun is known as the ecliptic. The plane in which the ecliptic lies is known as the ecliptic-plane. Each of the planets has a plane of its orbit, which is not parallel to the ecliptic-plane. The angle between the plane of the orbit of the planet and the ecliptic-plane is known as the obliquity of the orbit of the planet. Table 3 displays also the values of the obliquity of the orbit of the planets.

1.14 Obliquity of Rotation

Rotation (spin) axis (and thus the equatorial plane) of a planet is inclined with the ecliptic-plane. The angle between the equatorial plane of the planet and the ecliptic plane is known as the obliquity of the rotation. Values of the obliquity of the rotation of planets are displayed in Table 3.

1.15 Roche Limit

Around a celestial body, the Roche limit, also called the Roche radius, is the distance within which a second celestial body disintegrates. It happens because the tidal forces due to the first body exceed the gravitational self-attraction of second body. The disintegrated material of second body orbits around the first body and disperses to form the rings, whereas the material outside the Roche limit tends to coalesce. This theoretical limit around a celestial body, was first calculated by Edouard Roche in 1848, and it depends on the radius of the first body and on the ratio of the densities of the two bodies.

$$d \approx 1.26R\left(\frac{\rho_M}{\rho_m}\right)^{1/3}$$

This relation may be derived in the following manner.

The gravitational force F_G on the mass u towards a satellite having mass m and radius r can be expressed according to Newton's law of gravitation

$$F_G = -\frac{Gmu}{r^2}$$

The tidal force F_T on the mass u towards the primary with radius R and mass M, at a distance d between the centres of the two bodies, can be expressed approximately as

$$F_T = -\frac{2GMur}{d^3}$$

This expression can be obtained by taking the difference in the primary's gravitational pull on the centre of the satellite and on the edge of the satellite closest to the primary:

$$F_T = -\frac{GMu}{(d-r)^2} - \left(-\frac{GMu}{d^2}\right)$$

$$F_T = -GMu\frac{d^2-(d-r)^2}{d^2(d-r)^2}$$

or

$$F_T = -GMu\frac{2dr-r^2}{d^2(d-r)^2}$$

In the approximation where $r \ll R$ and $R < d$, it can be said that the r^2 the numerator and every term with r in the denominator goes to zero, which gives us:

$$F_T = -GMu\frac{2dr}{d^4}$$

or

$$F_T = -GMu\frac{2r}{d^3}$$

The Roche limit is reached when the gravitational force and the tidal force balance each other out

$$F_G = F_T$$

or

$$-\frac{Gmu}{r^2} = -GMu\frac{2r}{d^3}$$

It gives the Roche limit, d, as

$$d = r\left(2\frac{M}{m}\right)^{1/3}$$

The radius of the satellite should not appear in the expression for the limit, so it is re-written in terms of densities.

For the spheres the masses M and m, we can write

$$M = \frac{4\pi\rho_M R^3}{3}$$

and

$$m = \frac{4\pi\rho_m R^3}{3}$$

where ρ_M and ρ_m are densities, and R and r are radii, respectively. Substituting for the masses in the equation for the Roche limit, we get

$$d = R\left(2\frac{\rho_M}{\rho_m}\right)^{1/3}$$

or
$$d \approx 1.26R\left(\frac{\rho_M}{\rho_m}\right)^{1/3}$$

This is the required expression for the Roche limit.

1.16 Circulation Patterns

A large-scale movement of air is the atmospheric circulation. This air circulation along with the ocean circulation is a source of thermal energy redistributed on the surface of the Earth. The atmospheric circulation of the Earth though varies from year to year, but the large-scale structure of its circulation remains fairly constant.

The weather on the Earth is a consequence of its illumination by the Sun, and the laws of thermodynamics. The atmospheric circulation can be viewed as a heat engine driven by the Sun's energy. The work done by the engine produces the motion of the masses of air and in that process, it redistributes the energy absorbed by the Earth's surface near the tropics to the latitudes nearer the poles, and thence to space.

2. Mercury

Mercury (Fig. 1) is the innermost and smallest planet in our solar system, orbiting around the Sun in 87.969 days. It is almost spherical in shape. Since the Mercury is very close to the Sun, it is quite difficult to observe it from the Earth. The maximum angular separation of Mercury relative to the Sun is 28.3°. So, the Mercury is seen only for about one and half hour under the favourable conditions, either before the sunrise, or after the sunset.

Fig. 1: Mercury.
(*Courtesy:* Google website)

When viewed from the Earth, the apparent magnitude of the Mercury varies from 2.3 to 5.7. Since the orbit of the Mercury lies inside the orbit of the Earth, it shows Moon-like phases. It does not contain any atmosphere, because of its lower surface gravity. Mercury is similar to the Moon in the appearance. Its surface is full of craters produced either by volcanoes, or by the impacts of small foreign debris. Mercury has a large iron core and is found to have a bipolar magnetic field of strength about 200-300 nT (1 nT = 10^{-9} Tesla). It has no natural satellites. Surface temperature of Mercury varies from about 90 to 700 K.

3. Venus

Venus (Fig. 2) is the second planet in our solar system, revolving around the Sun in 224.7 days. It is almost spherical in shape. Venus is in many respect an Earth-like planet, particularly in its size, mass and retention of the atmosphere. Because of this, it is sometimes called the Earth's sister-planet. Owing to its exquisite brilliance and appearance in the morning and in the evening sky alternately, Venus is imagined as a goddess of love and beauty.

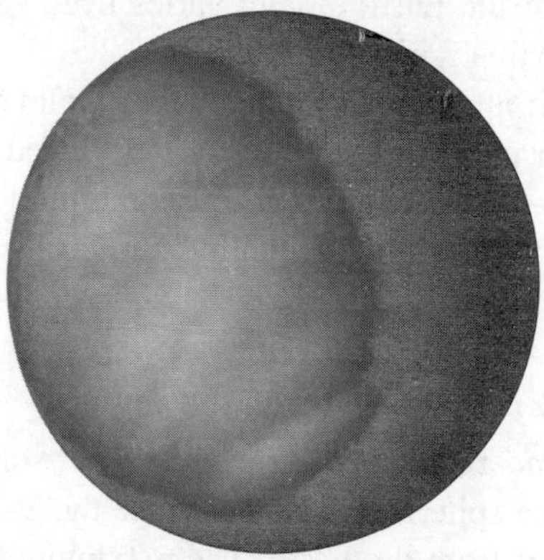

Fig. 2: Venus.
(*Courtesy:* Google website)

As seen from the Earth, Venus is the most brilliant object next to the Moon in the night sky. In fact, this is the only planet that can be seen in the broad daylight with the naked eyes. Its apparent magnitude reaches up to – 4.6. Its maximum angular separation from the Sun is 47.8°. Venus can be seen up to three hours under the favourable circumstances, either before the sunrise or after the sunset. It reaches its maximum brightness shortly before the sunrise or shortly after the sunset. For this reason, it is often called the Morning Star or the Evening Star.

The telescopic images of the Venus are rather dull and featureless, although its appearance in different Moon-like phases is quite attractive. Incidentally, it was the characteristic phase pattern of Venus that led Galileo Galilee to produce a convincing argument against the geocentric concept of the solar system.

The Venus is totally surrounded by its own cloud system. Clouds reflect light quite efficiently. In the case of the Venus, more than 70% of incident sunlight is reflected back by its topmost thin layer of clouds. Since its surface cannot be seen from outside, its rotation period cannot be ascertained directly.

Venus has a thick atmosphere containing ~96.6% of CO_2, 3.2% of N_2 and no water vapour. Possibly all the water molecules have been dissociated into H_2 and O_2, and escaped from the atmosphere of the Venus. CO_2 has a property that it allows visible light to pass through but absorbs the infrared radiations that the surface returns or re-emits. So, the sunlight can get in, and after being thermalized, it is radiated in the infrared radiations which are efficiently trapped due to the presence of CO_2 in the atmosphere. This effect, known as the **green house effect**, is responsible for raising the surface temperature of the Venus to a value as high as 650 K. The atmospheric pressure at the planet's surface is believed to be 92 times that of the Earth.

The Venus contains a lot of sulphur which finally forms the sulphuric acid and the clouds of the Venus are made up of the droplets of the sulphuric acid. Venus also has no natural satellites.

4. Earth

The Earth (Fig. 3) is the densest planet in our solar system. It makes one round around the Sun in 365.25 days and is sometimes referred to as the World, the Blue Planet, or by its Latin name, Terra. As we are housed on the Earth's surface, we must therefore be knowing a lot about it. The most reliable method that is used today for its age determination is the method of radioactive chronology. The Earth is estimated to be formed 4.54 billion years ago, and the life appeared on its surface within a billion years. Millions of species, including the human, live on its surface. Earth is currently the only place in the universe where life is known to exist.

Fig. 3: Earth.
(*Courtesy:* Google website)

The most important thing about the Earth's atmosphere is that it contains a lot of oxygen and nitrogen which are usually not found on other planets. They contain either too much H_2, He, CO_2 or methane, ammonia, etc. It is also important that the Earth's atmosphere contains ozone in sufficient amount which readily absorbs the ultraviolet radiations coming from the outer atmosphere and protects the life on the Earth's surface.

The Earth's magnetic field also shunts out most of the lethal solar cosmic rays from the equatorial regions which otherwise would have been extremely harmful to the life. Interaction of the Earth's magnetic field with the solar cosmic rays makes the latter channelled to the polar regions of the Earth. There they can penetrate sufficiently deep into the atmosphere and produce grand display of the auroral lights as a result of excitation and spontaneous de-excitation of the atmospheric nitrogen and oxygen.

The Earth has a sufficiently strong magnetic field of 32 μT (1 μT = 10^{-6} Tesla). It means that some part of the Earth's interior must be in the molten state. Detailed seismological studies have revealed that at a depth of about 2900 km from the surface, the matter changes its state, from the solid to the liquid state. At a depth about 5100 km from the surface there is another transition from the liquid to the solid state again. The central density and temperature of the Earth reaches as high as 1.5×10^4 kg/m^3 and 6900 K, respectively.

71% of the Earth's surface is covered with water having an average depth of about 3 km. Earth's axis of rotation is tilted 23.4° away from the perpendicular to its orbital plane, producing the seasonal variations on the Earth's surface.

Most part of the electromagnetic radiations from the outer space is blocked by the Earth's atmosphere. Radiations reaching on the surface of Earth, however, mainly lie in two bands, known as the optical window and the radio window. The optical window ranging from 3000 Å to 10,000 Å has been exploited for several centuries in order to get information/knowledge about the cosmic objects as well as about our solar system. Since the radiations of mm and cm wavelengths cannot be seen by the human eyes directly, therefore, they could not be noticed until 1932, when Karl G. Jansky accidentally discovered the radio window, which ranges from a few mm to a few decameters in wavelength.

The Earth has only one natural satellite, called the Moon.

4.1 Radioactive Dating

A technique used to date materials such as rocks or carbon, is knwon as the radioactive dating, radiometric dating, or radioisotope dating. In this technique, in the materials trace radioactive impurities were selectively incorporated when they were formed. This technique compares the abundance of a naturally occurring radioactive isotope within the material to the abundance of its decay products, which form at a known constant rate of decay. In the modern era, the use of radiometric dating is the principal source of information about the absolute age of rocks and other geological features, including the age of fossilized life forms or the age of the Earth itself. It can also be used to date a wide range of natural and man-made materials.

A mathematical expression relating the radioactive decay to geologic time is

$$D^* = D_0 + N(t)(e^{\lambda t} - 1)$$

where

- t is the age of the sample
- D^* is number of atoms of the radiogenic daughter isotope in the sample
- D_0 is number of atoms of the daughter isotope in the original or initial composition
- $N(t)$ is number of atoms of the parent isotope in the sample at time t (the present), given by $N(t) = N_0\, e^{-\lambda t}$, and
- λ is the decay constant of the parent isotope, equal to the inverse of the radioactive half-life of the parent isotope times the natural logarithm of 2.

For calculation of the age, it is assumed that the system is closed (i.e., neither parent nor daughter isotopes have been lost from system), D_0 must be either negligible or can be accurately estimated, λ is known to a high accuracy, and one has accurate and precise measurements of D^* and $N(t)$.

This equation makes the use of information on the composition of parent and daughter isotopes at the time the material being tested cooled below its closure temperature.

5. Moon

The Earth has only one natural satellite called the Moon (Fig. 4). Its radius is about 27% that of the Earth, mass being 1.23% that of the Earth, and density about 3.34 $\times 10^3$ kg/m^3. Besides its revolution around the Earth, the Moon spins about its own axis. The time-period of the revolution is exactly equal to the time-period of its spin motion. It is 27.32166 days. So, the Moon keeps its one face permanently turned towards the Earth. That is why we always see the same part of the surface of the Moon. On the surface of the Moon, there are craters and mountains.

Fig. 4: Moon.
(*Courtesy:* Google website)

The Moon's surface is found to be highly cratered, interspersed by the darkish plains, called Maria. The craters are mostly impact type with some of them having distinct ray patterns. Since the Moon has very little magnetic field, its interior must be geologically dead, but the surface must retain the craters of all ages, as it is practically devoid of atmosphere (about 10^{12} atoms/m^3). The heights of the lunar mountains can be measured by measuring the length of shadows they cast during the morning's or the late afternoon's low-lying Sun.

Lunar rocks that have been brought to the Earth by the Apollo astronauts have provided many interesting clues regarding the formation of the Moon, which is now believed to have either taken place simultaneously with the Earth or have been captured in very early times.

6. Solar Eclipse

As seen from the Earth, a solar eclipse occurs when the Moon passes between the Sun and the Earth, and the Moon blocks the Sun fully (Fig. 5) or partially. This can happen only at the new Moon day. For a total solar eclipse, the disk of the Sun is fully obscured by the Moon. In a partial or annular eclipse, only a part of Sun is obscured.

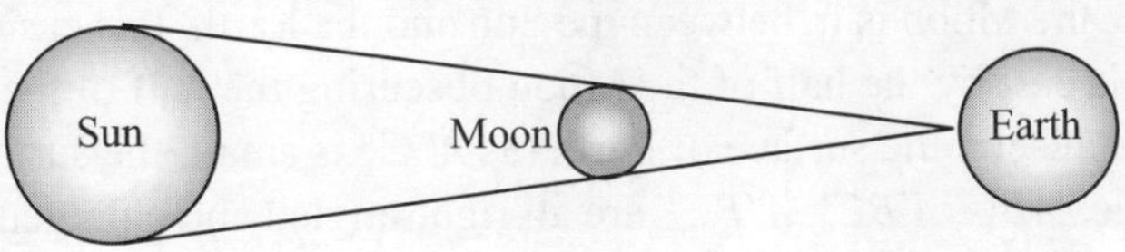

Fig. 5: During the total solar eclipse, the Moon covers the Sun.

Since the duration of a total solar eclipse is very small, it shows that the angular sizes of both the Moon and the Sun are nearly equal, when seen from the Earth.

If the Moon were in a perfectly circular orbit and in the same orbital plane, there have been total solar eclipses every month. However, the Moon's orbit is inclined (tilted) at more than 5° to the ecliptic, so its shadow at the new Moon usually misses the Earth.

Some dates when total solar eclipse will take place in future are: 8 April 2024, 12 August 2026, 2 August 2027 and 22 July 2028.

Some dates when the total solar eclipse occurred in the recent past are: 4 December 2021, 14 December 2020, 2 July 2019, 21 August 2017, 9 March 2016, 20 March 2015, 13 November 2012, 11 July 2010, 22 July 2009, 1 August 2008, 29 March 2006.

7. Lunar Eclipse

A lunar eclipse occurs when the Moon passes directly behind the Earth into its shadow (Fig. 6). This can occur only when the Sun, Earth and Moon are aligned exactly, or very closely so, with the Earth in the middle. Hence, a lunar eclipse can occur only at a night of full Moon.

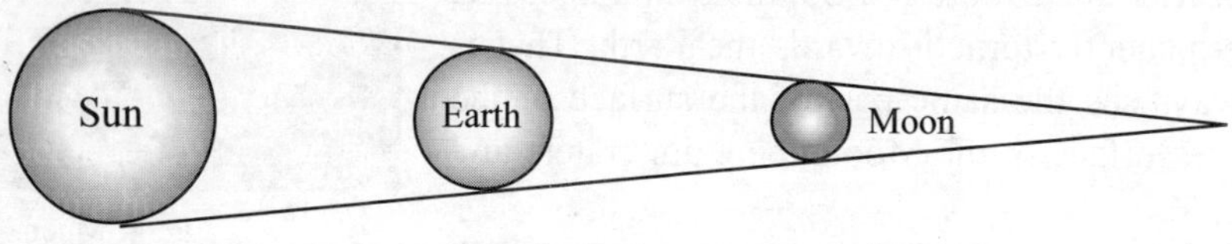

Fig. 6: During the total lunar eclipse, the Moon moves through the shadow of the Earth.

The type and length of a lunar eclipse depends upon the Moon's location relative to its orbital nodes[3].

Some dates when the total lunar eclipse will take place in future are: 14 March 2025, 7 September 2025, 3 March 2026 and 31 December 2028.

Some dates when the total lunar eclipse occurred in the recent past are: 26 May 2021, 21 January 2019, 27 July 2018, 31 January 2018, 28 September 2015, 4 August 2015, 8 October 2014, 15 August 2014, 20 December 2011, 15 June 2011, 21 December 2010, 21 February 2008, 28 August 2007, 3 March 2007, 28 October 2004.

Example 3.

Suppose, a solar eclipse takes place when the Sun and the Moon are separated from the Earth by 1.53×10^8 km and 3.79×10^5 km, respectively. Calculate the width of a strip on the Earth's surface from which a total solar eclipse is observed. The diameters of the Sun, Earth and Moon are 1.392×10^6 km, 3480 km, 12800 km, respectively.

Solution:

For the solar eclipse, the Moon is in between the Sun and the Earth. It is easier to work with the half shadow which is cast by the half of the Moon obscuring the half of the Sun and produces a shadow of length $A''C''$ on the surface of Earth (as $A''C''$ is small, the surface can be taken as a plane.) The triangles ABC, $A'BC'$, $A''BC''$ are all right-angled and enlargements of each other so that they all subtend the same angle, θ, at B. The position of B is initially unknown but can be calculated from the information available from the triangles ABC and $A'BC'$:

[3]The lunar orbital nodes are the points where the orbit of Moon crosses the orbit of the Earth. The ascending or north node is where the Moon crosses from the south of ecliptic to the north. In Hindu mythology, it is known as the Rahu. The descending or south node is where it crosses from the north of ecliptic to the south. In Hindu mythology, it is known as the Ketu.

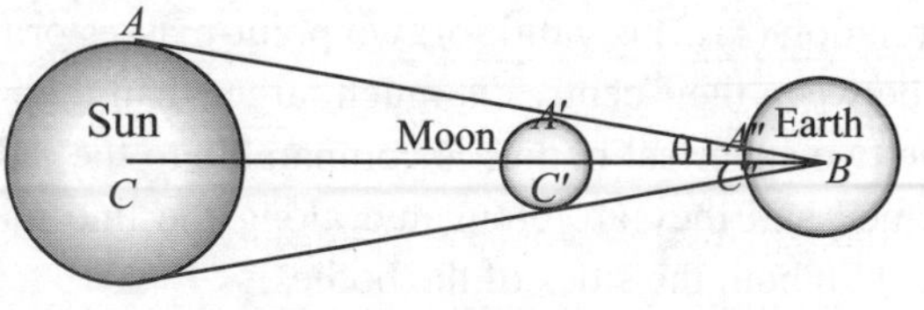

$$AC = BC \tan\theta$$

and $$A'C' = BC' \tan\theta$$

AC is the Sun's radius = $1.392 \times 10^6/2$

= 6.96×10^5 km

$A'C'$ is the Moon's radius = 3480/2

= 1740 km

CC'' is the Earth-Sun separation = 1.53×10^8 km

The Earth-Moon distance = 3.79×10^5 km

Thus, $CC' = 1.53 \times 10^8 - 3.79 \times 10^5$

= 1.526×108 km

The distance CC' is equal to $BC - BC'$. Three equation are now available:

$$6.96 \times 10^5 = BC \tan\theta$$

$$1740 = BC' \tan\theta$$

$$BC - BC' = 1.526 \times 10^8$$

As there are only three unknowns, the equations can be solved to give $BC' = 3.83 \times 10^5$ km and $\tan\theta = 4.55 \times 10^{-3}$ (*BC* is not required). This means that *B* is within the Earth's volume, slightly further away from the Moon than the Earth's centre. Its separation from the Earth's centre is ($3.83 \times 10^5 - 3.79 \times 10^5$ =) 4000 km. The distance BC'' is therefore 4000 km plus the Earth's radius (6400 km) which is 10,400 km.

The size of half-shadow on the Earth's surface is

$$A''C'' = BC'' \tan\theta$$

$$= 10400 \times (4.55 \times 10^{-3})$$

$$= 47 \text{ km}$$

The path of the total eclipse is thus only about 100 km across. It should be noted that this calculation is subject to quite large rounding errors as the scales involved vary from thousands of km to hundreds of millions of km.

8. Tides

We are familiar with the relation

$$F = -G\frac{m_1 m_2}{r^2} \tag{11.3}$$

expressing the gravitational force between the two point-masses m_1 and m_2, separated by a distance *r*. Here, *G* is the universal gravitational constant and the negative sign shows that the force is attractive. The relation (11.3) is valid for two point-masses or for two spherical bodies where the separation r between their centres is much larger than their radii (sizes). When the separation r between the two spherical bodies is comparable to the radii of bodies, the shapes of the bodies are deformed and they are elongated along the line joining their centres. The elongation is symmetrical on both the sides of the bodies.

These effects are clearly observable on the Earth's surface in the form of the tides, which may be understood by considering the pairs: (i) Earth and Moon, and (ii) Earth and Sun.

8.1 Tidal Force on the Earth due to the Moon

In order to understand how the tides take place on the surface of the Earth, let us consider the Earth with its centre at O and the Moon with its centre at O', separated by a distance r (Fig. 7). For simplicity, we assume that both the Earth and the Moon are not spinning. Let R be the radius of the Earth which is about 3.7 times that of the Moon. The mass of the Earth is about 81.3 times that of the Moon. Since we are interested in the tidal effects observed on the surface of the Earth, it is convenient to consider that the mass of the Moon is situated at its centre O'.

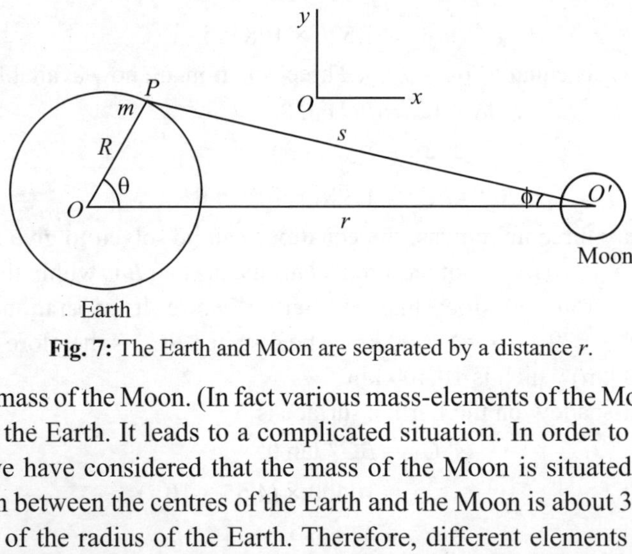

Fig. 7: The Earth and Moon are separated by a distance r.

Let M be the mass of the Moon. (In fact various mass-elements of the Moon attract different mass-elements of the Earth. It leads to a complicated situation. In order to get the reasonably reliable results, we have considered that the mass of the Moon is situated at its centre.) The average separation between the centres of the Earth and the Moon is about 3.82×10^8 m, which is about 60 times of the radius of the Earth. Therefore, different elements of the body of the Earth experience different gravitational forces due to the Moon, as the separation between an element of the Earth and the Moon varies. Further, the effects would be symmetrical about the line joining the centres O and O'. For convenience, we consider the forces in the xy-plane (lying in the plane of the page) only, where the x-axis is parallel to OO'. Suppose, a mass m is situated at O. The x and y components for the force between the mass m at O and the Moon are

$$F_{Ox} = -G\frac{Mm}{r^2},$$

$$F_{Oy} = 0$$

Now, suppose the mass m is situated at a point P on the surface of the Earth, which is at a distance s from O'. The x and y components for the force at P are

$$F_{Px} = -G\frac{Mm}{s^2}\cos\phi,$$

$$F_{Py} = G\frac{Mm}{s^2}\sin\phi$$

The difference between the forces at *P* and at *O* is

$$\Delta F = (F_{Px}\hat{i} + FP_y\hat{j}) - (F_{Ox}\hat{i} + F_{Oy}\hat{j})$$

$$= GMm\left(-\frac{\cos\phi}{s^2} + \frac{1}{r^2}\right)\hat{i} + \frac{GMm}{s^2}\sin\phi\,\hat{j} \tag{11.4}$$

From Fig. 7, we can write

$$s^2 = (r - R\cos\theta)^2 + (R\sin\theta)^2$$

$$= r^2\left[1 + \frac{R^2}{r^2} - \frac{2R\cos\theta}{r}\right] \approx r^2\left[1 - \frac{2R\cos\theta}{r}\right] \tag{11.5}$$

where the term $R^2/r^2 << 1$ is neglected. On using equation (11.5) in (11.4), we get

$$\Delta F = GMm\left[-\frac{\cos\phi}{(r^2 - 2Rr\cos\theta)} + \frac{1}{r^2}\right]\hat{i} + \frac{GMm\sin\phi}{(r^2 - 2Rr\cos\theta)}\hat{j} \tag{11.6}$$

On neglecting the terms having second and higher orders of R/r in equation (11.6), we get

$$\Delta F = \frac{GMm}{r^2}\left[-\cos\phi\left(1 + \frac{2R\cos\theta}{1}\right) + 1\right]\hat{i} + \frac{GMm\sin\phi}{r^2}\left[1 + \frac{2R\cos\theta}{r}\right]\hat{j} \tag{11.7}$$

On using the first order relations, $\cos\phi \approx 1$ and $\sin\phi \approx (R\sin\theta/r)$, in equation (11.7), and neglecting the terms containing R_2/r^2, we get

$$\Delta F = -\frac{GMmR}{r^3}(2\cos\theta\hat{i} - \sin\theta\hat{j}) \tag{11.8}$$

It shows a differential force on the surface of Earth relative to that at *O*.

Though all the forces on various elements of the Earth are directed towards the centre of the Moon, the forces relative to the force at *O* (equation 11.8) are directed outward along the *x*-axis and inward along the *y*-axis. The factor 2 in equation (11.8) shows that the magnitude of the outward force is two times larger than that of the inward one. Consequently, the Earth is elongated along the line *OO'*, and hence the tidal bulges are pronounced along the Earth-Moon line. The bulges are on both sides of the Earth, as the shape of the Earth is oblate spheroidal with larger radius along the line joining the Earth and Moon. (Similar effects, about 20 times more pronounced than those on the Earth, would be observable on the surface of the Moon. However, here, we are interested in the effects observed on the surface of Earth only.)

Owing to the revolution of the Moon around the Earth, the position of tidal bulge changes smoothly on the surface of the Earth. This phenomenon is periodic and one observes two high tides in a period of 24 h 53 m. This period known as the lunar day is the average time the interval between the two successive transits of the Moon through the observer's meridian.

8.2 Consideration of the Rotational Motion of the Earth

Let us see the effect of consideration of rotational motion of the Earth. (Since the Moon is being considered a point-mass, we are not accounting for the rotational motion of the Moon.)

If the time-period of Earth's rotation about its own axis is equal to the period of revolution of the Moon around the Earth, then the position of the tidal bulges would have remained fixed on the surface of Earth. But, in fact the time-period of the Earth's rotation is shorter than that of the Moon's revolution around the Earth by about 53 m. Therefore, the frictional forces on the surface of the Earth drag the bulge axis ahead of the Earth-Moon line. The bulge is pulling the Moon forward, speeding it up and causing it to move farther out. Since a frictional force is always dissipative, the Earth's rotational kinetic energy is being lost regularly, and hence, the Earth's spin rate is decreasing regularly. So, the Earth's rotational period is increasing regularly. Estimates show that the Earth's rotational period is, at present, increasing at a rate of 0.0016 s/century. The angular momentum of the Earth is obviously decreasing with time.

Following the conservation of angular momentum for the Earth-Moon system, the angular momentum lost by the Earth is being transferred to the orbital motion of the Moon, and therefore, the angular momentum of the Moon is increasing regularly. With the increase of the angular momentum of the Moon, the distance of the Moon from the Earth is increasing regularly. Estimates show that at present, the Moon is drifting away from the Earth at the rate of 3 to 4 cm/year.

8.3 Tidal Force on the Earth Due to the Sun

Another celestial body, after the Moon, which produces observable tidal effects on the Earth's surface is the Sun. Similar calculations can be performed for the Earth- Sun system. The ratio of the effects on the Earth's surface due to the Moon and due to the Sun may be written as (using equation 11.8).

$$\frac{(\Delta F)_{\text{Moon}}}{(\Delta F)_{\text{Sun}}} = \frac{(M/r^2)_{\text{Moon}}}{(M/r^2)_{\text{Sun}}}$$

$$= \frac{7.35\times10^{22}/(3.844\times10^{8})^{3}}{1.989\times10^{30}/1.496\times10^{11})^{3}} = 2.2$$

Obviously, the tidal effect of the Moon is 2.2 times larger than that of the Sun. The tidal effect of the Sun also is periodic. There are two high tides due to the Sun (other than those due to the Moon) in a period of one solar day, which is equal to 24 h. Hence, we have two sets of the periodic tidal bulges, one due to the Moon and the other due to the Sun. The time periods of two sets are different.

8.4 Superposition of the Tidal Forces

These two sets of the tidal bulges show a phenomenon of superposition. When the Sun, Earth and Moon are aligned along a straight line, at the time of full Moon or a new Moon, the tidal effects are added to create the largest possible bulges on the Earth's surface. These situations are known as the **spring tides**. The spring tide occurs twice each lunar month all year long without regard to the season.

At the first and third quarters of a lunar month, the Sun, Earth and Moon form a right-angle. In this situation, the tides produced by the Moon and by the Sun reduce the effects of each other to the minimum possible. These situations are known as the neap tides.

8.5 Future Scenario

Equation (11.8) shows that with the increase of *r*, the tidal effects would decrease rather rapidly, as the power of *r* is 3. Since the Moon is receding away from the Earth, the tidal effects due to the Moon would not be observable on the surface of the Earth in far future. Since the distance between the Sun and Earth is not changing, the Earth would always observe the tidal effects due to the Sun. Hence, in the far future, the tidal effect would not depend on the Moon calendar, and the intensity of the effect would remain about one-third of the observed maximum effect at present.

8.6 Tidal Forces in the Context of Other Planets

Equation (11.8) shows that the tidal effects become more pronounced with the decrease of distance *r* from the planet, generating the tidal effect. The effect may be so severe that a body coming into the range of the tidal force may break down into small pieces. This is an explanation for the material found in the ring(s) around some planets. The phenomenon would obviously be more intense in case of the massive planets, for example, Jupiter and Saturn. Jupiter is the most massive planet in our solar system. Thus, if some comet comes close to the surface of Jupiter (i.e., within the Roche limit), it would break into the pieces. Such event has been witnessed by a large section of people that the nucleus of the comet Shoemaker-Levy-9 was broken into 22 fragments when it passed through the perijove (nearest point in orbit of Jupiter) on 8 July 1992.

9. Mars

Mars (Fig. 8) appears markedly reddish, and therefore, it is sometimes called the red planet. Mars has been the first planet whose actual distance from the Earth was measured using the methods of trigonometric parallax by Cassini in 1673. However, in 1659, Huygens noticed a dark triangular mark on the disc of the Mars, which he found to rotate regularly once every 24.5 hours, giving the first measurement of rotation period of Mars. In 1781, Hershel showed that the axis of rotation of Mars was inclined by almost the same amount as that of the Earth. So, the seasons on the Mars must follow the terrestrial pattern, namely, alternately the opposite seasons in the opposite Martian hemispheres.

Mars has a thin atmosphere with a pressure about 1/200th that of the Earth, and with the composition of 95.3% CO_2, 2.7% N_2, 1.6% Ar, 0.3% O_2, 0.07% CO, 0.03% H_2O and very little Ne, Kr, Xe and O_3. The Mars' atmosphere is very dry because all the water is frozen into the ice, and the maximum daytime temperature is about –30°C. Even the CO_2 is frozen into the dry ice for most of the time.

The planet has polar caps composed of both water ice and dry ice. The extent of the northern cap is found to vary with the season, but not the southern cap. The surface as seen through a telescope is reddish, and this colour is due to the presence of iron oxide in the Marian limonites.

The question of existence of life on the Mars has been discussed from time to time. But, no concrete evidence for it has been established so far.

Mars has been found to have two natural satellites, named Phobos and Deimos. The Phobos is nearer as compared to the Deimos. The Phobos revolves with the time-period of 0.318 day, and the Deimos revolves with the time-period of 1.262 days. Interesting to note that both the satellites are rotating synchronously with their respective periods of revolution. That is, they keep always their same face turned towards the Mars. They are believed to have been captured from outside.

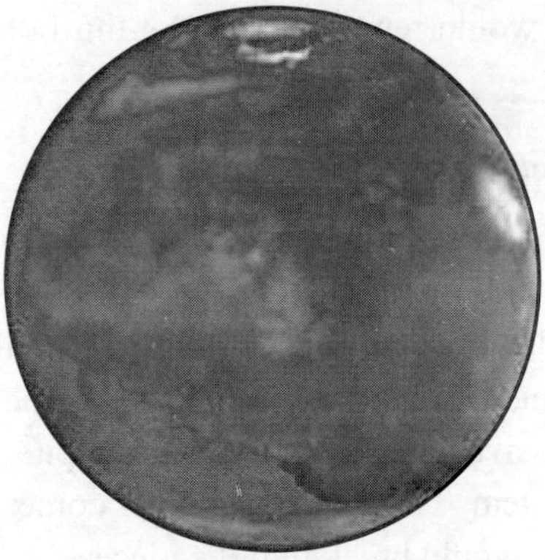

Fig. 8: Mars.
(*Courtesy:* Google website)

It is found that the Mars has very weak magnetic field of about 60 nT; a very weak ionosphere containing the plasma of singly ionized oxygen duly captured from the solar wind.

10. Jupiter

Jupiter (Fig. 9) is the largest planet of our solar system. It occupies ~ 0.1% of the mass of the Sun and a radius ~ 1/10th of the Sun. Its average density is almost equal to that of the Sun. Had it been 8 times more massive, it could have become a star by itself. From the Earth, it appears as the second brightest planet, next to Venus (however not always, because the Mars at its closest position appears somewhat brighter than the Jupiter).

With the help of a big telescope, it appears as a nebulous object with a sharp elliptical boundary. It is because the Jupiter rotates very fast about its own axis, and has therefore assumed the shape of an oblate spheroid with its polar diameter, ~ 6% smaller. A point on its equator has to move with a speed of about 45,000 km/hr in order to complete a full axial rotation in only 9.85 hr. Jupiter displays a number of dark and light colourful bands round the body. They are seen to be perpendicular to the rotation axis. Since it is a giant gaseous planet, we can see only the atmosphere and the cloud system from the outside.

There is a permanent oval-shaped spot in the southern hemisphere of the planet, which is visible since its discovery in 1664. It was dramatically described as the Great Red Spot in 1878. The Great Red Spot is ~ 14,000 km wide and 30,000 - 40,000 km long, and deep brownish red in the colour. It swifts very slowly with time as the rotational period of the Jupiter is $9^h\ 55^m\ 37.6^s$. There are however other smaller, red and white spots with the lifetime of several years. Also seen on its surface are the belts and zones which are stable, coloured formation, though the width and colour do vary with time. The belts are reddish and brownish in the colour. The motion of the gas inside a belt is downward.

Fig. 9: Jupiter.
(*Courtesy:* Google website)

The Jovian atmosphere is primarily made up of H and He in almost the same proportion as found in the interstellar clouds or the Sun.

Like the Sun, the Jupiter shows differential rotation. Its equatorial region rotates with a period of $9^h\ 50^m\ 30^s$. So, it is very difficult to determine the actual period of rotation of the rigid part of the planet. Fortunately, the Jupiter has a strong bipolar magnetic field which suggests that its inner core must be in a fluid state, supporting the electric currents. The differential rotation is also responsible for the stratification of the atmosphere parallel to the rotation axis.

The solid and liquid parts of the Jupiter are buried inside its thick atmosphere of depth about 1000 km. It is believed that possibly it has a solid iron, nickel and silicate core of few tens of the Earth masses, surrounded by a thick layer of the metallic liquid hydrogen and overlying thick layer of liquid molecular hydrogen. The central temperature of the Jupiter exceeds about 3×10^4 K with a central pressure $\sim 3 \times 10^6$ times the pressure on the the surface of the Earth.

One of the greatest puzzling discoveries made by the Pioneer probes in 1979 was that the Jupiter emits twice the radiation that it receives from the Sun, indicating that the Jupiter has an internal source of heat. Evolutionary calculations show that the Jupiter is still cooling; He^4 is still gradually settling down towards the centre, releasing tremendous amount of gravitational energy. This heat is being transported outwards by the convection currents inducing flows in the metallic hydrogen, responsible for the origin of the Jovian magnetic field.

Jupiter has the largest number of satellites. At least 67 satellites of the Jupiter are known. The four prime satellites of Jupiter are the Io, Europa, Ganymede and Callisto, discovered in 1610 by Galileo. The motion of inner three satellites, namely, Io, Europa and Ganymede is not independent of each other; in fact their orbital periods around the Jupiter are locked in the perfect tidal resonances with a ratio of the orbital periods 1 : 2 : 4. The satellite Io, a little bigger than our Moon, has its surface spotted with the numerous volcanoes but without any mountains. There is also no trace of impact craters on the surface of Io, and no trace of water either.

Since the time of Galileo, the tally of the Jupiter satellites has been steadily increasing from 4 to at least 67. Rings of Jupiter have also been discovered in 1979, which are sparse and narrow having thickness of smaller than 1 km and being composed of small particles, few micrometer (μm) in size.

Europa, slightly smaller than our Moon, appears appreciably brighter than Io because of its high reflectivity. The surface is extremely smooth and covered with ice bearing no sign of impact craters at all. Europa's surface is renewed continuously by small fountains of fresh water that comes through the icy crust.

Ganymede, the biggest of all the satellites in the solar system, is about 1.5 times bigger than our Moon. It is also bigger than the Pluto and Mercury. Ganymede is composed of the water or ice and the silicate rocks in almost equal proportion. Its surface bears the marks of impact craters.

Callisto, the outermost of Galilean satellites has a rather dark complexion and is still quite big in size, only a little smaller than the Ganymede. From the texture of surface, it seems that the Callisto is made mostly of stone and ice, and bears marks of impact craters.

The other satellites of Jupiter are considerably small in size.

11. Saturn

Saturn (Fig. 10) is the second largest planet of our solar system, after the Jupiter, having beautiful ring system. In the beginning, the planet appeared to have ears. These ears later on turned out to be a beautiful system of rings surrounding the planet. Saturn is the only planet in the solar system whose average mass-density is lower than that of the water. One can say that the Saturn can float on water like a piece of wood.

Fig. 10: Saturn.
(*Courtesy:* Google website)

Like Jupiter, Saturn is also found to radiate about 1.8 times the energy it receives from the Sun. One reason why the Saturn is so featureless is that the velocity of wind in the equatorial band is about 1800 km/hr. So, every feature is finely striated parallel to the Saturn's equator. The outflow of the Saturn's own internal energy becomes the driving force behind achieving such high velocities of the wind.

Like Jupiter, the Saturn has also got a very strong magnetic field of its own (about 40 μT near its equator). The Saturn has a rocky core of about 3 Earth masses and a temperature of 1.5×10^4 K. Beneath the clouds, there is a thick layer of molecular liquid hydrogen and metallic liquid hydrogen, the latter starting at a depth of 30,000 km. Although the conducting mantle of the Saturn rotates with a period of $10^h\ 39^m\ 24^s$, its equatorial atmosphere rotates somewhat slowly with a period of $10^h\ 10^m$.

The rings of the Saturn are made up of normal water-ice, having particle sizes widely varying from μm to metre. The thickness of rings is extremely small, ~ 100 m at the most, whereas the width of the major ring system is ~ 1.3×10^5 km. The reason why Galileo missed the identification of the ears of the Saturn as its ring was that he could no longer see those ears in 1612 which he saw in 1610-1611.

In fact the Saturn's ring system disappears, when viewed from the Earth, once every 15 years, and in the intervening period one alternately sees the northern and southern faces of the ring with a maximum opening angle of about 27°. That the ring is not continuous across its breadth was first noted by Cassini in 1675. He thought that there had been two rings separated by a dark portion, now known as the Cassini's division.

The outer ring has been designated as *A* and the inner *B*. In 1850, yet a smaller region inner than *B* was reported and was duly designated as *C*. Thanks to the wonderful space missions, today we know that there are about 7 prominent rings around the Saturn, named alphabetically *D*, *C*, *B*, *A*, *F*, *G* and arranged sequentially from the innermost to outermost.

Roche demonstrated that any satellite in orbit within ~ 2.44 times the radius of the planet finally breaks up into small pieces and forms a ring.

More than sixty-two known moons orbit the planet of which Titan is the largest. In size, Titan is slightly bigger than the Moon. It is the only satellite in the solar system that has an atmosphere. Its atmosphere contains 99% of nitrogen, the temperature is about 90 K, and the atmospheric pressure is 1.522 bars. Other prime satellites of the Saturn are the Mimas, Enceladus, Tethys, Dione, Rhea, Iapetus.

Except for Phoebe, the first 11 satellites of the Saturn are synchronous, that is, axial periods of rotation are the same as those of their orbital periods of the revolution around the Saturn. Compared to the Titan, Tethys, Dione and Rhea are the three next bigger satellites composed of ice and rock. All the other satellites are just small irregular fragments of ice and rock.

12. Uranus

The Voyager 2 passed by the Uranus on 24 January 1986. Before that, very little was known about the Uranus as its angular size seen from the Earth never exceeds 4″. Uranus (Fig. 11) was the first planet discovered using a telescope by Herschel in 1781 (the planets up to the Saturn are naked eye planets), and its orbit lies beyond that of the Saturn. It was later on named after the god of the sky and the father of the Saturn. When the planet comes closest to the Earth, that is, during an opposition, it can barely be seen with the naked eye and its total visual magnitude becomes +5.5.

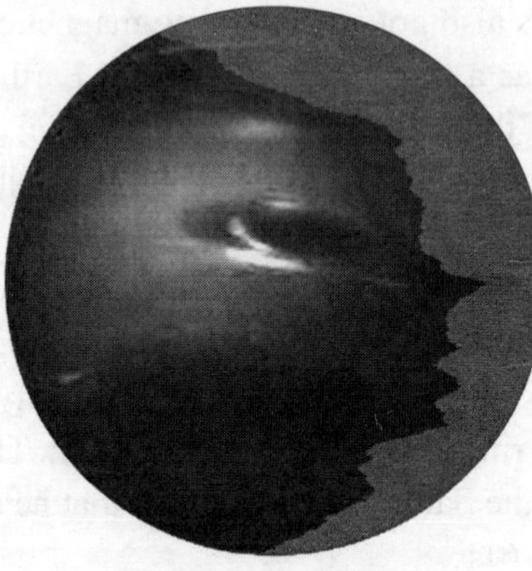

Fig. 11: Uranus.
(*Courtesy:* Google website)

The Uranus is surrounded by a system of eleven rings of which five were discovered in 1977 by observing the occultation of a star by the disc of the Uranus. The last two rings were discovered by Voyager 2. The Uranian system of the rings are found to be confined within the Roche limit of the Uranus.

Since the axis of rotation of Uranus is tilted by 98° with respect to the pole of the ecliptic, Uranus revolves around the Sun keeping its axis turned to a side. Even though the Uranus rotates very fast about its own axis, its outer surface shows very little oblateness. Thus, it is expected that the Uranus should have a rocky core about the size of the Earth. The envelope is a gaseous mixture of H, He, H_2O, NH_3 and CH_4, instead of a liquid mantle.

Due to a large tilt of its rotation axis, Uranus' north pole faces the Sun for 42 years (one half of the period of its orbital revolution) and during the next 42 years the south pole faces the Sun. So, the poles get alternately warmed up during the planet's complete revolution around the Sun.

The atmosphere is composed mostly of hydrogen (85% by mass) with about 12% of helium by mass and very small amounts of methane and other hydrocarbons. The Uranus appears blue in colour because its atmospheric methane differentially absorbs red light from the Sun. The visible surface is a layer of haze containing C_2H_2 and C_2H_6, formed due to the photo-dissociation of CH_4.

The average temperature at the cloud top is 52 K. Uranus reflects ~50% of the incident sunlight from its cloud top. There are intense radiation belts and radio emissions because it has a sufficiently strong magnetic field (23 μT on its magnetic equator). The axis of rotation and that of the magnetic field are however not aligned. They subtend an angle of 60° with each other.

The narrow rings of the Uranus appear to be very dark and are possibly composed of the metre-sized icy chunks. The thickness of the rings are less than 12 km, except the epsilon ring, which is 22-93 km.

More than 27 moons of the Uranus revolve synchronously with the axial rotation of Uranus. The biggest ones, the Titania, Oberon, Ariel, Umbriel and Miranda, can be spotted as tiny bright dots with the help of a large telescope. They are made up of dark grey ice and rock. The Ariel appears the brightest containing many features and extensive flows of the icy material.

13. Neptune

Neptune (Fig. 12) is the eighth and the farthest planet in our solar system. It is slightly smaller in size, yet more massive than the Uranus. It is the first planet of the solar system whose existence was predicted by the theory of gravitation, and which was subsequently discovered in 1846. After the discovery of Uranus in 1781, it was found that Uranus was not exactly following its predicted orbit. So, the problem was taken up and the conclusion drawn was that some other, yet unknown planet, must be perturbing the orbit of the Uranus. The discovery of Neptune marked the greatest triumph of the Newton's theory of gravitation.

Fig. 12: Neptune.
(*Courtesy:* Google website)

Neptune has so far completed one revolution around the Sun since its discovery. It appears as a point of light when seen through a pair of binoculars and has a magnitude at opposition about 7.8. With the help of the latest telescopes, Neptune appears as a tiny bluish disc of size 2.3″. The bluish colour is once again due to the presence of methane in its atmosphere.

A ring system around the Neptune was reported from the stellar occultation observations in 1985. Then the Voyager 2 flew past the Neptune on August 25, 1989, which provided the wealth of information about the planet, its ring system and satellites.

Neptune has got two complete thin rings at the distances of 62.9 and 52.3 thousand km from its centre. There are also evidences of 3 sheets or incomplete rings. Rings are once again made up of the micron-sized dust particles.

Neptune appears in many respects similar to the Uranus and the Jupiter. There is a Great Dark Spot on the Neptune almost at the same latitude of 22°S as the Great Red Spot on the Jupiter. Even though Neptune's period of axial rotation is $16^h\ 3^m$, its cloud system is slightly slower in the same direction with a period of ~ 18 hr, whereas on the Jupiter, Saturn and Uranus, the clouds revolve around the planet faster than the planets' axial rotation.

The temperature at the cloud top near the equator as well as at the poles is about 60 K, whereas at the mid-latitudes, it is only 50 K. Neptune has a rocky core surrounded by a waterly mantle and its thick atmosphere is primarily composed of H and He with an admixture of methane. Ionized water molecules in the mantle conduct electric current thereby giving the planet its magnetic field strength, 13 μT. The magnetic axis of the Neptune is tilted by 47° with respect to its rotation axis and the centre of the magnetic dipole seems to be greatly displaced from the centre of the Neptune towards south. The core contains rocky material and is believed to be cold and inert.

At least 14 satellites of the Neptune are known. However, only one satellite, Triton, is prime. Triton is slightly bigger than our Moon and revolves synchronously with the axial rotation, but in a retrograde sense with a period of 5.88 days. The Triton, having also a highly inclined orbit, was perhaps captured by the Neptune. The pinkish eyes that constitute Triton's south polar cap, is probably a layer of frozen methane and nitrogen, and the nitrogen is believed to have been a product of ammonia. The other satellites of the Neptune are quite small in size.

14. Asteroids

As stated earlier that every body revolving around the Sun is not a planet. For a body to be called a planet, it must have size (mass) more than the value specified by the International Astronomical Union (IAU). This value can be modified. For example, the Pluto was in the list of planets up to 24 August 2006, and after that it has been removed from the list, as the value of specified mass was modified. There are large number of bodies revolving around the Sun having small size. They are generally called the minor planets. The asteroids (Fig. 13) or minor planets are the bodies orbiting around the Sun, mainly in between the Mars and the Jupiter.

14.1 Titius-Bode Law

The Titius-Bode law first predicted a missing planet between the Mars and Jupiter. In order to have the idea about the Titius-Bode law, take the following series:

$$0, 1, 2, 4, 8, 16, 32, ...$$

Now, follow the following steps on each of the terms of the series:

(i) multiply each term by 3, to get
0, 3, 6, 12, 24, 48, 96, ...

(ii) add 4 to each term, to get
4, 7, 10, 16, 28, 52, 100, ...

(iii) divide each term by 10, obtaining
0.4, 0.7, 1.0, 1.6, 2.8, 5.2, 10.0, ...

These final numbers when taken in AU, they correspond to the distances of the planets from the Sun: 0.4 AU corresponds to the Mercury, 0.7 AU to the Venus, 1.0 AU to Earth, 1.6 AU to Mars, 5.2 AU to Jupiter, 10.0 AU to Saturn and so on. It obviously shows that no known planet has been assigned to 2.8 AU. Thus, the Titius-Bode law predicted a missing planet at a distance of 2.8 AU from the Sun.

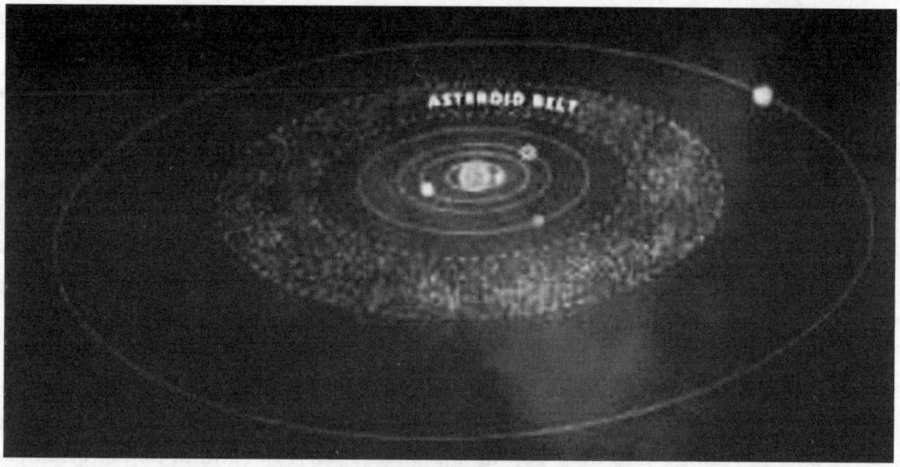

Fig. 13: Asteroids.

(*Courtesy:* Google website)

Table 4. Parameters of the planets

Planet	Distance (AU)	Bode No.
Mercury	0.39	0.4
Venus	0.72	0.7
Earth	1.00	1.0
Mars	1.52	1.6
Missing planet		2.8
Jupiter	5.20	5.2
Saturn	9.54	10.0

14.2 Search for Missing Planet

Investigations for the search of missing planet were made during the last decades of 18th century. On 1 January 1801, an object having magnitude 8 (very dim) was noticed moving across the sky. The same was found on the successive nights. The body was recognized as a new planet and it was named the Ceres. Very soon in 1802, another similar object was discovered and named the Pallas. Then the Juno was discovered in 1804 and the Vesta in 1807. Information about these four bodies is summarized in Table 5.

Table 5. Discovery of asteroids

Year	Asteroid	Diameter (km)
1801	Ceres	960
1802	Pallas	570
1804	Juno	233
1807	Vesta	530

By that time, the Uranus was discovered. After the discovery of the Vesta, from 1807 to 1845, the planetary system has been as the following:

Mercury, Venus, Earth, Mars, Vesta, Juno, Ceres, Pallas, Jupiter, Saturn, Uranus

All these asteroids are regarded as the minor planets. Because of their faintness and smallness, they were found to twinkle like the stars.

The orbital characteristics of these newly discovered minor planets seemed to fill the gap. In the Titius-Bode relation for the mean distance from the Sun about 2.8 AU, placing them between the orbits of the Mars and the Jupiter. As on 1 September 2017, there are more than 200 asteroids having diameter more than 100 km; between 1.1 and 1.9 million satellites having diameter larger than 1 km.

The asteroids were mostly named after the mythological gods and demons which were soon exhausted, and the list continued with the names of wives, friends, cities, colleges of discoverers. However, officially the minor planets, also called the asteroids, are designated by the serial numbers in addition to their common names, if any. For example, the first masculine name for an asteroid was given to 433 Eros. Examples for these, named after the countries are 136 Austria, 1125 China, 1279 Uganda and so on.

The asteroid number 4130 has been discovered by the astronomers working in the Indian Institute of Astrophysics (IIA), Bangalore, India. It is named as 'Ramanujan', in the honour of great mathematician from India. Pandit Jasraj becomes the fist Indian musician to have minor planet (2006 VP32; No. 300128) named after him. This minor was discovered on November 11, 2006. The number '300128' denotes the vocalist's date of birth, January 28, 1930.

The total mass of all the known asteroids does not exceed one-tenth of the mass of our Moon. The orbits of the asteroids were found to form two distinct groups: one group having length of semi-major axis between 2.3 and 2.8 AU and the other group with the length of semi-major axis between 3 and 3.2 AU. There are practically no asteroids having the length of semi-major axis 2.52 AU, 2.82 AU or 2.90 AU. These gaps are collectively known as the Kirkwood gaps, first predicted by Daniel Kirkwood in 1857 and confirmed in 1866.

Because of the large orbital eccentricities, some of the asteroids are found to come regularly close to the Earth. The shape of an asteroid can be spherical to irregular depending on its size. In fact, the biggest few asteroids, namely, the Ceres, Pallas, Vesta and Hygiea are quite spherical in shape. Most of the other asteroids remain, however, extremely irregular. By observing the light curve of the asteroids, the rotation period as well as the shape can be figured out.

Most asteroids have a rotation period of only a few hours. For example, the period of the Eros is $\sim 5^h\ 17^m$, the Juno has a period of $7^h\ 13^m$, and so on. Asteroids can have their satellites too. The first such satellite was discovered in 1973. The asteroid 532 Herculina is found to have a tiny satellite. This asteroid has a diameter of 217 km, whereas the satellite has a size of 50 km only.

Note:

It is now clarified that every body revolving around the Sun is not a planet. Up to the year 1800, a planet was defined as an object having a fairly circular orbit around the Sun and that did not show any cometary activity such as a coma or a tail. After discovery of the Ceres, Pallas, Vesta, etc., a point was raised to put a limit on the size of the body in order to distinguish between a planet and an asteroid. However, no clear demarcation was decided.

15. Small Planets Beyond Neptune

Up to 24 August 2006, Pluto has been in the list of planets in our solar system. Since the size of Pluto is about 18% of Earth, calling it as a planet had been objected by some astronomers. On the other side, search for additional planets in the Kuiper belt (away from Neptune) is progressing. There are about 800 known Kuiper belt objects; most of them are only 100 km in diameter. The Kuiper belt objects of diameter larger than 1200 km are given in Table 6. Here, we have included Pluto also in order to compare it with the other objects.

Table 6. Parameters of bodies

Body	Distance (AU)	Diameter (km)	Year of discovery
Pluto	39.53	2302	1930
Quaor	42	1250	2002
2004 DW	55	1600	2004
Sedna	90	1800	2004
2003 UB_{313}	97	3100	2005

An object, tentatively named 2003 UB_{313}, bigger than Pluto was found in the Kuiper belt. Like Pluto, it has a satellite. With the discovery of the 2003 UB_{313}, the point under consideration was that if the Pluto was continued as a planet, then 2003 UB_{313} was to be recognized as the tenth planet. In case the 2003 UB_{313} to be recognized as a small planet, then the Pluto also was to be recognized as a small planet. On 24 August 2006, the International Astronomical Union (IAU) decided to exclude the Pluto from the list of the planets. Likewise Quaor, 2004 DW, Sedna, the Pluto became a body revolving around the Sun beyond the Neptune.

The Pluto is found to have at least 5 satellites. The prime satellite of the Pluto is the Charon.

16. Satellites

Satellite is an object which revolves around a planet. Moon is the natural satellite of the Earth. Besides the Moon, there are artificial satellites revolving around the Earth. The mass M of the Earth is very large as compared to that of a satellite. Though the orbit of a satellite is elliptical in shape, but for the practical purposes, we can take it circular (Fig. 14). Suppose, a satellite of mass m is revolving around the Earth at a distance r from the centre of Earth. The force of attraction between the Earth and the satellite is (Newton's law of gravitation).

$$F = -\frac{GMm}{r^2}$$

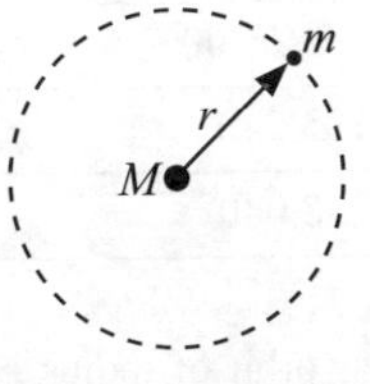

Fig. 14: A satellite of mass m revolving around the Earth of mass M in a circular orbit of radius r.

where G is the gravitational constant. This force is balanced by the centrifugal force[4] acting on the satellite in the outward direction due to its angular motion. If v is the velocity of satellite, the centrifugal force F' acting in the outward direction is

$$F' = \frac{mv^2}{r}$$

In the equilibrium position, the two opposite forces F and F' are equal in magnitude and we have

$$\frac{GM_m}{r^2} = \frac{mv^2}{r}$$

or
$$v = \sqrt{\frac{GM}{r}}$$

It is interesting to note that the velocity of the satellite does not depend on its mass. The period of revolution (T) of a satellite moving around a planet is

$$T = \frac{2\pi r}{v}$$

$$= \frac{2\pi}{\sqrt{GM}} r^{3/2}$$

Thus, we have $T^2 \propto r^3$. It is the Kepler's third law of planetary motion. If R is the radius of the Earth, the altitude h of satellite above the surface of the Earth is

$$h = r - R$$

Some man-made bodies have been put for revolving around the Earth, for example. Thus, the satellites are of two types: (i) natural satellites, and (ii) artificial satellites. Here, we are concerned with the natural satellites.

Table 7. Information about prime satellites

Satellite	Mean radius (km)	Semi-major axis (km)	Discovery year	Planet
Moon	1737.10	384,399	prehistoric	Earth
Phobos	11.1 ± 0.15	9,380	1877	Mars
Deimos	6.2 ± 0.18	23,460	1877	Mars
Io	1818.1 ± 0.1	421,800	1610	Jupiter
Europa	1560.7 ± 0.7	671,100	1610	Jupiter
Ganymede	2634.1 ± 0.3	1,070,400	1610	Jupiter
Callisto	2408.4 ± 0.3	1,882,700	1610	Jupiter
Mimas	198.2 ± 0.4	185,540	1789	Saturn
Enceladus	252.3 ± 0.6	238,040	1789	Saturn

[4]When a body of mass m moves in a circular orbit of radius r with velocity v, a fictitious force, called the centrifugal force of magnitude mv^2/r is developed in the direction away from the centre of the curved path.

Tethys	536.3 ± 1.5	294,670	1684	Saturn
Dione	562.5 ± 1.5	377,420	1684	Saturn
Rhea	764.5 ± 2.0	527,070	1672	Saturn
Titan	2575.5 ± 2.0	1,221,870	1655	Saturn
Iapetus	734.5 ± 4.0	3,560,840	1671	Saturn
Ariel	578.9 ± 0.6	190,900	1851	Uranus
Umbriel	584.7 ± 2.8	266,000	1851	Uranus
Titania	788.9 ± 1.8	436,300	1787	Uranus
Oberon	761.4 ± 2.6	583,500	1787	Uranus
Miranda	235.8 ± 0.7	129,900	1948	Uranus
Triton	1353.4 ± 0.9	354,800	1846	Neptune
Charon	603.6 ± 1.4	17,536	1978	Pluto*

*Pluto is no more a planet.

As we discussed, the Moon is the only natural satellite of the Earth. Other planets also have natural satellites. However, the Mercury and Venus do not have any satellites. Here, we do not consider for smaller bodies revolving around a planet. Even then, some planets, e.g., the Jupiter and Saturn, have large number of satellites. However, some of them are the prime ones. The prime satellites of planets are given in Table 7.

The planet Mars has two satellites, Phobos and Deimos. Jupiter has the largest number of satellites. At least 67 satellites of the Jupiter are known. The four prime satellites of the Jupiter are Io, Europa, Ganymede and Callisto. It is interesting to note that these four satellites were discovered in 1610. The planet Saturn is the next planet to have large number of satellites. At least 62 satellites of the Saturn are known. The seven prime satellites of the Saturn are the Mimas, Enceladus, Tethys, Dione, Rhea, Titan, Iapetus.

The planet Uranus has at least 27 satellites. The five prime satellites of the Uranus are the Ariel, Umbriel, Titania, Oberon, Miranda. At least 14 satellites of the Neptune are known. However, only one satellite, the Triton is prime. The Pluto is though not a satellite after August 2006, but it has at least 5 satellites and the prime satellite is the Charon.

16.1 Geostationary Satellite

For a geostationary satellite, the time period of revolution around the Earth is equal to the time-period of the spin motion of the Earth. It moves from west towards east – the direction of the spin motion of the Earth, so that it is always situated above the country to which it belongs to. The Earth completes one rotation about its own axis in 24 hours[5]. The distance r of satellite from the centre of the Earth is

$$r = \left(\frac{GM}{4\pi^2}\right)^{1/3} T^{2/3}$$

[5]Actual time-period is 23 hr 56 m 4.1 s. But, for the practical purposes, it is taken as 24 hours.

Using $T = 24$ hr = 86400 s, $G = 6.67 \times 10^{-11}$ kg^{-1} m^3 s^{-2} and mass of the Earth $M = 6 \times 10^{24}$ kg, we have

$$r = \left(\frac{6.67 \times 10^{-11} \times 6 \times 10^{24}}{4\pi^2}\right)^{1/3} (86400)^{2/3} = 42.3 \times 10^6 \text{ m}$$

Thus, the distance of the satellite from the centre of the Earth is about 42.3×10^6 m. For the radius of the Earth $R = 6.4 \times 10^6$ m, the distance of the satellite from the Earth's surface is about 35.9×10^6 m.

Example 4.

Calculate the altitude of a geostationary satellite. Given, $G = 6.67 \times 10^{-11}$ kg^{-1} m^3 s^{-2}, mass of the Earth $M = 6 \times 10^{24}$ kg, radius of Earth $R = 6.4 \times 10^6$ m.

Solution:

The time-period of revolution of geostationary satellite around the Earth is $T = 24$ hr = 86400 s. The distance r of the satellite from the centre of Earth is expressed as

$$r = \left(\frac{GM}{4\pi^2}\right)^{1/3} T^{2/3}$$

Putting the values, we get

$$r = \left(\frac{6.67 \times 10^{-11} \times 6 \times 10^{24}}{4\pi^2}\right)^{1/3} (86400)^{2/3}$$

$$= 42.3 \times 10^6 \text{ m}$$

The altitude of geostationary satellite is

$$h = r - R$$

$$= 42.3 \times 10^6 - 6.4 \times 10^6 = 35.9 \times 10^6 \text{ m}$$

17. Comets

Unlike a planet which moves across the sky with great regularity, there are some strangers which visit the inner part of the solar system quite unpredictably. As, in general, they have hair-like tail and therefore have been named the comets (Fig. 15) (the word 'comet' has been derived from the coma-like appearance).

Fig. 15: Halley's comet

It appears that all the comets are members of the solar system, having size up to a few kilometres. However, when they come very close to the Sun, a tail is developed which might extend up to millions of kilometres. Some comets, for example, the Halley's comet, have been known to the mankind for more than 2000 years. Occasionally, once or twice in a decade some comets become so bright as can be seen with naked eyes.

When a comet is away from the Sun, its size is of a few kilometres. But, when it approaches the Sun, it becomes brighter and develops a tenuous halo around it, known as the coma of the comet. The actual cometary body, now called the nucleus of comet, remains deeply buried inside the head, consists of nucleus and coma.

The tails of comet are generally of two types – one is straight, long and somewhat longitudinal. This is usually made up of ionized atoms and molecules, and is known as the ion-tail or the tail of Type I. The other type of tail, the so called the tail of Type II, is more diffused, more thick, shorter and curved. They were probably composed of dust particles and hence known also as the dust tail. Occasionally, a Sun-ward small spike-like tail is also seen in some comets. It is known as the anomalous anti-tail.

Almost all the comets are endowed with the tails as and when they approach the Sun inward of the Mars' orbit. The comets are distinguished from the asteroids by the characteristic developments of coma and tail. Obviously, the comets must be composed of some volatile substances, at least partially, not rock entirely.

In fact, the comets are mostly composed of water-ice, ammonia and methaneice before they come to the vicinity of the Sun. Dust grains of various sizes are embedded in them, making them dirty-ice-balls. At a large distance from the Sun, this cometary body serves as the bare nucleus of comet. It shines by the reflected sunlight. Therefore, its brightness is a function of its distance from the Sun and the Earth.

The spectrum of a comet is quite complex. There is a continuum superposed by the emission lines and molecular bands. The continuum results from the direct reflection of the sunlight from the grains. From the emission lines and bands, it is found that the coma is composed of CN, C_2, C_3, CH, NH, NH_2, OH and others, which are mostly the dissociation products of H_2O, CO, CH_4, CO_2, H_2CO and NH_3. The spectrum of a comet obviously varies with its distance from the Sun. When a comet occasionally comes within 0.1 AU of the Sun, the lines of Fe, Cr and Ni are also seen. The size of coma with the diluted gas and ions evaporated from the nucleus by the action of sunlight assumes a dimension as big as 10^5 km.

The tail of a comet usually begins to form when the comet comes within 1.5 to 2 AU of the Sun and it stems out from the coma in a direction opposite to the direction of the Sun. The speed of particles in the solar wind is about 430 km/s. With their high flux of momentum, they are able to streamline the flow of the ions and to duct them suitably along the magnetic field lines of the Sun that they must follow. The interaction of the magnetic field with the ambient plasma of the cometary ion tail is extremely complex.

The type II tails are mostly made up of dust particles of micron size. Their motions are affected most conspicuously by the pressure of solar radiation. The dust grains absorb the sunlight preferentially from the direction of the Sun. In this process, the momentum of the absorbed photons produces a recoil on the dust grain in a direction opposite to that of the Sun. However, when the dust grains re-radiate in the infrared, they do so isotropically in all directions, producing no net recoil on the dust grain. This is how the radiation pressure of the

Sun is able to drive the dust grains and channelize them in a direction opposite to that of the Sun.

Since the magnitudes of these accelerations are not very large as compared to those produced by the solar wind on the gas molecules and ions, the dust tails are produced at a slower rate. Since the comet follows a curved path, the dust tail appears to be curved, as the distant-most part of the dust tail remembers the direction of the Sun with respect to the former location of the comet and the nearer portion of the tail follows the latest direction of the Sun at the pressure position of the comet.

A comet must lose mass because of evaporation, though at a variable rate depending on the distance from the Sun. The rate of evaporation from a bright comet becomes about 10^{29}-10^{30} molecules/s, when it is inside the Earth's orbit. This is how both the coma and tail are sustained for a long time. During a fly-by, a comet can lose ~1/1000th of its mass. The process of ejection of the gas and dust particles from the comet's surface takes place preferentially from that part of its surface facing the Sun. Thus, the ejected gas also produces a recoil on the comet which tries to slow it down during the incoming and accelerates while outgoing from the Sun. A comet is also found to rotate about its own axis.

Every year, a dozen or two of comets are observed, of which 4 or 5 may be new. All the observed comets are usually named after the discoverers as well as after the specific year and their alphabetical order of discovery/recovery in a given calendar year. Only a fraction of comets are found to be periodic and they are further designated by the letter *P*. The periodic comets are classified as short and long periodic ones, the dividing line standing at $P = 200$ yrs. So, the Halley having an average period of 76 years is regarded as a short periodic comet. Some extremely short periodic ones have periodicity of less than 10 years.

18. Meteors, Meteoroids and Meteorites

The meteors (Fig. 16) are often known as the shooting stars. Almost every 5-10 min one finds a shooting star popping up from somewhere in the night sky. They appear as a bright streak of light across the sky. It is this burning state of matter that they are referred to as the meteors. The objects which become meteors due to the friction with the atmosphere are usually small particles coming from somewhere in the solar system. These particles are generally classified as the meteoroids.

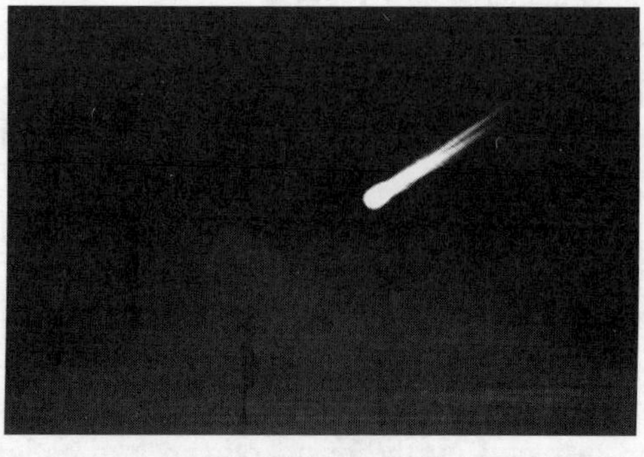

Fig. 16: Meteor.
(*Courtesy:* Google website)

Occasionally, a meteoroid can be so big that even after becoming the meteors they are not completely burnt out and a sizable remnant hits the surface of the Earth. The extraterrestrial matter recovered this way is referred to as a meteorite (Fig. 17).

Fortunately, the meteors can also be recorded during the daytime. This is possible because the meteors leave a hot and ionized column of the atmosphere behind their velocity track. In this column, the atoms of air are collisionally excited and even ionized. Before they recombine, they form a column of high density electron plasma in the atmosphere along the track.

An average meteor that is seen with the naked eye in a clear night has its track originating usually at a height of 100 km above the ground. Brighter meteors are caused by the cm-sized particles. The most bright ones called the 'fire balls', can be anything between a pebble and a cobble. The density of these objects varies from 3×10^2 kg/m^3 to 3×10^3 kg/m^3. The naked eye meteors have masses in the range form 10^{-4} kg to 10^{-2} kg with the density of about 3×10^2 kg/m^3.

Fig. 17: Meteorite.
(*Courtesy:* Google website)

The meteors which are recorded photographically but not seen with the naked eye, have masses of about 10^{-5} kg and have similar densities. The meteors detected by the radars are usually much smaller and less massive, having mass in the range from 10^{-9} kg to 10^{-7} kg and they have the mean density of about 8×10^2 kg/m^3.

There are occasionally bursts of the meteors, called the meteor showers, apparently found to originate in a constellation and seen for over few hours to few days. The hourly rate of such meteors is between 15 and 100. The meteor showers are usually named after the constellation in which they seem to originate. For example, the meteor shower named the Perseids originates in the constellation Perseus.

The fire-balls which finally hit the surface of the Earth are called the meteorites.

19. Zodiacal Light

The Zodiacal light (Fig. 18) is a faint, roughly triangular, whitish glow seen in the night sky which appears to extend up from the vicinity of the Sun along the ecliptic or zodiac. This phenomenon was first investigated by Cassini in 1683 and first explained by Fatio de Duillier in 1684. When the plane of the ecliptic passes through the zenith of a local observer, the ecliptic intercepts with the horizon at a right angle. Under such configuration, small particles that are concentrated mainly along the ecliptic scatter the sunlight in the forward direction.

Usually, about 3/2 hour after the local sunset during January and February or about 3/2 hour before the sunrise in September, the ecliptic passes through the zenith of an observer located in the latitude belt of 15°N-25°N. A conical glow of sunlight is visible for about 1/2 hour along the plane of the ecliptic projected away from the Sun, and extending up to the zenith. Such glows are known as the Zodiacal light. Thousands of years ago, Egyptians had made a record of these lights. It is called the zodiacal light because the glow extends along the ecliptic which runs through the zodiacal constellations.

Fig. 18. Zodiacal.

(*Courtesy:* Google website)

20. Oort Cloud

The Oort cloud (Fig. 19) is a hypothesized spherical cloud of comets which may lie roughly 50,000 AU, or nearly a light year away from the Sun. In 1950, Jan Oort made an interesting investigation of the orbits of very long period comets. Some of them having the semi-major axis longer than 2×10^4 AU. It meant, according to Oort, that all the comets are not only bound to the solar system but many of them are strongly populated at a distance of about 2×10^4 AU from the Sun. He therefore hypothesized the existence of a huge population of comets, later on called the Oort cloud, encircling the solar system at a distance of about 2×10^4 AU - 10×10^4 AU.

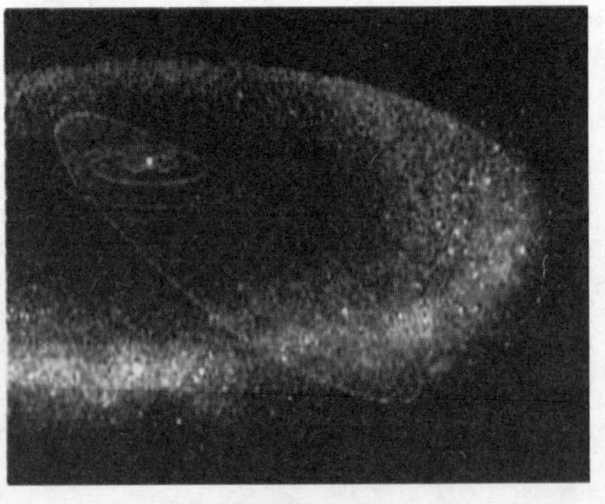

Fig. 19. Oort cloud.

(*Courtesy:* Google website)

The Oort cloud is thought to have two separate regions: a spherical outer Oort cloud and a disc-shaped inner Oort cloud. Objects in the Oort cloud are largely made up of ices, such as water, ammonia, and methane. The outer Oort cloud is loosely bound to the solar system, and this is easily affected by the gravitational pull both of passing stars and of the Milky Way galaxy itself. It may divert a large number of comets from their orbits within the cloud and send them towards the Sun.

A question arises how could the Oort cloud stay in its original position for over 4.5 billion years. In order to remove this problem, Gerard Kuiper suggested the existence of a belt of comets at a distance far closer than the Oort cloud. The Kuiper belt is hypothesized in the form of a flattened belt of comets beyond the orbit of the Neptune but within 150 AU of the Sun, with a pool of comets numbering 10^8-10^{10}.

The orbital characteristics of most of the short period comets are consistent with their presumed origin rather in the Kuiper belt than in the Oort cloud. Also hypothesized is the existence of an inner Oort cloud at a distance of 2×10^4 AU, having a bigger pool of comets, numbering about 6×10^{12}. Kuiper belt and the farther scattered disc have been observed and mapped by the scientists.

21. Origin of Solar System

The formation and evolution of the solar system is estimated to have begun 4.55 to 4.56 billion years ago with the gravitational collapse of a small part of a giant molecular cloud. There are a large number of different theories about the origin of the solar system, none of which seems to explain all the known peculiarities of the system. Nevertheless, it is one of the most important issues that relates to the origin of the Earth, Moon, planets and Sun, and therefore of the stars, in general, in the galaxy.

The solar system is believed to have evolved considerably since the formation of planets and their respective satellites. Some satellites might have been formed independently and later on been captured by their planets. Some satellites, as such the Earth's satellite the Moon, might have been the result of giant collisions between the bodies.

21.1 Early Theories of Solar System Formation

Emmanuel Kant, towards the middle of 18th century, proposed his famous theory about the origin of the solar system. He thought that the Sun was a product of self gravitational condensation of a cosmic nebula, made up of gas and dust particles. The nebula was initially rotating and because of its rotation, it was flattened. The planets were the nuclei of the matter in this flattened nebula, where the concentrations took place. Thus, he explained why the planets lie in the equatorial plane of the Sun.

Simon Laplace, towards the end of 18th century, expanded the idea of Kant. He imagined that the spinning gas cloud, out of which the solar system was born, was gradually shrinking due to cooling. Because of the contraction, size and hence the moments of inertia of cloud about its rotation axis kept on decreasing, and as a result of that the axial spinning was accelerated. As the angular speed of rotation increased, the gas cloud became more flattened and assumed a lenticular shape with the sharp edge. At some stage, the centrifugal force at the rim balanced the gravitational force of attraction due to the inner parts of the cloud.

Perhaps this is how the rim got separated into a number of equatorial rings. The series of rings thus released, finally condensed into the planets. Perhaps, the planets themselves also were initially in the form of the rotating clouds or the cloudlets which, on further contraction, led similarly to the formation of satellites around them. In this model, the Keplerian distribution of angular velocity among the planets and satellites were duly explained. Some of the modern theories of the solar system formation still support the basic ideas of Kant and Laplace.

The major criticism about the Kant-Laplace model is its inability to explain the distribution of sizes of planets. It is found that the largest planet Jupiter occupies a central position in the system of planets.

Comte de Buffon was the first to suggest in 1745 (that is even before Kant) that the planets were formed through the impact of a comet which collided with the Sun on its way through the space. The idea was revived in first two decades of twentieth century where the collision of comet was replaced by that of another star with the Sun. The collision of the Sun with a foreign star led to the ejection of matter from both the stars. These ejected materials perhaps formed a gaseous disc around one of these stars, which finally fragmented into the planets.

Very soon it became clear that collision of two stars in the sky has practically zero probability. Stars are literally distributed in otherwise free space like any two marbles at an average separation of about 10 km. Any two such objects moving however irregularly, would find very little chance of meeting in a head-on collision; not even once in 10 billion years. Therefore, it was suggested that instead of a head-on collision, two wandering stars might have interacted gravitationally at perhaps a close enough range, a range which may be as small as a few hundred to thousand times their radii.

For such a gravitational interaction of any two objects of finite size at a close range of separation, both the stars are attracted slightly more strongly as compared to their rear parts, making either or both of the stars assume the shape of a prolate spheroid, the elongated sides being often referred to as the tidal bulges. For a close enough interaction, the tidal bulges can be appreciably large and in the extreme cases can even separate from the respective parent bodies. Such a tidally ejected trail of hot gas was believed to have finally cooled down to form the small asteroid-like objects, called the planetesimals, which finally conglomerated to form the bodies of planets and satellites.

Sir James Jeans (1916) believed in the tidal theory for the formation of solar system, but deviated from the idea of forming planetesimals from a tidally ejected gas. He thought and in fact showed that the ejection would take place in the form of a curved cigar because of the relative angular momentum of two stars about each other. This cigar-like ejection finally splits up into big knots, and the rest of the material fell back on the Sun. Since the originally ejected material had already developed a sense of rotation, this fell-back material provided the Sun a rotation in the same sense as of the planetary revolutions. This theory of Jeans became quite popular in the twenties of twentieth century and remained so till the end of the second world war.

Meanwhile, Raymond Lyttleton in 1940 and Fred Hoyle in 1944 suggested another theory for the solar system formation. Lyttleton thought that the Sun was originally a part of a double star system. In the course of time, the pair interacted with a third star, which due to its tidal interaction with the pair took away one star of the pair and left the other, namely our Sun, behind with an envelope of hot gas cloud. This cloud finally condensed into the planets. Hoyle added a grain of salt to this whole hypothesis by saying that the original double star system had one component which was slightly more massive than the other, and exploded as a supernova leaving our Sun behind. So, he did not have to invoke the interaction of the third star for the want of ejecta, which he could obtain from the supernova outburst of the companion. The gaseous ejecta were captured by the solar system.

Later on, the theories of Carl von Weizäcker and Gerard Kuiper revived the idea of Kant and Laplace, suggesting that the initial disc of the gas developed semipermanent eddies and the local condensations were caused by the turbulence.

Let us discuss the specific features of the solar system that must be addressed to be any reasonable theory of the solar system formation.

21.2 Specific Features of the Solar System Related to its Origin

Here, we mention some outstanding features of the solar system that must be explained by a reasonably good theory of its origin.

(i) The total mass of planets, asteroids and comets constitutes less than 0.2% of the mass of the Sun or hence of the solar system.

(ii) The total angular momentum of all planets due to their revolution around the Sun constitutes about 98% of the total angular momentum of the solar system. Thus, the material containing about 0.2% of the total mass of the solar system shares about 98% of the total angular momentum of the solar system.

(iii) All planets revolve around the Sun in the same sense as the rotation of the Sun.

(iv) Planets are mostly confined within about 25° of the equatorial plane of the Sun.

(v) Chemical composition of the giant planets, particularly the Jupiter and Saturn, is almost identical to that of the Sun itself. The chemical composition of the Sun, other nearby stars and the interstellar medium out of which the stars are formed are also almost identical. This way, the local interstellar medium, the Sun and the gaseous giant planets all share the same average cosmic abundance of various elements.

(vi) The inner and the outer planets have widely different chemical compositions. The inner planets are absolutely devoid of the volatiles and of cosmically most abundant elements, namely, hydrogen and helium.

(vii) The chemical composition of the asteroids and the meteorites are quite similar. The meteorites are the oldest known objects in the solar system with an estimated age of 4.55 to 4.60 billion years.

(viii) The Moon and the Mercury bear an irrefutable evidence of heavy meteoritic bombardment, that took place about 3.8 billion years ago. The age of the crust of the inner planets and the Moon seems to be about 3.8-4.3 billion years.

(ix) The crystalline structure of Ni and Fe chondrules inside chondritic meteorites suggests that they must have been formed at high temperatures with a cooling rate not exceeding 1 or 2°C per million years. Such a low cooling rate is possible only when these meteorites were crystallized inside an object of the size not less than 200 km. This is often regarded as the evidence for the existence of planetesimals, before the formation of inner planets.

Based on these well established genesis related features of the solar system, the following facts about the origin of the solar system have substantially been agreed upon by about 1972.

1. The Sun and the nebula from which the solar system was born, called the solar nebula, were identical in the composition. This conclusion is based on the similarity in the composition of the solar photosphere and carbonaceous chondrites for the relative abundances of all the heavy elements. It is noted that the mantle of the Sun has lost substantial amount of Li^7 and Be^9 due to the convective mixing up to a certain depth at which the temperature becomes just too high to induce the nuclear reactions responsible for the destruction of Li and Be. The chondrites on the other hand, have lost all the volatiles namely, hydrogen and helium.
2. The solar nebula was essentially homogeneous in composition and hence had been well mixed. This is based on the similarity of all the non-radiogenic isotope ratios in the terrestrial, the lunar and the meteoritic rocks, subject to their availability.
3. The nucleosynthesis of the solar material seems to have taken place about 10^8 years prior to the onset of the condensation in the solar nebula. That is, any local supernova explosion that triggered the early condensation and brought about the change in the ratios of the radio isotopes had occurred within about 10^8 years before the condensation of the solar nebula began. This conclusion is based on the products of I^{129} and PU^{244} decay among various radiogenic isotopes of Xe.
4. The Earth, Moon and stony meteoritic parent bodies became separate entities at least 4.5 billion years ago. This is based on the radio-chronological studies of Rb/Sr and Pb/Pb isotopes systematics.

Review Questions

1. For the solar planets, describe the following terms:
   (i) Mass, (ii) Density, (iii) Mean distance from the Sun, (iv) Orbital period, (v) Orbital eccentricity, (vi) Rotation period, (vii) Polar obliquity, (viii) Polar oblateness (ix) Equatorial gravity, (x) Surface temperature, (xi) Albedo, (xii) Obliquity of the orbit, and (xiii) Obliquity of the rotation axis.
2. Show that a planet closer to the Sun revolves with higher speed.
3. Show that a planet closer to the Sun has smaller period of revolution.
4. Two planets *A* and *B*, at distances r_1 and r_2 from the Sun are revolving with velocities v_1 and v_2, respectively. If $r_1 < r_2$, show that $v_1 > v_2$.
5. Describe the solar eclipse.
6. Describe the lunar eclipse.
7. Describe the tides found on the Earth.
8. What is Titius-Bode relation? What has been the outcome of that?
9. Show that the speed of revolution of the Earth is faster than that of the Jupiter.
10. What is orbital period for a planet revolving around the Sun? Derive a relation between the speed of revolution and distance of a planet from the Sun.
11. Write specific features of the solar system which should be addressed by a theory for its origin.
12. Describe the Laplace theory for the formation of rings around a planet.
13. Explain why the Pluto is no more a planet.

14. What is Roche limit? Obtain expression for it.
15. Write notes on the following:
    (i) Escape velocity
    (ii) Mercury
    (iii) Venus
    (iv) Earth
    (v) Moon
    (vi) Solar eclipse
    (vii) Lunar eclipse
    (viii) Tides
    (ix) Mars
    (x) Jupiter
    (xi) Saturn
    (xii) Uranus
    (xiii) Neptune
    (xiv) Asteroids
    (xv) Titius-Bode law
    (xvi) Pluto
    (xvii) Satellites
    (xviii) Comets
    (xix) Meteors, meteorites and meteoroids
    (xx) Zodiacal light
    (xxi) Oort cloud
    (xxii) Roche limit

Multiple-Choice Questions

1. Conventionally, the retrograde motion is from
   A. west to east B. east to west C. north to south D. south to north
   Ans. A
2. Solar planet having the largest mass is
   A. Earth B. Mars C. Jupiter D. Saturn **Ans. C**
3. Solar planet closest to the Sun is
   A. Mercury B. Venus C. Earth D. Mars **Ans. A**
4. Solar planet having the smallest period of revolution is
   A. Mercury B. Venus C. Earth D. Neptune **Ans. A**
5. Solar planet having the largest speed of revolution is
   A. Mercury B. Venus C. Earth D. Neptune **Ans. A**
6. Solar planet having the lowest orbital eccentricity is
   A. Mercury B. Venus C. Earth D. Neptune **Ans. B**
7. Solar planet having the retrograde rotation about its own axis is
   A. Mercury B. Venus C. Earth D. Jupiter **Ans. B**

8. Which of the following is not a planet?
   A. Mercury B. Venus C. Earth D. Moon **Ans. D**
9. Mean distance of asteroids from the Sun is
   A. 1 AU B. 0.4 AU C. 2.8 AU D. 5.2 AU **Ans. C**
10. From the Earth's surface, the same face of Moon is seen always because
   A. The Earth revolves around the Sun in 1 year.
   B. The Earth spins about its own axis in 1 day.
   C. The Moon revolves around the Earth.
   D. The rotation and revolution periods of the Moon are equal. **Ans. D**
11. The number of satellites of Jupiter discovered by Galileo in 1610 is
   A. 2 B. 3 C. 4 D. 8 **Ans. C**
12. The Roche limit for a ring system around a planet is how many times the radius of the planet.
   A. 1.00 B. 1.44 C. 2.4 D. 4.2 **Ans. C**
13. Titan is a satellite of
   A. Jupiter B. Saturn C. Uranus D. Neptune **Ans. B**
14. Which satellite of the Jupiter has the largest size?
   A. Io B. Europa C. Ganymede D. Callisto **Ans. C**
15. Imagine that four planets, the Earth, Mars, Jupiter and Saturn are put in a big pond having water. Which one would float in the water?
   A. Earth B. Mars C. Jupiter D. Saturn **Ans. D**
16. Which of the following radiates more energy than it receives?
   A. Earth B. Mercury C. Mars D. Jupiter **Ans. D**
17. Which of the following radiates more energy than it receives?
   A. Earth B. Mars C. Saturn D. Uranus **Ans. C**
18. Which satellite of the Saturn has the largest size?
   A. Iapetus B. Enceladus C. Dione D. Titan **Ans. D**
19. Which of the following has the largest size?
   A. Ariel B. Umbriel C. Titania D. Miranda **Ans. C**
20. Which of the following appears blue?
   A. Mars B. Jupiter C. Uranus D. Saturn **Ans. C**
21. Which of the following appears red?
   A. Mars B. Jupiter C. Uranus D. Saturn **Ans. A**
22. Which of the following appears blue?
   A. Mars B. Jupiter C. Neptune D. Saturn **Ans. C**
23. Who explained that planets lie in the equatorial plane of the Sun?
   A. Newton B. Galileo C. Kant D. Einstein **Ans. C**
24. In the past, which of the followings was a planet?
   A. Ceres B. Pluto C. Pallas D. Vesta **Ans. B**

25. A planet revolves around the Sun in a circular orbit. Its orbital velocity does not depend on
    A. mass of the Sun B. distance from the Sun
    C. mass of the planet D. gravitational constant **Ans. C**
26. A planet revolves around the Sun in a circular orbit. Its time-period of revolution does not depend on
    A. mass of the Sun B. distance from the Sun
    C. mass of the planet D. gravitational constant **Ans. C**
27. The orbits of three satellites, Io, Europa and Ganymede of Jupiter are locked in the perfect tidal resonances with a ratio of the orbital periods, respectively,
    A. 1 : 2 : 3 B. 1 : 3 : 5 C. 1 : 2 : 4 D. 1 : 4 : 9 **Ans. C**
28. In water which of the planets would float?
    A. Earth B. Mercury C. Saturn D. Neptune **Ans. C**
29. Saturn's ring system disappears, when viewed from the Earth, once every
    A. 11 years B. 15 years C. 22 years D. 42 years **Ans. B**
30. Each of the north and south poles of the Uranus face the Sun for how many years?
    A. 11 years B. 15 years C. 22 years D. 42 years **Ans. D**
31. Since its discovery, which of the followings has completed one revolution around the Sun.
    A. Saturn B. Neptune C. Jupiter D. Uranus **Ans. B**

12 Nucleosynthesis

The baryonic matter, where the elements are made out of protons, neutrons and electrons, is a small part of the content of universe. The cosmological models support the idea that the total visible mass in the form of stars, planets, satellites, living beings, etc. (visible baryonic matter) plus the mass contributed by the energy, is only 4.6%. The rest of the mass is considered to be in the form of dark matter (23%) and dark energy. However, there are no direct observational evidences for these forms of matter and energy, and is a subject of research.

The matter found in the stars and in the interstellar clouds is in the form of atoms or ions (plasma). In some regions, such as high densities inside the white dwarfs and neutron stars, the matter may be found in the degenerate forms.

In the entire known universe, the hydrogen (atomic number 1) is found to be the most abundant element; the helium (atomic number 2) is found second. However, later on, the rank of abundance does not follow the correspondence with the atomic number. For example, the oxygen (atomic number 8) has abundance rank 3, and carbon (atomic number 6) has abundance rank 4. The nuclear reactions are responsible for the abundance of various elements. Answer to a question, how carbon, iron and other heavy elements are found inside the Earth, may be that the Earth is made out of the recycled material thrown out during the death of the stars. In the present chapter, we have discussed various elements and nuclear reactions prevailing inside the stars.

1. Stellar Nucleosynthesis

Stellar nucleosynthesis is a process through which the natural abundance of elements change within a star due to the nuclear fusion reactions in the cores and overlying mantles of the star. The stars are said to evolve (age) with the changes in the abundance of the elements within. The fusion process in the core increases the atomic weight of its gaseous elements, causing pressure loss and contraction accompanied by the increase of temperature. Structural changes of the star (evolution) become necessary to stabilize it. The stars lose most of their mass when they are ejected late in their stellar lifetimes, thereby increasing the abundance of the elements heavier than the helium in the interstellar medium.

The fusion of nuclei in a star, starting from its initial hydrogen and helium abundance, provides energy and synthesizes new nuclei as a byproduct of the fusion process.

2. Hydrogen Fusion

As discussed earlier, the hydrogen burning is a process of nuclear fusion of four protons to form one nucleus of $2He^4$. There are two predominant processes through which the hydrogen burning in a star takes place.

As discussed earlier, in lower-mass stars, such as the Sun, the dominant process of hydrogen burning is the proton-proton chain (pp-chain) reaction, whereas in the high-mass stars, the dominant process is the Carbon-Nitrogen-Oxygen (CNO) cycle. Which type of hydrogen burning process dominates inside a star is decided on the basis of the temperature therein. The pp-chain reaction starts at the temperatures around 4×10^6 K, making it the dominant mechanism in the low-mass stars. The CNO chain requires a higher temperature of approximately 15×10^6 K, but thereafter it increases more rapidly in efficiency than the pp-chain reaction as the temperature grows. Above approximately 17×10^6, the CNO cycle becomes the dominant source of the energy.

This temperature is achieved in the cores of the main sequence stars with at least 1.3 times the mass of the Sun. The Sun itself has a core temperature of around 15.7×10^6 K and only 0.8% of energy being produced in the Sun comes from the CNO cycle.

2.1 Deuterium Burning

The deuterium ${}_1D^2$ is a stable isotope of hydrogen with a mass approximately double that of a proton. It consists of one proton and one neutron. Deuterium burning is a nuclear fusion reaction that occurs in the stars and some sub-stellar objects, in which one deuterium nucleus and one proton combine to form one ${}_2He^3$ nucleus. It takes place as the second stage of the p-p chain reaction, in which a deuterium nucleus formed from two protons fuses with a further proton.

Stellar models suggest that if there were no deuterium burning, then there should be no stars with masses more than about two or three times the mass of the Sun in the pre-main sequence phase because the hydrogen burning would occur while the object was still accreting matter.

2.2 Proton-Proton Chain Reaction

The proton-proton chain reaction is one of two nuclear fusion reactions, along with the CNO cycle, by which the stars convert hydrogen to helium and which dominates in the stars of the size of Sun or smaller.

In general, the proton-proton fusion can occur only if the temperature (i.e., kinetic energy) of protons is high enough to overcome their mutual electrostatic or Coulomb repulsion. The first step involves the fusion of two ${}_1H^1$ nuclei (protons) into the deuterium, releasing a positron and a neutrino as one proton changes into a neutron.

$${}_1H^1 + {}_1H^1 \rightarrow {}_1D^2 + e^+ + \nu_e + \gamma$$

This first step is extremely slow. The half-life for a successful p-p fusion in the core of the Sun is estimated to be a billion years, even at the extreme density and temperatures found there. So produced positron is almost immediately annihilated with an electron, and their mass-energy and kinetic energy, is carried off by two gamma-ray photons.

$$e^- + e^+ \rightarrow 2\gamma$$

The deuterium produced in the first stage can fuse with another proton to produce an isotope of helium ${}_2He^3$:

$${}_1D^2 + {}_1H^1 \rightarrow {}_2He^3 + \gamma$$

This process is extremely fast in comparison to the first step. It is estimated that, under the conditions in the Sun's core, a newly-created deuterium nucleus exists for only about 4 seconds before it is converted to ${}_2He^3$. In the Sun, ${}_2He^3$ produced in these reactions exists for only about 400 years before it is converted into ${}_2He^4$. At this stage, there are four possible paths, pp I, pp II, pp III and pp IV to generate ${}_2He^4$. The frequencies of the first three paths are, respectively, 86%, 14% and 0.11%. There is also an extremely rare pp IV branch. The pp I branch is dominant at the temperatures of 10 to 14 MK and is as the following.

$$ {}_2He^3 + {}_2He^3 \rightarrow {}_2He^4 + 2\,{}_1H^1 $$

The pp II branch is dominant at the temperatures of 14 to 23 MK and is as the following.

$$ {}_2He^3 + {}_2He^4 \rightarrow {}_4Be^7 + \gamma $$
$$ {}_4Be^7 + e^- \rightarrow {}_3Li^7 + \nu_e $$
$$ {}_3Li^7 + {}_1H^1 \rightarrow 2\,{}_2He^4 $$

The pp III branch is dominant if the temperature exceeds 23 MK and is as the following.

$$ {}_2He^3 + {}_2He^4 \rightarrow {}_4Be^7 + \gamma $$
$$ {}_4Be^7 + {}_1H^1 \rightarrow {}_5B^8 + \gamma $$
$$ {}_5B^8 + {}_1H^1 \rightarrow {}_4Be^8 + e^+ + \nu_e $$
$$ {}_4Be^8 \rightarrow 2\,{}_2He^4 $$

The pp III chain is not a major source of energy in the Sun (only 0.11%), but was very important in the solar neutrino problem because it generates very high energy neutrinos (up to 14.06 MeV). The pp IV branch, as the following, is predicted but has never been observed due to its rarity.

$$ {}_2He^3 + {}_1H^1 \rightarrow {}_2He^4 + e^+ + \nu_e $$

Comparing the mass of the final ${}_2He^4$ atom with the masses of four protons reveals that 0.007 or 0.7% of the mass of original protons has been lost. This mass has been converted into the energy, in the form of gamma-rays and neutrinos released during each of the individual reactions.

The energy released as the gamma-rays interacts with electrons and protons and heats the interior of the Sun. Neutrinos do not interact significantly with any matter and therefore do not help to support the Sun against the gravitational collapse. Their energy is lost and the neutrinos in the pp I, pp II and pp III branches carry away 2.0%, 4.0%, and 28.3% of energy in those reactions, respectively.

2.3 CNO Cycle

Besides the proton-proton chain reaction, the CNO (carbon-nitrogen-oxygen) cycle is another set of the fusion reactions by which the stars convert hydrogen into helium. Unlike the pp-chain, the CNO cycle is a catalytic cycle. Theoretical models suggest that the CNO cycle is a dominant source of energy in the stars – more massive than about 1.3 times the mass of Sun. The proton-proton chain is more important in the stars having mass equal to that of the Sun or less. This difference stems from temperature dependency differences between the two reactions. The pp-chain reactions start occurring at temperatures around 4×10^6 K, making it the dominant energy source in the smaller stars. A CNO cycle starts occurring at approximately 15×10^6 K, but its energy output rises much more rapidly with the increasing temperature. At approximately 17×10^6 K, the CNO cycle starts becoming the dominant source of energy. The Sun has a core temperature of around 15.7×10^6 K, and only 1.7% of He^4 nuclei are produced in the Sun through the CNO cycle. Various CNO cycles are proposed by the scientists. The CNO-I proposed by Carl von Weizsäcker and Hans Bethe is as the following.

$$_6C^{12} + {}_1H^1 \rightarrow {}_7N^{13} + \gamma$$
$$_7N^{13} \rightarrow {}_6C^{13} + e^+ + \nu_e$$
$$_6C^{13} + {}_1H^1 \rightarrow {}_7N^{14} + \gamma$$
$$_7N^{14} + {}_1H^1 \rightarrow {}_8O^{15} + \gamma$$
$$_8O^{15} \rightarrow {}_7N^{15} + e^+ + \nu_e$$
$$_7N^{15} + {}_1H^1 \rightarrow {}_6C^{12} + {}_2He^4$$

The positrons almost instantly annihilate with electrons, releasing energy in the form of two gamma radiations. The neutrinos escape from the star carrying away some energy. One nucleus goes to become carbon, nitrogen, and oxygen isotopes through a number of transformations in an endless loop.

In the CNO-II cycle, instead of the production of a $_6C^{12}$ and an alpha particle, a $_8O^{16}$ and a photon are produced. Like the carbon, nitrogen, and oxygen involved in the main branch (CNO-I), the fluorine produced in the minor branch is merely an intermediate product in the CNO-II, and the sequence of the reactions is as the following.

$$_7N^{15} + {}_1H^1 \rightarrow {}_8O^{16} + \gamma$$
$$_8O^{16} + {}_1H^1 \rightarrow {}_9F^{17} + \gamma$$
$$_9F^{17} \rightarrow {}_8O^{17} + e^+ + \nu_e$$
$$_8O^{17} + {}_1H^1 \rightarrow {}_7N^{14} + {}_2He^4$$
$$_7N^{14} + {}_1H^1 \rightarrow {}_8O^{15} + \gamma$$
$$_8O^{15} \rightarrow {}_7N^{15} + e^+ + \nu_e$$

The CNO-III cycle is a sub-dominant branch only for the massive stars where the reactions are started when one of the reactions in the CNO-II results in a $_9F^{18}$ and a gamma radiation, instead of a $_7N^{14}$ and an alpha particle, and the sequence of reactions is as the following.

$$_8O^{17} + {}_1H^1 \rightarrow {}_9F^{18} + \gamma$$
$$_9F^{18} \rightarrow {}_8O^{18} + e^+ + \nu_e$$
$$_8O^{18} + {}_1H^1 \rightarrow {}_7N^{15} + {}_2He^4$$
$$_7N^{15} + {}_1H^1 \rightarrow {}_8O^{16} + \gamma$$
$$_8O^{16} + {}_1H^1 \rightarrow {}_9F^{17} + \gamma$$
$$_9F^{17} \rightarrow {}_8O^{17} + e^+ + \nu_e$$

Like the CNO-III, the CNO-IV is also only significant in massive stars. The reactions are started when one of the reactions in the CNO-III results in a $_9F^{19}$ and a gamma radiation, instead of $_7N^{15}$ and an alpha particle, and the sequence of reactions is as the following.

$$_9F^{19} + {_1H^1} \rightarrow {_8O^{16}} + {_2He^4}$$
$$_8O^{16} + {_1H^1} \rightarrow {_9F^{17}} + \gamma$$
$$_9F^{17} \rightarrow {_8O^{17}} + e^+ + \nu_e$$
$$_8O^{17} + {_1H^1} \rightarrow {_9F^{18}} + \gamma$$
$$_9F^{18} \rightarrow {_8O^{18}} + e^+ + \nu_e$$
$$_8O^{18} + {_1H^1} \rightarrow {_9F^{19}} + \gamma$$

2.4 Hot CNO Cycle

Under the conditions of higher temperature and pressure, such as those found in the novae and x-ray bursts, the catalytic cycles are typically referred to as the hot CNO (HCNO) cycles. The difference between the HCNO-I cycle and the CNO-I cycle is that $_7N^{13}$ captures a proton instead of decaying, leading to the sequence of reactions as the following.

$$_6C^{12} + {_1H^1} \rightarrow {_7N^{13}} + \gamma$$
$$_7N^{13} + {_1H^1} \rightarrow {_8O^{14}} + \gamma$$
$$_8O^{14} \rightarrow {_7N^{14}} + e^+ + \nu_e$$
$$_7N^{14} + {_1H^1} \rightarrow {_8O^{15}} + \gamma$$
$$_8O^{15} \rightarrow {_7N^{15}} + e^+ + \nu_e$$
$$_7N^{15} + {_1H^1} \rightarrow {_6C^{12}} + {_2He^4}$$

The notable difference between the CNO-II cycle and the HCNO-II cycle is that $_9F^{17}$ captures a proton instead of decaying, and neon is produced in a subsequent reaction on $_9F^{18}$, leading to the sequence of reactions as the following.

$$_7N^{15} + {_1H^1} \rightarrow {_8O^{16}} + \gamma$$
$$_8O^{16} + {_1H^1} \rightarrow {_9F^{17}} + \gamma$$
$$_9F^{17} + {_1H^1} \rightarrow {_{10}Ne^{18}} + \gamma$$
$$_{10}Ne^{18} \rightarrow {_9F^{18}} + e^+ + \nu_e$$
$$_9F^{18} + {_1H^1} \rightarrow {_8O^{15}} + {_2He^4}$$
$$_8O^{15} \rightarrow {_7N^{15}} + e^+ + \nu_e$$

The HCNO-III cycle is an alternative to the HCNO-II cycle where a $_9F^{18}$ captures a proton moving towards higher mass and using the same helium production mechanism as in the CNO-IV cycle, as the following.

$$_9F^{18} + {_1H^1} \rightarrow {_{10}Ne^{19}} + \gamma$$
$$_{10}Ne^{19} \rightarrow {_9F^{19}} + e^+ + \nu_e$$
$$_9F^{19} + {_1H^1} \rightarrow {_8O^{16}} + {_2He^4}$$
$$_8O^{16} + {_1H^1} \rightarrow {_9F^{17}} + \gamma$$
$$_9F^{17} + {_1H^1} \rightarrow {_{10}Ne^{18}} + \gamma$$
$$_{10}Ne^{18} \rightarrow {_9F^{18}} + e^+ + \nu_e$$

3. Helium Burning

Older stars start to accumulate helium produced by the proton-proton chain reaction and the carbon-nitrogen-oxygen cycle in their cores. The products of further nuclear fusion reactions of helium with hydrogen or another helium nucleus produce the lithium-5 and the beryllium-8 respectively, both of which are highly unstable and decay almost instantly back into the smaller nuclei. When the star starts to run out of the hydrogen to fuse, the core of the star begins to collapse until the central temperature rises to 10^8 K. At this point the helium nuclei are fusing together faster than their product, the beryllium-8 decays back into two helium nuclei.

3.1 Triple-alpha Process

The triple-alpha process is a set of nuclear fusion reactions by which three $_2He^4$ nuclei (alpha particles) are transformed into carbon.

Once a beryllium-8 is produced, a little faster than it decays, the number of beryllium-8 nuclei in the stellar core increases to a large number. Then in its core there are many beryllium-8 nuclei that can fuse with another helium nucleus to form carbon-12, which is stable:

$$_2He^4 + {_2He^4} \rightarrow {_4Be^8}$$

$$_4Be^8 + {_2He^4} \rightarrow {_6C^{12}}$$

Because the triple-alpha process is unlikely, it needs a long time to produce much carbon. One consequence of this is that no significant amount of carbon was produced in the Big Bang, because within minutes after the Big Bang, the temperature fell below the critical point for nuclear fusion.

Ordinarily, the probability of the triple alpha process is extremely small. However, the beryllium-8 ground state has almost exactly the energy of two alpha particles. In the second step, $_4Be^8 + {_2He^4}$ has almost exactly the energy of an excited state of $_6C^{12}$. These resonances greatly increase the probability that an incoming alpha particle combines with the beryllium-8 to form the carbon.

As a side effect of the process, some carbon nuclei can fuse with additional helium to produce a stable isotope of oxygen.

$$_6C^{12} + {_2He^4} \rightarrow {_8O^{16}}$$

3.2 Alpha Process

The alpha process, also known as the alpha-ladder, is one of two classes of the nuclear fusion reactions by which the stars convert helium into the heavier elements, the other being the triple-alpha process. While the triple-alpha process only requires helium, once some carbon is present, other reactions that consume the helium are possible:

$$_6C^{12} + {_2He^4} \rightarrow {_8O^{16}} + \gamma$$

$$_8O^{16} + {_2He^4} \rightarrow {_{10}Ne^{20}} + \gamma$$

$$_{10}Ne^{20} + {_2He^4} \rightarrow {_{12}Mg^{24}} + \gamma$$

$$_{12}Mg^{24} + {_2He^4} \rightarrow {_{14}Si^{28}} + \gamma$$

$$_{14}Si^{28} + {_2He^4} \rightarrow {_{16}S^{32}} + \gamma$$

$$_{16}S^{32} + {}_2He^4 \rightarrow {}_{18}Ar^{36} + \gamma$$
$$_{18}Ar^{36} + {}_2He^4 \rightarrow {}_{20}Ca^{40} + \gamma$$
$$_{20}Ca^{40} + {}_2He^4 \rightarrow {}_{22}Ti^{44} + \gamma$$
$$_{22}Ti^{44} + {}_2He^4 \rightarrow {}_{24}Cr^{48} + \gamma$$
$$_{24}Cr^{48} + {}_2He^4 \rightarrow {}_{26}Fe^{52} + \gamma$$
$$_{26}Fe^{52} + {}_2He^4 \rightarrow {}_{28}Ni^{56} + \gamma$$
$$_{28}Ni^{56} + {}_2He^4 + \gamma \rightarrow {}_{30}Zn^{60}$$

All these reactions have a very low rate at temperatures and densities found in the stars and therefore do not contribute significantly to a star's energy production; with the elements heavier than neon (atomic number > 10), they occur even less easily due to the increasing Coulomb barrier.

Alpha process elements (or alpha elements) are so-called since their most abundant isotopes are integer multiples of four, the mass of helium nucleus (alpha particle). Stable alpha elements are: C, O, Ne, Mg, Si, S, Ar, Ca.

4. Burning of Heavy Elements

4.1 Lithium Burning

Lithium is generally present in the brown dwarfs and not in the low-mass stars. The stars, which by definition must achieve high temperature (2.5×10^6 K) necessary for fusing hydrogen, rapidly deplete their lithium. This occurs by a collision of lithium-7 and a proton producing two helium-4 nuclei. The temperature necessary for this reaction is just below the temperature necessary for the hydrogen fusion. From the study of lithium abundances in 53 T Tauri stars, it has been found that the lithium depletion varies strongly with the size, suggesting the lithium burning by the pp-chain. The pp-chain for the lithium burning is as the following.

$$p + {}_3Li^6 \rightarrow {}_4Be^7 \text{ (unstable)}$$
$$_4Be^7 + e^- \rightarrow {}_3Li^7 + \nu$$
$$p + {}_3Li^7 \rightarrow {}_4Be^8 \text{ (unstable)}$$
$$_4Be^8 \rightarrow 2\,{}_2He^4 + \text{energy}$$

It does not occur in the stars less than sixty times the mass of Jupiter.

4.2 Carbon Burning

The carbon burning process or carbon fusion is a set of nuclear fusion reactions that take place in the massive stars (at least $8M_\odot$ at birth) that have used up the lighter elements in their cores. It requires high temperatures ($> 5 \times 10^8$ K) and densities ($> 3 \times 10^9$ kg/m^3). The principal reactions are:

$$_6C^{12} + {}_6C^{12} \rightarrow {}_{10}Ne^{20} + {}_2He^4$$
$$_6C^{12} + {}_6C^{12} \rightarrow {}_{11}Na^{23} + {}_1H^1$$
$$_6C^{12} + {}_6C^{12} \rightarrow {}_{12}Mg^{24} + n$$
$$_6C^{12} + {}_6C^{12} \rightarrow {}_8O^{16} + 2\,{}_2He^4$$

4.3 Neon Burning

The neon burning process is a set of nuclear fusion reactions that take place in the massive stars (at least 8 solar masses). Neon burning requires high temperature and density (around 1.2×10^9 K and 4×10^9 kg/m^3). At such high temperature, the photo-disintegration becomes a significant effect, so that some neon nuclei decompose, releasing alpha particles:

$$_{10}Ne^{20} + \gamma \rightarrow {}_{8}O^{16} + {}_{2}He^{4}$$

$$_{10}Ne^{20} + {}_{2}He^{4} \rightarrow {}_{12}Mg^{24} + \gamma$$

$$_{10}Ne^{20} + n \rightarrow {}_{10}Ne^{21} + \gamma$$

$$_{10}Ne^{20} + {}_{2}He^{4} \rightarrow {}_{12}Mg^{24} + n$$

where the neutron consumed in the first step is regenerated in the second step.

4.4 Oxygen Burning

The oxygen burning process is a set of nuclear fusion reactions that take place in the massive stars that have used up the lighter elements in their cores. It occurs at temperature around 1.5×10^9 K and density of 10^{10} kg/m^3. The principal reactions are:

$$_{8}O^{16} + {}_{8}O^{16} \rightarrow {}_{14}Si^{26} + {}_{2}He^{4}$$

$$_{8}O^{16} + {}_{8}O^{16} \rightarrow {}_{15}P^{31} + {}_{1}H^{1}$$

$$_{8}O^{16} + {}_{8}O^{16} \rightarrow {}_{16}S^{31} + n$$

$$_{8}O^{16} + {}_{8}O^{16} \rightarrow {}_{14}Si^{30} + 2\,{}_{1}H^{1}$$

$$_{8}O^{16} + {}_{8}O^{16} \rightarrow {}_{15}P^{30} + {}_{1}D^{2}$$

$$_{8}O^{16} + {}_{8}O^{16} \rightarrow {}_{16}S^{32} + \gamma$$

$$_{8}O^{16} + {}_{8}O^{16} \rightarrow {}_{12}Mg^{24} + 2\,{}_{2}He^{4}$$

The oxygen burning process is the last nuclear reaction in the star's core which does not proceed via the alpha process.

4.5 Silicon Burning

In Astrophysics, silicon burning is a very brief sequence of nuclear fusion reactions that occur in the massive stars with a minimum of about 8-11 solar mass. Silicon burning is the final stage of fusion for the massive stars that have run out of the fuels that power them for their long lives in the main sequence on the Hertzsprung-Russell diagram. It follows the previous stages of hydrogen, helium, carbon, neon and oxygen burning processes. Silicon burning begins when the gravitational contraction raises the star's core temperature to $(2.7\text{-}3.5) \times 10^9$ K. The exact temperature depends on the mass. When a star has completed the silicon burning phase, no further fusion is possible.

After a star completes the oxygen burning process, its core is composed primarily of silicon and sulphur. If it has sufficiently high mass, it further contracts until its core reaches temperature in the range of $(2.7\text{-}3.5) \times 10^9$ K. At these temperatures, silicon and other elements can photo-disintegrate, emitting a proton or alpha particle. Silicon burning entails the alpha process, which creates new elements by adding one of these alpha particles per step in the following sequence:

$$_{14}Si^{28} + {}_{2}He^{4} \rightarrow {}_{16}S^{32}$$
$$_{16}S^{32} + {}_{2}He^{4} \rightarrow {}_{18}Ar^{36}$$
$$_{18}Ar^{36} + {}_{2}He^{4} \rightarrow {}_{20}Ca^{40}$$
$$_{20}Ca^{40} + {}_{2}He^{4} \rightarrow {}_{22}Ti^{44}$$
$$_{22}Ti^{44} + {}_{2}He^{4} \rightarrow {}_{24}Cr^{48}$$
$$_{24}Cr^{48} + {}_{2}He^{4} \rightarrow {}_{26}Fe^{52}$$
$$_{26}Fe^{52} + {}_{2}He^{4} \rightarrow {}_{28}Ni^{56}$$
$$_{28}Ni^{56} + {}_{2}He^{4} \rightarrow {}_{30}Zn^{60}$$

The entire silicon burning sequence lasts about one day and stops when the nickel-56 is produced. The star can no longer release energy via nuclear fusion because a nucleus with 56 nucleons has the lowest mass per nucleon of all the elements in the alpha process sequence. Although the iron-58 and the nickel-62 have slightly higher binding energies per nucleon than the iron-56, the next step in the alpha process would be the zinc-60, which has slightly more mass per nucleon and thus, is less thermodynamically favourable. The nickel-56 (which has 28 protons) has a half-life of 6.02 days and decays via the β+ decay to the cobalt-56 (27 protons), which in turn has a half-life of 77.3 days as it decays to the iron-56 (26 protons).

During this phase of contraction, the potential energy of the gravitational contraction heats the interior to 5×10^9 K and this opposes and delays the contraction. However, since no additional heat energy can be generated via the new fusion reactions, the final unopposed contraction rapidly accelerates into a collapse lasting only a few seconds. The central portion of the star is now crushed into either a neutron star or, if the star is massive enough, a black hole. The outer layers of star are blown off in an explosion known as a Type II supernova that lasts from days to months.

5. Production of Elements Heavier than Iron

It is sill possible to produce the elements heavier than iron. It can be done through the neutron capture and proton capture.

5.1 *r*-process

The *r*-process (or rapid-process) is a nucleosynthesis process that occurs in the core-collapse supernovae and is responsible for the creation of approximately half of neutron-rich atomic nuclei heavier than iron. The process entails a succession of rapid neutron captures (hence the name *r*-process) by heavy seed nuclei, typically the Fe^{56} or other more neutron-rich heavy isotopes.

The most probable candidate sites for the *r*-process has long been suggested to be the core-collapse supernovae, which may provide necessary physical conditions for *r*-process.

5.2 *s*-process

The *s*-process (or slow-process) is another nucleosynthesis process that occurs by means of the slow captures of neutrons, primarily occurring in the AGB stars. It occurs at relatively low neutron density and intermediate temperature conditions in the stars. Under these conditions heavier nuclei are created by the neutron capture, increasing the atomic weight of the nucleus by one. The s-process is secondary, meaning that it requires preexisting heavy isotopes as seed nuclei to be converted into other heavy nuclei.

The s-process is believed to occur mostly in asymptotic giant branch stars. In contrast to the r-process which is believed to occur over time scales of seconds in the explosive environments, the s-process is believed to occur over time scales of thousands of years, passing decades between neutron captures.

The main neutron source reactions are:

$$_6C^{13} + {}_2He^4 \rightarrow {}_8O^{16} + n$$

$$_{10}Ne^{22} + {}_2He^4 \rightarrow {}_{12}Mg^{25} + n$$

The cycle that terminates the *s*-process is:

$$_{83}Bi^{209} + n \rightarrow {}_{83}Bi^{210} + \gamma$$

$$_{83}Bi^{210} \rightarrow {}_{84}Po^{210} + e^- + \overline{\nu}_e$$

$$_{84}Po^{210} \rightarrow {}_{82}Pb^{206} + {}_2He^4$$

$$_{82}Pb^{206} + 3n \rightarrow {}_{82}Pb^{209}$$

$$_{82}Pb^{209} \rightarrow {}_{83}Bi^{209} + e^- + \overline{\nu}_e$$

The net result of this cycle therefore is that 4 neutrons are converted into one alpha particle, two electrons, two anti-electron neutrinos and gamma radiation:

$$4n \rightarrow {}_2He^4 + 2e^- + 2\overline{\nu}_e$$

The process thus terminates in the bismuth, the heaviest stable element, and polonium, the first non-primordial element after bismuth.

These *r*- and *s*-processes account for a majority of galactic chemical evolution of elements heavier than iron.

5.3 *rp*-process

The *rp*-process (rapid proton capture process) consists of consecutive proton captures onto seed nuclei to produce the heavier elements. It is a nucleosynthesis process and, along with the s-process and the *r*-process, may be responsible for the generation of many of heavy elements present in the universe.

The process has to occur in very high temperature environment (above 10^9 K) so that the protons can overcome large Coulomb barrier for charged particle reactions. A hydrogen-rich environment is also a prerequisite due to large proton flux needed.

Sites suggested for the rp-process are accreting binary systems where one star is a neutron star. In these systems the donor star is accreting material onto its compact partner star. The accreted material is usually rich in hydrogen and helium because of its origin from the surface layers of the donor star.

5.4 *p*-process

The *p*-process is a proton capture process which is the source of certain, naturally occurring, proton-rich isotopes of elements from the selenium to the mercury. These nuclides are called the p-nuclei. Proton-rich nuclides can be produced by sequentially adding one or more protons to an atomic nucleus. Such a nuclear reaction of type (p, γ) is called the proton capture reaction. By adding a proton to a nucleus, the element is changed because the chemical element is defined by the proton number of a nucleus. At the same time the ratio of the protons to the neutrons is changed, resulting in a proton-richer isotope of the next element.

Such a proton captures on stable nuclides (or nearly stable), however, are not very efficient in producing a p-nuclei, especially the heavier ones, because the electric charge increases with each added proton, leading to an increased repulsion of the next proton to be added, according to the Coulomb's law.

Review Questions

1. Discuss the stellar nucleosynthesis.
2. Write notes on the following:
   (i) Hydrogen fusion
   (ii) Deuterium burning
   (iii) Proton-proton chain reaction
   (iv) CNO cycle
   (v) Helium burning
   (vi) Triple-alpha process
   (vii) Lithium burning
   (viii) Carbon burning
   (ix) Neon burning
   (x) Oxygen burning
   (xi) Silicon burning
   (xii) Production of elements heavier than iron
   (xiii) *r*-process
   (xiv) *s*-process
   (xv) *rp*-process
   (xvi) *p*-process

Multiple-Choice Questions

1. Which is catalytic process for production of energy in the core of a star?
   A. pp-chain B. both pp-chain and CNO cycle
   C. CNO cycle D. neither pp-chain nor CNO cycle **Ans. C**
2. In a CNO cycle, the isotopes of which element act a catalyst?
   A. C only B. N only C. C and N only D. C, N, O all **Ans. D**
3. In nucleosynthesis process in stars, the heaviest element produced is
   A. He B. C C. Ca D. Fe **Ans. D**

13 Galaxies and Universe

Though the boundaries of universe are not known, scientists believe that the universe consists of several hundred billions of galaxies. These galaxies have various shapes, sizes and ways of their evolution. Our galaxy, Milky Way is one of them. As we live inside our galaxy, it is not possible to have direct information about its shape. In a night we see a dense stream of stars, which has very large number of stars in the summer nights as compared to those in the winter nights. We are in the northern hemisphere on the Earth and are situated in the disc at a distance of about 30,000 ly from the centre of the Milky Way. Hence, in the summer nights, we look towards the centre whereas in the winter night we see away from the centre. Though the galaxies are very large in number, but on the basis of their finite number of shapes, they are classified into some groups. Some galaxies are found active. Based on the gravitational effects, it is concluded that there is a large amount in the form of the dark matter as compared to the normal one. The recession of the galaxies obtained from the redshift provides information about the expanding universe. In this chapter, we have discussed about various galaxies.

1. Messier Catalog

Charles Messier prepared a document regarding 110 astronomical objects. As Messier was interested in finding only comets, he prepared a list of those non-comet objects that created problems in his hunt for them. The compilation of this document, in collaboration with his assistant Pierre Mechain, is known as the Messier catalogue. This catalogue of objects is one of the most famous lists of astronomical objects, and many of them are still referenced by their Messier number. The catalogue includes most of the astronomical deep-sky objects that can easily be observed from Earth's Northern Hemisphere.

2. Sharply-Curtis Debate

The ShapleyCurtis Debate, also known as The Great Debate, took place on 26 April 1920 at the Smithsonian Museum of Natural History, between two renowned astronomers Harlow Shapley and Heber Curtis. This debate is about the nature of so-called spiral nebulae and the size of the universe. Shapley was of the opinion that distant nebulae were relatively small and lay within the outskirts of the Milky Way galaxy (at that time, the Milky Way galaxy was thought to be the entire universe), whereas Curtis had opinion that they were in fact independent galaxies, implying that they were exceedingly large and distant.

3. Doppler Shift in Emission Lines Observed by Humason, Slipher and Hubble

An American astronomer, Vesto Melvin Slipher (November 11, 1875-November 8, 1969), is given credit to perform the first measurements of radial velocities for galaxies. He is considered the first astronomer who discovered that distant galaxies are redshifted, thus providing the first empirical basis for the expansion of the universe. Since he was also the first to relate these redshifts to the velocity of object, he may be given very large share of the credit for the discovery of the expanding universe.

Milton La Salle Humason (August 19, 1891-June 18, 1972) was an American astronomer. At the age of 14, he dropped out of school and had no formal education. In 1917, he became a caretaker at the Mount Wilson observatory, and out of his interest in astronomy, he volunteered to be a night assistant at the observatory. His technical skill and quiet manner made him a favourite on the mountain. Recognizing his talent, in 1919, George Ellery Hale appointed him as a staff member at Mt. Wilson observatory. This was a unique example, as Humason did not have a Ph.D., or even a high school diploma. He soon proved that Hale's judgment was correct, as he made several key observational discoveries. He was known as a meticulous observer, obtaining photographs and difficult spectrograms of faint galaxies. His contribution is sig-nificant and played important role in the development of physical cosmology, including assisting Edwin Hubble in formulating the Hubble's law. In 1950, he got a D.Sc. from Lund University. He discovered Comet C/1961 R1 (Humason), notable for its large perihelion distance.

Edwin Powell Hubble (November 20, 1889-September 28, 1953) played a crucial role in establishing the fields of extragalactic astronomy and observational cosmology. He showed that many objects previously considered to be clouds of dust and gas and classified as "nebulae" were actually galaxies beyond the Milky Way. He measured galactic and extragalactic distances with the help of strong direct relationship between a classical Cepheid variable's luminosity and pulsation period.

Hubble provided evidence that the recessional velocity of a galaxy increases with its distance from the Earth, now known as the "Hubble's law". However, it is found that it had been proposed and demonstrated observationally two years earlier by Georges Lematre; so the law is now named as the Hubble-Lematre law. This law implies that the universe is expanding. A decade before, an American astronomer Vesto Slipher had provided the first evidence that the light from many of these nebulae was strongly redshifted. It was the first evidence for the expansion of the universe.

4. Surface Brightness of Galaxies

In astronomy, for a spatially extended object, such as a galaxy or nebula, or of the night sky background, the apparent brightness or flux density per unit angular area, is expressed as the surface brightness. The surface brightness of an object depends on its surface luminosity density, i.e., its luminosity emitted per unit surface area. In visible and infrared astronomy, surface brightness is often expressed in terms magnitudes. The measurement of surface brightnesses of celestial objects is called surface photometry.

5. Star Count Analysis

In fact, a star count is a survey (or analysis of existing survey data) counting stars according to some criteria. A motivation behind this procedure has been to determine the density of stars, through classifying them by apparent magnitude and making adjustments to accommodate biases that affect the counts. The counting and analysis of stars is used for other astronomical phenomena, e.g., galaxies, supernovae, and extra-solar planets. Counts of the former two are considered revealing regarding current ideas in cosmology.

6. Milky Way Galaxy

Our galaxy, named Milky Way, is one of the several hundred billions galaxies in the universe. It contains over 100 billion stars. In a dark night, one can even casually notice in the sky, an almost continuous band of light, extending from one end of the horizon to other, inclined to some 62.6° with respect to the celestial equator (Fig. 1). This band of light is a part of our Milky Way. Galileo was the first who realized this band having a large collection of stars.

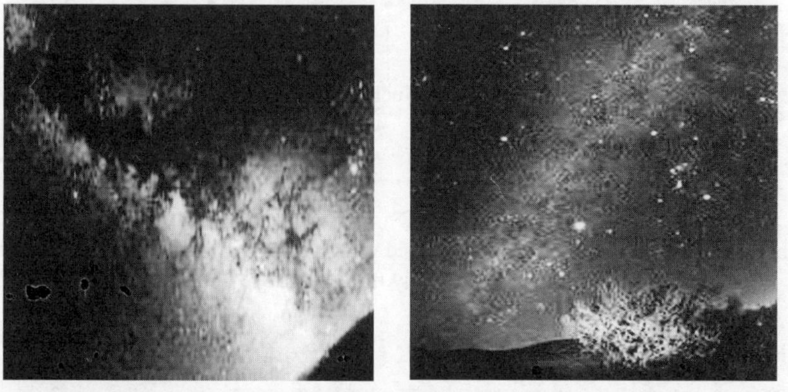

Fig. 1: Milky Way in summer (left) and in winter (right), seen in the northern hemisphere.
(*Courtesy:* Google website)

6.1 Shape and Size

Since we cannot go outside the Milky Way to see it from there, it is only possible to prepare a schematic drawing of the Milky Way, based on the observations made from the Earth in various directions. A schematic drawing of the Milky Way is shown in Fig. 2. The Milky Way may be divided into three major components: (i) a thin disc of thickness about 2000 ly and containing stars, gas, and dust, (ii) a central bulge and (iii) a halo which is roughly spherical and contains roughly 200 globular clusters. Our Sun is situated in the disc of Milky Way at a distance of about 30,000 ly from the centre.

As discussed earlier, because of the difference between the sidereal day and mean solar day by 3 min and 56 sec, stars in the summer nights observed at a place are different from those in the winter nights. On the Earth, we are situated in the northern hemisphere on the Earth and in the summer nights, we see towards the galactic centre (centre of the Milky Way) whereas in the winter nights, we see opposite to the galactic centre. Thus, to an observer in the northern hemisphere, our Milky Way appears more prominent during the summer nights than that during the winter nights. In the southern hemisphere, the case is the reverse and the Milky Way appears more prominent during the winter nights than that during the summer nights.

6.2 Stellar Populations

The stars found in the disc are mostly Population I (with high abundance of heavy-elements) stars. The central bulge has a mixture of Population I and Population II (with low abundance of heavy-elements) stars. The halo of the galaxy virtually contains Population II stars.

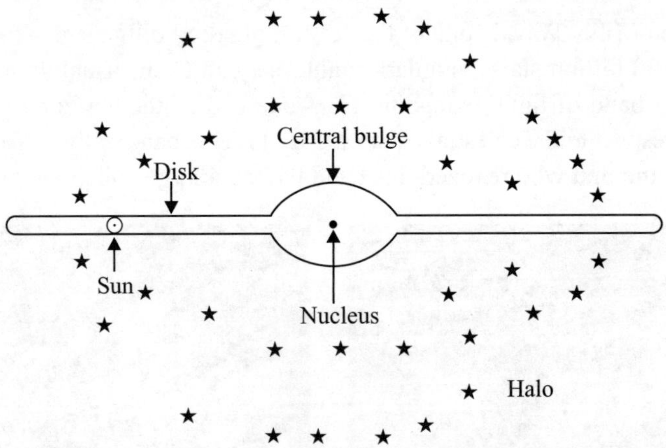

Fig. 2: A schematic drawing of the Milky Way viewed wedge-on by an observer outside the galaxy. Our Sun is at a distance of about 30,000 ly from the centre of the Milky Way.

6.3 Stellar Motions

Our Milky Way is a bound stellar system, where in order to balance the gravitational force by the centrifugal force, the stars must go in a rotational motion. In the disc, the rotational speeds of stars are larger as compared to the small random motions. This high value of rotational speed makes the disc highly flattened. In the central bulge, the rotational speeds of the stars are not very large and therefore they move in a random manner. Thus, these stars form a roughly spherical distribution. In the galactic halo, the stars have very high random velocities and they are less tightly bound to the galaxy as compared to those in the central bulge. It is not known if there is a smooth transition from the stars in the central bulge to the stars in the halo.

6.4 Missing Mass

From the measured gravitational field, it is possible to calculate the mass of the material required locally to provide the observed gravitational force. On the other side, the mass of the visible stars, gas, and dust can be estimated, and is found about one-half of the mass required for the observed gravitational field. This is known as the missing mass problem of the Milky Way. This is one of the major unresolved problems in the modern astronomy. The missing mass problem exists in all the galaxies and thus there is a question:

- Does there exist in the universe a large amount of non-luminous matter which is undetected (and perhaps is undetectable) except for its gravitational influence?
- Scientists have an opinion that there is dark matter in our Milky Way whose amount is more than that of the visible one.

Zwicky

Fritz Zwicky (February 14, 1898-February 8, 1974) was an astronomer, who made many important contributions in theoretical and observational astronomy. In 1933, Zwicky was the first to use the virial theorem to infer the existence of unseen dark matter.

6.5 Local Mass-to-light Ratio

From the star counts, in the solar neighbourhood, we can estimate the mean surface brightness of the galactic disc. If we divide the surface mass density (mass per unit area, denoted by μ) by the surface brightness (luminosity per unit area, denoted by L), we can get the local mass-to-light ratio μ/L. In the solar units, the local mass-to-light ratio is found to be 5. That is,

$$\frac{\mu}{L} = 5\frac{M_{\odot}}{L_{\odot}}$$

Fig. 3: Fritz Zwicky (February 14, 1898 - February 8, 1974).

(*Coutesy:* Google website)

This value is typical for the visible part of the disc galaxies in general. The mass-to-light ratio for our Sun is 1 in solar units. The fact that the ratio is 5 in the immediate vicinity of the Sun has some interesting implications.

(i) It can be interpreted that the mean gravitating mass in the galaxy has a lower efficiency for producing light per unit mass than in the Sun. This conclusion is consistent with the fact that most of the stars in the galaxy are faint, *K* or *M* dwarfs. Such stars have low mass and further low luminosities.

(ii) The second interpretation is concerned with the production of helium in the stars. The Sun with the mass-to-light ratio 1 could convert only ~ 10% of its hydrogen-mass into the helium-mass through the nuclear processes in ~ 10^{10} years. Thus, the matter with the mass-to-light ratio 5 could manage to convert only 2% of its hydrogen mass into the helium mass through the nuclear processes in ~ 10^{10} years, which is roughly the age of the Milky Way.

6.6 Age-metallicity Relation

The age-metallicity relation is a fundamental observational constraint which helps for understanding about the formation of Galactic disc and its chemical evolution with time. To our knowledge, it appears that there no simple relation between age and metallicity for cosmic objects, of even one type.

For example, the ages and metallicities of dwarf stars in the solar neighbourhood provide a unique record about the progressive chemical enrichment of the interstellar medium forming stars. With the passage of time, the interstellar medium gets more and more enriched in heavy elements. Thus, it may be expected that the more recently formed stars have a higher metallicity as compared to those formed at an earlier time.

6.7 Mass of Halo

The observed mass of all the luminous stars in the halo of the Milky Way is estimated to be less than a few percent of the mass of the disc. It is believed that the halo may contain a very large and very massive invisible component. Astronomers believe that the total mass of the halo is about 10 times the mass of the disc. This again leads to the situation of missing mass in the Milky Way, which is still an unresolved problem.

6.8 Differential Rotation

Out of the three components of the Milky Way, the disc is the most studied one. A body gets flattened when it spins about some axis. Thus, the flattened shape of the disc provides a clue that the stars as well as the interstellar gas clouds in the disc rotate about the centre of galaxy. Random motions of the stars as well as of the interstellar gas clouds are superimposed on the respective circular motions. The flattening of the disc however does not provide any clue whether the rotation is uniform or non-uniform. It was found that the rotation of the disc is not uniform but differential, i.e., the disc does not rotate like a solid body. The inner parts of the disc rotate faster than the outer ones.

Our Sun being part of the disc also rotates around the centre of the Milky Way. The mean motion of the disc's material in the neighbourhood of the Sun is defined as the local standard of rest (LSR). Though the Sun has a slight motion relative to the LSR, but the non-circular component of the Sun's motion is negligibly small and we can take the motion of Sun to be LSR. Stars outside the solar circle appear as in a slower lane and the stars inside the solar circle appear as in a faster lane than the stars in the solar circle. Individual disc stars also possess small excesses or deficits of velocities as compared with the mean of all stars in their neighbourhood.

In the solar neighbourhood, there are few stars which have large velocity relative to the LSR. Since these high-velocity stars move in a random manner, they exhibit an asymmetry.

6.9 Spiral Structure

In the universe, many galaxies having disc are found to have spiral structure. Since the Milky Way is considered to have a disc, it is quite reasonable to assume that it also has a spiral structure (Fig. 4). The galaxy has a beautiful spiral pattern having two trailing spiral arms. The spiral arms have brilliant young stars looking like beads in a string. Hence, the spiral arms show a beautiful scenario. The beads are actually the giant H ɪɪ regions where the embedded OB stars cause them to shine with a large power output.

We are aware of the fact that the main sequence, the lifetime of *OB* stars is a few million years whereas the age of the galaxy is about 10 billion years. Since we find the *OB* stars in the spiral arms, it is only possible when the process of formation of such stars is going on regularly. The newly formed stars become visible whereas the old ones disappear and produce the dark matter.

Fig. 4: Spiral galaxy.

(*Courtesy:* Google website)

The main question now arises: Why new stars are born in a grand spiral pattern?

The spiral arms delineated by the strongest H ii regions are often narrow and extend to a size of more than 100 ly from one end to another. What could be the mechanism responsible for the simultaneous formation of stars throughout such a narrow and long front?

The OB stars form a spiral pattern as the gas and dust, which is raw material for the star formation, are always found in the spiral arms. In order to answer about the spiral structure, differential rotation of the disc was suggested to be responsible. This concept cannot be acceptable because of the reason which can be understood with the help of Fig. 5.

For understanding the proposed winding process in a galaxy, consider the gas clouds lined up in the form of a long string stretched radially in the galaxy at time t_1. We are aware of the differential rotation of the disc of the galaxy. The inner layers rotate faster than the outer ones. Suppose, at time t_2, when the innermost layer completes one-half circle, the line of the clouds forms the trailing spiral arc. By the time t_3, when the innermost layers complete one full circle, the outermost layers complete only a fraction of a circle.

It gives an impression of the formation of spiral structure. The time-period of the rotation of the innermost layers is generally less than 10^8 years. Typical age of a galaxy is about 10^{10} years. It shows that in the lifetime of a galaxy, the innermost layers would have completed about 100 rounds. But, in most of the galaxies, we do not have such large number of turns in the spiral arms. It is a rare event when more than one or two turns are available in a spiral galaxy.

Hence, the differential rotation cannot be a mechanism for the spiral structure. A way out for this dilemma was suggested in the form of the Density-Wave theory.

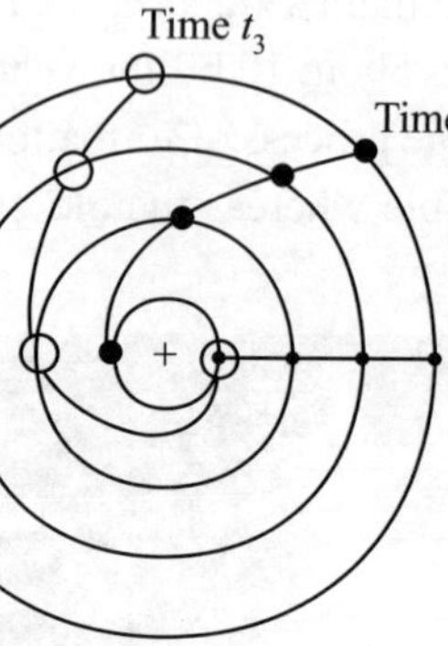

Fig. 5: Concept of winding process in a galaxy.

6.10 Magnetic Fields

Since most of the visible matter, in the Universe, is in the ionized state, it is easy to generate cosmic magnetic fields, and due to the lack of magnetic monopoles, it is hard to destroy it. Magnetic fields have been found practically in or around all celestial objects, either by in-situ measurements of spacecraft or by the electromagnetic radiation of embedded cosmic rays, gas, or dust. Significant strengths of magnetic fields have been found in the Earth, Sun, solar planets, stars, pulsars, Milky Way, nearby galaxies, more distant (radio) galaxies, quasars, and intergalactic space in clusters of galaxies. Information about the existence of cosmic magnetic fields has increased tremendously due to the rapid development of observational techniques, especially in radio astronomy.

Magnetic fields are a major agent in the interstellar medium (ISM) of spiral, barred, irregular and dwarf galaxies. They have significant contribution to the total pressure which balances the ISM against gravity. For the onset of star formation, the magnetic fields are essential, as it plays important role for the removal of angular momentum from the protostellar cloud during its collapse.

6.11 Faber-Jackson Relation

In 1976, for elliptical galaxy, two astronomers, S.M. Faber and R.E. Jackson derived the first empirical power-law relation between the luminosity L and the central stellar velocity dispersion σ as:

$$L \propto \sigma^{\gamma}$$

where the index γ is approximately equal to 4. Derivation of this relation may be understood as the following.

For a spherical body of mass M and radius R, and having uniform density, the gravitational binding energy U is expressed as:

$$U = -\frac{3GM^2}{5R}$$

where G is the gravitational constant. For a mass M distributed over a radius R, the gravitational potential may be expressed as:

$$U = -\alpha \frac{GM^2}{R}$$

where α is a constant depending, e.g., on the density profile of the system. For a constant density, $\alpha = 3/5$. The kinetic energy is:

$$K = \frac{1}{2}MV^2 = \frac{2}{3}M\sigma^2$$

where σ is the one-dimensional velocity dispersion, so that $3\sigma^2 = V^2$.[1] Using the virial theorem, $2K + U = 0$, we have

$$\sigma^2 = \frac{\alpha}{3}\frac{GM}{R}$$

Assuming that the mass-to-light ratio, M/L is constant, i.e., $M \propto L$, we can use this and the above expression to obtain a relation between R and σ^2

$$R \propto \frac{LG}{\sigma^2}$$

They considered the surface brightness, $B = L/(4\pi R^2$ and assumed it a constant (which from a fundamental theoretical point of view, is a totally unjustified assumption) to get

$$L = 4\pi R^2 B$$

Using this and combining it with the relation between R and L, this results in

$$L \propto 4\pi\left(\frac{LG}{\sigma^2}\right)^2 B$$

This relation can be rewritten as the following.

$$L \propto \frac{\sigma^4}{4\pi G^2 B}$$

That is,

$$L \propto \sigma^4$$

7. Oort Constants

By analyzing the local rotational properties of the Milky Way, Jan Oort derived empirically two parameters, generally denoted as A and B, and known as the Oort constants, expressed as

$$A = \frac{1}{2}\left(\frac{V_0}{R_0} - \left.\frac{dv}{dr}\right|_{R_0}\right)$$

[1]For a group of astronomical objects, such as an open cluster, globular cluster, galaxy, galaxy cluster, or supercluster, in astronomy, the velocity dispersion (σ) is expressed as the statistical dispersion of velocities about the mean velocity. By measuring the radial velocities of the members of group through astronomical spectroscopy, the velocity dispersion of that group can be estimated and used to derive the group's mass using the virial theorem.

and
$$B = -\frac{1}{2}\left(\frac{V_0}{R_0} + \left.\frac{\mathrm{d}v}{\mathrm{d}r}\right|_{R_0}\right)$$

where V_0 and R_0 are the rotational velocity and distance from the Galactic centre, respectively, measured at the position of the Sun. These constants depend only on the motions and positions of stars in the solar neighbourhood. As of 1997, the derived most accurate values of these constants are $A = 14.82 \pm 0.84$ km/s/kpc and $B = -12.37 \pm 0.64$ km/s/kpc. From the Oort constants, it is possible to determine the orbital properties of the Sun, such as the orbital velocity and period, and to infer the local properties of the Galactic disk, such as the mass-density and how the rotational velocity changes as a function of radius measured from the Galactic centre.

8. Galaxy Rotation Curve

For a disc-galaxy, the rotation curve (also known as velocity curve) is a plot of the magnitude of orbital velocities (i.e., the speeds) of the visible stars or gas in that galaxy versus their radial distance measured from the centre of that galaxy.

So far, the galaxy rotation curves are found to show a general behaviour that up to the distance from the galactic centre, the measurements can be made, with the increase of distance from the galactic centre, the orbital speed of the stars and gas increases or remains almost constant. That is, over the distances measured from the centre of a disc-galaxy, the measurements could be made, the stars and gas are found to revolve around the centre of the galaxy at a velocity that increases or remains almost constant. If the mass distribution in the disc-galaxies is similar to that observed from the distribution of stars and gas, then on increasing the distance, the orbital speed would always decrease in the same manner as in other systems where most of their mass is concentrated in the centre, such as the Solar System or the moons of the Jupiter.

For the spiral galaxies, the rotation curves are known to be asymmetric. Then the observational data from each side of a galaxy are averaged. The asymmetry in the rotation curve is a normal feature rather than the exceptional.

The theoretical calculations based on the assumption of centrally dominated mass associated with the observed luminous material for the galaxy rotation curve differs much from the observations.

The calculations of mass profiles of galaxies from the luminosity profiles and mass-to-light ratios in the stellar disks do not match with the masses derived from the observations of rotation curves and the law of gravity. This discrepancy can be considered by postulating a large amount of dark matter that permeates the galaxy and extends into the galaxy's halo.

8.1 Halo Density Profiles

To accommodate a flat rotation curve, a density profile for a galactic environ must be different from that for a centrally concentrated. According to the Newton's version, for the Kepler's third law, the radial density profile $\rho(r)$ should be as the following.

$$\rho(r) = \frac{3v(r)^2}{4\pi G r^2}\left(1 + 2\frac{\mathrm{d}\log_{10} v(r)}{\mathrm{d}\log_{10} r}\right)$$

where $v(r)$ is the radial orbital velocity at a distance r and G the gravitational constant. Observations did not match with such a simple profile. Then, the famous Navarro-Frenk-White profile, which is consistent with the N-body simulations and observations both, was derived as the following.

$$\rho(r) = \frac{\rho_0 R_s}{r(1 + r/R_s)^2}$$

where the central density ρ_0, and the scale radius R_s, are the parameters which vary from halo to halo. Here, the slope of the density profile diverges at the centre, and therefore, other alternative profiles were proposed.

9. Dark Matter

The dark matter is a hypothetical form of matter, which cannot be seen with a telescope but is considered to account for most of the matter in the universe. It is one of the greatest mysteries in modern astrophysics. The existence and properties of the dark matter are derived mainly from its gravitational effects observed on the visible matter, radiation, and the large-scale structure of the universe.

The dark matter neither emits nor absorbs any kind of electromagnetic radiation at any significant level. The observations of the Planck mission, for the standard model of cosmology, have been concluded that the total mass-energy of the known universe contains 4.9% ordinary matter, 26.8% dark matter and 68.3% dark energy. Hence, the invisible matter is estimated to constitute 95.1% of the total mass-energy content of the universe.

The first person to interpret the evidence and infer the presence of dark matter was Dutch astronomer Jan Oort, a pioneer in radio astronomy, in 1932. Later on. much of the evidence for dark matter comes from the study of the motions of galaxies. Many of them appear to be fairly uniform, and according to the Virial theorem, the total kinetic energy should be half the total gravitational binding energy of the galaxies. From the observations, however, the total kinetic energy is found to be much larger, when assumed that the gravitational mass is due to only the visible matter of galaxy. The stars far from the centre of the galaxies have much higher velocities than predicted by the Virial theorem.

Gravitational lensing observations of the galaxy clusters allow direct estimates of the gravitational mass based on its effect on the light from the background galaxies, since large collections of matter (dark or otherwise) deflect the light due to gravitational effect.

10. Classification of Galaxies

Our universe is made up of several hundred billions of galaxies. Hubble studied a large number of galaxies and noticed that though the number of galaxies is very large, but their geometrical shapes, which are the projections on the sky-plane, can be classified into a limited number of categories. The first point he noticed that a majority of galaxies had a regular shape whereas a few percent of them had an irregular shape. Based on this criterion, he classified the galaxies into two categories: (i) those having regular shape, called regular galaxies, and (ii) those having irregular shape, called the irregular galaxies.

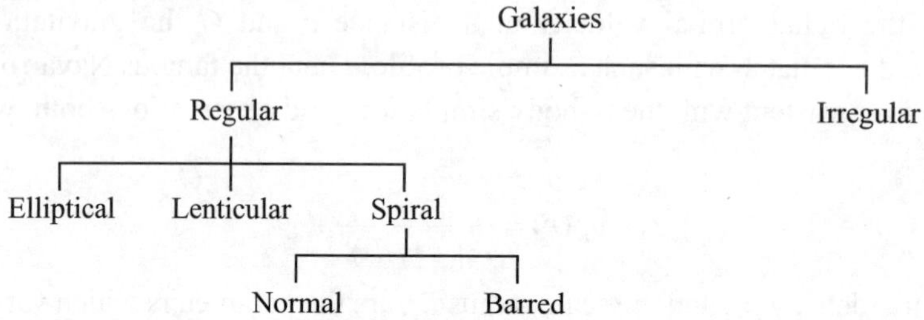

Fig. 6: Classification of galaxies.

On the basis of their shapes, the regular galaxies were further classified into two categories: (a) elliptical galaxies, also called the ellipticals, having a shape of an ellipse, and (b) spiral galaxies, also called the discs or spirals, having spiral arms. The spiral galaxies were again divided into two kinds: the ordinary spirals also called the normal spirals, and the barred spirals. Classification of the galaxies is shown in Fig. 6. Hubble's morphological classification of galaxies is shown in Fig. 7, which is of the form of a tuning fork. Hence, this diagram is known as the Hubble's tuning fork diagram.

A large majority of galaxies, about 70%, are spirals (ordinary plus barred); irregular galaxies are a few percent; and the rest are elliptical galaxies. Thus, a bulk of the mass of the visible universe is contained in the regular galaxies.

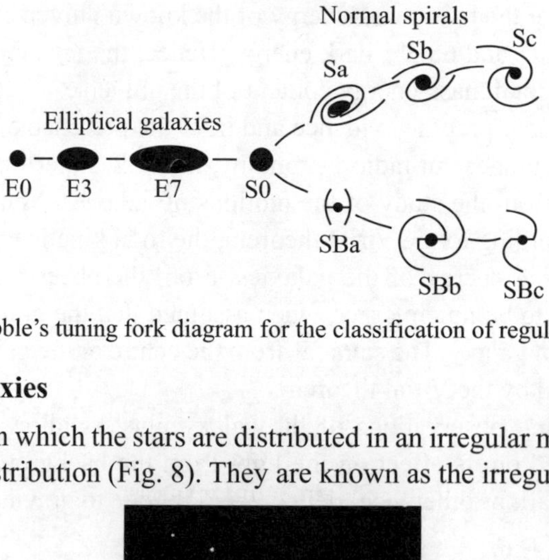

Fig. 7: Hubble's tuning fork diagram for the classification of regular galaxies.

10.1 Irregular Galaxies

These are the galaxies in which the stars are distributed in an irregular manner without showing any regularity in the distribution (Fig. 8). They are known as the irregular galaxies.

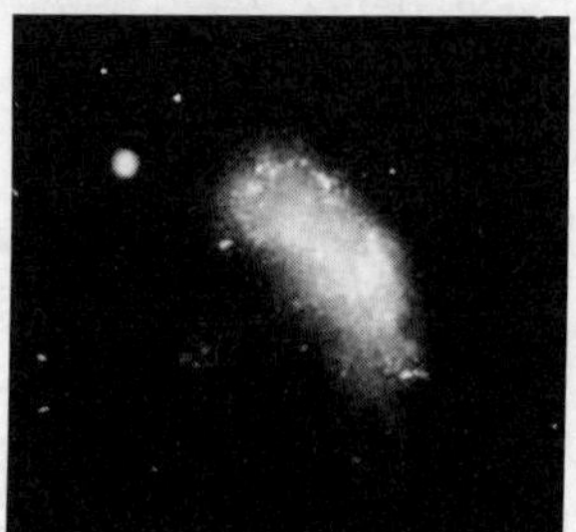

Fig. 8: Irregular galaxy.

(*Courtesy:* Google website)

10.2 Elliptical Galaxies

These are the galaxies which look like an ellipse when projected on the sky-plane (Fig. 9). Depending on the flatness of the ellipse of the galaxy, they are classified into eight groups denoted by from E0 to E7, where E0 denotes a circular form (zero value of eccentricity) and E7 denotes the available most flattened form (the largest available value of the eccentricity). Equation of an ellipse with semi-major axis a and semi-minor axis b is

$$\frac{x^2}{a^2}+\frac{y^2}{b^2}=1$$

and eccentricity e of the ellipse is $e = \sqrt{1-b^2/a^2}$. For an elliptical galaxy denoted by En, the integer n is expressed as

$$n=10\left(1-\frac{b}{a}\right)$$

Earlier, the reason for the flatness of an elliptical galaxy was assumed to be its rotation about its own axis. The speed of rotation was assumed to increase when we go from E0 to E7.

Fig. 9: Elliptical galaxy.
(*Courtesy:* Google website)

However, later on the observations found that the reason for the flattening was not the rotation, but it was due to that the large random velocities of the constituent stars are anisotropic (i.e., the velocities of the constituent stars have different values in different directions). From the model calculations of the stellar dynamics, this effect of the flattening of a galaxy due to anisotropy in velocities was known for a long time, but it was not accepted earlier.

When we neglect the disc of a spiral galaxy, the bulge of the galaxy shows a resemblance with an elliptical galaxy. About the dynamics of stars in an elliptical galaxy, it can be said that the stars are moving in a more or less random manner under their mutual gravitational force. Almost no gas, which is the raw material for the star formation, has been found in the elliptical galaxies.

Example 1.

For an elliptical galaxy En, derive an expression for the eccentricity. Calculate the eccentricity of the galaxy E6.

Solution:

For an ellipse with the semi-major axis a and the semi-minor axis b, we have

$$n = 10\left(1 - \frac{b}{a}\right)$$

or

$$\frac{b}{a} = \frac{10 - n}{10}$$

The eccentricity e is

$$e = \sqrt{1 - \frac{b^2}{a^2}}$$

$$= \sqrt{1 - \left(\frac{10 - n}{10}\right)^2} = \frac{\sqrt{n(20 - n)}}{10}$$

For the galaxy $E6$, we have $n = 6$, and thus

$$6 = 10\left(1 - \frac{b}{a}\right)$$

or

$$\frac{b}{a} = 0.4$$

Thus, eccentricity e of the ellipse is

$$e = \sqrt{1 - b^2/a^2}$$

$$= \sqrt{1 - 0.16} = 0.917$$

10.3 Spiral Galaxies

Spiral galaxies have spiral arms. Our galaxy Milky Way also is assumed to be a spiral galaxy. The spiral galaxies can be classified into two categories: (i) normal spirals, denoted by S, and (ii) barred spirals, denoted by SB (Fig. 10). The normal spirals, besides the spiral arms, have a disc whereas the barred spirals, besides the spiral arms, have a bar. Each of these types is further divided into three sub-categories: for the normal spirals, we have Sa, Sb and Sc; for the barred spirals, we have SBa, SBb and SBc.

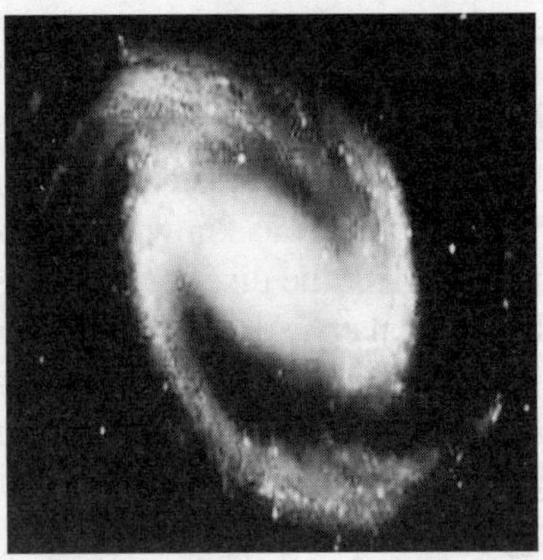

Fig. 10: Spiral (left) and barred (right) spiral galaxy.

(*Courtesy:* Google website)

In each of the sub-category, the spiral arms are increasingly more loosely wound. The tightness of the spiral winding correlates closely with the relative size of the central part: the bigger the central part, the more tightly bound are the spiral arms. Thus, Sa and SBa galaxies have larger central part than the Sc and SBc galaxies, respectively. However, some observations show that this correlation may be more complicated.

It has been suggested that the spirals may be losing their spiral arms and turning to S0 and SB0 type galaxies. This suggestion is however not accepted by most of the scientists. Although not all present S0 and SB0 galaxies might have emerged from the spirals, all spirals which avoid merging with other galaxies apparently one day would become S0 and SB0 type galaxies.

It is universally accepted that the amount of gas in the spiral galaxies is gradually decreasing in the formation of new stars. Though during the death of the stars, some of the gas may return to the interstellar medium, but in each cycle, a net amount of gas is locked in the end stages (white dwarfs, neutron stars and black holes) of the stars. Confinement of the birth of stars to the spiral arms suggests that most of the spiral galaxies have become so diffuse that only in the regions where the gas is compressed, the formation of new stars takes place.

Though astronomers have an outline of theory for the evolution of galaxies, but they are still very far from making a definite calculations explaining the scenario of the galaxies.

Tully-Fisher relation

In Astronomy, the Tully-Fisher relation (TFR) is an empirical relationship between the mass or intrinsic luminosity of a spiral galaxy and its angular velocity or emission line width. The luminosity is calculated by multiplying the galaxy's apparent brightness by $4\pi d^2$, where d is its distance from us, and the spectral-line width is measured using long-slit spectroscopy.

Several different forms of the TFR exist, depending on which precise measures of the mass, luminosity or rotation velocity one takes it to relate. Its modified form, known as the Baryonic Tully-Fisher relation (BTFR) states that the baryonic mass is proportional to the velocity to the power of roughly 3.5-4.

The TFR can be used to estimate the distance to the spiral galaxies by allowing the luminosity of a galaxy to be derived from its directly measurable line width.

10.4 Lenticular Galaxies

The galaxies at the transition between the elliptical and spiral galaxies come under the class of the lenticular galaxies, denoted by S0 as well as SB0 (Fig. 11).

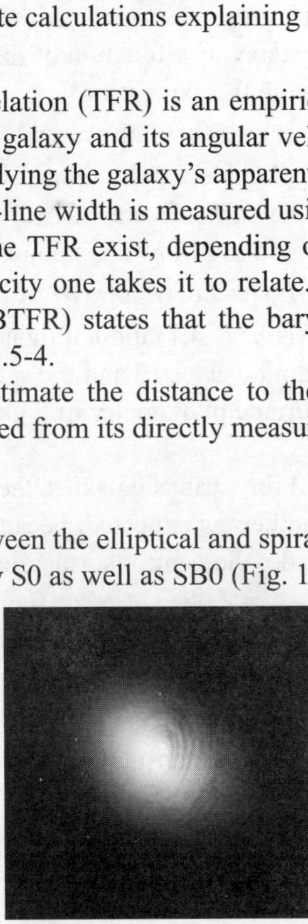

Fig. 11: Lenticular galaxy.
(*Courtesy:* Google website)

10.5 Major differences between the Spirals and Ellipticals

The spirals and ellipticals have the following major differences:

1. In the elliptical galaxies, the ratio of the random velocities to the rotational velocities is much larger than that in the spiral galaxies.
2. Elliptical galaxies have almost no gas and no dust. On the other side, most of the spirals have gas as well as dust.
3. Since the gas and dust, which are raw materials for the formation of new stars, are absent in the ellipticals, there are neither young stars nor the formation of stars is going on. On the other side, young stars are being regularly formed in the spirals as the required raw material is present there in a large amount.
4. There are no spiral arms in the elliptical galaxies. On the other side, almost all the spirals have the spiral arms.

10.6 Various Distributions in Regular Galaxies

The regular galaxies have been found to show some distributions.

(i) Distribution of light: Following the observations, the surface brightness L (luminosity per unit area) of an elliptical galaxy as a function of distance r from the centre along the major-axis could be fitted through the expression

$$L(r) = L(0)\exp[-(r/r_0)^{1/4}] \tag{13.1}$$

This equation (13.1) is also applicable for the light-distribution of the central bulge of a spiral galaxy. However, when the light is averaged over the circles, the light distribution in the disc of a spiral galaxy is found to follow an exponential law

$$L(\mathrm{r}) = L(0)\exp[-(r/r_0)] \tag{13.2}$$

Equations (13.1) and (13.2) give the distribution of light in the regular galaxies. Numerical values of the central surface luminosity $L(0)$ and the scale length r_0 are adjusted to obtain the best fit for the observed surface photometry of a galaxy (or various components of a galaxy).

When such fits were performed for various galaxies, the scale length r_0 was found to vary greatly from one galaxy to another, as expected, because the galaxies are found to have different sizes. On the other side, the central surface luminosity $L(0)$ was found to have smaller variation in its value.

(ii) Distribution of mass: For a galaxy if v denotes either kind of velocity, then the mass $M(r)$ contained in a sphere of radius r can be roughly expressed through the relation

$$M(r) = rv^2/G \tag{13.3}$$

where G is the universal gravitational constant. Equation (13.3) gives the distribution of mass in a regular galaxy.

(iii) The $L \propto v^4$ law for the ellipticals and spirals: Because of very large distances of galaxies, angular size of most of them is too small and the detailed kinematics of the structure of a galaxy cannot be investigated. From the line profile of radiation from the visible part of a galaxy, it is possible to measure the total spread v of the velocity. In the ellipticals, the velocity v results from the random velocities of the stars whereas in the spirals, it is from the rotational velocities of the stars and gas. It was interesting to find that in the optical spectrum, the velocity spread v of an elliptical galaxy or a spiral galaxy is related to its luminosity L through the relation

$$L \propto v^4$$

11. Active Galaxies

So far we have discussed about such galaxies which have very little activity. These galaxies may be described as the quiet galaxies. A small percentage of the observed galaxies show violent activity much beyond the norms shown by the quiet galaxies. These galaxies are known as the active galaxies. The active galaxies may be classified into various categories and some of them are as the following.

1. Seyfert galaxies or Seyferts
2. Radio galaxies
3. Quasars
4. QSOs

11.1 Seyfert Galaxies

Carl Seyfert was the first person who discovered anomalies in a class of galaxies with very bright centres. These galaxies are now known as the Seyfert galaxies (Fig. 12). In many of such galaxies, the bright nucleus outshines the rest part of the galaxy put together. In these galaxies not only the nucleus is very bright, but the energy output from the nucleus can often vary by more than twice in the span of less than one year. It is interesting to find that the size of the nucleus of a Seyfert galaxy is less than the average separation between the two stars in a typical galaxy, and its energy output is equal to that of a system having 100 billion stars.

Fig. 12: Seyfert galaxy.
(*Courtesy:* Google website)

Further, the light coming from the nucleus of a Seyfert galaxy is quite different in nature from that emitted by a typical galaxy. For a typical galaxy, the optical light is mostly star-like which generally consists of a thermal continuum plus absorption lines, and the same is true for the light combined from a large number of stars in a typical galaxy. On the other side, a Seyfert nucleus exhibits strong emission lines like those from H II regions and planetary nebulae (recombination spectra) and a continuum which looks non-thermal (in that $\lambda^2 f_\lambda$ increases with increasing λ). It gives an impression that the basic energy source of nucleus of a Seyfert galaxy probably does not derive from the thermonuclear power of ordinary stars. This point got a support from the observations that the Seyfert galaxies emit also strongly in the radio, infrared and x-rays.

From the observations, it is found that the nuclei of Seyfert galaxies are surrounded by a large amount of dust and gas, which would obscure a direct view of the nucleus of the Seyfert galaxy. Thus, it is believed that the real size of the central power house in the Seyfert galaxies may be considerably less than one light year.

11.2 Radio Galaxies

The interest in radioactive galaxies was stimulated by the fact that a large number of such galaxies emit strongly in the radio continuum. Besides our Sun, the Cygnus A was the first source found to have radio emission. Deep photographs show the Seyfert galaxies to have extended optical images, but their radio appearance is essentially point-like. On the other side, some of the strongest extragalactic sources have extended radio structures. This structure of radio galaxies typically consists of two radio lobes which are the extended parts of the central galaxy whose nucleus may or may not be abnormally bright in the radio and in the visible. The radio lobes typically lie at a distance of millions of light years on the diametrically opposite sides of the central galaxy, with the sizes of lobes being about a third to fifth of the spacing. The central part of the galaxy is generally a giant elliptical or a super giant elliptical, but never a spiral.

Jets and lobes

Radio jet is the material spewing from the centres of some galaxies at close to the speed of light and emitting strong radio waves. The most powerful extragalactic sources of radio waves are double-lobed sources in which two large regions of radio emission are situated in a line on diametrically opposite sides of galaxy (Fig. 13). The parent galaxy is usually a giant elliptical.

11.3 Active Galactic Nuclei

An active galactic nucleus (AGN) is a compact region at the centre of a galaxy that has a much-higher-than-normal luminosity over at least some portion of the electromagnetic spectrum with characteristics indicating that the luminosity is not produced by stars. Such excess non-stellar emission has been observed in the radio, microwave, infrared, optical, ultraviolet, x-ray and gamma ray wavebands. A galaxy hosting an AGN is called an "active galaxy". The non-stellar radiation from an AGN is theorized to result from the accretion of matter by a supermassive black hole at the centre of its host galaxy.

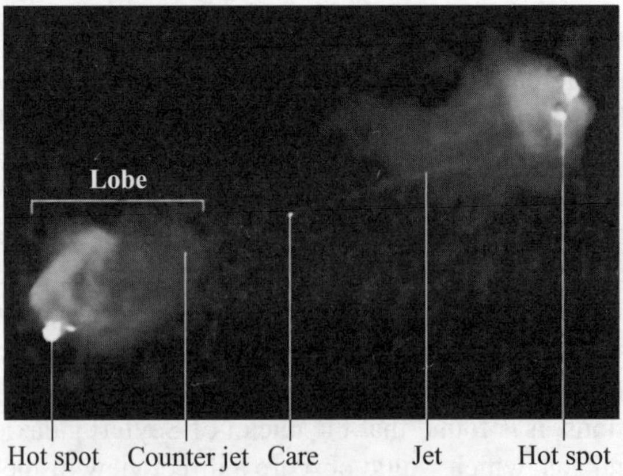

Fig. 13: Jets and lobes of a galaxy.

(*Courtesy:* Google website)

Active galactic nuclei are the most luminous persistent sources of electromagnetic radiation in the universe, and as such can be used as a means of discovering distant objects; their evolution as a function of cosmic time also puts constraints on models of the cosmos.

The observed characteristics of an AGN depend on several properties such as the mass of the central black hole, the rate of gas accretion onto the black hole, the orientation of the accretion disk, the degree of obscuration of the nucleus by dust, and presence or absence of jets.

Numerous subclasses of AGN have been defined based on their observed characteristics; the most powerful AGN are classified as quasars. A blazar is an AGN with a jet pointed towards the Earth, in which radiation from the jet is enhanced by relativistic beaming.

Radio source counts

From a radio-astronomical survey, the source counts distribution of radio-sources is the cumulative distribution of the number of sources (N) brighter than a given flux density (S). Since it is usually plotted on a log-log scale, its distribution is known as the log N-log S plot. It is one of several cosmological tests that were considered in the 1930s to check the viability of and compare new cosmological models.

11.4 Star Burst Galaxies

A galaxy having an exceptionally high rate of star formation, as compared to most other galaxies having average rate of star formation, is known as the star burst galaxy. For example, in the Milky Way galaxy, the star formation rate is approximately $3M_{\odot}$/yr, whereas in the starburst galaxies, the star formation rates may be 33 times larger than that of the Milky Way. In a star burst galaxy, the star formation is so large that the galaxy consumes all of its gas reservoir, on a much shorter timescale. Starburst galaxies include M82, NGC 4038/NGC 4039, and IC 10.

11.5 Quasars

The word quasar stands for a quasi-stellar radio source. Owing to poor resolution in radio Astronomy, quasars were first detected as unresolved radio sources. After the development of the very long baseline interferometry (VLBI) technique, radio images of a few quasars were resolved (Fig. 14). The quasars were occasionally found to show a classic double-lobed structure, similar to that of the strong radio galaxies. However, the major apparent difference between the quasars and the radio galaxies is that the central optical image of a quasar is not obviously a galaxy. It may be a galaxy, but it is not obviously a galaxy. Optical spectrum of the central source of a quasar contains a continuum with a non-thermal slope superimposed with broad emission lines and narrow absorption lines.

Fig. 14: Quasar.
(*Courtesy:* Google website)

Astronomers have found about 20 radio-quiet star-like objects whose optical properties are similar to those of the quasars. These objects are known the QSOs (quasi-stellar objects). Both the QSOs and the quasars are believed to be similar kinds of the systems except that QSOs do not generate appreciable non-thermal radio emission to give large amount of the optical, infrared and x-ray radiation. The resemblance of the quasars to the radio galaxies and of the QSO's to the nuclei of Seyfert galaxies, is so strong that the most (but not all) astronomers believe that the quasars and the QSO's arise from nuclear activity in the galaxies.

11.6 LINERS

A low-ionization nuclear emission-line region (LINER) is a kind of galactic nucleus which is characterized by its spectral line emission. The spectra typically include line emission from singly ionized ions or neutral atoms, such as O, O^+, N^+, and S^+. Further, the spectral line emission from doubly ionized atoms, such as O^{++}, Ne^{++}, and He^+, is relatively weak.

11.7 BL Lac Objects

A BL Lacertae object or BL Lac object (Fig. 15) is a type of active galaxy with an active galactic nucleus (AGN). In contrast to other types of the AGNs, the BL Lacs are characterized by the rapid and large-amplitude flux variability and significant optical polarization. Due to these properties, the prototype of the class (BL Lac) was originally believed to be a variable star. When compared to the more luminous active nuclei (quasars) with strong emission lines, the BL Lac objects have spectra dominated by a featureless non-thermal continuum.

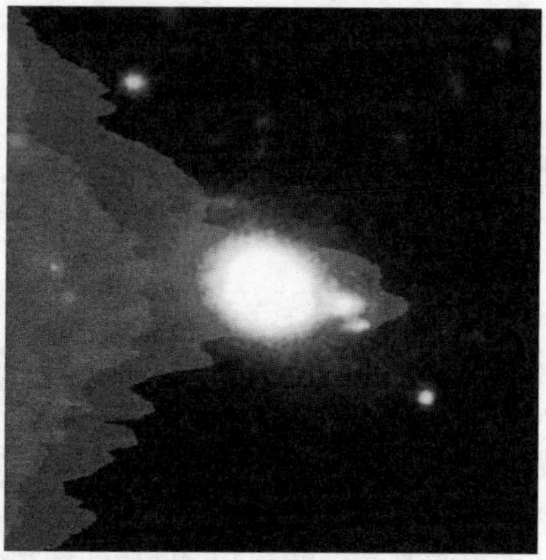

Fig. 15: BL Lac object.
(*Courtesy:* Google website)

Some examples of BL Lac objects are BL Lacertae itself, OJ 287, AP Librae, PKS 2155-304, PKS 0521-365, Markarian 421, 3C 371, W Com, ON 325 and Markarian 501.

11.8 Photoionization Models

Photoionization plays an important role in the mechanism of positive streamer discharges in air.

11.9 Continuum Radiation Processes

There are several processes which produce radiation. Radiation having all wavelengths are known as the continuum radiation. The emission mechanism of continuum radiation are black-body radiation, thermal bremsstrahlung, synchrotron radiation, and Compton scattering.

11.10 Optically Violent Variables

An optically violent variable quasar (often abbreviated as OVV quasar) is a type of highly variable quasar. It is a subtype of blazar that consists of a few rare, bright radio galaxies, whose visible light output can change by 50% in a day. They are similar in appearance to BL Lacs but generally have a stronger broad emission line, and tend to have higher redshift components.

12. Interaction of Galaxies

So far, we have discussed an isolated galaxy which had no interaction between them. In fact, in the universe, various galaxies interact with one another, as was the case of the interaction among various stars in a star cluster. Thus, like the star clusters, there are clusters of galaxies. Let us look into the interacting galaxies.

12.1 Interacting Binary Galaxies

Though a large percentage of galaxies behave like isolated galaxies, but a small percentage of the galaxies are found having strongly interacting pairs of the galaxies. These interacting pairs of the galaxies produce intriguing structures which are very different from those of the single galaxies. With the advent of the computers, numerical simulations for the galactic systems can be carried out and it has advanced greatly our theoretical understanding about such systems.

If the masses of two interacting galaxies are very different from each other, the smaller galaxy can pull out the material from the near side of the bigger one, forming a bridge between the two galaxies. When the two interacting galaxies have nearly equal masses, each of the galaxies develops a long tail that generally extends away from the system.

12.2 Mergers

When two galaxies encounter very closely, internal motions of each of them are modified. Following the law of conservation of energy, the energy required to make such changes obviously must come from the orbital motions in the galaxies. In a system of the collection of gravitating stars, the kinetic energy of motion cannot be dissipated in the form of radiation. The energy would only be shared among the stars. During the encounter between the two galaxies, individual stars suffer gravitational deflections by the entire collection of stars.

A close encounter of two bound spiral galaxies, having a lot of energy in the form of orbital and angular motions, must tend eventually to produce a merger into a single pile of stars, having much more energy in the disordered form of random motions. Such merger process involves a form of 'violent relaxation', in which violently changing gravitational fields help to produce a relaxed (final) smooth distribution of stars. Except in having relatively great amounts of interstellar gas and dust, the collection of stars strongly look like an elliptical galaxy. The NGC 7252 has been found an excellent example for such a merger process. When the gas and dust are completely used up to form more stars, the final form of the system would probably look like an ordinary elliptical galaxy.

It was also speculated that perhaps all the elliptical galaxies were formed in this manner. It is discussed that mergers have been more common in the past than now, or is more common in the clusters than in the general field.

12.3 Clusters of Galaxies

Likewise the stars, in addition to doublets, galaxies are also found in small groups and rich clusters. A small group of galaxies may have about one dozen to two dozen galaxies extending over a radius of about a million light years. An example of such small group is the Local Group whose list of members includes our galaxy Milky Way as well as Andromeda galaxy. On the other side, rich clusters of galaxies may contain several thousand galaxies extending over a radius of about ten million light years.

On even at a larger scale, say, extending over a radius of about hundred million light years, another level of this hierarchical clustering, called a supercluster, has been discovered. A supercluster contains about one hundred clusters. It is not yet known whether any cluster of superclusters exists or not.

12.4 Expansion of Universe

Cosmological interpretation for redshifts of spectral lines from the galaxies is that the universe is expanding and the velocity of the recession of a galaxy is proportional to its distance from an observer. Thus, the larger the redshift of a galaxy, the large the distance of it from the observer. The redshift provides the velocity of the galaxy.

13. Hubble's Law

It is remarkable to have a note that observed wavelengths of all radiations coming from the space are longer than the wavelengths of the corresponding transitions measured in the laboratories. If λ_0 is wavelength of a radiation (emission line) measured in a laboratory and if the observed wavelength of the corresponding line coming from a cosmic object is λ, then the redshift z of the object is defined as

$$z = \frac{\lambda - \lambda_0}{\lambda_0}$$

According to the Doppler effect, the wavelength of a radiation emitted by a cosmic object receding away from an observer with a velocity v is found to increase as compared to its wavelength measured in a laboratory. The relation between the redshift z of a cosmic object and its receding velocity v is

$$z + 1 = \frac{\sqrt{1 + v/c}}{\sqrt{1 - v/c}}$$

For the non-relativistic case ($v \ll c$), we have $z = v/c$.

Since all the cosmic objects show the redshift (no object shows the blueshift) and if the Doppler effect is valid, it shows that all the objects are moving away from us, i.e., the universe is expanding.

For some galaxies, it is possible to estimate their distances from us. Hubble realized that there was a relation between the distance of a galaxy and its redshift. He found that the velocity of recession v of a galaxy was proportional to its distance r from the Earth. Thus, the Hubble's law can be stated as

$$v = H_0 r$$

where the proportionality constant H_0 is known as the Hubble's constant. When it was calculated first time by Hubble, its value was found to be 536 (km/s)/Mpc. Various scientists from time to time have obtained different values of the Hubble's constant. The latest value measured is 70.1 ± 1.3 (km/s)/Mpc.

13.1 Determination of Hubble Constant

Example 2.

A galaxy has a redshift $z = 0.02$. What is its distance if we adopt the Hubble's constant H_0 to be 20 km s^{-1}/Mly.

Solution:

We have

$$v = H_0 r$$
$$0.02c = 20r$$
$$0.02 \times 3 \times 10^5 = 20r$$
$$r = 300 \text{ Mly}$$

Example 3.

Two galaxies A and B, moving with velocities v_A and v_B, respectively, relative to the observer O, are at the distances d_A and d_B, respectively, from the observer as shown below. The galaxies A and B follow the Hubble's law when observed from O. Show that the galaxy B follows the Hubble's law when observed from A.

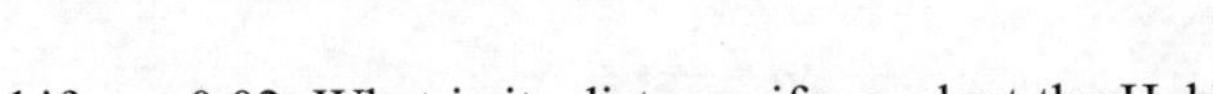

Solution:

The distance of B from A is $d' = d_B - d_A$ and velocity of B relative to A is $v' = v_B - v_A$. As the galaxies A and B follow the Hubble's law when observed from O; we have

$$v_A = H_0\, d_A$$

and

$$v_B = H_0\, d_B$$

On subtracting the first equation from the second, we have

$$(v_B - v_A) = H_0(d_B - d_A)$$

or

$$v' = H_0 d'$$

It shows that the galaxy B follows the Hubble's law when observed from A.

Example 4.

When a quasar has redshift of 4.5 how fast it is moving away and how far away it is. Hubble's constant $H = 6.5 \times 10^4$ m/s.

Solution:

A redshift of 4.5 implies that $\Delta\lambda/\lambda$ is 4.5. This is the case of relativistic Doppler effect. We have

$$\frac{\Delta\lambda}{\lambda} = 4.5 = \left(\frac{1 + v/c}{1 - v/c}\right)^{1/2} - 1$$

$$5.5^2 = \frac{1 + v/c}{1 - v/c}$$

$$\frac{v}{c} = 0.94$$

The quasar is receding at a speed of 94% of the speed of light or 2.8×10^8 m/s. For this value, we have

$$d = \frac{v_r}{H}$$

$$= \frac{2.8 \times 10^8}{6.5 \times 10^4} = 4300 \text{ Mpc}$$

The uncertainty in the Hubble's constant leads to an actual distance for the quasar to be between 3 billion parsecs and 6 billion parsecs.

14. Cosmic Microwave Background

In the scenario of expanding universe, since we are not at the outermost surface of the universe, there should be a background around us. Around 1950, George Gamow proposed that the present-day cosmos must be filled with the black-body radiations at a kinetic temperature of 10 K. Hence, there should be cosmic microwave background (CMB) around us. At that time, radio-astronomical technology was not efficient enough to record these radiations, corresponding to a low temperature of the order of 10 K.

Scientists, however, got interested in recording these black-body radiations (now known as the CMB radiations) corresponding to a low temperature of the order of 10 K. Efforts were put to improve the technology for radio astronomical observations. Improved calculations for the temperature of the CMB were performed, and it was found to be around 4 K. The CMB radiation was discovered coincidentally in 1965 by A.A. Penzias and R.W. Wilson. They detected weak radio frequency radiation at $\lambda = 7.35$ cm (4.08 GHz) and were honoured by the Noble Prize in Physics in 1978.

They found the isotropic unpolarised radiations coming from all directions which were independent of the time of day or of the season. It was later on confirmed that these radiations were in fact the remnant of the Big Bang. It was interesting to notice that, in the entire accessible wavelength region, all the radiations follow the Planck's law at the temperature, $T = 2.73 \pm 0.05$ K.

14.1 Big Bang Model

The Big Bang theory is the prevailing cosmological model that explains the formation and existence of the observable universe from the earliest known periods through its subsequent large-scale evolution. The model explains how the universe expanded from an initial state of high density and temperature, and offers a comprehensive explanation for a broad range of observed phenomena, including the abundance of light elements, the cosmic microwave background (CMB) radiation, and large-scale structure. Detailed measurements of the expansion rate of the universe place the Big Bang singularity at around 13.8 billion years ago, which is thus considered the age of the universe.

15. Formation and Evolution of Galaxies and Clusters

Matter is distributed throughout the universe in an inhomogeneous manner, and a large part of the universe is empty. But, embedded in these empty regions, there are concentrations of gases, concentrations of stars, and concentrations of galaxies. There are neutral and ionized clouds of gas, groups of stars, clusters of stars, small and big galaxies, small groups of galaxies, and clusters of galaxies. Evidently, a number of different physical processes compete in the determination of the characteristics and makeup of these different types of objects. The physical properties which lead to the substantial differences that distinguish the globular clusters from the galaxies, and the individual galaxies from the clusters of galaxies, are completely unknown.

It is not known why there should be distinctions when we switch from a scale of 10^{20} cm to 10^{23} cm, and finally to 10^{25} cm. These are the dimensions of the globular clusters, galaxies and clusters of galaxies, respectively. Larger yet, by a factor of $\sim 10^3$, is the scale of the entire universe whose horizon is roughly 10^{28} cm distant.

Before talking about the differences due to the sizes, let us make sure that they really exist. For example, we can think of very small galaxies whose characteristics are similar to those of the largest globular clusters. However, since the galaxies would be very faint, we would not see them at large distances, and cannot tell about their existence.

In order to resolve this problem, we can undertake an exhaustive study of all discoverable clusters and groupings in the immediate neighbourhood of our own galaxy, where we may be able to determine the properties of the galaxies and the clusters in some details. In particular, we would like to have answers to the questions:

(a) If there are small galaxies whose sizes are intermediate between those of the globular clusters and of the large galaxies such as the Andromeda Nebula, our nearest spiral galactic neighbour in space.

(b) If yes, whether such galaxies show the same kind of structure as a globular cluster, or whether there are more like galaxies in that they actually contain a number of subordinate globular clusters as part of their makeup. Further, if they contain the interstellar gas like the larger galaxies have.

(c) If the globular clusters sometimes behave like independent galaxies, in the sense that they are not gravitationally bound to any galaxy, but exist in space by themselves. Whether such clusters would have their own quota of the interstellar gas.

Such types of queries can be fortunately answered. For the Galaxy and Andromeda nebula do not form a pair of galaxies completely isolated from all others. A careful search has shown that there is a Local Group of objects containing some 21 identified members. There are some such objects which are obscured by the absorbing matter within the Galaxy, and probably many more that simply have been too faint to be observed by the present day techniques.

It has been discussed that the Local Group is the beginning to throw light on some of the questions posed at the beginning. Further, there is no absolutely sharp cut-off distinguishing galaxies from the globular clusters. We sense that there are differences; but the criteria we choose do not always provide a clear distinction in borderline cases.

Let us ask how the galaxy formation may take place. We know that the galaxies exist in all parts of the universe. At the largest distances accessible with the modern techniques, the galaxies can still be found in apparently unchanged form and undiminished numbers.

The normal approach of Astrophysicists has been to postulate that the galaxies are the condensations formed out of previously tenuous material. Parallel of this, there are other approaches taken in the theories of the star formation. In both cases, there are difficulties.

For the galaxy formation, the prime difficulty is the cosmic expansion. For understanding the galaxy formation, we would like to think of the gravitation, or some other force, acting to concentrate matter into a small volume. This action, however, is always countered by the continuing recession of any two or more elements of the matter, as the expansion of the universe forces them further apart. There results a tug-of-war between the cosmic expansion and the gravitation. The gravitation evidently wins but we do not observe the process anywhere and so we cannot be certain about it.

Let us now re-examine our main concept: How sure we are about the condensation process through which the galaxies are formed. Not very sure. There is an alternate approach given by Ambartsumian. He noticed that there are many pairs of galaxies observed in the sky and that the differences in the radial recession velocity sometimes is so great that the galaxies are escaping from one another. Their mutual gravitational attraction is too weak to hold them together, unless there is some large unobserved mass that makes the total aggregate much more massive than we find from the brightness of the individual galaxies alone.

A prevailing view proposes that much of the mass in the universe is in the form of dark matter. As there is some evidence for low rates of galaxy formation at the present epoch, we believe that most of the galaxies were formed at a much earlier stage, and that contemporary galaxy formation could only represent the closing stages of a much more vigorous period.

For evolving the cosmological models, the galaxy formation from an intergalactic medium could be most easily understood in a Lemaitre universe. This is a cosmological model in which a rapid expansion following the birth of the cosmos is followed by a slowing down, and a quiescent stage of zero expansion. Thereafter, the expansion picks up again. During the quiescent period, the universe is roughly described by a static state in which the galaxies might well be able to form, as there is no cosmic expansion to interfere with the condensation process. When the instabilities leading to the clumping of the extragalactic gas occur at all, this is the kind of universe in which the instabilities would be most likely to set in.

It is not clear if it is a sufficient reason to support the Lemaitre models. Most modern theories are concerned not just with the formation of the galaxies, but also of galaxy clusters and large intergalactic voids. The formation process in all such models therefore have to be theoretically analyzed not only with a view towards the understanding how the condensation can arise at all, but also with the idea of determining whether the condensations of the right-size range are formed. So far, this problem has not been successfully solved.

While the galaxy formation then constitutes a serious problem in Astrophysics, it does not have the great promise of leading to greater insights into the processes on a very large scale in cosmology. If we can understand how the galaxies are formed, we can also understand a great deal more about the origin-provided there ever was one-of the universe and about the dynamic laws that hold on a scale some 1010 times larger than the scale of the solar system on which our current laws of dynamics are tested.

16. Intergalactic Medium

The warmhot intergalactic medium (WHIM) is the sparse, warm-to-hot (10^5 to 10^7 K) plasma that cosmologists believe to exist in the spaces between galaxies and to contain 40-50% of the baryonic 'normal matter' in the universe at the current epoch. The WHIM can be described as a web of hot, diffuse gas stretching between galaxies, and consists of plasma, as well as atoms and molecules, in contrast to dark matter. The WHIM is a proposed solution to the missing baryon problem, where the observed amount of baryonic matter does not match theoretical predictions from cosmology.

16.1 Sunyaev-Zeldovich Effect

The Sunyaev-Zeldovich effect, abbreviated as the SZ effect, named after Rashid Sunyaev and Yakov B. Zeldovich, is the spectral distortion of the cosmic microwave background (CMB). It occurs through inverse Compton scattering by high-energy electrons in galaxy clusters, where the low-energy CMB photons receive an average energy boost during collision with the high-energy cluster electrons. The disturbance of density in the universe may be detected with the help of the observed distortions of the cosmic microwave background spectrum. Dense clusters of galaxies have been observed with the help of Sunyaev-Zeldovich effect.

17. Filaments

Galaxy filaments are considered the largest known structures in the universe. They are consists of walls of gravitationally bound galaxy superclusters. The size of these massive, thread-like structures can be of the order of 160 to 260 million light years and they form the boundaries between the large voids.

18. Voids

Cosmic voids are vast spaces between the galaxy. They contain very few or no galaxies, and they look almost empty for an observer, even the emptiest regions of voids have more than ~15% of the average matter density of the Universe. Voids typically have a diameter of 10 to 100 megaparsecs (30 to 300 million light years). Large voids, defined by the absence of rich superclusters, are sometimes called supervoids.

19. Faraday Rotation

The Faraday rotation, also known as the Faraday effect or magneto-optic Faraday effect, is a physical magneto-optical phenomenon. It was the first experimental evidence that light and electromagnetism are related. The Faraday effect produces a polarization rotation which is proportional to the projection of the magnetic field along the direction of propagation of light.

For understanding the Faraday rotation, the linear polarized light can be regarded as consisting of the superposition of two components, a right-circularly polarized beam and another left-circularly polarized beam. Now, we consider each component, which is circularly polarized beam.

The direction of the electric field, in a circularly polarized light, rotates at the frequency of the light, either clockwise or counterclockwise. While propagating through the material, this rotating electric field produces a force on the charged particles in the material. The largest effect is on the electrons, because of their low mass. Further, the effect will be circular, and these circularly moving charges create their own (magnetic) field in addition to the external magnetic field. Consequently, the created fields are in two opposite directions, for two circularly polarized light. One field is parallel to the external magnetic field, whereas the other field is opposite to the external magnetic field. Thus, the net magnetic field is enhanced in one direction and diminished in the opposite direction. This affects the dynamics of the interaction for each beam and one of the beams is slowed more than the other, producing a phase difference between the left- and right-polarized beams. After this phase shift, on addition of two circularly polarized beams, the result is again a linearly polarized beam, but with a rotation of the polarization vector.

The direction of polarization rotation depends on the properties of the material through which the linear polarized light propagates.

20. Unified Model for AGNs

For Active Galactic Nuclei (AGN), the unified model states that all the AGN are same phenomenon and the difference between them is due to orientation effects with respect to the line of sight to the observer. That is all the observed differences between different AGNs are due to directions they are observed.

21. Gravitational Lenses

A gravitational lens is a distribution of matter (such as a cluster of galaxies) between a distant light source and an observer, that is capable of bending the light from the source as the light travels toward the observer. This effect is known as gravitational lensing, and the amount of bending is one of the predictions of Albert Einstein's general theory of relativity. Treating light as corpuscles travelling at the speed of light, Newtonian physics also predicts the bending of light, but only half of that predicted by general relativity.

Review Questions

1. Discuss various parts of the Milky Way, stellar populations and motions of stars in the Milky Way.
2. Discuss the differential rotation and spiral structure of the Milky Way.
3. Discuss the problem of dark matter on our Milky Way.
4. Describe in brief the classification of galaxies.
5. Discuss the spiral galaxies and elliptical galaxies. Describe the major differences between them.
6. Discuss the distribution of light and distribution of mass in a regular galaxy.
7. Discuss the active galaxies.
8. Discuss the interaction of galaxies.

9. For the expanding universe, describe the Hubble's law.
10. Write notes on the following:
    (i) Milky Way galaxy
    (ii) Dark matter
    (iii) Differential rotation of Milky Way
    (iv) Oort constants
    (v) Galaxy rotation curve
    (vi) Classification of galaxies
    (vii) Hubble's tuning fork diagram
    (viii) Irregular galaxies
    (ix) Elliptical galaxies
    (x) Spiral galaxies
    (xi) Tully-Fisher relation
    (xii) Lenticular galaxies
    (xiii) Active galaxies
    (xiv) Seyfert galaxies
    (xv) Radio galaxies
    (xvi) Quasars
    (xvii) BL Lac objects
    (xviii) Interaction of galaxies
    (xix) Clusters of galaxies
    (xx) Expansion of universe
    (xxi) Hubble's law
    (xxii) Cosmic microwave background

Multiple-Choice Questions

1. A person in the north hemisphere on the Earth sees towards the centre of Milky Way in the night of
   A. Summer B. Autumn C. Winter D. Spring **Ans. A**
2. A person in the north hemisphere on the Earth sees opposite to the centre of Milky Way in the night of
   A. Summer B. Autumn C. Winter D. Spring **Ans. C**
3. A person in the south hemisphere on the Earth sees towards the centre of Milky Way in the night of
   A. Summer B. Autumn C. Winter D. Spring **Ans. C**
4. A person in the south hemisphere on the Earth sees opposite to the centre of Milky Way in the night of
   A. Summer B. Autumn C. Winter D. Spring **Ans. A**
5. Typical size of a globular cluster is
   A. 10^{15} cm B. 10^{20} cm C. 10^{25} cm D. 10^{28} cm **Ans. B**
6. Typical size of a galaxy is
   A. 10^{20} cm B. 10^{23} cm C. 10^{25} cm D. 10^{28} cm **Ans. B**

7. Estimated size of the entire universe is
   A. 10^{20} cm B. 10^{23} cm C. 10^{25} cm D. 10^{28} cm **Ans. D**
8. Dark matter
   A. emits radiations in the visible range.
   B. is experienced by the gravitational attraction.
   C. emits radio waves having long wavelength.
   D. is experienced by Coulomb attraction. **Ans. B**
9. The ratio of the local mass-to-light ratio for Milky Way to that of the Sun is 5. If in the period of 10^{10} years, in the Sun only 10% of its hydrogen mass is converted into the helium mass through the nuclear processes.In the Milky Way, the percent of hydrogen mass converted into the helium mass through the nuclear processes in the same period is
   A. 10% B. 5% C. 2% D. 50% **Ans. C**
10. The class of stars, the disc of Milky Way contains mainly is
    A. Population I B. Population I & II both
    C. Population II D. Neither Population I nor II **Ans. A**
11. The class of stars, the central bulge of Milky Way contains is
    A. Population I B. Population I & II both
    C. Population II D. Neither Population I nor II **Ans. B**
12. The class of stars, the halo of Milky Way contains mainly is
    A. Population I B. Population I & II both
    C. Population II D. Neither Population I nor II **Ans. C**
13. In the disc of Milky Way, the rotational speed and random speed, respectively, are
    A. large, large B. large, small C. small, large D. small, small
    Ans. B
14. In the central bulge of Milky Way, the rotational speed and random speed, respectively, are
    A. large, large B. large, small C. small, large D. small, small
    Ans. D
15. In the halo of Milky Way, the rotational speed and random speed, respectively, are
    A. large, large B. large, small C. small, large D. small, small
    Ans. C

14 Cosmic Distance Ladder

By looking the sky, we can measure angular separation between two given objects, situated on the celestial sphere, but cannot say which one of them is closer and which one is far away. Measurement of distance of an cosmic object, in general, is a difficult task, particularly when we go to the large distances. For the nearby objects, parallax method can be used for the determination of distance of an object. For the distant objects, various methods are used. In the present chapter, we have discussed about various methods used for the measurement of distance of cosmic objects.

1. Size of Solar System

In order to establish a cosmic distance scale, the first requirement is to measure the distances within our solar system, with accuracy. The basic step in this direction is to measure the distance of Venus from the Earth. Being very close to the Sun, detection of Mercury from the Earth is a bit difficult task. The most reliable and precise way to obtain the distance of Venus is with the help of radar techniques. In this method, a radar pulse is communicated towards the Venus and is received back after its reflection from the Venus surface. The time interval between the communication of the pulse and its reception back is measured. As the time measurements can be made to a large accuracy, the distance of the Venus and dimensions of its orbit can be calculated within an accuracy of one kilometre.

Example 1.

A radar pulse sent to the Venus was received back after 26 min 40 sec. Calculate the distance of the Venus from the Earth. Speed of light is 3×10^8 m/s.

Solution:

Time spent by the pulse for one way is 13 min 20 sec = 800 sec.

$$\begin{aligned}\text{Distance} &= \text{Speed of light} \times \text{Time}\\ &= (3 \times 10^8) \times 800\\ &= 2.4 \times 10^{11} \text{ m.}\end{aligned}$$

The distance of the Venus from the Earth at that time is 2.4×10^8 km.

1.1 Parallax Method

The distance of the Venus from the Earth may be calculated by measuring the angle on the star-background, when the measurements are taken from two different positions on the Earth separated by a known distance. Two examples are as the following.

(i) Two observers are separated by a distance of 10^4 km along a line perpendicular 469 to the direction of the Venus from the Earth, find the position of the Venus on the star-background to differ by 49″. Calculate the distance of the Venus from the Earth.

Let us consider two observers A and B on the surface of the Earth. The separation between A and B is 10^4 km. The angle subtended by the two observers at the Venus V is $\angle AVB$. The distance of Venus from the Earth is much larger than the separation between A and B. Thus, $\angle AVB$ is very small. It is 49″. On expressing the angle in radians, we have

$$\text{Angle} = \frac{\text{Separation}}{\text{Distance}}$$

or

$$\frac{49 \times \pi}{60 \times 60 \times 180} = \frac{10^4}{d}$$

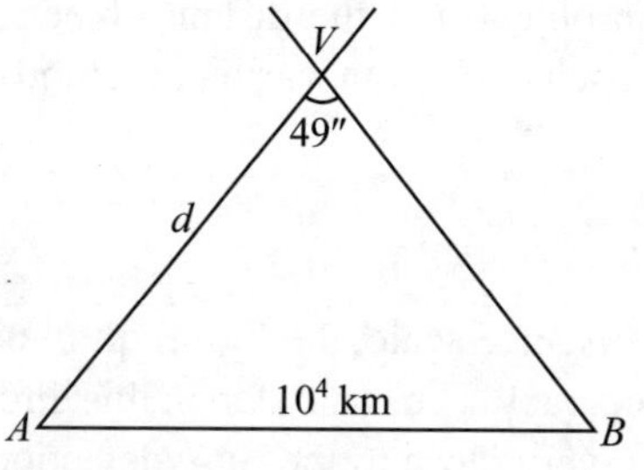

On solving, the value of d is 4.21×10^7 km. Thus, the distance d of the Venus from the Earth is 4.21×10^7 km.

(ii) Consider that both the Venus and the Earth are revolving in their circular orbits around the Sun. At the closest approach of the Venus to the Earth, its angular diameter is 64″, and at the greatest separation of the Venus from the Earth, its angular diameter is 10″. Calculate the radii of the orbits of the Venus and the Earth. Diameter of the Venus is 1.21×10^4 km.

The distance of the Venus from the Earth is much larger than its diameter. Thus, we have

$$\text{Angle} = \frac{\text{Diameter}}{\text{Distance}}$$

For the closest distance a and the farthest distance b, we have

$$\frac{64 \times \pi}{60 \times 60 \times 180} = \frac{1.21 \times 10^4}{a}$$

and

$$\frac{10 \times \pi}{60 \times 60 \times 180} = \frac{1.21 \times 10^4}{b}$$

On solving, we get $a = 3.9 \times 10^7$ km and $b = 2.5 \times 10^8$ km. The radius of the orbit of Earth is

$$R_E = \frac{b + a}{2} = 1.44 \times 10^8 \text{ km}$$

The radius of the orbit of Venus is

$$R_V = \frac{b - a}{2} = 1.05 \times 10^8 \text{ km}$$

The distance of the Venus at the closest approach a and at the most distant separation b can be thus measured (Fig. 1). These measurements are repeated over a number of times for several years. These data can be used for the determination of the diameter and eccentricity of both the Earth and the Venus. The mean distance between the Earth and the Sun is $(a + b)/2$.

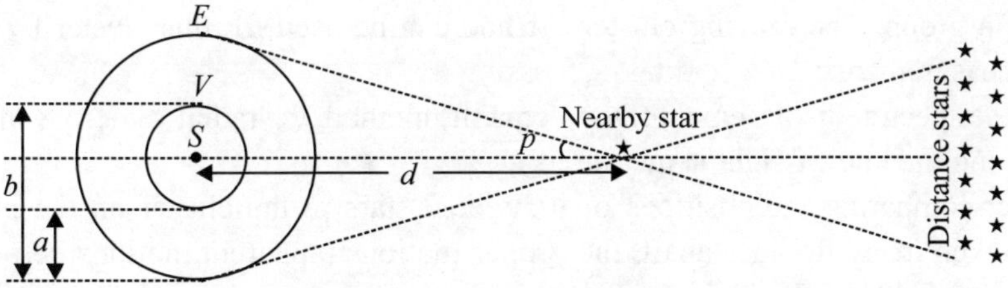

Fig. 1: Here, S denotes the Sun. The Venus (V) and the Earth (E) are revolving around the Sun in the circular orbits with the common centre at the centre of the Sun.

Example 2.

The closest and the farthest distances of the Venus from the Earth are 4×10^7 km and 2.6×10^8 km, respectively. Calculate the distance of the Earth from the Sun.

Solution:

The distance d of the Earth from the Sun is

$$d = \frac{4\times10^7 + 2.6\times10^8}{2}$$

$$= 1.5 \times 10^8 \text{ km}$$

2. Trigonometric Parallax

This is same as the parallax, and the observations are made from the opposite extremes in the Earth's orbit about the Sun. It is already discussed earlier. The nearby stars appear to be displaced relative to the more distant stars in the same part of the sky. The parallax, p is defined as the half the apparent angular displacement measured in this manner. The distance d of the star from the Earth is

$$d = \frac{\text{Astronomical unit}}{\tan p}$$

or

$$d = \frac{1.5\times10^{13}}{\tan p} \text{ cm}$$

2.1 Limitation

The trigonometric parallax can be reliable for the determination of distance up to about 50 pc, where the parallax is 0.002″. The parallax less than 0.002″ cannot be measured with accuracy.

3. Spectroscopic Parallax

After measuring the distance to nearby stars, the absolute brightness with the spectral type can be correlated. The bright stars of recognizable spectral type are then used as the distance indicators across a large distances, where only the brightest stars are individually recognized and the trigonometric parallax cannot be used.

4. Moving Cluster Method

In the Hyades star cluster, the members are close together and they move through space in the form of a group. The moving cluster method can be used for this cluster by making the following measurements:

(a) Using the spectral displacement (Doppler shift) method, the radial velocity v of the cluster-stars along the line-of-sight is measured.

(b) By inter-comparing the positions of individual stars in the cluster on the photographic plates taken many decades apart, the proper motions (apparent motions perpendicular to the line-of-sight) of the individual stars in the cluster are measured. This motion has the form of an apparent contraction of the cluster and provides fractional rate of increase of the cluster's distance from the Sun, as shown in Fig. 2.

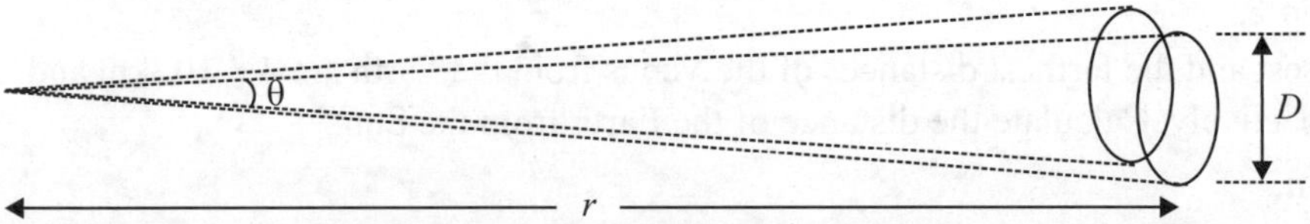

Fig. 2: The Hyades cluster at a distance r recedes away from us with a radial velocity component v. The cluster diameter D remains constant, and therefore the angle θ decreases.

Here, the cluster's distance r can be expressed in terms of the angular size θ, rate of change of the angular size, θ and receding velocity v of the cluster in the following manner. When the angle θ is small and is expressed in radians, we have

$$\theta = \frac{D}{r}$$

where D denotes the diameter of the cluster. Considering D as constant, the differentiation of this relation with respect to time gives

$$\dot{\theta} = -\frac{D\dot{r}}{r^2}$$

The differentiation of quantity is denoted by putting a dot over the quantity. Using $\dot{r} = v$ and $D/r = \theta$, we have

$$\dot{\theta} = -\frac{Dv}{r^2} = -\frac{\theta v}{r}$$

or

$$r = -\frac{\theta v}{\dot{\theta}}$$

The distance r of the Hyades cluster can be measured in terms of three quantities, which can be measured directly. This simple method is restricted to its use for the Hyades stars, because no other cluster is near enough to obtain accurate proper motions. The use of this method obviously depends on the dynamical stability of the Hyades group, as the stars are gravitationally bound.

5. Superposition of Main Sequences

In this method, one assumes that in all the galactic clusters, the main sequence stars have identical properties. That is, for all such clusters, the slop of the main sequence is the same and the main sequence stars of a given spectral type or colour have the same absolute brightness in all the clusters. On the basis of this assumption, one can compare the brightness of the main sequence of the Hyades and any other galactic cluster. The vertical shift necessary to bring the two main sequences into their superposition provides the relative distances of the clusters.

Suppose, a shift in the apparent magnitudes of galactic cluster and of Hyades is $\Delta m = m_{GC} - m_{Hya}$, the relative distances r_{GC} of the galactic cluster and r_{Hya} of the Hyades are related as

$$\Delta m = 5 \log_{10}\left(\frac{r_{GC}}{r_{Hya}}\right) + A' \tag{14.1}$$

where A' is a correction for the difference in the interstellar reddening of the galactic cluster and the Hyades. It can be determined with the help of the stellar line spectra.

For finding the distances of globular clusters, one can proceed on one of the three different assumptions:

(a) The Hertzsprung-Russell diagram of the globular cluster has a segment that runs essentially parallel to the galactic cluster main sequence. One can assume that this segment coincides with the main sequence of the Hyades cluster. Then, one can use equation (14.1) to calculate the distance of the globular cluster.

(b) One can assume that the segment of the H-R diagram of the globular cluster coincides with the main sequence, defined by a group of the dwarf stars in the Sun's immediate neighbourhood. The distance to these dwarf stars can be determined with the help of the trigonometric parallax.

(c) One can assume that the mean absolute magnitude for short-period variables (RR Lyrae variables) is the same as for the globular clusters and in the solar neighbourhood.

None of these three choices is safe in itself. However, when applied to the globular cluster M3 (the third object in the Messier catalog), all the three methods are found to give distances in fair agreement with each other. This verifies that the main sequences of different groupings of the stars coincide reasonably well and can be used as a distance indicator.

6. Brightness of RR Lyrae Variables

It has been observed that the apparent brightness of all the RR Lyrae variables in a given globular cluster is the same, irrespective of the variable period. As these stars are intrinsically bright, and their short periods make them stand out, they can serve as the ideal distance indicators. One can assume that the absolute brightness of these stars is the same not only within a given cluster, but in all the clusters. Thus, equation (14.1) can be used to calculate the relative distance of two clusters.

7. Brightness of Cepheid Variables (Henrietta Leavitt)

An important contribution of Henrietta Leavitt in Astronomy is the discovery of about 2,400 variable stars during the period from 1907 to 1921 (when she died). She found that some of these stars, called the Cepheid variables, have consistent brightness irrespective of their location in the space. These Cepheid variables can be used as a good measuring stick for astronomical distances. The intrinsic brightness of Cepheid variables is strongly related to their period. This famous period-luminosity relationship was discovered by Henrietta Leavitt.

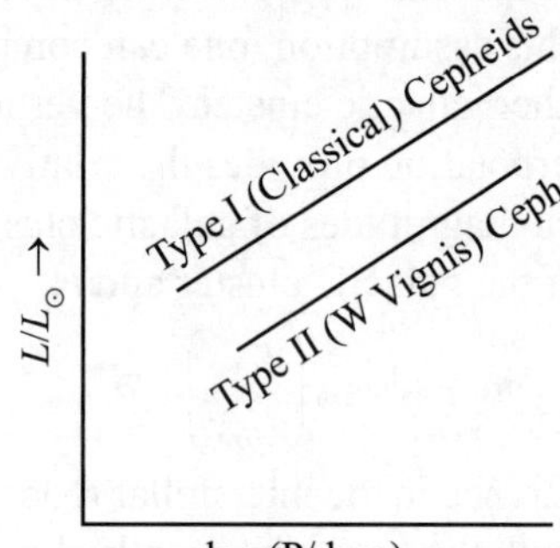

Fig. 3: Typical period-luminosity relationship for the Cepheids of Types I and II.

Period-luminosity relationship for the Cepheids of Types I and II is shown in Fig. 3. Both relations are almost straight lines.

We now discuss how this relationship can be used to determine the distance to a Cepheid variable. Let us take the case of Type I (Classical) Cepheids (and the same procedure can be applied for Type II (W Virginis) Cepheid stars.

1. We record the apparent magnitude of a given Cepheid using naked eye estimates, photographic plates, or photoelectric CCD images.
2. We plot the apparent magnitude values from the observations at different times results in a light curve.
3. From the light curve, two values can be determined: the average apparent magnitude, m, of the star and its period *P* of intensity variation in days.
4. Using the period-luminosity plot (such plot for Cepheids within the Milky Way is given in Fig. 4), for the known value of the period P_0 of the Cepheid we can now determine its mean absolute magnitude, M_0. The figure is the variation of *M* versus log *P*, where *P* is the period of variation in days.

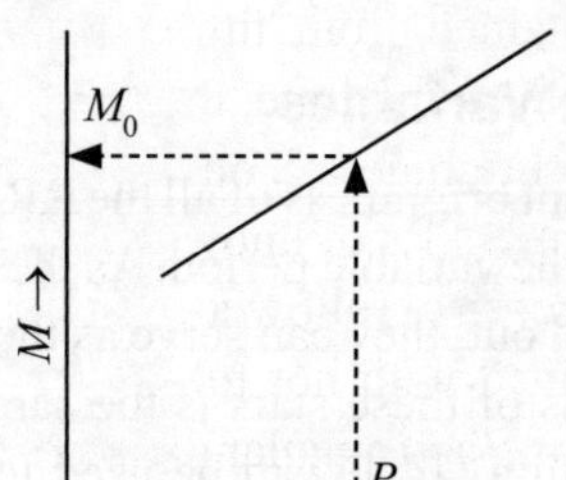

Fig. 4: Corresponding to the period P_0, the absolute magnitude M_0 is obtained.

5. Once both the apparent magnitude, *m*, and the absolute magnitude, M_0 are known we can simply substitute in to the distance-modulus formula (16.4) and rework it to give a value for *d*, the distance to the Cepheid.

$$m - M_0 = 5 \log(d/10) \tag{14.2}$$

As you know

$$\log d = (m - M + 5)/5$$

For example, for a Cepheid in the LMC has the mean apparent magnitude $m = 15.56$ and period $P = 4.76$ days. $\log P = 0.68$ and from the period-Luminosity plot for the Cepheids, the mean absolute magnitude, M_0, corresponding to $\log P = 0.68$ is about -3.6. Now, substituting the values, we have

$$\log d = (15.57 - (-3.6) + 5)/5$$

$$\log d = 24.17/5$$

$$\log d = 4.834$$

$$d = 68{,}230 \text{ parsecs}$$

This means that the Cepheid in the LMC is about 68.2 kpc (or about 222,000 light years away).

In practice astronomers would try and observe as many Cepheids as possible in another galaxy in order to determine a more accurate distance. With the increase of the number of stars observed, the uncertainties involved in the calculations for individual stars reduce.

8. Brightness of Novae and H II Regions

Novae have an absolute brightness which is related to the decay rate of the brightness, after an outburst. The great intrinsic brightness of a nova may be used as an useful indicator for the distance of nearby galaxies. The diameters of bright H II (ionized hydrogen) regions also form a good yardsticks through which the distances of such galaxies can be estimated.

9. Σ-*D* Relation

Clark and Casewell in 1976 found surface brightness, denoted by Σ and measured at 408 MHz frequency, of a supernova remnant (SNR) is related to the linear diameter D of the SNR as

$$\Sigma = A\, D^{\beta}$$

where A and β are parameters obtained from fitting, and this is Σ-*D* relation. Using the information about the SNRs available at that time, the β was found in the range from -4.54 to -3.0. The dependence of the surface brightness Σ on the diameter D of SNRs has been used to determine the distance to them. Using SNRs of known distance, we may plot their distribution in the Σ-*D* plane to determine whether they follow a common evolutionary track.

If an evolutionary track is defined with not too much scatter, then this may be used in conjunction with the measurements of Σ and angular diameter to determine the distance to SNRs for which no other distance information is available. Many known SNRs have no other distance measurement, the method is very valuable. In order to construct the Σ-*D*-dependence, various authors have used different calibrator remnants with independently established distances.

Thus, the investigation will adopt the following Σ-*D* relationships. For (i) $\Sigma_{408} > 3 \times 10^{-20} Wm^{-2}\, Hz^{-1}\, sr^{-1}$, we have

$$\Sigma_{408} = 10^{-15} D_{pc}^{-3}$$

so that $$d_{kpc} = 64.6(S_{408}\ \theta)^{-1/3}$$

whereas for (ii) $\Sigma_{408} < 3 \times 10^{-20} Wm^{-2} Hz^{-1} sr^{-1}$, we have

$$\Sigma_{408} = 3.6 \times 10^{-5} D_{pc}^{-10}$$

so that $$d_{kpc} = \frac{94.2}{\theta}\left(\frac{S_{408}}{\theta^2}\right)^{-1/10}$$

Here S stands for the flux density, the θ is the angular diameter of the SNR.

10. Seeliger's Theorem

Regarding the estimation of the distance of stars, the Seeliger's theorem may be expressed in the following manner. Consider a set of emitting objects is distributed homogeneously in space, then the ratio N_m/N_{m-1} of the number of objects having apparent magnitude less than m to the number of those having apparent magnitude is less than $m - 1$ is 3.98.

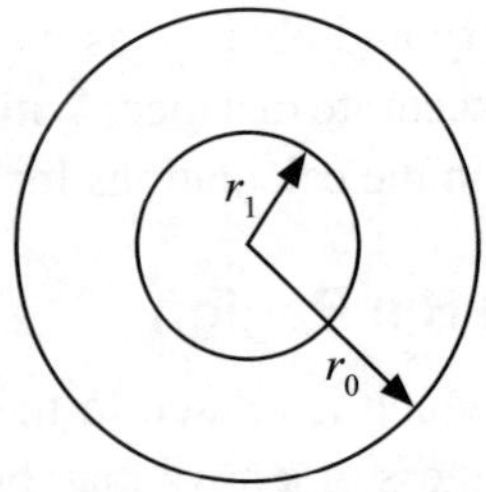

Fig. 5: When a star moves away from a distance r_1 to r_0, its apparent magnitude changes from $m - 1$ to m.

The proof of the theorem can be given in the following manner. Consider a star at a distance r_1 from an observer having apparent magnitude $m - 1$, as shown in Fig. 5. Suppose, on moving, the star to a distance r_0, its apparent magnitude becomes m. The radiant fluxes F_1 and F_0 received from the distances r_1 and r_0, respectively are given as (star is the same)

$$\frac{F_0}{F_1} = \frac{r_1^2}{r_0^2}$$

The fluxes are related to the magnitudes as

$$\frac{F_0}{F_1} = 10^{0.4(m_1 - m_0)}$$

Using $m_1 = m - 1$ and $m_0 = m$, these two equations can be rearranged as

$$\frac{r_1^2}{r_0^2} = 10^{-0.4}$$

or $$r_0 = (2.512)^{1/2}\ r_1$$

If the stars have a fixed brightness and are distributed uniformly in space, they appear brighter than the apparent magnitude m out to a distance r_0, but brighter than $m - 1$ only out to a distance r_1. The ratio of the number of stars, brighter than a certain magnitude, N_m/N_{m-1}, is proportional to the volume occupied, by the stars of a certain magnitude. The ratio of the number of stars, N_m/N_{m-1}, is proportional to the volume occupied, and we have

$$\frac{N_m}{N_{m-1}} = \frac{4\pi r_0^3/3}{4\pi r_1^3/3}$$

$$= \frac{r_0^3}{r_1^3}$$

Using the value of r_0/r_1, we have

$$\frac{N_m}{N_{m-1}} = (2.512)^{3/2}$$

$$= 3.98$$

This is the proof of the theorem.

Example 3.

Consider a space where identical stars are distributed homogeneously throughout. Suppose, a star has magnitude m_1 when it is at a distance r_1 from an observer. On moving the star to a distance r_2, its magnitude becomes m_2. Calculate the ratio of the number of stars having magnitude less than m_1 to in the number of those having magnitude less than m_2.

Solution:

When a star is moved from a distance r_1 to r_2, its corresponding radiant fluxes F_1 and F_2 are expressed as

$$\frac{F_2}{F_1} = \frac{r_1^2}{r_2^2}$$

In terms of the magnitudes m_1 and m_2, the fluxes are related as

$$\frac{F_2}{F_1} = 10^{0.4(m_1 - m_2)}$$

These two equations can be rearranged as

$$\frac{r_1^2}{r_2^2} = 10^{0.4(m_1 - m_2)}$$

or

$$\frac{r_1}{r_2} = 10^{0.2(m_1 - m_2)}$$

The stars within the sphere of radius r_1 have the magnitude m_1 or less and the stars within the sphere of radius r_2 have the magnitude m_2 or less. The ratio of the number of stars brighter than a certain magnitude, N_1/N_2, is proportional to the volume occupied, by the stars of a certain magnitude, and we have

$$\frac{N_1}{N_2} = \frac{4\pi r_1^3/3}{4\pi r_2^3/3}$$

$$= \frac{r_1^3}{r_2^3}$$

Using the value of r_1/r_2, we have

$$\frac{N_1}{N_2} = 10^{0.6(m_1 - m_2)}$$

Review Questions

1. A radar pulse sent from the Earth to thc Vcnus was received back after 6 min 40 sec. Calculate the distance of the Venus from the Earth. Given, $c = 3 \times 10^8$ m/s.
2. Discuss the parallax method for determination of distance of the Venus from the Earth.
3. Discuss the trigonometric parallax for determination of the distance of nearby stars and write its limitation.
4. Describe the moving cluster method used for determination of the distance of the Hyades star cluster.
5. Describe the method for the determination of the distance with the help of the superposition of the main sequences.
6. Describe the method for determination of the distance of the RR Lyrae variables.
7. State the Seeliger's theorem and prove it.
8. Write notes on the following:
   (i) Trigonometric parallax method
   (ii) Spectroscopic parallax
   (iii) Superposition of the main sequences method
   (iv) Moving cluster method
   (v) Distance of the RR Lyrae variables
   (vi) Distance of the Cepheid variables
   (vii) Distance of the novae and the H II regions
   (viii) Σ-D relation
   (ix) Seeliger's theorem

Multiple-Choice Questions

1. The limitation of the trigonometric parallax is about
   A. 10″ B. 1″ C. 0.01″ D. 0.002″ **Ans. D**
2. The ratio N_m/N_{m-1} of the number of objects having apparent magnitude less than m to the number of those having apparent magnitude less than $m - 1$ is
   A. 10 B. 2.512 C. 3.98 D. 1 **Ans. C**
3. The closest and farthest distances of the Venus from the Earth are 4×10^7 km and 2.6×10^8 km, respectively. The distance of the Earth from the Sun is
   A. 3×10^8 km B. 2.2×10^8 km C. 1.1×10^8 km D. 1.5×10^8 km **Ans. D**
4. A radar pulse sent to the Venus was received back after 26 min 40 sec. Speed of light is 3×10^8 m/s. The distance of the Venus from the Earth is
   A. 6×10^8 m B. 4.8×10^8 km C. 3×10^8 m D. 2.4×10^8 km **Ans. D**

15 Cosmology

The word **cosmos** is derived from the Greek word **kosmos**, which means the **world**. The cosmology is a branch of science which deals with the study of the origin of the universe, its evolution, and its eventual fate. So, under the cosmology, scientists study about the origin and evolution of the universe, large-scale structures and their dynamics, and the ultimate fate of the universe. It deals with the scientific laws also which govern these realities. There is a difference between the **Cosmology** and the **Astronomy** and **Astrophysics**. The cosmology deals with the universe as a whole whereas the Astronomy and Astrophysics deal with the individual celestial objects in the universe. The modern cosmology is based on the Big Bang theory, where the universe is considered emerged out of the Big Bang, which occurred about 13.8 billion years ago. The cosmology assumes the homogeneous and isotropic universe, which is justified on the scales of larger than 100 Mpc. Thus, apart from the local irregularities, the universe looks the same from all positions in the space. For the evolution of the universe, various models are given by scientists from time to time. This chapter deals with cosmology.

1. Steady State Theory of Universe

Though the Big Bang theory has now been accepted by most cosmologists, but this has not always been the case in the past when the **Steady State theory** was very popular. This theory was developed in 1948 by Fred Hoyle, Herman Bondi and Thomas Gold as an alternative to the Big Bang theory, to explain the origin and the expansion of the Universe, and its basis is the **Perfect Cosmological Principle**. This principle considers that the universe is infinitely large and infinitely old. As a whole, the universe is the same in all directions (isotropic universe) and it does not evolve or change over the time. That is, it is the same as was in the past and will remain the same in future.

The theory however accepts that on a smaller scale, the change takes place. Suppose, we take a small segment of the universe, such as the neighbourhood of the Sun, it does change over time as individual stars burn up their fuel and die, eventually becoming objects such as black dwarfs, neutrons stars and black holes. The Steady State state theory proposes the formation of new stars continuously with the time at the rate needed to replace the stars which have used up their fuel and have stopped shining. So, if we take a large enough region of space, having size of tens of millions of light years across, the average amount of light emitted does not change over time.

2. Hoyle-Narlikar's Steady State Model

In the early twentieth century, as the universe was probed by powerful telescopes, it was clear that the universe appeared to be isotropic (i.e., the same in all directions) on the large average.

Coupling this isotropy with the reasonable conjecture that our place was not a special place in the universe; in other words, the observed isotropy should be around every cosmic observer, motivated what is called the Cosmological Principle. It simply states that the universe possesses large scale homogeniety and isotropy.

Space-time intervals in a background that obeys cosmological principle are described by the Freidman Robertson Metric.

$$ds^2 = c^2 dt^2 - a(t)^2 \left[\frac{dr^2}{1 - kr^2} + r^2 d\theta^2 + r^2 \sin^2 \theta \, d\phi^2 \right]$$

The dynamics is thus described by a curvature index: $k = \pm 1$, and a scale factor $a(t)$ that is only a function of the cosmic time t. The evolution of this scale factor is determined by requiring this metric to be a solution of the Einstein equations:

$$G_{\mu\nu} + \Lambda g_{\mu\nu} = \kappa \, T_{\mu\nu}$$

where $G_{\mu\nu}$ is the Einstein tensor, $g_{\mu\nu}$ is the metric tensor, $T_{\mu\nu}$ is the stress energy tensor, Λ is the cosmological constant and κ is the Einstein gravitational constant.

The Einstein gravitational constant is defined as

$$\kappa = \frac{8\pi G}{c^4} \approx 2.077 \times 10^{-43}\ \mathrm{N}^{-1}$$

The above narrative forms the basis of the Standard Model. The model provides for a template that is used to accommodate and describe the observed expansion of the universe along with classical cosmologic observations (namely: the observed magnitude-redshift Hubble Diagram; the angular diameter-redshift curve; and the number count variation with redshift), the Hot Big Bang and nucleosynthesis, and Cosmic Microwave Background.

In contrast, Bondi, Gold, and later Fred Hoyle and Jayant Narlikar toyed with a Steady State model with the idea of a Perfect Cosmological Principle that could account for a universe that, besides being homogeneous and isotropic, looks the same at all times on large scales. In other words, a universe that looks the same from any place and also at any time. At the very outset, while in the standard model the total amount of matter remained constant in coordinates that were co-moving with the expanding universe, in the Steady State model the total amount of matter was required to be fixed in a given volume. With the observed large scale outflow from a fixed volume, Hoyle proposed creation of matter to maintain a steady state inside the given fixed volume.

To determine the required rate of matter creation, consider an observer surrounded by a sphere of a fixed radius r. Requiring the density of the observed matter to be constant ρ, the amount of matter in a shell of thickness dr, that travels with the Hubble flow velocity v, would imply a rate of loss of mass in the sphere to be simply $4\pi r^2 \rho v$. This has to be compensated by an average creation of matter density ($\equiv \bar{R}$)

$$4\pi r^2 \rho v = \frac{4\pi}{3} r^3 \bar{R}$$

Assuming a uniform Hubble flow $v = Hr$, gives the required rate of matter creation as $\bar{R} = 3\rho H$. Taking the currently preferred value of the Hubble parameter from observations as roughly 75 km/sec/Mpc gives $H^{-1} \approx 4 \times 10^{17}$ sec, and the observed density $\rho \approx 2 \times 10^{-31}$ gm/

cm^3. This would amount to roughly some 10^{-25} hydrogen atoms per sec per cc-which would be far below the threshold of any experimental verification.

Formally, Hoyle and Narlikar accounted for matter creation at the expense of a "C" field. The details are available in their original articles. For the purpose of this note, it would suffice to mention that the entire idea fell into disfavour due to numerous reasons. Firstly, it was pointed out by Stephen Hawking in his PhD thesis that the ad hoc "C" field was an unstable field with a negative energy and thus the idea was not theoretically realizable. Secondly, careful observations by Malcolm Longair and Martin Rayle demonstrated that the model was not consistent with the classical cosmological observational tests. Finally, the model did not offer any account for nucleosynthesis and Cosmic Microwave Background. This was because the steady state implied a constant temperature while the nucleosynthesis in the early universe requires a high enough temperature in the past.

3. Radio Source Counts: Setback to the Steady State Theory

From a radio-astronomical survey, the source counts distribution of radio-sources is the cumulative distribution of the number of sources (N) brighter than a given flux density (S). As it is usually plotted on a log-log scale its distribution is known as the log N-log S plot. It is one of several cosmological tests that were conceived in the 1930s to check the viability of and compare new cosmological models.

Early work to catalogue radio sources aimed to determine the source count distribution as a discriminating test of different cosmological models. For example, a uniform distribution of radio sources at low redshift, such as might be found in a 'steady-state Euclidean universe,' would produce a slope of –1.5 in the cumulative distribution of log(N) versus log(S).

Data from the early Cambridge 2C survey (published 1955) apparently implied a (log(N), log(S)) slope of nearly –3.0. This appeared to invalidate the steady state theory of Fred Hoyle, Hermann Bondi and Thomas Gold. Unfortunately, many of these weaker sources were subsequently found to be due to 'confusion' the blending of several weak sources in the side-lobes of the interferometer, producing a stronger response.

Later radio survey data have shown a complex picture – the 3C and 4C claims appear to hold up, while at fainter levels the source counts flatten substantially below a slope of –1.5. This is now understood to reflect the effects of both density and luminosity evolution of the principal radio sources over cosmic timescales.

4. Angular Resolution

For an image-forming device such as a radio or optical telescope, a microscope, a camera, or an eye, the angular resolution describes its ability to distinguish small details of an object. In the optics, it is used for the light waves, in the antenna theory, it is used for the radio waves, and in the acoustics, it is used for the sound waves. The colloquial use of the term "resolution" is rather confusing. For example, for a good quality of image, a camera is said to have a high resolution, but the camera actually has a low angular resolution. According to the Rayleigh criterion, the minimum angular spread which can be resolved by an image forming system is limited by diffraction to the ratio of the wavelength of the waves to the aperture width of that the system used. Thus, for high angular resolution, a telescope need to have large aperture.

5. Olbers Paradox

The Olbers' paradox, also known as **dark night-sky paradox**, is an argument that darkness of night-sky conflicts with the assumption of an infinite and eternal static universe. Thus, the darkness of night sky is an evidence for a dynamic universe, such as the Big Bang model.

The paradox may be stated that a static, infinitely large universe having infinite number of stars distributed uniformly would be bright rather than dark. In order to show this, let us consider universe which is static, homogeneous at a large scale, and is populated by an infinite number of stars. Now, we divide the universe into a series of concentric spherical shells, each of thickness of 1 ly. The volume of a spherical shell of radius r and thickness dr is $4\pi r^2 dr$. Suppose, ρ denotes the uniform density of stars in the universe. The number of stars in the shell lying between n ly and $(n + 1)$ ly (n is very large number, say 10^9) is

$$N_1 = 4\pi r^2 dr\rho$$
$$= 4\pi n^2\rho$$

where the distance is measured in ly and $dr = 1$ ly. The number of stars in the shell lying between $2n$ ly and $(2n + 1)$ ly is

$$N_2 = 4\pi r^2\, dr\rho$$
$$= 4\pi(2n)^2\rho$$

Thus, the ratio of the number of stars in two shells is

$$\frac{N_2}{N_1} = 4$$

Suppose, the stars are identical and I is light-energy emitted by each star. The amount of energy received by observer, situated at the common centre of shells, per unit area from each star in the first shell is

$$E_1 = \frac{I}{4\pi n^2}$$

where n is the radius of first shell. The amount of energy received per unit area from each star in the second shell is

$$E_2 = \frac{I}{4\pi(2n)^2}$$

where $(2n)$ is the radius of second shell. Thus, the ratio of light received from a star in two shells is

$$\frac{E_2}{E_1} = \frac{1}{4}$$

It shows that $E_1N_1 = E_2N_2$, i.e., the total energy received by the observer from each shell is the same. Thus each shell of a given thickness provides the same amount of light regardless of how far away it is from observer. That is, the light of each shell adds to the total amount received by observer. Thus, the more shells, the more light; and with infinitely many shells, there would be a bright night sky.

While dark clouds could obstruct the light, these clouds would heat up, until they were as hot as the stars, and then radiate the same amount of light. We have dark nights and it contradicts the existence of static, infinitely large universe having infinite number of stars distributed uniformly. It may be concluded that the universe is dynamic and not static.

6. Age of Universe

The estimates about the ages of the Earth, Sun and star clusters are important cosmological observations and they do not depend on the specific cosmological models. From the decay of radioactive isotopes, the age of the Earth is estimated as 4.6 billion years. The age of the Sun is thought to be slightly more than the age of the Earth. The ages of the oldest star clusters in Milky Way are estimated to be from 10 billion years to the age of universe.

Example 1.

The speed of expansion of particles in a planetary nebula is measured spectroscopically to be 20 km/s. The planetary nebula has a diameter of 0.1 pc. Calculate when it started to form.

Solution:

For solving the problem, we assume that the particles are moving with the same speed throughout their journey. It is then simple matter to calculate how long ago they began. As the diameter of nebula is 0.1 pc then particles have travelled 0.05 pc (the distance from centre). Their speed is 20 km/s. The time elapsed is

$$\text{Time} = \frac{\text{Distance}}{\text{Speed}} = \frac{0.05 \text{ pc}}{20 \text{ km/s}}$$

$$= \frac{0.05 \times 9.5 \times 10^{15}}{20 \times 10^{3}}$$

$$= 2.4 \times 10^{10} \approx 800 \text{ years}$$

7. Relative Helium Abundance

The Big Bang theory is treated as a natural way to explain the present abundance of elements. After about 2 to 3 minutes of Big Bang, the expanding universe had cooled to below about 10^9 K so that protons and neutrons could fuse to make stable deuterium nuclei that would not be broken by 'energetic photons'. Remember that the deuterium was produced in the fusion chain process in the core of stars, for the fusion of hydrogen nuclei to make a helium nucleus. But, because of the abundant free neutrons in the early universe, the fusion chain process in early universe was slightly different as compared to what occurs in the stars.

However, the general process is the same as the following. Two protons combine to from a deuteron.

$${}_1H^1 + {}_1H^1 \rightarrow {}_1H^2$$

The deuteron combines with a proton to form helium isotope.

$${}_1H^2 + {}_1H^1 \rightarrow {}_2He^3$$

Two helium isotopes obtained from two sets of reactions combine to form a helium nucleus.

$$_2He^3 + {}_2He^3 \rightarrow {}_2He^4 + {}_1H^1 + {}_1H^1$$

These fusion chain reactions could not take place until the universe had cooled down enough. The exact value of temperature depends sensitively on the density of protons and neutrons at that time.

Beside $_2He^4$, extremely small amounts of $_3Li^7$ were also produced during the early universe nucleosynthesis process and no other elements heavier than helium were produced. The elements heavier than helium were produced in the cores of stars. After about 15 minutes from Big Bang, the universe had expanded and cooled so much that there was no more fusion process. The composition of universe was 10% helium and 90% hydrogen in terms of numbers or 25% helium and 75% hydrogen in terms of mass. Stars also produce some of the helium visible today, but not most of it.

In this context, the deuterium nucleus $_1H^2$ is of great importance because its production is sensitive to the density of protons and neutrons, and temperature in the early universe. The number of deuterium nuclei which later on participate in the fusion reaction to make $_2He^3$ nuclei also depends sensitively on the temperature, and density of protons and neutrons. A denser universe would have had more deuterium fused to form helium. A less dense universe would have had more deuterium remaining. The amount of final $_2He^4$ product is not as sensitive to the ordinary matter density of early universe, and therefore the amount of remaining deuterium seen today is used as a probe of early density.

8. Cosmological Principle

Before development of a theory about universe, we have to have some postulates granted about it. These postulates must be such that they are not only internally consistent but their consequences must explain the observed facts. One of the postulates about the universe is **Cosmological principle**, which can be expressed in a number of forms. The cosmology deals at large scales and the universe is assumed to be homogeneous (no change during linear motion) and isotropic (no change during angular motion) at any instant of time. This assumption is a form of Cosmological principle. It can be justified on the scales of larger than 100 Mpc. Thus, following the Cosmological Principle, apart from the local irregularities, the universe looks the same from all positions in space.

The **Perfect Cosmological Principle** may be described that for an observer situated at an arbitrary position in space, at an arbitrary time in the history of the universe, the cosmos gives exactly the same results as those obtained by some other observer situated at some other position in space, at the same or other time. This principle has far reaching consequences and it directly leads to the steady state model of universe. In this scenario, the average properties of universe become well defined. The average density of universe is constant.

9. Homogeneous and Isotropic Universe

Results of observations made so far of outer space tell that there are no preferred directions or unusually dense regions in the universe. The data are nearly same as for a homogeneous and isotropic universe. Hence, in the universe, there are no preferred locations or directions. An observer situated at a location in the universe would find the distant galaxies redshifted, showing recession away from observer, irrespective of the direction of observation. That is, the galaxies situated in all directions of an observer appear to be receding away from observer.

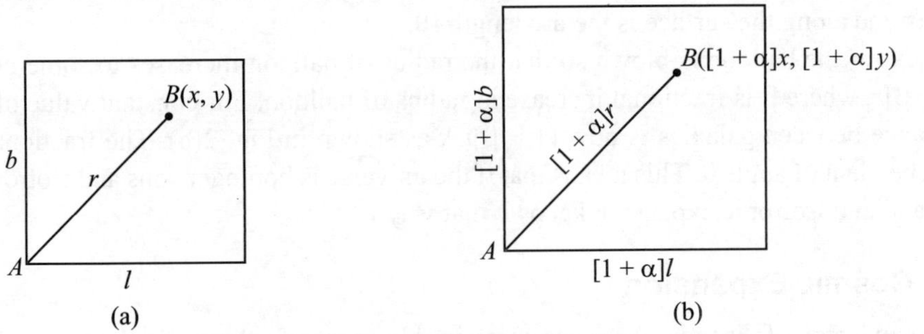

Fig. 1: Two-dimensional model of universe.

For having a model of such universe, we assume that the redshift indicates a genuine expansion indeed. Let us attempt to understand a model of universe in which an observer situated at a point sees all other observers in distant galaxies receding away from him/her.

A simple model in two dimensions may be a rubber sheet of length l (along x-axis) and breadth b (along y-axis) as shown in Fig. 1(a). Let the spots be painted on this sheet in a random manner. The two spots A and B at the positions (0, 0) and (x, y), respectively, are separated by the distance r, expressed as

$$r = \sqrt{x^2 + y^2}$$

Now, the sheet is stretched in length and breadth by a fractional amount α, so that the sides of sheet are $[1 + \alpha]l$ and $[1 + \alpha]b$, respectively, as shown in Fig. 1(b). Moreover, the coordinates of the spots A and B are (0, 0) and $([1 + \alpha]x, [1 + \alpha]y)$, respectively, and the separation between them is $[1 + \alpha]r$, such that

$$([1 + \alpha]r)^2 = ([1 + \alpha]x)^2 + ([1 + \alpha]y)^2$$

which is equivalent to

$$r = \sqrt{x^2 + y^2}$$

showing each part is stretched by the same fractional amount.

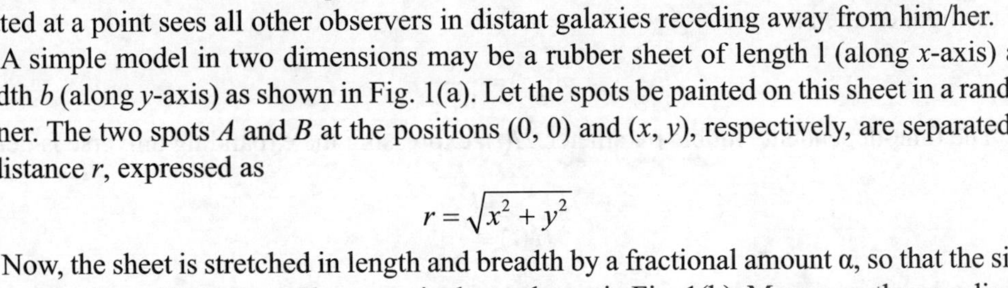

Fig. 2: Three-dimensional model of universe.

A flat rubber sheet is not the only two-dimensional model for an expanding homogeneous isotropic universe. It can be a three dimensional also. Let us have a rubber balloon and paint spots on it in a random manner to represent galaxies. Suppose, the radius of balloon is r at a given instant, as shown in Fig. 2(a). Let the angle subtended by two galaxies A and B on the surface of balloon be θ at the centre of balloon. The distance between the galaxies A and B measured along the surface is the arc length $r\theta$.

Now, the balloon is blown so that the radius of balloon increases to some new value say $[1 + \beta]r$, where β is fractional increase in radius of balloon. For constant value of angle θ, the distance between galaxies is now $[1 + \beta]r\theta$, as shown in Fig. 2(b). The fractional increase is independent of angle θ. This means that if the universe is homogeneous and isotropic at a given instant, and isotropic expansion keeps it that way.

10. Cosmic Expansion

The expansion of the universe is parameterized in terms of a dimensionless scale factor $a = a(t)$ (with time t counted from the birth of the universe), defined relative to the present day, so $a_0 = a(t_0) = 1$. The usual convention in cosmology is that the subscript 0 denotes present-day values, so t_0 is the current age of the universe. The scale factor is related to the observed redshift z of the light emitted at time t_{em} by

$$a(t_{em}) = \frac{1}{1+z}$$

The time-dependent Hubble parameter, $H(t)$, expressing the expanding universe is defined as

$$H(t) \equiv \frac{\dot{a}}{a}$$

where $\dot{a}$ denotes the time-derivative of the scale factor. The first Friedmann equation gives the expansion rate in terms of the matter + radiation density ρ, the curvature k, and the cosmological constant Λ,

$$\begin{aligned} H^2 &= \left(\frac{\dot{a}}{a}\right)^2 \\ &= \frac{8\pi G}{3}\rho - \frac{kc^2}{a^2} + \frac{\Lambda c^2}{3} \end{aligned}$$

where c is the speed of light and G the gravitational constant. A critical density ρ_c is the present-day density, which corresponds to $k = 0$, assuming the cosmological constant $\Lambda = 0$, regardless of its actual value. Substituting these conditions to the Friedmann equation gives

$$\begin{aligned} \rho_c &= \frac{3H_0^2}{8\pi G} \\ &= 1.878\,47(23) \times 10^{-26}\, h^2 \text{ kg m}^{-3} \end{aligned}$$

where $h \equiv H_0/(100 \text{ km s}^{-1} Mpc^{-1})$ is the reduced Hubble constant. For the Lambda-CDM model with a positive cosmological constant (as observed), the universe is predicted to expand forever regardless of whether the total density is slightly above or below the critical density. Other outcomes are possible in extended models where the dark energy is not constant but actually time-dependent.

It is standard to define the present-day density parameter Ω_x for various species as the dimensionless ratio

$$\Omega_x \equiv \frac{\rho_x(t = t_0)}{\rho_c}$$

$$= \frac{8\pi G \rho_x(t = t_0)}{3H_0^2}$$

where the subscript x is one of the b for baryons, c for cold dark matter, r for radiation (photons plus relativistic neutrinos), and Λ for dark energy.

Since the densities of various species scale as different powers of a, e.g., a^{-3} for matter, etc., the Friedmann equation can be conveniently expressed in terms of the various density parameters as

$$H(a) \equiv \frac{\dot{a}}{a}$$

$$= H_0\sqrt{(\Omega_c + \Omega_b)a^{-3} + \Omega_r a^{-4} + \Omega_k a^{-2} + \Omega_\Lambda a^{-3(1+w)}}$$

where w is parameter for the equation of state of dark energy, and assuming negligible neutrino mass (significant neutrino mass requires a more complex treatment).

In the minimal 6-parameter Lambda-CDM model, it is assumed that curvature Ω_k is zero and $w = -1$, so this simplifies to

$$H(a) = H_0\sqrt{\Omega_m a^{-3} + \Omega_r a^{-4} + \Omega_\Lambda}$$

Observations show that the radiation density is very small today, $\Omega_r \sim 10^{-4}$. On neglecting this term, the equaition has an analytic solution

$$a(t) = (\Omega_m/\Omega_\Lambda)^{1/3} \sinh^{2/3}(t/t_\Lambda)$$

where $t_\Lambda \equiv 2/(3H_0\sqrt{\Omega_\Lambda})$. This is fairly accurate for $a > 0.01$ or $t > 10$ million years. Solving for $a(t) = 1$ gives the present age of the universe t_0 in terms of the other parameters.

11. Some Simple Models of Universe

For understanding universe, some simple models proposed are the following.

11.1 Steady State Universe

In this model, described by Bondi and Gold, and by Hoyle, the universe is flat and provides the same aspect at all places and at all times. Its expansion rate is uniform in space and time.

11.2 Static Universe of Einstein

Before the discovery of expanding universe described by Hubble, Einstein in 1917 proposed a cosmic model based on his general relativistic field equation expressed as:

$$R_{\mu\nu} - Rg_{\mu\nu} + \Lambda g_{\mu\nu} = \frac{8\pi G}{c^4} T_{\mu\nu}$$

where $R_{\mu\nu}$ is Ricci curvature tensor, R is scalar curvature, $g_{\mu\nu}$ is metric tensor, Λ is cosmological constant, G is gravitational constant, c is speed of light in vacuum and $T_{\mu\nu}$ is stress-energy tensor. This model was static (not expanding). The Einstein universe is spherical and has constant radius of curvature a. It was later found that if Einstein had not taken Λ constant, he would have found expanding universe.

In 1930, Lemaitre and Eddington discovered that the Einstein universe is unstable. A small deviation from the perfect conditions taken by Einstein, the universe results in either a continuing expansion or an accelerating collapse.

11.3 de Sitter Universe

Soon after the proposal of static model by Einstein, de Sitter found that the general relativistic field equations permitted the description of a second model, one that existed in a flat space. In this case, the universe was found expanding. In the beginning, it had no more than an academic interest. But, in the late 1920s, Hubble's discovery of cosmic expansion revived interest in this model. The main drawback found in this model is that the density must be zero.

11.4 Eddington Universe

Starting of this universe is from the Einstein state. Then an initially uniform distribution of gas becomes perturbed by the processes involving the formation of galaxies and then goes over into a uniform expansion. One difficulty in this model is the uncertainty whether the formation of galaxies should not lead to an instability that would give rise to contraction rather than expansion. This model is found interesting as it focused attention on the fact that cosmology is not just a matter of geometry. A model in principle must also be able to account for the physical state of matter available in the universe. Galaxies are considered to have condensed out of an initially uniform gas. In case this gas was in rapid expansion, how it could be possible to counteract the expansion in order to force the gas to contract into galaxies? The answer is not known. But, Eddington and Lemaitre tried to make a plausible guess.

11.5 Lemaitre Universe

Following Lemaitre, the universe starts out in a highly contracted state which initially expands at a rapid rate. Later on, the expansion is slowed down and brought to a halt in a state. Galaxies are formed at this stage and give rise to a new expanding phase that continues indefinitely.

A noticeable feature of Lemaitre's model is high initial density of universe, very similar to Big Bang. We can compute the temperature and pressure that must have existed at the initial time, and thus establish the nuclear reactions that should have occurred. We obtain information about the chemical composition of matter at early stages of universe before the formation of galaxies.

This gives the importance of chemical characteristics of universe. A cosmological model must be able to simulate not only the overall density and pressure of universe, or the existence of galaxies, it must also be able to predict in some detail the chemical composition of matter for the formation of earliest stars. The chemical composition of evolved stars need not be predicted in this way, because nuclear reactions take place in the interior of stars so that early generation of stars can produce heavier chemical elements that later are distributed in the interstellar space through ejection or explosion. Subsequent generations of stars can incorporate this newly formed matter and their atmospheres contain these chemical elements.

11.6 Friedmann Universe

From the Einstein field equations, Friedmann derived two equations for homogeneous and isotropic universe as the following.

$$\frac{\dot{a}+kc^2}{a^2}=\frac{8\pi G_\rho+\Lambda c^2}{3}$$

and
$$\frac{\dot{a}}{a}=-\frac{4\pi G}{3}\left(\rho+\frac{3p}{c^2}\right)+\frac{\Lambda c^2}{3}$$

where $a(t)$, a function of time t, is the scale factor, $\dot{a}$ and ä are first and second derivatives, respectively, of a with respect to time t. The Hubble parameter (constant) H is expressed as $H = \dot{a}/a$. The G, Λ and c are, respectively, the universal gravitational constants, cosmological constant, and speed of light in vacuum. For a particular case, the intrinsic curvature k remains constant, but may vary from one case to another. Here, a, H, ρ and p are functions of time t. The ratio k/a^2 is spatial curvature in any time slice of universe. The solution of equations is independent of coordinate system chosen for the spatial slices. There are two commonly used choices for a and k which describe the universe.

- The spatial curvature k is considered to have three integer values: +1, 0, −1.
  For $k = +1$, the universe is a closed three-dimensional sphere. Here, a denotes the radius of curvature of universe (sphere).
  For $k = 0$, the universe is flat (Euclidean space). Here, a may be given any arbitrary positive number at one particular time.
  For $k = -1$, the universe is an open three-dimensional hyperboloid. Here, ia denotes the radius of curvature of universe.
- The parameter a is the scale factor which is taken at the present time. If R_t is the radius of curvature of hypothetical universe at a time t (R_0 in the present day radius of curvature), the $a = R_t/R_0$.
  These equations may be simplified by replacing

$$\rho \to \rho - \frac{\Lambda c^2}{8\pi G}$$

and
$$p \to p + \frac{\Lambda c^2}{8\pi G}$$

to give
$$H^2 = \left(\frac{\dot{a}}{a}\right)^2 = \frac{8\pi G}{3}\rho - \frac{kc^2}{a^2}$$

and
$$\dot{H} + H^2 = \frac{\ddot{a}}{a} = -\frac{4\pi G}{3}\left(\rho + \frac{3p}{c^2}\right)$$

The Hubble parameter can change over time when other parts of equation are time dependent (in particular, the mass density, vacuum energy, or spatial curvature). Evaluation of Hubble parameter at the present time gives the Hubble's constant, which is the proportionality constant of Hubble's law. Applied to a fluid with a given equation of state, the Friedmann equations yield the time evolution and geometry of universe as a function of fluid density.

Some cosmologists call the second of these two equations as the Friedmann acceleration equation and the first equation as the Friedmann equation.

11.7 Density Parameter

In the Friedmann universe, the density parameter Ω is expressed as the ratio of actual (or observed) density ρ to the critical density ρ_c. This parameter determines the overall geometry of universe. When $\Omega = 1$, the universe is flat (Euclidean).

To date, the critical density is estimated to be approximately five atoms (of mono-atomic hydrogen) per cubic metre, whereas the average density of ordinary matter in the universe is believed to be 0.2 atoms/m^3. A large contribution to density comes from unidentified dark matter. Both the ordinary matter and dark matter lead to the contraction of universe. The largest contribution to density comes from the so-called dark energy, which accounts for the cosmological constant term. Although the total density is equal to the critical density (exactly up to the measurement error), the dark energy does not lead to the contraction of universe, but rather accelerates its expansion. Therefore, the universe expands forever.

An expression for the critical density is found by assuming Λ to be zero and setting the normalized spatial curvature k equal to zero. Substitution of these values in Friedmann's first equation gives

$$\rho_c = \frac{3H^2}{8\pi G}$$

The density parameter is thus defined as

$$\Omega = \frac{8\pi G_\rho}{3H^2}$$

Assuming a zero vacuum energy density, when Ω is larger than unity, the space sections of universe are closed; the universe eventually stops expanding, then collapses. When Ω is less than unity, they are open, and the universe expands forever.

11.8 Mathematical Forms of Models

It is interesting to note that all the models have similar mathematical forms: Lemaitre, Robertson and Walker, independently, found that in an isotropic, homogeneous space, the field equations of Einstein reduce to two simple differential equations in the radius of curvature or scale factor a. The first of these gives the mass density, including all radiation energy, in terms of the mass density of vacuum, the expansion or contraction kinetic energy, and a curvature energy:

$$\kappa\rho = -\Lambda + 3\left(\frac{\dot{a} + \kappa c^2}{c^2 a^2}\right)$$

The second gives the change in this density on expansion or contraction, including work done against the pressure of fluid. This is the energy conservation equation.

$$\frac{d\rho}{dt} + 3\left(\rho + \frac{P}{c^2}\right) = 0$$

or

$$\frac{\kappa P}{c^2} = \Lambda - \left(\frac{2a\ddot{a} + \dot{a}^2 + \kappa c^2}{c^2 a^2}\right)$$

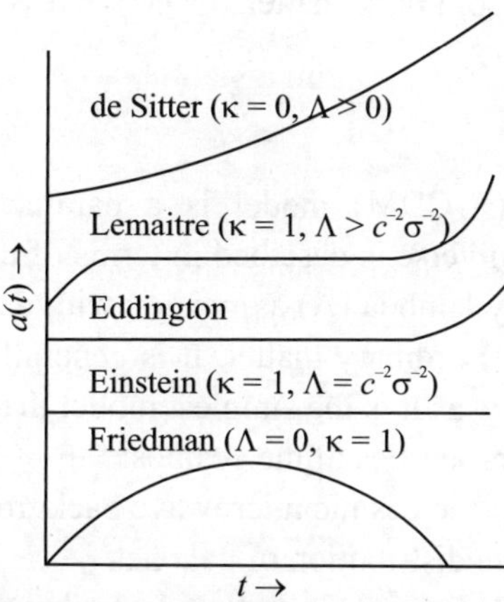

Fig. 3. Some cosmological models. The scales of $a(t)$ and of t are not the same for different curves. These are qualitative natures of variation of $a(t)$ versus t in different models.

Here, κ is $8\pi G/c^2 = 1.86 \times 10^{-27}$ cm/g. κ is sometimes called the Einstein gravitational constant. Λ is cosmological constant, and ρ and P are density and pressure of universe. Dots represent differentiation with respect to world time. A qualitative nature of the variation of $a(t)$ versus time t for various models is shown in Fig. 3.

12. Type Ia Supernovae and Accelerating Universe

The outcomes of two independent projects, the Supernova Cosmology Project and the High-Z Supernova Search Team, in 1998, found that the universe is expanding with an accelerated speed, i.e., the velocity at which a distant galaxy recedes from the observer is continuously increasing with time. This finding has been contrary to the expectation of the cosmologists that due to the gravitational attraction of the matter in the universe, the recession velocity would always be decelerating, i.e., the velocity would be decreasing with time. In both the projects, the teams used type Ia supernovae. The idea behind the use of type Ia supernovae was that they have almost the same intrinsic brightness (a standard candle), and since objects that are further away appear dimmer, and one can use the observed brightness of these supernovae to measure the distance to them.

The distance of Supernova can be compared to the supernovae's cosmological redshift, which measures how much the universe has expanded since the occurrence of the supernova. The result was unexpected, as both teams found that objects in the universe are moving away from one another at an accelerated rate. For this discovery of the accelerating universe, three members of these two groups have subsequently been awarded Nobel Prize of Physics. It is believed that the accelerated expansion of the universe has started since the universe entered its dark-energy-dominated era, roughly 4 billion years ago.

In the general relativity model, an accelerated expansion can be accommodated by considering a positive value of the cosmological constant Λ. It is equivalent to the presence of a positive vacuum energy, dubbed dark energy. While there are alternative possible explanations, the description of assumption of dark energy (positive Λ) is used in the standard model of cosmology, which also includes cold dark matter (CDM) and is known as the lambda-CDM model.

12.1 Lambda-CDM Model

The lambda-cold dark matter (Λ-CDM) model is a parameterization of the Big Bang cosmological model where the universe is descibed in terms of three major components: (i) a cosmological constant denoted by lambda (Λ) associated with dark energy, (ii) the postulated cold dark matter (CDM), and (iii) ordinary matter. It is generally referred to as the standard model of the Big Bang cosmology, as it is the simplest model that provides a reasonably good information about the following properties of the cosmos:

- the existence and structure of the cosmic microwave background
- the large-scale structure in the distribution of galaxies
- the observed abundances of hydrogen (including deuterium), helium, and lithium
- the accelerating expansion of the universe observed in the light from distant galaxies and supernovae

The model assumes that general relativity is the correct theory of gravity on cosmological scales. This model can be extended by adding cosmological inflation, quintessence and other elements which are current areas of speculation and research in cosmology. Some alternative models challenge the assumptions of the Λ-CDM model.

13. Puzzle of Dark Energy

The dark energy is a form of energy whose source is unknown so far and it affects the universe on the largest scales. The first observational evidence for its existence came from the measurements of the supernovae, where it is found that the universe does not expand at a constant speed, but its speed is continuously increasing with time, i.e., the universe is accelerating. For understanding the evolution of the universe, one needs information about its initial conditions and its composition. This finding has been contrary to the expectation of the cosmologists that due to the gravitational attraction of the matter in the universe, the recession velocity would always be decelerating, i.e., the velocity would be decreasing with time.

Existence of the cosmic microwave background suggests that the universe began in a hot Big Bang, from which the general relativity explains its evolution and the subsequent large-scale motion. For understanding the accelerating universe, there is need to introduce a new form of energy, called the dark energy. Since the 1990s, scientists have acepted the concept of dark energy, and are trying to understand its source. However, understanding the source of dark enery has so far remained a puzzle.

Considering the lambda-CDM model of cosmology, it has been estimated that the contribution of dark energy is 68% of the total energy in the present-day observable universe.

14. Eddington Limit

Under hydrostatic equilibrium in a star, there is balance between the radiation force acting outward and the gravitational force acting inward. In the state of hydrostatic equilibrium, the **Eddington luminosity**, also known as the **Eddington limit**, is the maximum luminosity a body (such as a star) can achieve. The Eddington limit is invoked to explain the observed luminosity of accreting black holes such as quasars. A star initiates a very intense radiation-driven stellar wind from its outer layers, when a star exceeds the Eddington luminosity. As most massive stars have luminosities far below the Eddington luminosity, their winds are mostly driven by the less intense line absorption.

In the original work, Sir Arthur Eddington considered only the electron scattering for calculating this limit. That is now is known as the classical Eddington limit. Nowadays, for the modified Eddington limit, other radiation processes such as bound-free and free-free radiation interaction, also have been considered.

14.1 Derivation

The limit is derived under the hydrostatic equilibrium where the outward radiation pressure is set equal to the inward gravitational force. As both the forces vary as inverse square laws, and therefore, once equality is reached, the hydrodynamic flow remains the same throughout the star.

From Euler's equation in hydrostatic equilibrium, the mean acceleration is zero, and thus, we have

$$\frac{du}{dt} = \frac{\nabla p}{\rho} - \nabla\Phi = 0$$

where u is the velocity, p the pressure, ρ the density, and Φ the gravitational potential. If the pressure is dominated by radiation pressure associated with a radiation flux F_{rad}, we have

$$-\frac{\nabla p}{\rho} = \frac{\kappa}{c} F_{\text{rad}} \tag{15.1}$$

where κ is the opacity of the stellar material defined as the fraction of radiation energy flux absorbed by the medium per unit density and per unit length. For ionized hydrogen $\kappa = \sigma_T / m_p$, where σ_T denotes the Thomson scattering crosssection for electron and m_p is the mass of a proton.

The energy flux over a surface F_{rad} is expressed as

$$F_{\text{rad}} = \frac{d^2 E}{dA dt}$$

It can be expressed with the momentum flux using $E = pc$ for radiation. Therefore, the rate of momentum transfer from the radiation to the gaseous medium per unit density is $\kappa F_{\text{rad}} / c$, which explains the right side of equation (15.1).

The luminosity of a source bounded by a surface S may be expressed with these relations as

$$L = \int_S F_{\text{rad}} \cdot dS$$

$$= \int_S \frac{c}{\kappa} \nabla\Phi \cdot dS.$$

Now assuming that opacity is a constant, it can be brought outside of the integral. Use of Gauss's theorem and Poisson's equation gives

$$L = \frac{c}{\kappa} \int_S \nabla\Phi \cdot dS = \frac{c}{\kappa} \int_V \nabla^2 \Phi dV$$

$$= \frac{4\pi Gc}{\kappa} \int_V \rho \, dV = \frac{4\pi GMc}{\kappa}$$

where M is the mass of the central object. This is called the Eddington luminosity. For pure ionized hydrogen

$$L_{\text{Edd}} = \frac{4\pi GMm_p c}{\sigma_T} \cong 1.26 \times 10^{31} \left(\frac{M}{M_\odot}\right) W$$

$$= 1.26 \times 10^{38} \left(\frac{M}{M_\odot}\right) \text{erg/s}$$

$$= 3.2 \times 10^4 \left(\frac{M}{M_\odot}\right) L_\odot$$

where $M_\odot$ is the mass of the Sun and $L_\odot$ the luminosity of the Sun.

15. Unification Models

Everyone wants to have a simple throry of everyting that works at all energies, at all scales, and for all times. We want to have a single theory, because a single theory is more beautiful than many theories. We want that all the laws should be different parts of a single pattern. Grand Unified Theory (GUT) is an attempt to describe all four fundamental forces and the relationships between the elementary particles in terms of a single throretical framework. There are four fundamental forces:

(i) Strong nuclear force
(ii) Electromagnetic force
(iii) Weak nuclear force
(iv) Gravitational force

The strong nuclear force is the strongest one, and the strength of forces decreases in sequence, and thus, the gravitational force is the weakest among them. The strong nuclear force holds the nuclei of atoms together. Though the force is very strong, but has very short distance range. The electromagnetic force causes electric and magnetic effects, such as two like charges repel each other, two opposite charges attract each other, and two magnets interact with each other. Through the electromagnetic force is much weaker, but has very long range. The weak nuclear force is responsible for nuclear decay. This force is weak and has very short distance range. The gravitational force is always attractive and the weakest of all the fundamental forces, but acts over very long distances. It acts between any two pieces of matter in the universe, and is very improtant in explaining the structure of the universe.

Attempts for unification of forces have always been made. The experiments of Faraday have contributed greatly to the understanding of electromagnetism (Faraday's law of electromagnetic induction). Maxwell formulated electromagnetic theory, and incorporated light and wave phenomenon into electromagnetism (Four Maxwell's equations, where variations of electric and magnetic fields are related). In 1905, Einstein's special theory of relativity proved that electricity and magnetism are linked and Maxwell was correct, and the single force of electromagnetism emerged (unification of electric and magnetic phenomena).

Each of the Sheldon Glashow, Abdus Salam, and Steven Weinberg, independently found that the weak nuclear force could be unified with the electromagnetic force. At extremely high energies, 10^{15} GeV, the electromagnetic force and weak force merge with each other. The relationship is based on the exchange of four particles: the photon of electromagnetic interactions, two charged W particles ($W^{\pm}$ particles), and a neutral Z particle (Z^0 particle). For their contributions, they were awarded the 1979 Nobel Prize in Physics.

Newton showed that both the astronomical and terrestrial motions could be explained with gravitation. Einstein's theory of general relativity was written in 1915 where the concept of Anti Gravity & Black Hole was given. Einstein worked for 30 years to unify electromagnetism with gravity, and failed.

16. Pressureless Dust Model

In general relativity, a dust solution is an important special case of fluid solution, a type of exact solution of the Einstein field equation, where the gravitational field is produced entirely by the mass, momentum, and stress density of a perfect fluid that has positive mass density but vanishing pressure.

A pressureless perfect fluid can be considered a model of a configuration of dust particles that locally move and interact with each other only gravitationally. Because of that the dust models are often employed in cosmology as the models of a toy universe, in which the dust particles are considered highly idealized models of galaxies, clusters, or superclusters.

The dust models, in astrophysics, have been employed as models of gravitational collapse. Dust solutions can also be used to model finite rotating disks of dust grains. If superimposed somehow on a stellar model comprising a ball of fluid surrounded by vacuum, a dust solution could be used to model an accretion disk around a massive object; however, no such exact solutions that model rotating accretion disks are yet known due to the extreme mathematical difficulty of their construction.

17. Critical Density

The critical density is the average density of matter required for the universe to just halt its expansion, but only after an infinite time. That is, when the density of universe becomes equal to the critical density, its expansion would stop. A universe with the critical density is said to be flat.

18. Lookback Time

The lookback time is the time elapsed between when we detect the light here on the Earth and when it was originally emitted by the source. The more distant an object is from us, the further back in time we are looking. The light left some of the more distant galaxies in the Hubble Ultra-Deep Field (HUDF) over 10 billion years ago.

19. Deceleration Parameter

In cosmology, the deceleration parameter q is a dimensionless measure of the cosmic acceleration of the expansion of space in a Friedmann-Lematre-Robertson-Walker universe. It is defined as

$$q \stackrel{\text{def}}{=} -\frac{\ddot{a}a}{\dot{a}^2}$$

where a is the scale factor of the universe and the dots indicate derivatives by proper time. The expansion of the universe is considered accelerating if $\ddot{a} > 0$ (as has been suggested in the recent measurements), and in this case the deceleration parameter is negative. At the time of definition, $\ddot{a}$ was expected to be negative, so a minus sign was introduced in the definition to make the q positive in that case. Since the evidence for the accelerating universe, it is now believed that $\ddot{a}$ is positive therefore the present-day value q_0 is negative (though q was positive in the past before dark energy became dominant).

The Friedmann acceleration equation can be written as

$$\frac{\ddot{a}}{a} = -\frac{4\pi G}{3}\sum_i \left(\rho_i + \frac{3p_i}{c^2}\right)$$
$$= -\frac{4\pi G}{3}\sum_i \rho_i(1+3w_i)$$

where the sum i extends over the different components, matter, radiation and dark energy, ρ_i is the equivalent mass density of each component, p_i is its pressure, and $w_i = p_i/(\rho_i c^2)$ is the equation of state for each component. The value of w_i is 0 for non-relativistic matter (baryons and dark matter), 1/3 for radiation, and –1 for a cosmological constant; for more general dark energy it may differ from –1, in which case it is denoted w_{DE} or simply w.

Defining the critical density as

$$\rho_c = \frac{3H^2}{8\pi G}$$

the density parameters $\Omega_i \equiv \rho_i/\rho_c$, and substituting $\rho_i = \Omega_i\,\rho_c$ in the acceleration equation, we get

$$q = \frac{1}{2}\sum \Omega_i(1+3w_i)$$
$$= \Omega_{\text{rad}}(z) + \frac{1}{2}\Omega_m(z) + \frac{1+3w_{DE}}{2}\Omega_{DE}(z)$$

where the density parameters are at the relevant cosmic epoch. At the present-day $\Omega_{\text{rad}} \sim 10^{-4}$ is negligible, and if $w_{DE} = -1$ (cosmological constant) this simplifies to

$$q_0 = \frac{1}{2}\Omega_m - \Omega_\Lambda$$

where the density parameters are present-day values; with $\Omega_\Lambda + \Omega_m \approx 1$, and $\Omega_\Lambda = 0.7$ and then $\Omega_m = 0.3$, this evaluates to $q_0 \approx -0.55$ for the parameters estimated from the Planck spacecraft data.

Review Questions

1. Describe the Hoyle-Narlikars Steady State model of the universe.
2. Describe the Steady State Theory of the universe.
3. State and describe Olbers paradox.
4. Describe cosmological principle.
5. Describe homogeneous and isotropic universe.
6. Discuss some simple models of universe.
7. For Einstein's universe a = constant and $k = 1$.
   (i) Show that
   $$\Lambda = \frac{1}{a^2} + \frac{\kappa P}{c^2}$$
   and that the density of universe is fixed at
   $$\rho = \frac{2}{a^2 \kappa} - \frac{P}{c^2}$$
   We know that $P/c^2 << \rho$, at present. If we lived in an Einstein universe, Λ would have to have a value $\sim a^{-2}$, and $\rho \sim 2(a^2\kappa)^{-1}$.
   (ii) Show that if $\kappa = 0$ and $P = 0$, a static universe would require $\Lambda = 0$ and $\rho = 0$, leaving a undefined.
8. De Sitter universe is flat and empty, $\kappa = \rho = P = 0$. Show that the scale factor a of the xpanding universe obeys
   $$a = a_0 e^{(\Lambda c^2/3)^{1/3} t}$$
   and the age of universe judged from the Hubble's constant H is
   $$\frac{1}{H} = \sqrt{\frac{3}{\Lambda c^2}}$$
9. From the field equations with cosmological constant $\Lambda = 0$ and with zero pressure, obtain the relations
   $$\frac{\ddot{a}}{a} = -\frac{4\pi G\rho}{3}$$
   and
   $$H^2 q_0 = \frac{4\pi G\rho}{3}$$
10. Prove the instability of Einstein universe for $P = 0$.
11. Prove for Friedman universe ($\Lambda = 0$) that
    (a) with $\kappa = 1$ and an initially dense universe for which $P/c^2 = \rho/3$, the solution has the parametric form
    $$a = b_0 c \sin x, \quad t = b_0(1 - \cos x)$$
    where x is a parameter.
    (b) For late stages of the universe, with $\kappa = 1$, $P = 0$
    $$a = a_0 c\,(1 - \cos x), \quad t = a_o(x - \sin x)$$
    (c) For the dense stage of a hyperbolic universe ($\kappa = -1$)
    $$a = b_0 c \sinh x, \quad t = b_o(\cosh x - 1)$$

(d) For late stage of a hyperbolic universe

$$a = a_0 c\,(\cosh x - 1), \quad t = a_o(\sinh x - x)$$

12. Write notes on the following:
   (i) Radio source counts
   (ii) Steady State Theory of universe
   (iii) Angular resolution
   (iv) Olbers paradox
   (v) Age of universe
   (vi) Relative helium abundance in the universe
   (vii) Cosmological principle
   (viii) Homogeneous and isotropic universe
   (ix) Steady state universe
   (x) Static universe of Einstein
   (xi) de Sitter model for universe
   (xii) Eddington model for universe
   (xiii) Lemaitre model for universe
   (xiv) Friedmann models for universe
   (xv) Type Ia supernovae and accelerating universe
   (xvi) Lambda-CDM model
   (xvii) Puzzle of dark energy
   (xviii) Eddington limit
   (xix) Unification models
   (xx) Pressureless dust model
   (xxi) Critical density
   (xxii) Lookback time
   (xxiii) Deceleration parameter

Multiple-Choice Questions

1. Which aspect is not dealt with under Cosmology?
   A. origin of universe B. evolution of universe
   C. evolution of the Sun D. ultimate fate of universe **Ans. C**
2. The age of the Earth is considered about
   A. 4.6×10^3 years B. 4.6×10^6 years C. 4.6×10^9 years D. 4.6×10^{12} years
   Ans. C
3. The density (ρ) of a flat universe is related to the critical density ρ_c as
   A. $\rho < \rho_c$ B. $\rho = \rho_c$ C. $\rho > \rho_c$ **Ans. B**

16 Life in Universe

Our Earth is the only place in the universe where the existence of life is known so far. Scientists have always been interested in the search of life in the universe, other than that we know on the Earth's surface. This motto of scientists may be rephrased as: "Are we alone in the universe?". An Earth-based mission to explore, understand and explain the origin, nature and prevalence of life outside the Earth in the universe is the SETI (Search for Extraterrestrial Intelligence). In the SETI mission, researchers study various aspects related to the existence of life, its origin, environment in which a life develops, and its ultimate fate. Because of the interest, this topic has been widely studied by the scientists worldwide.

The very first question in this context may arise about the existence of life on the surface of the Earth. In our solar system, the planet Earth is situated at such an advantageous position so that it has an atmosphere having constituents, which are necessary for the formation of life and for its protection from the intense and hazardous radiations coming from the outer space. Keeping in view our Earth, the scientists have opinion that some other stars may have planetary system and some of the planets may be at an advantageous position, similar to our Earth. Consequently, the search of exoplanets (planets in other solar systems) is going on extensively worldwide. Some scientists have the opinion that the life may be found in some other parts in our solar system. Thus, the mission for the SETI may be classified into two parts:

(i) Search of life in our solar system
(ii) Search of life on the exoplanets

1. Search of Life in Our Solar System

Our solar system is the only place where the life has been found so far in the universe. Some scientists believe that, besides on the surface of Earth, the life may be found in some parts of our solar system. Keeping in view the physical conditions prevailing there, the places where the life is being searched are: (i) the planet Mars, and (ii) the satellite Titan.

1.1 Missions for Mars

The Viking mission was launched by NASA in 1975 to scan the Martian surface for its structure and also the atmospheric composition in order to search some evidence for the existence of life there. This mission was composed of two spacecraft, namely, Viking 1 and Viking 2. The Viking 1 was launched on August 20, 1975 and arrived at Mars on June 19, 1976, whereas the Viking 2 was launched on September 9, 1975 and it entered the Mars orbit on August 7, 1976. The mission revealed many features of Mars which have been studied in detail by the **Spirit** (Mars exploration Rover-A) and the **Opportunity** (Mars exploration Rover-B) launched in

2003. Later on, other missions were launched, but so far no evidence for the existence of life on Mars has been witnessed.

1.2 Mission for Titan

Titan (the largest moon of the planet Saturn) being the only natural satellite in our solar system with atmosphere and adequate density, has been studied thoroughly by the NASA with the help of **Cassini-Huygens mission** since 2004. The duration of the mission has been from 15 October 1997 to 15 September 2017. Being rich in organic compounds, Titan has been studied for the traces of life. **Miller experiment** in 2010 confirmed the possibility of nucleotides and amino acids with similar conditions as on the Earth in a laboratory. In April 2013, NASA reported that complex organic molecule could arise on the Titan, on the basis of the simulation of the atmosphere of Titan. In June 2013, polycyclic aromatic hydrocarbons (PAHs) were detected in the upper atmosphere of the Titan.

Similar study has been done for the **Europa** (the satellite of Jupiter) where a vast reservoir of water is predicted by the scientists.

1.3 Voyager Mission

The Voyager mission was launched in 1979. The Voyager 1 and Voyager 2 spacecraft crossed our solar system in July 2013. They were launched with the sole motto of searching of extraterrestrial life in the deep space.

2. Discovery of Exoplanets

In 1584, Catholic Monk Giordano Bruno asserted that there are countless Suns and countless Earths, and all these Earths are revolving around their respective Suns. When he said this in public, he was removed from the Church by charging that he was against the religion. But, this fact is now found true, as we have information about a large number of exoplanets revolving around their respective Suns.

The extrasolar planets (also known as the exoplanets) are the bodies orbiting around the respective stars, outside our solar system. A total of 3758 such planets (in 2808 planetary systems, including 627 multiple planetary systems) have been identified as of April 1, 2018. This number is increasing day by day and the latest information can be obtained on websites. The discovery of the first two exoplanets was announced on 9 January 1992 as orbiting the pulsar PSR 1257 + 12. Direct detection of exoplanets is very difficult because they are so much dimmer than the host star they are orbiting. For example, our Sun is billion (10^9) times brighter than the Jupiter and 10 billion (10^{10}) times brighter than the Earth. The nearest known exoplanet is at a distance of 12 light year (ly) from the Earth, in the habitable zone of Proxima Centauri.

There has been **Kepler mission** with the aim to discover the Earth-size extrasolar planets orbiting around other stars in or near the habitable zone and to estimate how many of the billions of stars in our galaxy have such planets. The Kepler spacecraft was launched on March 7, 2009. On February, 2011, the Kepler space observatory mission team released a list of 1235 extrasolar planets, including 54 that may be in the *habitable zone*. Some of these candidates were the Earth-sized and super Earth-sized (defined as less than or equal to the twice of the Earth radii). Six of these candidates were in the **habitable zone**.

For the detection of exoplanets, the following methods are generally used.

1. Radial velocity method
2. Transit method
3. Gravitational microlensing
4. Astrometry
5. Direct imaging
6. Pulsar timing

2.1 Radial Velocity Method

This method is based on the fact that if a star has a planet (or planets) revolving around it, it is not strictly correct to say that the planet orbits around the star. Instead, the planet and the star orbit together about their common centre-of-mass. When the star is so much more massive as compared to the planet, the centre-of-mass is within the star and the star appears to wobble slightly on the travelling of planet around it. Astronomers can measure this wobble with the help of spectroscopic methods. If a star is travelling towards us, its light appears as blueshifted, and if it is travelling away from us, its light appears as redshifted. Spectroscopy can show this change in colour from a star as it moves towards and away from us while orbiting around the centre-of-mass of the star-planet system.

Many of the planets have been discovered using this method, when they are very large and very close to their stars. These planets, called the super-Jupiters, are much larger than the Jupiter, and have their period of revolution of a few days. Such a large planet results a large wobble, which is one of the features along with the short orbit-time, to detect them rather than a smaller planet or planets which is farther from its star.

2.2 Transit Method

This method only works for the star-planet systems which have orbits aligned in such a way that, as seen from the Earth, the planet travels between the observer and the star, and consequently blocks some of the light from the star. When a planet crosses (or transits) in front of its parent star-disk, then the observed brightness of the star drops by a small amount. The amount by which the star dims depends on the size of the planet. This has been the second most productive method of detection of exoplanets.

The transit method tells us about the radius of the planet, and it has the benefit that it sometimes allows a planet's atmosphere to be investigated through the spectroscopy. A planet does not block much light from a star, but that is detectable. Similar to the radial velocity method, this method has a bias towards the discovery of a large number of planets orbiting closer to their stars, because the larger planets block more light and transit more frequently so that they are easier to detect. This method does not work for all the systems because only 10% of the super-Jupiters are aligned in such a way, so that we see their transit.

Smaller planets in larger orbits are even less likely to be aligned in such a way that we can observe transits.

2.2.1 Transit timing variation (TTV)

When more than one planet (multiple planets) are revolving around a common star, each one of them slightly perturbs the orbits of the other planets. Small variations in the times of transit for one planet can thus indicate the presence of other planet, which itself may or may not transit.

2.2.2 Transit duration variation (TDV)

When a planet orbits multiple stars or the planet has moons, its transit time can significantly vary from transit to transit. Although no new planets or moons are discovered with this method, it has been used for successful confirmation of many transiting circumbinary planets.

2.3 Gravitational Microlensing

Microlensig occurs when the gravitational field of a star acts like a lens, magnifying the light of a distant background star. Planets orbiting the lensing star can cause detectable anomalies in the magnification as it varies over time. Gravitational microlensing relies on chance events where one star passes in front of another. When the farther star is a bright one and the nearer one is a normal one, we could not ordinarily see from the Earth. When the nearer star passes in front of the farther star, its gravity causes the light from the farther star to bend and the star is magnified, as seen from the Earth. If, during the event, the background star appears to be magnified even more for a short time that means a planet orbiting the smaller star is increasing the effect of magnification. This method has resulted in the detection of some exoplanets but it has the advantage of being sensitive to planets at large separations from their parent stars specifically.

2.4 Astrometry

This method gives accurate measurements of position of a star in the sky and the changes in its position over a period of time. The movement of a star due to the gravitational effect of a planet may be noticeable. But, as the motion is so small, this method is not yet so effective. It has predicted only a few disputed detection. However, it has been successfully used for the determination of the properties of the planets found with the help of other methods. Astrometry is found to be more sensitive to the planets having large orbits. This is an advantage over other methods. No extrasolar planet discoveries by this method have been confirmed by other methods, but the field is quite promising.

2.5 Direct Imaging

Direct imaging method uses the infrared wavelengths for observing the planets. This method works because at the infrared wavelengths, a Sun-like star is only 100 times brighter than the Jupiter. This method works for the planets which are very far from their stars, and therefore, a period of hundreds or thousands of years may be required for a planet to be discovered with the help of this method. Astronomers have to ascertain that the planet and star move simultaneously across the space to show that the planet revolving around the star, since to observe the complete orbit takes much longer period. This method does not permit astronomers to compute the mass of a planet instantly, but they can use the spectrum and luminosity to obtain the data regarding its surface temperature and diameter.

2.6 Pulsar Timing

A pulsar emits radio waves very frequently as it rotates. A planet orbiting a pulsar causes a bit abnormality in the timing of its observed radio pulses.

3. Drake Equation

Formation of stars is a regular process in galaxies. Some stars may have planets orbiting around them and out of them some planets may have possibility of the existence of life there. Drake in 1961 suggested an equation, known as the Drake equation, which is a probabilistic argument used to estimate the number of active, communicative, extraterrestrial civilizations in the Milky Way galaxy. The number of such civilizations N may be expressed in the form of the product of various parameters:

$$N = R_* \cdot f_p \cdot n_e \cdot f_l \cdot f_i \cdot f_c \cdot L$$

where

R_* denotes the average rate of the star-formation in our galaxy.

f_p is the fraction of the formed stars that have planets.

n_e is the average number of planets (for stars that have planets) that can potentially support life.

f_l is the fraction of those planets that actually develop life.

f_i is the fraction of planets bearing the life on which intelligent, civilized life has developed.

f_c is the fraction of these civilizations that have developed the communication.

L is the length of time over which such civilizations release the detectable signals.

The equation was written in 1961 by Frank Drake, not for the purposes of quantifying the number of civilizations, but as a way to stimulate the scientific dialogue at the first scientific meeting on the SETI.

Criticism made on the Drake equation focused not on the equation itself, but on the fact that the estimated values for several of its factors are highly conjectural. The combined effect being that the uncertainty associated with any derived value is so large that the equation cannot be used to draw any firm conclusions.

There is considerable disagreement on the values of above parameters used in the Drake equation. However, the educated guesses used by Drake and his colleagues in 1961 were as the following.

$R_* = 1\ \text{yr}^{-1}$ (1 star formed per year, on the average over the life of galaxy)

$f_p = 0.2$ to 0.5 (one fifth to one half of all stars formed have planets)

$n_e = 1$ to 5 (stars with the planets have between 1 and 5 planets capable of developing life)

$f_l = 1$ (100% of these planets develop life)

$f_i = 1$ (100% of which develop intelligent life)

$f_c = 0.1$ to 0.2 (10-20% of which is able to communicate)

$L = 10^3$ to 10^8 years

On inserting the minimum values in the Drake equation, we get

$N = 1\ (0.2)\ 1\ (1)\ 1\ (0.1)\ 10^3 = 20$

On inserting the maximum values in the Drake equation, we get
$N = 1\ (0.5)\ 1\ (1)\ 1\ (0.2)\ 10^8 = 5 \times 10^7$
Drake stated that, given the uncertainties, the original meeting concluded that $N = L$, and there were probably between 10^3 and 10^8 civilizations in the Milky Way galaxy.

4. Earth Similarity Index

Similarity index is a mathematical tool that can be applied for a set of data. This method is a measure of deviation from a reference system, usually on a scale lying between zero and one. The Earth similarity index (ESI) is a quantitative measure of the Earth-likeness. One way to understand the ESI may be as the following. Consider a physical quantity having value x_0 on the surface of the Earth. Suppose, the value of this quantity varies from x_a to x_b, such that $x_a < x_0 < x_b$, on other Earth. Then, the percentage variations p and q are expressed as

$$p = \frac{x_0 - x_a}{x_0} \times 100$$

and

$$q = \frac{x_b - x_0}{x_0} \times 100$$

Now, we define a threshold value T as

$$T = \left[1 - \left(\frac{x_0 - x_a}{x_0 + x_a}\right)^m\right]^{w_a} \quad (16.1)$$

and

$$T = \left[1 - \left(\frac{x_b - x_0}{x_b + x_0}\right)^m\right]^{w_b} \quad (16.2)$$

where w_a and w_b denote the weight exponents and m is a free parameter. The threshold is the limiting value of ESI above which a planet is considered similar to the Earth. Further, the threshold value T can be expressed as

$$T = \left(1 - \frac{t}{100}\right)$$

where t can assume positive values between zero and 100, so that T is positive, having the value between zero and 1. Thus, we have taken T also as another free parameter. The parameter m may be considered to assume the values from 0.6 to 1.2.

From equations (16.1) and (16.2), we get

$$w_a = \frac{\ln T}{\ln\left[1 - \left(\frac{p}{200 - p}\right)^m\right]}$$

and

$$w_b = \frac{\ln T}{\ln\left[1 - \left(\frac{q}{200 + q}\right)^m\right]}$$

The weight exponent w_x may be taken as the geometrical mean of w_a and w_b. That is,

$$w_x = \sqrt{w_a \times w_b}$$

Here, the subscript x denotes a physical quantity. The basic ESI_x for a physical quantity x is expressed as

$$ESI_x = \left(1 - \left|\frac{x - x_0}{x + x_0}\right|^m\right)^x$$

For the planetary property x, the terrestrial reference is x_0. The ESI_x can assume a value between zero (no similarity with the Earth) and one (identical to the Earth). One can calculate ESI_r, ESI_ρ, ESI_e and ESI_T corresponding to the radius, density, escape velocity and temperature, respectively.

The ESI_r and ESI_ρ, for the mean radius r and bulk density ρ, respectively, are used to define the interior Earth similarity index ESI_I, expressed as

$$ESI_I = (ESI_r \times ESI_\rho)^{1/2} \tag{16.3}$$

Thus, the ESI_I is the geometrical mean of ESI_r and ESI_ρ. The ESI_e and ESI_T for escape velocity v and mean surface temperature T, respectively, are used to define the surface Earth similarity index ESI_S, expressed as

$$ESI_S = (ESI_e \times ESI_T)^{1/2} \tag{16.4}$$

Thus, ESI_S is the geometrical mean of ESI_e and ESI_T. The ESI (sometimes called global ESI) is expressed as

$$ESI = (ESI_I \times ESI_S)^{1/2} \tag{16.5}$$

Using equations (16.3) and (16.4) in (16.5), we have

$$ESI = (ESI_r \times ESI_\rho \times ESI_e \times ESI_T)^{1/4}$$

On considering the Earth as the most superior planet, the value of T can never be greater than 1. The variations of various parameters, relative to those of the mean values on the Earth, may be taken as the following. The definitional limits for the radius are from 0.5 to 1.5 times the Earth's radius. The range for the mass of an existing planet is from 0.1 to 10 times that of Earth. For the density, the definitional range is from 0.7 to 1.5 times the Earth's density. The temperature variation is taken from 273 K to 232 K. The range for the escape velocity is considered from 0.4 to 1.4 times that of the Earth. For these limits, one can calculate the Earth similarity indexes corresponding to the radius, density, escape velocity and temperature, and finally the Earth similarity index.

For given values of parameters for planets in our solar system, one can calculate the ESI, where m and T are considered free parameters. Table 1 has the values of physical parameters, interior-ESI, surface-ESI and ESI for 31 objects. The abbreviation EU stands for the Earth Unit. The ESI for some objects is within 10% limit from the threshold value, supporting the opinion for the existence of life in the universe.

5. Prebiotic Molecules

The amino acids are the constituents of life. A number of molecules pertaining to the life (responsible for the production of amino acids) have been identified in the interstellar medium. Some of the prebiotic molecules are:

Table 1. Values of various ESI for some solar planets and satellites and exoplanets with $m = 0.8$ and $T = 0.8$.

S. No.	Body	Radius (EU)	Density (EU)	Esc. Vel. (EU)	Temp K	ESI_I	ESI_S	ESI
1	Earth	1.000	1.000	1.000	288	1.000	1.000	1.00
2	Mars	0.532	0.713	0.449	227	0.864	0.739	0.80
3	Mercury	0.382	0.984	0.379	440	0.877	0.614	0.73
4	Moon	0.272	0.606	0.212	220	0.751	0.624	0.68
5	Venus	0.949	0.950	0.925	730	0.985	0.44	0.66
6	Io	0.285	0.639	0.228	130	0.766	0.394	0.55
7	Callisto	0.378	0.352	0.218	134	0.692	0.402	0.53
8	Jupiter	10.973	0.240	5.379	152	0.480	0.439	0.46
9	Ganymede	0.412	0.352	0.244	110	0.701	0.338	0.49
10	Ceres	0.074	0.376	0.045	167	0.523	0.345	0.42
11	Europa	0.244	0.546	0.180	102	0.719	0.293	0.46
12	Saturn	9.140	0.124	3.225	134	0.405	0.432	0.42
13	Titan	0.404	0.341	0.235	94	0.693	0.284	0.44
14	Uranus	3.980	0.230	1.910	76	0.573	0.260	0.39
15	Neptune	3.864	0.297	2.105	72	0.620	0.241	0.39
16	Titania	0.123	0.301	0.069	60	0.543	0.130	0.27
17	Enceladus	0.039	0.291	0.021	75	0.427	0.126	0.23
18	Pluto	0.180	0.371	0.109	40	0.618	0.088	0.23
19	Triton	0.212	0.379	0.130	38	0.640	0.086	0.23
20	HD69830 d	4.190	0.250	2.100	312	0.582	0.844	0.70
21	55 Cnc c	5.680	0.250	2.840	310	0.551	0.803	0.67
22	55 Cnc f	4.910	0.390	3.060	310	0.641	0.792	0.71
23	61 Vir d	3.680	0.460	2.500	375	0.703	0.709	0.71
24	HIP 57050 b	6.640	0.320	3.780	250	0.574	0.722	0.64
25	Mu Ara d	8.380	0.280	4.450	327	0.529	0.706	0.61
26	HD 142 b	11.120	0.240	5.430	286	0.479	0.740	0.60
27	HD 96167 b	9.360	0.260	4.810	334	0.507	0.683	0.59
28	HD 108874 b	12.490	0.220	5.890	294	0.456	0.720	0.57
29	HD 210277 b	11.980	0.230	5.720	275	0.466	0.711	0.58
30	HD 147513 b	10.990	0.240	5.380	263	0.480	0.697	0.58
31	HD 69830 c	2.820	0.540	2.070	549	0.764	0.525	0.63

EU stands for Earth Unit.

1. Formic acid ($HCOOH$)
2. Glycolaldehyde
3. Glycine (NH_2CH_2COOH) - the simplest amino acid.
4. Ethylene oxide (c-C_2H_4O)
5. 2-Aminopyridine ($H_2NC_5H_4N$)
6. 2-Aminoethanol ($NH_2CH_2CH_2OH$)
7. Alanine ($NH_2CH_2CH_2COOH$) - amino acid.
8. Amino acetonitrile (NH_2CH_2CN)
9. Methanimine (CH_2NH)

Some reactions pertaining to the molecules in interstellar medium are as the following.

The methanimine (CH_2NH) is a precursor of aminoacetonitrile (NH_2CH_2CN)

$$CH_2NH + HCN \rightarrow NH_2CH_2CN$$

The aminoacetonitrile can be hydrolysed to give the glycine (NH_2CH_2COOH).

$$NH_2CH_2CN + H_2O \rightarrow NH_2CH_2OH + HCN$$

$$NH_2CH_2OH + HCN + H_2O \rightarrow NH_2CH_2COOH + NH_3$$

The formic acid has great importance, as it is a precursor for the amino acids (Glycine and Alanine, for example) :

$$\left.\begin{array}{r}\text{Aminomethanol } (NH_2CH_2OH) \\ \text{Aminoethanol } (NH_2CH_2CH_2OH)\end{array}\right\} + \begin{array}{c}\text{Formic Acid} \\ (HCOOH)\end{array} \rightarrow \left\{\begin{array}{l}\text{Glycine} \\ \text{Alanine}\end{array}\right.$$

Review Questions

1. Describe the places where the life, other than that on the Earth, is being searched.
2. Describe in brief about the radial velocity method for searching the exoplanets.
3. Describe in brief about the transit method for searching the exoplanets.
4. Describe the Earth similarity index.
5. Write notes on the following:
   (i) Search of life in our solar system
   (ii) Missions for Mars
   (iii) Missions for Titan
   (iv) Voyager mission
   (v) Radial velocity method
   (vi) Transit method
   (vii) Gravitational microlensing
   (viii) Astrometry
   (ix) Direct imaging
   (x) Exoplanets
   (xi) Drake equation
   (xii) Earth similarity index
   (xiii) Prebiotic molecules

Multiple-Choice Questions

1. The place in our solar system where life is being searched is
   A. Moon B. Jupiter C. Mars D. Saturn **Ans. C**
2. The place in our solar system where life is being searched is
   A. Moon B. Titan C. Jupiter D. Saturn **Ans. B**

17 Waves in Fluid Plasma

1. Phase Velocity and Group Velocity

A wave travelling along positive *z*-direction may be expressed as

$$y = A\cos(kz - \omega t) \tag{17.1}$$

where A is amplitude, $\omega(= 2\pi\nu)$ is angular frequency and $k(= 2\pi/\lambda)$ is wavenumber. Here, ν is frequency and λ is wavelength of wave. **Phase velocity** of wave is defined as the ratio of angular frequency ω to wavenumber k. Thus, the phase velocity v_p is

$$v_p = \frac{\omega}{k}$$

The quantity $(kz - \omega t)$ in equation (17.1) is known as the phase of wave motion. The particle of constant phase travels such that $(kz - \omega t)$ = constant with time. Thus, we have

$$\frac{\mathrm{d}}{\mathrm{d}t}(kz - \omega t) = 0$$

or

$$k\frac{\mathrm{d}z}{\mathrm{d}t} - \omega = 0$$

Therefore, we get

$$\frac{\mathrm{d}z}{\mathrm{d}t} = \frac{\omega}{k} = v_p$$

Hence, the phase velocity of a wave is the same as the velocity of wave. We can thus write phase velocity

$$v_p = \nu\lambda \tag{17.2}$$

For electromagnetic wave

$$E = h\nu$$

or

$$\nu = \frac{E}{h} = \frac{mc^2}{h}$$

Here, we have used the energy-mass equivalence relation, $E = mc^2$. According to de Broglie hypothesis, we have

$$\lambda = \frac{h}{p} = \frac{h}{mv}$$

where v is velocity of particle. Using expressions for ν and λ in equation (17.2), we get

$$v_p = \frac{mc^2}{h}\frac{h}{mv} = \frac{c^2}{v} \tag{17.3}$$

Since the velocity of particle is always smaller than the speed of light ($v < c$), the phase velocity is always larger than the speed of light. Therefore, the phase velocity does not correspond to the velocity of a particle. This difficulty of phase velocity being larger than the speed of light can be overcome by assuming that each moving particle is associated with a group of waves or a wave packet rather than a single wave. Thus, a particle is represented by a wave packet and we have group velocity associated with a particle.

To understand the concept of group velocity, let us consider a combination of two waves having equal amplitude A as

$$y_1 = A\cos(k_1 z - \omega_1 t)$$

and

$$y_2 = A\cos(k_2 z - \omega_2 t)$$

where k_1 and k_2 are wavenumbers, and ω_1 and ω_2 are angular frequencies of waves. The phase velocities of waves are ω_1/k_1 and ω_2/k_2, respectively.

The superposition of waves gives

$$\begin{aligned} y &= y_1 + y_2 \\ &= A\cos(k_1 z - \omega_1 t) + A\cos(k_2 z - \omega_2 t) \end{aligned}$$

Using trigonometric relation

$$\cos C + \cos D = 2\cos\left[\frac{C-D}{2}\right]\cos\left[\frac{C+D}{2}\right]$$

we have

$$\begin{aligned} y &= 2A\cos\left[\frac{(k_1 z - \omega_1 t) - (k_2 z - \omega_2 t}{2}\right]\cos\left[\frac{(k_1 z - \omega_1 t) + (k_2 z - \omega_2 t}{2}\right] \\ &= 2A\cos\left[\frac{\Delta k}{2} z - \frac{\Delta\omega}{2} t\right]\cos(kz - \omega t) \end{aligned} \tag{17.4}$$

where we have

$$\omega = \frac{\omega_1 + \omega_2}{2}$$

$$k = \frac{k_1 + k_2}{2}$$

$$\Delta\omega = (\omega_1 - \omega_2)$$

$$\Delta k = (\mathrm{k}_1 - k_2)$$

The resultant wave (17.4) has two segments:

(i) A wave having angular frequency ω and wavenumber k so that its velocity (phase velocity) is

$$v_p = \frac{\omega}{k}$$

(ii) Another wave having angular frequency $\Delta\omega/2$ and wavenumber $\Delta k/2$, and corresponding velocity is

$$\begin{aligned} v_g &= \frac{\Delta\omega/2}{\Delta k/2} \\ &= \frac{\Delta\omega}{\Delta k} = \frac{d\omega}{dk} \end{aligned}$$

The v_g is known as the **group velocity**.

1.1 Relation between v_p and v_g

The phase velocity v_p and group velocity v_g are expressed as

$$v_p = \frac{\omega}{k}$$

and
$$v_g = \frac{d\omega}{dk}$$

Now, we have

$$\begin{aligned} v_g &= \frac{d\omega}{dk} = \frac{d(v_p k)}{dk} \\ &= v_p + k\frac{dv_p}{dk} \\ &= v_p + k\frac{dv_p}{d\lambda}\frac{d\lambda}{dk} \end{aligned}$$

We know

$$\lambda = \frac{2\pi}{k}$$

and
$$\frac{d\lambda}{dk} = -\frac{2\pi}{k^2}$$

Thus, we have

$$\begin{aligned} v_g &= v_p + k\frac{dv_p}{d\lambda}\left(-\frac{2\pi}{k^2}\right) \\ &= v_p - \frac{2\pi}{k}\frac{dv_p}{d\lambda} \\ &= v_p - \lambda\frac{dv_p}{d\lambda} \end{aligned}$$

It is the required relation.

Example 1.

Show that group velocity v_g of a wave packet is equal to the velocity of a material particle. Derive the relation $v_g v_p = c^2$, where v_p is phase velocity of a wave in the wave packet and c is speed of light.

Solution:

For the relations

$$E = \hbar\omega$$

and
$$p = \hbar k$$

we have
$$dE = \hbar d\omega$$

and
$$dp = \hbar dk$$

group velocity
$$v_g = \frac{d\omega}{dk} = \frac{dE}{dp}$$

(i) For a free non-relativistic particle, potential energy is zero and we have

$$E = \frac{p^2}{2m}$$

and thus,
$$\frac{dE}{dp} = \frac{p}{m} = v$$

(ii) For a relativistic particle, we have[1]

$$E^2 = c^2p^2 + m_0^2c^4$$

and thus,
$$\frac{dE}{dp} = \frac{c^2p}{E}$$
$$= \frac{c^2mv}{mc^2} = v$$

This shows that velocity of a particle and group velocity of corresponding wave packet are the same, i.e., $v_g = v$. The phase velocity expressed as

$$v_p = \frac{\omega}{k} = \frac{E}{p}$$
$$= \frac{mc^2}{mv} = \frac{c^2}{v}$$

has no physical significance and it is not a measurable quantity. The product of group velocity and phase velocity is

$$v_g v_p = v\frac{c^2}{v} = c^2$$

2. Fluid Equations

As a simple case, let us consider plasma comprising two species: positive ions and electrons. For more species, the extension of following equations is straightforward. When ions and electrons in a plasma have charges q_i and q_e, particle densities n_i and n_e, and are moving with velocities $\vec{v}_i$ and $\vec{v}_e$, respectively, the charge density ρ and current density $\vec{J}$ are

$$\rho = n_iq_i + n_eq_e$$

and
$$\vec{J} = n_iq_i\,\vec{u}_i + n_eq_e\,\vec{u}_e$$

[1]We have

$$E = mc^2$$

and
$$p = mc$$

Therefore,
$$E^2 - p^2c^2 = m^2c^4 - c^2m^2c^2$$
$$= m^2c^4\left(1 - \frac{c^2}{c^2}\right)$$

Using the relation

$$m = \frac{m_0}{\sqrt{1 - c^2/c^2}}$$

we have
$$E^2 - p^2c^2 = m_0^2c^4$$

In the plasma (fluid), we have a large number of particles. Hence, we no more consider motion of individual particle and therefore, we use the velocity $\vec{u}$ for fluid instead of $\vec{v}$ for individual particle. When we neglect collisions (i.e., fully ionized plasma)[2] and viscosity (i.e., isotropic plasma), a set of equations is as the following.

$$\nabla \cdot \vec{E} = \frac{1}{\epsilon_0}(n_i q_i + n_e q_e) \tag{17.5}$$

$$\nabla \cdot \vec{E} = -\frac{\partial \vec{B}}{\partial t} \tag{17.6}$$

$$\nabla \cdot \vec{B} = 0 \tag{17.7}$$

$$\nabla \cdot \vec{B} = \frac{1}{c^2}\frac{\partial \vec{E}}{\partial t} + \mu_0(n_i q_i \vec{u}_i + n_e q_e \vec{u}_e) \tag{17.8}$$

$$m_j n_j\left[\frac{\partial \vec{u}_j}{\partial t} + (\vec{u}_j \cdot \nabla)\vec{u}_j\right] = q_j n_j[\vec{E} + \vec{u}_j \times \vec{B}] - \nabla p_j \tag{17.9}$$

$$\frac{\partial n_j}{\partial t} + \nabla \cdot (n_j \vec{u}_j) = 0 \tag{17.10}$$

$$P_j = C_j n_j^{\gamma_j} \tag{17.11}$$

where $c = 1/\sqrt{\epsilon_0 \mu_0}$ is the velocity of electromagnetic wave propagating through a medium having permittivity ϵ_0 and permeability μ_0, and j denotes the ion i or electron e.

The equations (17.5), (17.6), (17.7), (17.8) are Maxwell's equations. In a plasma, the ions and electrons comprising the plasma are the equivalent of the bound charges and currents. Since these charges move in a complicated way, it is impractical to try to lump their effects into two constants, ϵ and μ. Consequently, in the plasma physics, one generally works with the vacuum equations, in which ρ and $\vec{J}$ include all the charges and currents, both external and internal.

Equation (17.9) is the equation of motion where electromagnetic field is included and p denotes the pressure. Equations (17.10) (17.11) are the continuity equation and equation of state, respectively. Equations (17.9), (17.10), (17.11) depend on the species (ion or electron). There are 16 unknowns: n_i, n_e, p_i, p_e, $\vec{u}_i$, $\vec{u}_e$, $\vec{E}$ and $\vec{H}$ and 18 equations when we account each vector equation as three scalar equations. Two Maxwell's equations are superfluous as (17.5) and (17.7) can be obtained from the divergence of (17.8) and (17.6)[3]. A simultaneous solution

[2]As the collisions between charged particles and neutral particles are quite frequent whereas the collisions between charged particles themselves are rare.

[3]Taking divergence of equation (17.6), we have

$$\nabla \cdot (\nabla \times \vec{E}) = 0 = -\nabla \cdot \frac{\partial \vec{B}}{\partial t}$$

$$\frac{\partial}{\partial t}(\nabla \cdot \vec{B}) = 0$$

Thus, $\nabla \cdot \vec{B} = 0$ when it is initially zero. This is equation (17.7). Taking divergence of equation (17.8), we have

$$\nabla \cdot (\nabla \times \vec{B}) = 0 = \mu_0[q_i \nabla \cdot (n_i \vec{u}_i) + q_e \nabla \cdot (n_e \vec{u}_e) + \frac{1}{c^2}\nabla \cdot \frac{\partial \vec{E}}{\partial t}$$

Contd.

of 16 equations in 16 unknowns gives a self-consistent values of fields and motions in the fluid approximation.

2.1 Equation of State of Plasma

A fluid where the exchange of energy from the outside is not allowed, the pressure p and density ρ of the fluid are related through the equation of state

$$p = C\rho^{\gamma} \tag{17.12}$$

where C is constant and γ the ratio of specific heats C_p/C_v of fluid. For the particle density n, equation (17.12) can be written as

$$\begin{aligned} p &= C(mn)^{\gamma} \\ &= C_1\, n^{\gamma} \end{aligned} \tag{17.13}$$

where m is mass of a particle, and $C_1(= Cm^{\gamma})$ is constant. Thus, we have

$$\nabla p = C_1\, \gamma\, n^{\gamma-1}\, \nabla n$$

and therefore,

$$\frac{\nabla p}{p} = \gamma \frac{\nabla n}{n} \tag{17.14}$$

For an isothermal system, we have $p = nKT$. Thus,

$$\nabla p = \nabla(nKT) = KT\, \nabla n$$

and therefore,

$$\frac{\nabla p}{p} = \frac{\nabla n}{n} \tag{17.15}$$

Comparison of equation (17.15) with (17.14) tells that γ is 1 for an isothermal fluid. For an adiabatic compression, T also changes and therefore, the value of γ is larger than 1. For a system with N degrees of freedom, the γ is

$$\gamma = 1 + \frac{2}{N}$$

For a fluid, we have $\nabla p = \gamma\, KT\, \nabla n$ and the equation of motion is expressed as

$$mn\left[\frac{\partial \vec{u}}{\partial t} + (\vec{u}\cdot\nabla)\vec{u}\right] = qn[\vec{E} + \vec{u}\times\vec{B}] - \gamma KT \nabla n$$

Now, for the equation of continuity, we have

$$\nabla\cdot(n_i\vec{u}_i) = -\frac{\partial n_i}{\partial t}$$

$$\nabla\cdot(n_e\vec{u}_e) = -\frac{\partial n_e}{\partial t}$$

Using these relations here, we get

$$\mu_0\left[-q_i\frac{\partial n_i}{\partial t} - q_e\frac{\partial n_e}{\partial t}\right] + \frac{1}{v^2}\nabla\cdot\frac{\partial \vec{E}}{\partial t} = 0$$

$$\frac{\partial}{\partial t}[\nabla\cdot\vec{E} - \mu_0 c^2(n_i q_i + n_e q_e)] = 0$$

Thus, $(\nabla\cdot\vec{E} - (n_i q_i + n_e q_e)/\epsilon_0 = 0$ when it is initially zero. This is equation (17.5).

3. Linearisation of Equations

Consider a plasma where a particle has mass m, charge q and average velocity is u in an electric field $\vec{E}$ and magnetic field $\vec{B}$. Let the plasma be homogeneous and isotropic with the particle density n and kinetic temperature T. We have given a set of equations. This set of equations can be solved with the help of the process of linearisation of equations.

Here, we have discussed the procedure of linearisation of some equations.

(a) Equation of motion of particle

$$mn\left[\frac{\partial \vec{u}}{\partial t} + (\vec{u}\cdot\nabla)\vec{u}\right] = qn[\vec{E} + \vec{u}\times\vec{B}] - \gamma KT\nabla n \tag{17.16}$$

where K is Boltzmann constant, and γ the ratio of specific heats at constant pressure and constant volume.

(b) Equation of continuity

$$\frac{\partial n}{\partial t} + \nabla\cdot(n\vec{u}) = 0 \tag{17.17}$$

(c) Poisson equation

$$\nabla\cdot\vec{E} = \frac{e(n_i - n_e)}{\epsilon_0} \tag{17.18}$$

where n_i and n_e are densities of positive ions and electrons, respectively, and an ion (proton and an electron carry equal magnitude e of charge.

The procedure of linearisation is applicable only when the amplitude of each oscillating quantity is small, so that the terms having square or higher degrees of amplitude can be neglected. For linearisation, the first step is to separate the variable parameters into two parts: (i) an equilibrium part, which represents the state of matter in absence of oscillations and is indicated by the subscript 0, and (ii) the perturbation part due to oscillations, indicated by the subscript 1:

$$\begin{aligned} n &= n_0 + n_1 \\ \vec{u} &= \vec{u}_0 + \vec{u}_1 \\ \phi &= \phi_0 + \phi_1 \\ \vec{E} &= \vec{E}_0 + \vec{E}_1 \\ \vec{B} &= \vec{B}_0 + \vec{B}_1 \end{aligned} \tag{17.19}$$

Before perturbations, the plasma is homogeneous and at rest, so that

$$\nabla n_0 = \nabla\phi_0 = \vec{u}_0 = \vec{E}_0 = 0$$

$$\frac{\partial n_0}{\partial t} = 0 \tag{17.20}$$

3.1 Linearisation of Equation of Motion

In equilibrium, we have

$$mn_0\left[\frac{\partial \vec{u}_0}{\partial t} + (\vec{u}_0\cdot\nabla)\vec{u}_0\right] = qn_0[\vec{E}_0 + \vec{u}_0\times\vec{B}_0] - \gamma KT\nabla n_0$$

Using equations (17.19) and (17.20) in (17.16), we get

$$m(n_0 + n_1)\left[\frac{\partial}{\partial t}(\vec{u}_0 + \vec{u}_1) + \{(\vec{u}_0 + \vec{u}_1)\cdot\nabla\}(\vec{u}_0 + \vec{u}_1)\right]$$

$$= q(n_0 + n_1)[(\vec{E}_0 + \vec{E}_1) + (\vec{u}_0 + \vec{u}_1) \times (\vec{B}_0 + \vec{B}_1)] - \gamma KT\,\nabla(n_0 + n_1)$$

so that

$$m(n_0 + n_1)\left[\frac{\partial \vec{u}_1}{\partial t} + (\vec{u}_1 \cdot \nabla)\vec{u}_1\right] = q(n_0 + n_1)[(\vec{E}_1 + \vec{u}_1 \times (\vec{B}_0 + \vec{B}_1) - \gamma KT\,\nabla n_1 \quad (17.21)$$

Neglecting the square and higher degree terms in equation (17.21), we have

$$mn_0 \frac{\partial \vec{u}_1}{\partial t} = qn_0(\vec{E}_1 + \vec{u}_1 \times \vec{B}_0) - \gamma KT\,\nabla n_1$$

3.2 Linearisation of Equation of Continuity

In equilibrium, we have

$$\frac{\partial n_0}{\partial t} + \nabla \cdot (n_0 \vec{u}_0) = 0$$

Using equations (17.19) and (17.20) in (17.17), we get

$$\frac{\partial}{\partial t}(n_0 + n_1) + \nabla \cdot [(n_0 + n_1)(\vec{u}_0 + \vec{u}_1)] = 0$$

$$\frac{\partial n_1}{\partial t} n_0 \nabla \cdot \vec{u}_1 = 0$$

3.3 Linearisation of Poisson equation

We have

$$\nabla \cdot \vec{E} = \frac{e(n_i - n_e)}{\epsilon_0}$$

Using equations (17.19) and (17.20) in (17.18), we get

$$\nabla \cdot (\vec{E}_0 + \vec{E}_1) = \frac{e[n_{i0} - (n_{e0} + n_{e1})]}{\epsilon_0}$$

$$\nabla \cdot \vec{E}_1 = -\frac{en_{e1}}{\epsilon_0}$$

where we have used $n_i = n_{i0}$ as ions are stationary, $n_e = n_{e0} + n_{e1}$, and $n_{i0} = n_{e0}$ (as in the equilibrium, the neutrality is maintained).

3.4 Linearisation of Electron Current Density

Electron current density is expressed as

$$\vec{J} = -n_e e \vec{u}_e$$

Expressing $\vec{J} = \vec{J}_0 + \vec{J}_1$, $\vec{u}_e = \vec{u}_{e0} + \vec{u}_{e1}$, $n_e = n_{e0} + n_{e1}$ and consider that for plasma, $\vec{u}_{e0} = 0$, linearized form of the equation of current density is

$$\vec{J} = -n_{e0} e \vec{u}_{e1}$$

3.5 Perturbed Quantities

Perturbed quantities are considered oscillating, behaving sinusoidally as $e^{i(\vec{k}\cdot\vec{r}-\omega t)}$. Then, the time derivative $(\partial/\partial t)$ can be replaced by $-i\omega$, and the gradient ∇ by $i\vec{k}$.

$$\frac{\partial}{\partial t}e^{i(\vec{k}\cdot\vec{r}-\omega t)} = -i\omega e^{i(\vec{k}\cdot\vec{r}-\omega t)}$$

$$\frac{\partial}{\partial t} \to -i\omega$$

We have $\vec{k} = k_x\hat{x} + k_y\hat{y} + k_z\hat{z}$ and $\vec{r} = x\hat{x} + y\hat{y} + z\hat{z}$, so that $\vec{k}\cdot\vec{r} = k_x x + k_y y + k_z z$.

$$\begin{aligned}\nabla e^{i(\vec{k}\cdot\vec{r}-\omega t)} &= \left(\frac{\partial}{\partial x}\hat{x} + \frac{\partial}{\partial y}\hat{y} + \frac{\partial}{\partial z}\hat{z}\right)e^{i(k_x x+k_y y+k_z z-\omega t)}\\ &= i(k_x\hat{x} + k_y\hat{y} + k_z\hat{z})e^{i(k_x x+k_y y+k_z z-\omega t)}\\ &= i\vec{k}\,e^{i(k_x x+k_y y+k_z z-\omega t)}\end{aligned}$$

$$\nabla \to i\vec{k}$$

$$\frac{\partial}{\partial x} \to ik_x$$

4. Plasma Oscillations

In equilibrium position, charged particles in plasma are uniformly distributed in such a manner that neutrality is maintained everywhere. When the electrons in plasma are displaced relative to uniform background of ions, an electric field is developed in such a direction that it tries to pull the electrons back to restore neutrality. While returning back, owing to inertia, the electrons overshoot and the electric field is developed in reverse direction which tries to pull back the electrons to their positions of equilibrium. Thus, the electrons oscillate about their equilibrium positions with a characteristic frequency, known as the **plasma frequency**.

Because of low mass of electrons, these oscillations are so fast that massive ions are not capable to respond the oscillating field, generated by the oscillations of electrons. Hence, the ions may be treated as fixed. Expression for plasma frequency ωp can be derived in a simple manner using following assumptions:

(i) The plasma has a quite large extent.

(ii) There is no magnetic field, so that from Maxwell's equation, we have

$$\nabla \times \vec{E} = -\frac{\partial \vec{B}}{\partial t} = 0$$

and hence, $\vec{E} = -\nabla\phi$

where ϕ is scalar quantity.

(iii) The ions are fixed and form uniform background in space.

(iv) There are no thermal motions.

(v) Due to electric field generated in plasma, motions of electrons are along one direction, say, the x direction, so that we have

$$\nabla = \frac{\partial}{\partial x}\hat{x}$$

and $\vec{E} = E\hat{x}$

The equations of motion and continuity for electron are

$$m_e n_e\left[\frac{\partial \vec{u}_e}{\partial t}+(\vec{u}_e\cdot\nabla)\vec{u}_e\right]=-en_e\vec{E} \tag{17.22}$$

and
$$\frac{\partial n_e}{\partial t}+\nabla\cdot(n_e\vec{u}_e)=0 \tag{17.23}$$

Here, we are assuming collision-less, homogeneous and isotropic plasma. These are high frequency oscillations, generated due to deviation from neutrality and therefore, we can use Poisson equation for one-dimensional motion

$$\frac{\partial E}{\partial x}=\frac{e(n_i-n_e)}{\epsilon_0} \tag{17.24}$$

where n_i and n_e are densities of ions and electrons, respectively. When amplitude of each oscillating quantity is small, the set of equations (17.22) – (17.24) can be solved by using the procedure of linearisation, where the terms having square and higher powers of amplitude are neglected. For this, let us first separate the variable parameters into two parts:

(i) an equilibrium part, which represents the state of plasma in absence of oscillations, and is indicated by subscript 0, and

(ii) perturbation part due to oscillations, indicated by subscript 1:

$$n_e=n_0+n_1$$
$$\vec{u}_e=\vec{u}_0+\vec{u}_1$$
$$\vec{E}=\vec{E}_0+\vec{E}_1 \tag{17.25}$$

Before displacement of electrons, the plasma is homogeneous and at rest, so that

$$\nabla n_0=\vec{u}_0=\vec{E}_0=0$$

and
$$\frac{\partial n_0}{\partial t}=0 \tag{17.26}$$

Using equations (17.25) and (17.26) in (17.22) – (17.24) and on linearisation, we get

$$m_e\frac{\partial \vec{u}_1}{\partial t}=-e\vec{E}_1;$$

$$\frac{\partial n_1}{\partial t}+n_0\nabla\cdot\vec{u}_1=0;$$

$$\frac{\partial E_1}{\partial x}=-\frac{en_1}{\epsilon_0} \tag{17.27}$$

For convenience, we consider one-dimensional case and assume that oscillating quantities behave sinusoidally as $e^{i(kx-\omega t)}$. Then time derivative $(\partial/\partial t)$ can be replaced by $-i\omega$, and gradient ∇ by ik. From equations (17.27), we get

$$-im_e\omega u_1=-eE_1 \tag{17.28}$$

$$-i\omega n_1=-n_0iku_1 \tag{17.29}$$

$$ikE_1=-\frac{en_1}{\epsilon_0} \tag{17.30}$$

Equations (17.28) – (17.30) can be rearranged as

$$-im_e\omega u_1 = -e\left(\frac{-e}{ik\epsilon_0}\right)n_1$$

$$= \left(\frac{e^2}{ik\epsilon_0}\right)\left(\frac{n_0 k}{\omega}\right)u_1$$

$$= -i\left(\frac{n_0 e^2}{\omega\epsilon_0}\right)u_1$$

so that

$$\left(m_e\omega - \frac{n_0 e^2}{\omega\epsilon_0}\right)u_1 = 0$$

Since u_1 is not zero, we have

$$\omega^2 = \frac{n_0 e^2}{m_e\epsilon_0}$$

Thus, plasma frequency ω_p is

$$\omega_p = \left(\frac{n_0 e^2}{m_e\epsilon_0}\right)^{1/2} \qquad (17.31)$$

The plasma frequency depending on plasma density is also a fundamental parameter for plasma. Since the mass of electron is very low, plasma frequency is very high. The frequency f_p is

$$f_p = \frac{\omega_p}{2\pi} = \left(\frac{n_0 e^2}{4\pi^2 m_e\epsilon_0}\right)^{1/2}$$

In the SI system, charge of electron $e = 1.6 \times 10^{-19}$ C, mass of electron $m_e = 9.1 \times 10^{-31}$ kg, and $(1/4\pi\epsilon_0) = 8.99 \times 10^9$ N m^2/C^2. Thus, we have

$$f_p = \frac{1.6\times10^{-19}\sqrt{8.99\times10^9}}{\sqrt{\pi 9.1\times10^{-31}}}\sqrt{n_0}$$

$$= 8.975\sqrt{n_0}\ \text{Hz}$$

where n_0 is electron density per m^3. For plasma having electron density $n_0 = 10^{18}$ m^{-3}, we have

$$f_p = 8.975(10^{18})^{1/2}$$

$$= 9 \times 10^9\ \text{Hz} = 9\ \text{GHz}$$

Hence, the plasma frequency generally lies in the microwave region. Equation (17.31) shows that ω_p does not depend on k, and therefore, the group velocity $d\omega_p/dk$ is zero. Thus, no information transmits through these waves from one region to another.

Electromagnetic radiation of frequency smaller than plasma frequency ($\omega < \omega_p$), on incidence on plasma are reflected back, whereas the radiation of frequency larger than plasma frequency ($\omega > \omega_p$) transmit through the plasma. This property of plasma in ionosphere around the earth has been exploited for communication purpose. Obviously, in order to communicate with geostationary satellites, frequency of signal must be larger than the plasma frequency of ionosphere.

5. Electron Plasma Waves

In the above, we have assumed no thermal motions and found that electrons oscillate about their equilibrium positions with frequency ω_p which is independent of k. Thus, the group velocity is zero, and hence, there is no transmission of information from one place to another. When thermal motions are included, they provide propagation of plasma oscillations. Electrons moving with their thermal velocities into adjacent layers of plasma can carry information regarding the happening in the previous oscillating region.

Propagation of such information is known as **electron plasma waves**. Such waves can be obtained through inclusion of term $-\nabla p_e$, due to thermal motions of electrons, in the equation of motion. If we take all assumptions of preceding section except the assumption (iv), (i.e., here we include thermal motions), the equations of motion and continuity for electron are

$$m_e n_e \left[\frac{\partial \vec{u}_e}{\partial t} + (\vec{u}_e \cdot \nabla)\vec{u}_e\right] = -en_e\vec{E} - \nabla p_e \tag{17.32}$$

and

$$\frac{\partial n_e}{\partial t} + \nabla \cdot (n_e \vec{u}_e) = 0 \tag{17.33}$$

Here, we are assuming collision-less, homogeneous and isotropic plasma. These are high frequency oscillations which are generated due to deviation from neutrality and therefore, we can use Poisson equation for one-dimensional motion

$$\frac{\partial E}{\partial x} = \frac{e(n_i - n_e)}{\epsilon_0} \tag{17.34}$$

where n_i and n_e are densities of ions and electrons, respectively. When amplitude of each oscillating quantity is small, the set of equations (17.32) − (17.34) can be solved by using the procedure of linearisation, where the terms having square and higher powers of amplitude are neglected. For this, let us first separate variable parameters into two parts (one-dimensional case): (i) an equilibrium part, which represents the state of plasma in absence of oscillations, and is indicated by subscript 0 and (ii) perturbation part due to oscillations, indicated by subscript 1:

$$\begin{aligned} n_e &= n_0 + n_1 \\ u_e &= u_0 + u_1 \\ E &= E_0 + E_1 \end{aligned} \tag{17.35}$$

Before displacement of electrons, plasma is homogeneous and at rest, so that

$$\frac{\partial n_0}{\partial x} = u_0 = E_0 = 0$$

and

$$\frac{\partial n_0}{\partial t} = 0 \tag{17.36}$$

The term ∇p_e for one-dimensional space can be written as

$$\frac{\partial p_e}{\partial x} = \frac{\gamma p_e}{n_e}\frac{\partial n_e}{\partial_x} = \frac{\gamma n_e K T_e}{n_e}\frac{\partial n_e}{\partial x}$$

$$= \gamma KT_e \frac{\partial}{\partial x}(n_0 + n_1)$$

$$= \gamma KT_e \frac{\partial n_1}{\partial x} \tag{17.37}$$

For one-dimensional space we have

$$\gamma = 1 + \frac{2}{N} = 1 + \frac{2}{1} = 3$$

Using equations (17.35) – (17.37) in (17.32) – (17.34) and on linearisation, we get

$$m_e n_0 \frac{\partial u_1}{\partial t} = -en_0 E_1 - 3KT_e \frac{\partial n_1}{\partial x} \tag{17.38}$$

$$\frac{\partial n_1}{\partial t} + n_0 \frac{\partial u_1}{\partial x} = 0 \tag{17.39}$$

$$\frac{\partial E_1}{\partial x} = -\frac{en_1}{\epsilon_0} \tag{17.40}$$

For convenience, we consider one-dimensional case and assume that oscillating quantities behave sinusoidally as $e^{i(kx-\omega t)}$. Then time derivative $(\partial/\partial t)$ can be replaced by $-i\omega$, and gradient $\partial/\partial x$ by ik. From the equations (17.38) – (17.40), we get

$$-im_e n_0 \omega u_1 = -en_0 E_1 - 3KT_e ikn_1 \tag{17.41}$$

$$-i\omega n_1 = -n_0 iku_1 \tag{17.42}$$

$$ikE_1 = -\frac{en_1}{\epsilon_0} \tag{17.43}$$

Equations (17.41) – (17.43) can be rearranged as

$$-im_e n_0 \omega u_1 = -en_0\left(\frac{-e}{ik\epsilon_0}\right)n_1 - 3KT_e ikn_1$$

$$= -i\left(\frac{n_0 e^2}{k\epsilon_0} + 3KT_e k\right)n_1$$

$$= -i\left(\frac{n_0 e^2}{k\epsilon_0} + 3KT_e k\right)\frac{n_0 k}{\omega}u_1$$

so that $\left[\omega^2 - \frac{n_0 e^2}{m_3 \epsilon_0} - \frac{3KT_e k^2}{m_e}\right]u_1 = 0$

Since u_1 is not zero, we have

$$\omega^2 = \frac{n_0 e^2}{m_e \epsilon_0} + \frac{3KT_e k^2}{m_e}$$

$$= \omega_p^2 + \frac{3}{2}k^2 u_{th}^2$$

where $\omega_p (= \sqrt{n_0 e^2 / m_e \epsilon_0})$ is the plasma frequency and $u_{th} (= \sqrt{2KT_e / m_e})$ is thermal velocity in one-dimensional space. Figure 1 shows a plot of the dispersion relation – a variation of ω versus k.

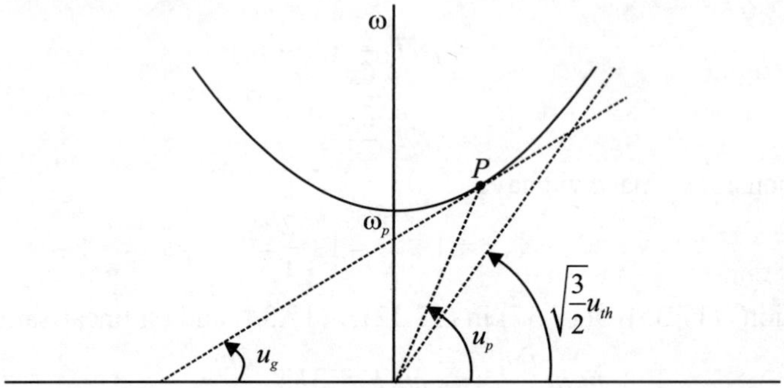

Fig. 1: Variation of ω versus *k* for electron plasma waves

At a point *P* on the curve, the slope of line joining the point and origin gives the phase velocity u_p, whereas tangent to the curve at *P* gives the group velocity u_g. We can write

$$\frac{\omega^2}{k^2} = \frac{\omega_p^2}{k^2} + \frac{3}{2}u_{th}^2$$

or

$$\frac{\omega}{k} = \sqrt{\frac{3}{2}}u_{th}\left(1 + \frac{2\omega_p^2}{3k^2u_{th}^2}\right)^{1/2} \equiv u_p$$

It shows that the phase velocity u_p is never smaller than $\sqrt{3/2}u_{th}$. When $k \to \infty$, the phase velocity $u_p \to \sqrt{3/2}u_{th}$. The group velocity u_g is obtained in the following manner

$$2\omega\,d\omega = \frac{3}{2}u_{th}^2 2k\,dk$$

Hence, the group velocity is

$$u_g = \frac{d\omega}{dk} = \frac{3}{2}\frac{k}{\omega}u_{th}^2$$

$$= \sqrt{\frac{3}{2}}u_{th}\left(1 + \frac{2\omega_p^2}{3k^2u_{th}^2}\right)^{-1/2}$$

It shows that group velocity u_g is never larger than $\sqrt{3/2}u_{th}$. When $k \to \infty$, group velocity $u_g \to \sqrt{3/2}\,u_{th}$.

When we move along the dispersion curve from a large value of *k* towards the point $k = 0$, the phase velocity varies from $\sqrt{3/2}\,u_{th}$ to infinite value whereas the group velocity varies from $\sqrt{3/2}\,u_{th}$ to zero value. Thus, the range of phase velocity is from $\sqrt{3/2}\,u_{th}$ to infinite whereas the range of group velocity is from zero to $\sqrt{3/2}\,u_{th}$. We can write

$$k = \sqrt{\frac{2}{3}}\frac{1}{u_{th}}(\omega^2 - \omega_p^2)^{1/2}$$

For $\omega < \omega_p$, *k* becomes imaginary. It shows that there is cut-off frequency $\omega = \omega_p$ above which an electron plasma wave can propagate. For imaginary *k*, the spatial dependence exp(*ikx*) shows an exponential attenuation. The depth δ through which an electron wave with $\omega < \omega_p$ can propagate can be obtained as

$$e^{ikx} = e^{-|k|x} = e^{-k/\delta}$$

or

$$\delta = |k|^{-1} = \sqrt{\frac{3}{2}}\frac{u_{th}}{(\omega_p^2 - \omega^2)^{1/2}}$$

6. Sound Waves

After discussion of electron waves, we want to discuss about ion waves. Before the discussion of ion waves, however, let us first discuss about the sound waves, which take place in an ordinary medium. Neglecting the viscosity, equation of motion is

$$\rho\left[\frac{\partial \vec{u}}{\partial t} + (\vec{u}\cdot\nabla)\vec{u}\right] = -\nabla p \tag{17.44}$$

and equations of continuity and state are

$$\frac{\partial \rho}{\partial t} + \nabla\cdot(\rho\vec{u}) = 0$$

and
$$p = C\rho^{\gamma} \tag{17.45}$$

where ρ is mass density. We have

$$\nabla p = \frac{\gamma p}{\rho}\nabla\rho \tag{17.46}$$

When amplitude of each oscillating quantity is small, the set of equations (17.44) – (17.46) can be solved by using the procedure of linearisation, where the terms having square and higher powers of amplitude are neglected. For this, let us first separate variable parameters into two parts: (i) an equilibrium part, which represents the state of matter in absence of oscillations, and is indicated by subscript 0 and (ii) perturbation part due to oscillations, indicated by subscript 1:

$$\rho = \rho_0 + \rho_1$$
$$\vec{u} = \vec{u}_0 + \vec{u}_1$$
$$p = p_0 + p_1 \tag{17.47}$$

Before displacement of particles, medium is homogeneous and at rest, so that

$$\nabla\rho_0 = \vec{u}_0 = \nabla p_0 = 0$$

and
$$\frac{\partial \rho_0}{\partial t} = \frac{\partial \vec{u}_0}{\partial t} = 0 \tag{17.48}$$

Using equations (17.46) – (17.48) in (17.44) and after linearisation, we get

$$\rho_0\frac{\partial \vec{u}_1}{\partial t} = -\frac{\gamma p_0}{\rho_0}\nabla\rho_1 \tag{17.49}$$

Using equations (17.47) – (17.48) in first of (17.45) and after linearisation, we get

$$\frac{\partial \rho_1}{\partial t} + \rho_0\nabla\cdot\vec{u}_1 = 0 \tag{17.50}$$

For convenience, we consider one-dimensional case and assume that oscillating quantities behave sinusoidally as $e^{i(kx-\omega t)}$. Then time derivative $(\partial/\partial t)$ can be replaced by $-i\omega$, and the gradient $(\partial/\partial x)$ by ik. Now, equations (17.49) and (17.50) become

$$-i\omega\rho_0 u_1 = -\frac{\gamma p_0}{\rho_0}ik\rho_1 \tag{17.51}$$

and
$$-\omega\rho_1 + \rho_0 k u_1 = 0 \tag{17.52}$$

On elimination of ρ, equations (17.51) and (17.52) can be rearranged as

$$\left[\omega^2 - \frac{\gamma p_0}{\rho_0}k^2\right]u_1 = 0$$

Since u_1 is not zero, we have

$$\omega^2 = \frac{\gamma p_0}{\rho_0} k^2$$

or
$$\frac{\omega}{k} = \left(\frac{\gamma p_0}{\rho_0}\right)^{1/2}$$

Using $p_0 = n_0 KT$ and $\rho_0 = n_0 m_p$, where m_p is mass of proton, we get

$$\frac{\omega}{k} = \left(\frac{\gamma KT}{m_p}\right)^{1/2} \equiv c_s$$

where c_s is velocity of sound wave in neutral gas. The waves are propagating from one layer to another through collisions among the gas molecules, and hence they are **pressure waves**. For one-dimensional case, $\gamma = 3$, and thus, the phase velocity v_p and group velocity v_g of wave are

$$v_p = \frac{\omega}{k} = \sqrt{\frac{3}{2}} u_{th}$$

and
$$v_g = \frac{d\omega}{dk} = \sqrt{\frac{3}{2}} u_{th}$$

where $u_{th} = \sqrt{2KT/m_p}$ is average thermal velocity of atoms in gas. It is interesting to find that in a fully ionized plasma (i.e., in absence of collisions), analogous waves, called *ion acoustic waves*, or simply, *ion waves* are found.

7. Ion Waves

In a fully ionized plasma, in absence of collisions, sound waves do not occur. Ions, being charged particles, can transmit vibrations to each other. Since the ions are massive particles, these vibrations are low frequency oscillations. In absence of magnetic field $\vec{B}$, from Maxwell's equation, we have

$$\nabla \times \vec{E} = -\frac{\partial \vec{B}}{\partial t} = 0$$

and hence,

$$\vec{E} = -\nabla \phi$$

where ϕ is scalar quantity. The equations of motion and continuity for an ion are

$$m_i n_i \left[\frac{\partial \vec{u}_i}{\partial t} + (\vec{u}_i \cdot \nabla)\vec{u}_i\right] = en_i \vec{E} - \nabla p_i \tag{17.53}$$

and
$$\frac{\partial n_i}{\partial t} + \nabla \cdot (n_i \vec{u}_i) = 0 \tag{17.54}$$

Using the equation of state $p_i = C\rho_i^{\gamma i}$, we have

$$\nabla p_i = \gamma_1 p_i \frac{\nabla n_i}{n_i}$$

$$= \frac{\gamma_i n_i KT_i}{n_i} \nabla n_i = \gamma_i KT_i \nabla n_i$$

Here, we have used $p_i = n_i KT_i$. Hence, equation (17.53) of motion can be written as

$$m_i n\left[\frac{\partial \vec{u}_i}{\partial t} + (\vec{u}_i \cdot \nabla)\vec{u}_i\right] = -en\nabla\phi - \gamma_i KT_i \nabla n \tag{17.55}$$

where we have used the plasma approximation $n_i = n_e = n$, and therefore, we do not use Poisson equation. When amplitude of each oscillating quantity is small, let us linearize equation (17.55) by neglecting the terms having square and higher powers of amplitude. For this, let us first separate variable parameters into two parts: (i) an equilibrium part, which represents the state of plasma in absence of oscillations, and is indicated by subscript 0 and (ii) perturbation part due to oscillations, indicated by subscript 1:

$$\begin{aligned} n &= n_0 + n_1 \\ \vec{u}_i &= \vec{u}_0 + \vec{u}_1 \\ \phi &= \phi_0 + \phi_1 \end{aligned} \tag{17.56}$$

Before displacement of electrons, plasma is homogeneous and at rest, so that

$$\nabla n_0 = \vec{u}_0 = \nabla\phi_0 = 0$$

and

$$\frac{\partial n_0}{\partial t} = 0 \tag{17.57}$$

Using equations (17.56) and (17.57) in (17.55) and on linearisation, we get

$$m_i n_0 \frac{\partial \vec{u}_1}{\partial t} = -en_0\,\nabla\phi_1 - \gamma_i KT\,\nabla n_1 \tag{17.58}$$

Fluid drift parallel to $\vec{B}$ is the same as in absence of $\vec{B}$. For electric field $\vec{E} = -\nabla\phi = -\nabla(\phi_0 + \phi_1) = -\nabla\phi_1$, density of mass-less electrons is (mass of electron may be considered negligible in comparison to that of proton)

$$\begin{aligned} n_e = n &= n_0 \exp\left(\frac{e\phi_1}{KT_e}\right) \\ &= n_0\left(1 + \frac{e\phi_1}{KT_e} + \ldots\right) \\ &= n_0\left(1 + \frac{e\phi_1}{KT_e}\right) \end{aligned}$$

We have neglected higher order terms, and we have

$$n - n_0 = n_0 \frac{e\phi_1}{KT_e}$$

or

$$n_1 = n_0 \frac{e\phi_1}{KT_e} \tag{17.59}$$

Equation (17.59) can be written as

$$m_i n_0 \frac{\partial \vec{u}_1}{\partial t} = -en_0\nabla\phi_1 - \frac{\gamma_i n_0 e T_i}{T_e}\nabla\phi_1 \tag{17.60}$$

Using equations (17.56) – (17.57) in (17.54) and on linearisation, we get

$$\frac{\partial n_1}{\partial t} + n_0 \tilde{N} \cdot \vec{u}_1 = 0 \tag{17.61}$$

For convenience, we consider one-dimensional case and assume that oscillating quantities behave sinusoidally as $e^{i(kx-\omega t)}$. Then time derivative $(\partial/\partial t)$ can be replaced by $-i\omega$, and gradient ∇ by ik. From equations (17.59) – (17.61), we get

$$n_1 = n_0 \frac{e\phi_1}{KT_e} \tag{17.62}$$

$$-im_i n_0 \omega u_1 = -en_0 ik\phi_1 - \frac{\gamma_i n_0 e T_i}{T_e} ik\phi_1 \tag{17.63}$$

$$-i\omega n_1 + n_0 iku_1 = 0 \tag{17.64}$$

Equations (17.62) – (17.64) can be rearranged as

$$\begin{aligned} -im_i n_0 \omega u_1 &= -iken_0\left(1 + \frac{\gamma_i T_i}{T_e}\right)\phi_1 \\ &= -iken_0\left(1 + \frac{\gamma_i T_i}{T_e}\right)\frac{KT_e}{n_0 e} n_1 \\ &= -ikK(T_e + \gamma_i T_i)\frac{n_0 k}{\omega} u_1 \end{aligned}$$

so that

$$\left[\omega^2 - \frac{k^2 K}{m_i}(T_e + \gamma_i T_i)\right] u_1 = 0$$

Since u_1 is not equal to zero, we have

$$\omega^2 = \frac{k^2 K}{m_i}(T_e + \gamma_i T_i)$$

or

$$\frac{\omega}{k} = \left(\frac{KT_e + \gamma_i KT_i}{m_i}\right)^{1/2} \equiv v_s \tag{17.65}$$

This is dispersion relation for ion acoustic waves. Here, v_s is sound speed in plasma. Obviously, v_s is equal to the phase velocity as well as group velocity of wave. When ion temperature T_i tends to zero, ion wave still exist and its velocity is

$$v_s = \left(\frac{KT_e}{m_i}\right)^{1/2}$$

This does happen in neutral gas.

8. Validity of Plasma Approximation

In preceding section, in derivation of the velocity of ion waves, we have used neutrality condition $n_i = n_e$, and have taken finite value for electric field $\vec{E}$. It is in violation of Poisson equation. For estimating error introduced in the process, let us assume $n_i \neq n_e$ and consider Poisson equation for hydrogen plasma

$$\nabla \cdot \vec{E} = \frac{e(n_i - n_e)}{\epsilon_0}$$

In absence of $\vec{B}$, we have $\vec{E} = -\nabla\phi$, and Poisson equation is

$$-\nabla^2\phi = \frac{e(n_i - n_e)}{\epsilon_0} \tag{17.66}$$

When amplitude of each oscillating quantity is small, let us linearize equation (17.66) by neglecting the terms having square and higher powers of amplitude. For this, let us first separate variable parameters into two parts: (i) an equilibrium part, which represents the state of plasma in absence of oscillations, and is indicated by subscript 0 and (ii) perturbation part due to oscillations, indicated by subscript 1:

$$\begin{aligned} n_i &= n_{i0} + n_{i1} \\ n_e &= n_{e0} + n_{e1} \\ \phi &= \phi_0 + \phi_1 \end{aligned} \tag{17.67}$$

Using equation (17.67) in (17.66), we have

$$-\nabla^2(\phi_0 + \phi_1) = \frac{e(n_{i0} + n_{i1} - n_{e0} - n_{e1})}{\epsilon_0} \tag{17.68}$$

Before displacement of electrons, plasma is homogeneous and at rest, so that

$$-\nabla^2\phi_0 = \frac{e(n_{i0} - n_{e0})}{\epsilon_0} \tag{17.69}$$

Subtracting equation (17.69) from (17.68), we get

$$-\nabla^2\phi_1 = \frac{e(n_{i1} - n_{e1})}{\epsilon_0} \tag{17.70}$$

Fluid drift parallel to $\vec{B}$ is the same as in absence of $\vec{B}$. The density of mass-less electrons is (mass of an electron may be considered as negligible in comparison to that of proton)

$$n_e = n_0 \exp\left(\frac{e\phi}{KT_e}\right)$$

$$n_{e0} + n_{e1} = n_0 \exp\left(\frac{e(\phi_0 + \phi_1)}{KT_e}\right) \tag{17.71}$$

We have

$$n_{e0} = n_0 \exp\left(\frac{e\phi_0}{KT_e}\right) \tag{17.72}$$

Subtracting equation (17.72) from (17.71), we get

$$\begin{aligned} n_{e1} &= n_0 \exp\left(\frac{e(\phi_0 + \phi_1)}{KT_e}\right) - n_0 \exp\left(\frac{e\phi_0}{KT_e}\right) \\ &= n_0 \exp\left(\frac{e\phi_0}{KT_e}\right)\left[1 + \frac{e\phi_1}{KT_e} + \ldots - 1\right] \\ &= n_{e0}\frac{e\phi_1}{KT_e} \end{aligned}$$

Using this expression in equation (17.70), we get

$$-\nabla^2\phi_1 + \frac{n_{e0}}{\epsilon_0}\frac{e^2\phi_1}{KT_e} = \frac{en_{i1}}{\epsilon_0} \tag{17.73}$$

For convenience, we consider one-dimensional case and assume that oscillating quantities behave sinusoidally as $e^{i(kx-\omega t)}$. Then time derivative $(\partial/\partial t)$ can be replaced by $-i\omega$, and gradient $\partial/\partial x$ by ik. From equation (17.73), we get

$$k^2\phi_1 + \frac{en_{e0}}{\epsilon_0}\frac{e\phi_1}{KT_e} = \frac{en_{i1}}{\epsilon_0}$$

$$\phi_1\left(k^2 + \frac{n_{e0}e^2}{KT_e\epsilon_0}\right) = \frac{en_{i1}}{\epsilon_0}$$

$$\phi_1\left(k^2 + \frac{1}{\lambda_D^2}\right) = \frac{en_{i1}}{\epsilon_0}$$

$$\phi_1(k^2\lambda_D^2 + 1) = \frac{en_{i1}\lambda_D^2}{\epsilon_0} \quad (17.74)$$

Equations (17.63), (17.64) and (17.74) can be rearranged as

$$-im_in_0\omega u_1 = -iken_0\left[1+\frac{\gamma_iT_i}{T_e}\right]\phi_1$$

$$= -iken_0\left[1+\frac{\gamma_iT_i}{T_e}\right]\left[\frac{e\lambda_D^2}{(1+k^2\lambda_D^2)\epsilon_0}\right]n_{i1}$$

$$= -iken_0\left[1+\frac{\gamma_iT_i}{T_e}\right]\left[\frac{e}{(1+k^2\lambda_D^2)\epsilon_0}\right]\left[\frac{KT_e\epsilon_0}{n_{e0}e^2}\right]\left[\frac{n_0k}{\omega}\right]u_1$$

so that $\left[\omega^2 - \frac{k^2K}{m_i}(T_e+\gamma_iT_i)\frac{1}{(1+k^2\lambda_D^2)}\right]u_1 = 0$

Since u_1 is not equal to zero, we have

$$\omega^2 = \frac{k^2K}{m_i}(T_e+\gamma_iT_i)\frac{1}{(1+k^2\lambda_D^2)}$$

or

$$\frac{\omega}{k} = \left(\frac{KT_e+\gamma_iKT_i}{m_i}\right)^{1/2}\frac{1}{\sqrt{(1+k^2\lambda_D^2)}} \quad (17.75)$$

Comparison of equations (17.75) and (17.65) shows that in our plasma approximation ($n_i = n_e$), an error of order of $k^2\lambda_D^2$ $(= \{2\pi\lambda_D/\lambda\}^2)$ is introduced. Since λD in comparison to λ is very small in most of the cases, plasma approximation is valid, except for the cases of shortest wavelengths.

9. Some Definitions for Waves

In the discussion so far, we did not account for the magnetic field $\vec{B}$. In presence of $\vec{B}$, several other types of waves are possible. Let us first introduce some terminology for waves.

(i) The words 'parallel' and 'perpendicular' are used to denote the direction of $\vec{k}$ relative to unperturbed magnetic field $\vec{B}_0$.

(ii) The words 'longitudinal' and 'transverse' are used to denote the direction of $\vec{k}$ relative to perturbed (oscillating) electric field $\vec{E}_1$.

(iii) The wave is termed 'electrostatic' when oscillating magnetic field $\vec{B}_1$ is zero.

(iv) The wave is termed 'electromagnetic' when oscillating magnetic field $\vec{B}_1$ is finite (non-zero)

Let us consider Maxwell's equation

$$\nabla \times \vec{E} = -\frac{\partial \vec{B}}{\partial t} \tag{17.76}$$

and resolve $\vec{E}$ and $\vec{B}$ into two parts: (i) an equilibrium part, which represents the state of plasma in absence of oscillations, and indicated by subscript 0 and (ii) perturbation part due to oscillations, indicated by subscript 1:

$$\vec{E} = \vec{E}_0 + \vec{E}_1$$

and

$$\vec{B} = \vec{B}_0 + \vec{B}_1 \tag{17.77}$$

Using equation (17.77) in (17.76), we have

$$\nabla \times (\vec{E}_0 + \vec{E}) = -\frac{\partial \vec{B}_0}{\partial t} - \frac{\partial \vec{B}_1}{\partial t}$$

Since the equilibrium parts do not depend on time and space coordinates, we have

$$\nabla \times \vec{E}_1 = -\frac{\partial \vec{B}_1}{\partial t} \tag{17.78}$$

Let us assume that perturbation parts behave sinusoidally as $e^{i(\vec{k}\cdot\vec{r}-\omega t)}$, then we replace time derivative $(\partial/\partial t)$ by $-i\omega$, and gradient ∇ by $i\vec{k}$. Thus, we have

$$i\vec{k} \times \vec{E}_1 = -(-i\omega)\vec{B}_1$$

or

$$\vec{k} \times \vec{E}_1 = \omega \vec{B}_1$$

When a wave is longitudinal, i.e., $\vec{k}$ is parallel to $\vec{E}_1$, we have $\vec{k} \times \vec{E}_1 = 0$, and $\vec{B}_1 = 0$ showing that the wave is electrostatic also. When a wave is transverse, i.e., $\vec{k}$ is perpendicular to $\vec{E}_1$, we have $\vec{k} \times \vec{E}_1 \neq 0$, and $\vec{B}_1 \neq 0$ showing that the wave is electromagnetic also.

In general, $\vec{k}$ may have an arbitrary angle to $\vec{B}_0$ and $\vec{E}_1$, then the wave has a combination of principal modes discussed above in (i) – (iv).

10. Electrostatic Electron Waves Perpendicular to $\vec{B}$

In equilibrium position, charged particles in plasma are uniformly distributed in such a manner that neutrality is maintained. When electrons in a plasma are displaced relative to uniform background of ions, an electric field is developed in such a direction that it tries to pull electrons back to restore the neutrality of plasma. While returning back, owing to inertia, the electrons overshoot and now electric field is developed in reverse direction which tries to pull back electrons to their equilibrium positions.

Thus, the electrons oscillate about their equilibrium positions. Because of low mass of electrons, their oscillations are so fast that massive ions are not capable to respond the oscillating field, generated by oscillations of electrons. Hence, the ions may be treated as fixed forming uniform background. Further, owing to the presence of magnetic field, the motions of electrons are not linear. We assume also that there are no thermal motions ($T_e = 0$). The equations of motion and continuity for electrons are

$$m_e n_e \left[\frac{\partial \vec{u}_e}{\partial t} + (\vec{u}_e \cdot \nabla)\vec{u}_e \right] = -en_e(\vec{E} + \vec{u}_e \times \vec{B}) \tag{17.79}$$

and
$$\frac{\partial n_e}{\partial t} + \nabla \cdot (n_e \vec{u}_e) = 0 \tag{17.80}$$

Here, we are assuming collision-less, homogeneous, isotropic and cold plasma. These are high frequency oscillations which are generated due to deviation from neutrality and therefore, we can use Poisson equation

$$\nabla \cdot \vec{E} = \frac{e(n_i - n_e)}{\epsilon_0} \tag{17.81}$$

where n_i and n_e are densities of ions and electrons, respectively. When amplitude of each oscillating quantity is small, the set of equations (17.79) – (17.81) can be solved by using the procedure of linearisation, where the terms having square and higher powers of amplitude are neglected. For this, let us first separate variable parameters into two parts: (i) an equilibrium part, which represents the state of plasma in absence of oscillations, and is indicated by subscript 0 and (ii) perturbation part due to oscillations, indicated by subscript 1:

$$n_e = n_0 + n_1$$
$$\vec{u}_e = \vec{u}_0 + \vec{u}_1$$
$$\vec{E} = \vec{E}_0 + \vec{E}_1 \tag{17.82}$$

Since we want to address an electrostatic wave, perturbation of $\vec{B}$ is not considered (i.e., $\vec{B} = \vec{B}_0$). Before displacement of electrons, plasma is homogeneous and at rest, so that

$$\nabla n_0 = \vec{u}_0 = \vec{E}_0 = 0 \text{ and } \frac{\partial n_0}{\partial t} = 0 \tag{17.83}$$

Using equations (17.82) and (17.83) in (17.79) – (17.81) and on linearisation, we get

$$m_e \frac{\partial \vec{u}_1}{\partial t} = -e(\vec{E}_1 + \vec{u}_1 \times \vec{B}_0) \tag{17.84}$$

$$\frac{\partial n_1}{\partial t} + n_0 \nabla \cdot \vec{u}_1 = 0 \tag{17.85}$$

$$\nabla \cdot \vec{E}_1 = -\frac{en_1}{\epsilon_0} \tag{17.86}$$

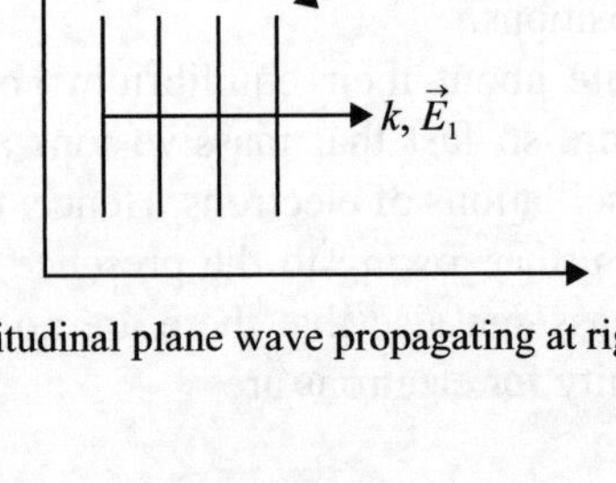

Fig. 2: A longitudinal plane wave propagating at right angle to $\vec{B}_0$

Here, we consider only longitudinal waves with $\vec{k}$ parallel $\vec{E}_1$. Let us consider x-axis along $\vec{k}$ and $\vec{E}_1$, and z-axis along $\vec{B}_0$ (Fig. 2). Hence, $\vec{k} = k\,\hat{x}$; $\vec{E}_1 = E_1\,\hat{x}$; $\vec{B}_0 = B_0\,\hat{z}$; $\vec{u}_1 = u_x\hat{x} + u_y\hat{y} + u_z\,\hat{z}$. Equation (17.84) can be expressed as

$$m_e \frac{\partial u_x}{\partial t} = -eE_1 - eu_y B_0 \tag{17.87}$$

and

$$m_e \frac{\partial u_y}{\partial t} = eu_x B_0 \tag{17.88}$$

We assume that oscillating quantities behave sinusoidally as $e^{i(\vec{k}\cdot\vec{r}-\omega t)}$. Then time derivative $(\partial/\partial t)$ can be replaced by $-i\omega$, and gradient ∇ by $i\vec{k}$. From equations (17.85) – (17.88), we get

$$-i\omega n_1 + n_0 iku_x = 0 \tag{17.89}$$

$$ikE_1 = -4\pi e n_1 \tag{17.90}$$

$$-im_e\omega u_x = -eE_1 - eu_y B_0 \tag{17.91}$$

$$-im_e\omega u_y = eu_x B_0 \tag{17.92}$$

Using u_y from equation (17.92) in (17.91), we get

$$-im_e\omega u_x = -eE_1 - eB_0\left(\frac{eB_0}{-im_e\omega}\right)u_x$$

Thus,

$$u_x = \frac{eE_1}{im_e\omega} + \frac{e^2B_0^2}{\omega^2 m_e^2}u_x$$

$$= \frac{eE_1}{im_e\omega} + \frac{\omega_0^2}{\omega^2}u_x$$

giving,

$$u_x\left(1 - \frac{\omega_c^2}{\omega^2}\right) = \frac{eE_1}{im_e\omega}$$

or

$$u_x = \frac{eE_1}{im_e\omega}\left(1 - \frac{\omega_c^2}{\omega^2}\right)^{-1} \tag{17.93}$$

where $\omega_c = eB_0/m_e$ is cyclotron frequency. The u_x obviously tends to infinite when $\omega \to \omega_c$. Equations (17.89), (17.90), (17.93) can be rearranged as

$$ikE_1 = -e\left(\frac{n_0 k}{\omega\epsilon_0}\right)u_x$$

$$= -\left(\frac{n_0 ek}{\omega\epsilon_0}\right)\frac{eE_1}{im_e\omega}\left(1 - \frac{\omega_c^2}{\omega^2}\right)^{-1}$$

so that

$$\left[1 - \frac{\omega_c^2}{\omega^2} - \frac{n_0 e^2}{m_e\epsilon_0}\frac{1}{\omega^2}\right]E_1 = 0$$

Since E_1 is not zero, we have

$$1 - \frac{\omega_c^2}{\omega^2} = \frac{4\pi n_0 e^2}{m_e}\frac{1}{\omega^2}$$

or

$$\omega^2 = \omega_p^2 + \omega_c^2 \equiv \omega_h^2 \tag{17.94}$$

where $\omega_p = (n_0 e^2/m_e \epsilon_0)^{1/2}$ is plasma frequency. The frequency ω_h is known as **upper hybrid frequency**. Electrostatic electron waves perpendicular to $\vec{B}$ have this frequency ω_h whereas electron waves parallel to $\vec{B}$ have frequency ω_p. The group velocity is zero, as the thermal motion is not considered. When magnetic field goes to zero, ω_c goes to zero and we have plasma oscillations with frequency ω_p. When electron density goes to zero, ω_p goes to zero and we have a simple Larmor gyration. Existence of upper hybrid frequency has been verified experimentally.

11. Electrostatic Ion Waves Perpendicular to $\vec{B}$

Let us investigate ion acoustic wave where $\vec{k}$ is perpendicular to $\vec{B}_0$. Here, we consider that angle between $\vec{k}$ and $\vec{B}_0$ is almost $\pi/2$ so that $\vec{k} \cdot \vec{B} \neq 0$. The geometry is shown in Fig. 3.

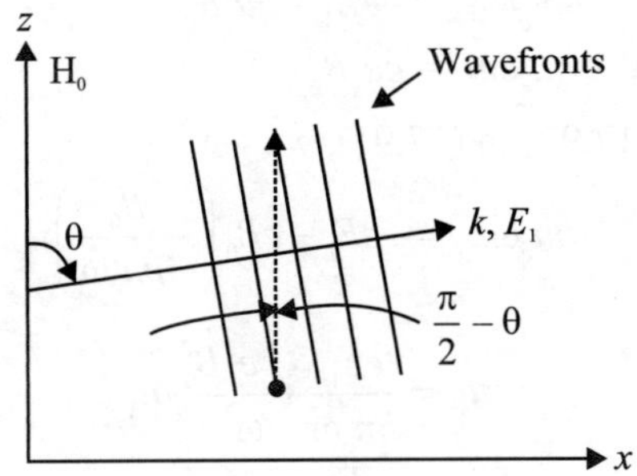

Fig. 3: Electrostatic ion cyclotron wave propagating nearly at right angle to $\vec{B}_0$

We consider electrostatic waves with $\vec{B}_1 = 0$, and therefore, $\vec{k} \times \vec{E}_1 = 0$. The wave is longitudinal. Further, $\nabla \times \vec{E}_1 = 0$ and therefore, $\vec{E}_1 = -\nabla\phi = -(\partial\phi/\partial x)\hat{x}$, where ϕ is scalar quantity. We assume also that there are no thermal motions ($T_i = 0$). The equations of motion and continuity for ions are

$$m_i n_i \left[\frac{\partial \vec{u}_i}{\partial t} + (\vec{u}_i \cdot \nabla)\vec{u}_i \right] = e n_i \left(-\frac{\partial \phi}{\partial x}\hat{x} + \vec{u}_i \times \vec{B}_0 \right) \tag{17.95}$$

and
$$\frac{\partial n_i}{\partial t} + \nabla \cdot (n_i \vec{u}_i) = 0 \tag{17.96}$$

Here, we have used plasma approximation $n_i = n_e = n$, and therefore, we do not use Poisson equation. When amplitude of each oscillating quantity is small, let us linearize equations by neglecting the terms having square and higher powers of amplitude. For this, let us first separate variable parameters into two parts: (i) an equilibrium part, which represents the state of plasma in absence of oscillations, and is indicated by subscript 0 and (ii) perturbation part due to oscillations, indicated by subscript 1:

$$\begin{aligned} \vec{u}_i &= \vec{u}_0 + \vec{u}_1 \\ \phi &= \phi_0 + \phi_1 \\ n_i &= n_0 + n_1 \end{aligned} \tag{17.97}$$

At equilibrium, we have

$$\nabla n_0 = \vec{u}_0 = \nabla \phi_0 = 0$$

and
$$\frac{\partial n_0}{\partial t} = 0 \tag{17.98}$$

Using equations (17.97) and (17.98) in (17.95) and on linearisation, we get

$$m_i \frac{\partial \vec{u}_1}{\partial t} = -e\frac{\partial \phi_1}{\partial x}\hat{x} + e\vec{u}_1 \times \vec{B}_0 \tag{17.99}$$

For $\vec{u}_1 = u_x\hat{x} + u_y\hat{y} + u_z\hat{z}$, equation (17.99) can be expressed as following.

$$m_i \frac{\partial \vec{u}_x}{\partial t} = -e\frac{\partial \phi_1}{\partial x} + eu_y B_0 \tag{17.100}$$

and
$$m_i \frac{\partial u_y}{\partial t} = -eu_x B_0 \tag{17.101}$$

For electric field $\vec{E} = -\nabla\phi = -\nabla(\phi_0 + \phi_1) = -\nabla\phi_1$, density of mass-less electrons is (mass of an electron may be treated as negligible in comparison to that of proton)

$$n_e = n = n_0 \exp\left(\frac{e\phi_1}{KT_e}\right)$$

$$= n_0\left(1 + \frac{e\phi_1}{KT_e} + ...\right)$$

$$= n_0\left(1 + \frac{e\phi_1}{KT_e}\right)$$

Neglecting higher order terms, we have

$$n - n_0 = n_0 \frac{e\phi_1}{KT_e}$$

or
$$n_{e1} = n_0 \frac{e\phi_1}{KT_e}$$

Under plasma approximation, we have $n_e = n_i$. Further, we have $n_{e0} = n_{i0}$ and therefore $n_{e1} = n_{i1}$. So, under plasma approximation, we get

$$n_1 = n_{i1} = n_0 \frac{e\phi_1}{KT_e} \tag{17.102}$$

Using equations (17.97) and (17.98) in (17.96) and on linearisation, we get

$$\frac{\partial n_1}{\partial t} + n_0 \nabla \cdot \vec{u}_1 = 0 \tag{17.103}$$

We consider only longitudinal waves with $\vec{k}$ parallel to $\vec{E}_1$. Let us consider x-axis along $\vec{k}$ and $\vec{E}_1$, and z-axis along $\vec{B}_0$. Hence, $\vec{k} = k\hat{x}$; $\vec{B}_0 = B_0\hat{z}$. We assume that oscillating quantities behave sinusoidally as $e^{i(\vec{k}\cdot\vec{r}-\omega t)}$. Then time derivative $(\partial/\partial t)$ can be replaced by $-i\omega$, and gradient ∇ by $i\vec{k}$. From equations (17.100) – (17.103), we get

$$-i\omega m_i u_x = -eik\phi_1 + eu_y B_0 \tag{17.104}$$

$$-i\omega m_i u_y = -eu_x B_0 \tag{17.105}$$

$$n_1 = n_0 \frac{e\phi_1}{KT_e} \tag{17.106}$$

$$-i\omega n_1 + n_0 iku_x = 0 \tag{17.107}$$

Using u_y from equation (17.105) in (17.104), we get

$$-i\omega m_i u_x = -eik\phi_1 + eB_0\left(\frac{eB_0}{i\omega m_i}\right)u_x$$

Hence,

$$u_x = \frac{ek\phi_1}{\omega m_i} + \left(\frac{e^2B_0^2}{\omega^2 m_i^2}\right)u_x$$

$$= \frac{ek\phi_1}{\omega m_i} + \frac{\Omega_c^2}{\omega^2}u_x$$

giving,

$$u_x\left(1 - \frac{\Omega_c^2}{\omega^2}\right) = \frac{ek\phi_1}{\omega m_i}$$

or

$$u_x = \frac{ek\phi_1}{\omega m_i}\left(1 - \frac{\Omega_c^2}{\omega^2}\right)^{-1} \tag{17.108}$$

where $\Omega_c = eB_0/m_i$ is ion cyclotron frequency. Equations (17.106) – (17.108) can be rearranged as

$$u_x = \frac{ek}{\omega m_i}\left(1 - \frac{\Omega_c^2}{\omega^2}\right)^{-1}\left(\frac{KT_e}{n_0 e}\right)n_1$$

$$u_x\left(1 - \frac{\Omega_c^2}{\omega^2}\right) = \frac{ek}{\omega m_i}\left(\frac{KT_e}{n_0 e}\right)\left(\frac{n_0 k}{\omega}\right)u_x$$

Since u_x is not zero, we have

$$\left(1 - \frac{\Omega_c^2}{\omega^2}\right) = \frac{k^2 KT_e}{\omega^2 m_i}$$

or

$$\omega^2 - \Omega_c^2 = k^2 v_s^2$$

where $v_s = \sqrt{KT_e/m_i}$ is velocity of ion waves. This is dispersion relation for electrostatic ion waves. Electrostatic ion waves, which also have been detected experimentally.

12. Lower Hybrid Frequency

In the preceding section, we have considered that angle between $\vec{k}$ and $\vec{B}_0$ is almost $\pi/2$ so that $\vec{k} \cdot \vec{B}_0$. When angle between $\vec{k}$ and $\vec{B}_0$ is exactly $\pi/2$, the electrons are not allowed to preserve neutrality by flowing along the lines of force, and hence they also obey the equation of motion. The equations of motion and continuity for ions and electrons are

$$m_i n_i\left[\frac{\partial \vec{u}_i}{\partial t} + (\vec{u}_i \cdot \nabla)\vec{u}_i\right] = en_i(-\nabla\varphi + \vec{u}_i \times \vec{B}_0) \tag{17.109}$$

$$\frac{\partial \vec{u}_1}{\partial t} + \nabla \cdot (n_i \vec{u}_i) = 0 \tag{17.110}$$

$$m_e n_e \left[\frac{\partial \vec{v}_e}{\partial t} + (\vec{v}_e \cdot \nabla)\vec{u}_e \right] = -en_e(-\nabla\phi + \vec{v}_e \times \vec{B}_0) \tag{17.111}$$

$$\frac{\partial n_e}{\partial t} + \nabla \cdot (n_e \vec{u}_e) = 0 \tag{17.112}$$

Equations (17.109) and (17.110) are solved in the preceding section and we have

$$u_{ix} = \frac{ek\phi_1}{\omega m_i}\left(1 - \frac{\Omega_c^2}{\omega^2}\right)^{-1}$$

and
$$n_{i1} = \frac{n_0 k}{\omega} u_{ix} \tag{17.113}$$

Solution of the equations (17.111) and (17.112) can be obtained in a similar manner by replacing e by $-e$, m_i by m_e, Ω_c by $-\omega_c$ in equation (17.113) as

$$u_{ex} = -\frac{ek\phi_1}{\omega m_e}\left(1 - \frac{\omega_c^2}{\omega^2}\right)^{-1}$$

and
$$n_{e1} = \frac{n_0 k}{\omega} u_{ex} \tag{17.114}$$

The plasma approximation $n_i = n_e = n_0$ requires $n_{i1} = n_{e1}$ and therefore, we have (using second equations of 17.113 and 17.114)

$$\frac{n_0 k}{\omega} u_{ix} = \frac{n_0 k}{\omega} u_{ex}$$

or
$$u_{ix} = u_{ex}$$

Using first equations of 17.113 and 17.114 here, we have

$$\frac{ek\phi_1}{\omega m_i}\left(1 - \frac{\Omega_c^2}{\omega^2}\right)^{-1} = -\frac{ek\phi_1}{\omega m_e}\left(1 - \frac{\omega_c^2}{\omega^2}\right)^{-1}$$

$$m_i\left(1 - \frac{\Omega_c^2}{\omega^2}\right) = -m_e\left(1 - \frac{\omega_c^2}{\omega^2}\right)$$

$$\omega^2(m_i + m_e) = m_e\omega_c^2 + m_i\Omega_c^2 = \frac{e^2 B_0^2}{c^2}\left(\frac{1}{m_i} + \frac{1}{m_e}\right)$$

giving,
$$\omega^2 = \frac{e^2 B_0^2}{c^2 m_i m_e} = \Omega_c \omega_c$$

or
$$\omega = \sqrt{\Omega_c \omega_c} \equiv \omega_l$$

This frequency ω_l is known as **lower hybrid frequency**.

13. Electromagnetic Waves with $\vec{B}_0 = 0$

Let us consider Maxwell's equations

$$\nabla \times \vec{E} = -\frac{\partial \vec{B}}{\partial t}$$

and

$$\nabla \times \vec{B} = \frac{1}{c^2}\frac{\partial \vec{E}}{\partial t} + \mu_0 \vec{J}$$

For linearization of these equations, let us first separate variable parameters into two parts: (i) an equilibrium part, which represents the state of plasma in absence of oscillations, and is indicated by subscript 0 and (ii) perturbation part due to oscillations, indicated by subscript 1:

$$\vec{B} = \vec{B}_0 + \vec{B}_1$$

$$\vec{E} = \vec{E}_0 + \vec{E}_1$$

$$\vec{J} = \vec{J}_0 + \vec{J}_1$$

In plasma in equilibrium state, we have $\vec{E}_0 = 0$ and $\vec{J}_0 = 0$. Thus, for $\vec{B}_0 = 0$, the Maxwell's equations reduce to

$$\nabla \times \vec{E}_1 = -\frac{\partial \vec{B}}{\partial t} \tag{17.115}$$

and

$$\nabla \times \vec{B}_1 = \frac{1}{c^2}\frac{\partial \vec{E}_1}{\partial t} + \mu_0 \vec{J}_1 \tag{17.116}$$

Differentiating equation (17.116) with respect to time, we have

$$\nabla \times \frac{\partial \vec{B}_1}{\partial t} = \frac{1}{c^2}\frac{\partial^2 \vec{E}_1}{\partial t^2} + \mu_0 \frac{\partial \vec{J}_1}{\partial t} \tag{17.117}$$

Taking curl of equation (17.115), we have

$$\nabla \times (\nabla \times \vec{E}_1) = -\nabla \times \frac{\partial \vec{B}_1}{\partial t}$$

$$\nabla(\nabla \cdot \vec{E}_1) - \nabla^2 \vec{E}_1 = -\nabla \times \frac{\partial \vec{B}_1}{\partial t} \tag{17.118}$$

Using equation (17.117) in (17.118), we have

$$\nabla(\nabla \cdot \vec{E}_1) - \nabla^2 \vec{E}_1 = -\mu_0 \frac{\partial \vec{J}_1}{\partial t} - \frac{1}{c^2}\frac{\partial^2 \vec{E}_1}{\partial t^2} \tag{17.119}$$

We assume that the oscillating quantities behave sinusoidally as $e^{i(\vec{k}\cdot\vec{r}-\omega t)}$. Then time derivative ($\partial/\partial t$) can be replaced by $-i\omega$, and gradient ∇ by $i\vec{k}$. From equation (17.119), we get

$$i\vec{k}(i\vec{k} \cdot \vec{E}_1) - (i\vec{k})^2 \vec{E}_1 = i\omega\mu_0 \vec{J}_1 - \frac{(-i\omega)^2}{c^2}\vec{E}_1$$

$$-\vec{k}(\vec{k} \cdot \vec{E}_1) + k^2 \vec{E}_1 = i\omega\mu_0 \vec{J}_1 + \frac{\omega^2}{c^2}\vec{E}_1$$

For transverse wave, we have $\vec{k} \cdot \vec{E}_1 = 0$. Thus,

$$k^2\vec{E}_1 = i\omega\mu_0\vec{J}_1 + \frac{\omega^2}{c^2}\vec{E}_1$$

or
$$(\omega^2 - c^2k^2)\vec{E}_1 = -i\omega\mu_0 c^2 \vec{J}_1 \tag{17.120}$$

For high frequency waves, ions may be treated as fixed. Then current density is due to the motion of electrons only

$$\vec{J}_1 = -n_0 e\, \vec{u}_{e1} \tag{17.121}$$

The equation of motion for electrons is

$$m_e n_e \left[\frac{\partial \vec{u}_e}{\partial t} + (\vec{u}_e \cdot \nabla)\vec{u}_e\right] = -en_e(\vec{E} + \vec{u}_e \times \vec{B})$$

Here, we are considering that electron temperature T_e is zero (cold plasma). We express as $\vec{u}_e = \vec{u}_0 + \vec{u}_1$. After linearisation with $\vec{u}_0 = 0$, $\vec{E}_0 = 0$ and $\vec{B}_0 = 0$, the equation of motion reduces to

$$m_e \frac{\partial \vec{u}_{e1}}{\partial t} = -e\vec{E}_1$$

We assume that oscillating quantities behave sinusoidally as $e^{i(\vec{k}\cdot\vec{r}-\omega t)}$. Then time derivative $(\partial/\partial t)$ can be replaced by $-i\omega$ and we get

$$m_e(-i\omega)\vec{u}_{e1} = -e\vec{E}_1$$

$$\vec{u}_{e1} = \left(\frac{e}{im_e\omega}\right)\vec{E}_1 \tag{17.122}$$

Using equation (17.122) in (17.121), we have

$$\vec{J}_1 = \left(\frac{n_0 e^2}{im_e\omega}\right)\vec{E}_1$$

Using $\vec{J}_1$ in equation (17.120), we get

$$(\omega^2 - c^2k^2)\vec{E}_1 = \omega_p^2\,\vec{E}_1$$

Since $\vec{E}_1 \neq 0$, we have

$$\omega^2 - c^2k^2 = \omega_p^2$$

$$\omega^2 = \omega_p^2 + c^2k^2$$

where $\omega_p = (n_0e^2/m_e\epsilon_0)^{1/2}$ is plasma frequency. This is dispersion relation for transverse electromagnetic waves propagating in plasma with $\vec{B}_0 = 0$. For electromagnetic waves in the medium, we have[4] $\omega^2 = c^2k^2$. It shows that there is additional term ω_p^2. The phase velocity is obtained as

---

[4]Maxwell's equations for electromagnetic waves in plasma are

$$\nabla \times \vec{E}_1 = -\frac{\partial \vec{B}_1}{\partial t} \tag{17.123}$$

$$\nabla \times \vec{B}_1 = \frac{1}{c^2}\frac{\partial \vec{E}_1}{\partial t} \tag{17.124}$$

Since in vacuum we have $\vec{J} = 0$. Differentiating equation (17.123) with respect to time t, we have

$$\nabla \times \frac{\partial \vec{E}_1}{\partial t} = -\frac{\partial^2 \vec{B}_1}{\partial t^2} \tag{17.125}$$

$$\frac{\omega^2}{k^2} = \frac{\omega_p^2}{k^2} + c^2$$

or
$$\frac{\omega}{k} = c\sqrt{1+\frac{\omega_p^2}{c^2k^2}} \equiv u_p$$

The phase velocity v_p is greater than c. The group velocity is obtained as

$$2\omega\, d\omega = 2c^2 k\, dk$$

$$\frac{d\omega}{dk} = \frac{c^2}{\omega/k} = c\left(1+\frac{\omega_p^2}{c^2k^2}\right)^{-1/2} \equiv v_g$$

The group velocity v_g is smaller than c. When $k \to \infty$, the phase velocity $v_p \to c$, and also the group velocity $v_g \to c$. The dispersion curve is shown in Fig. 4.

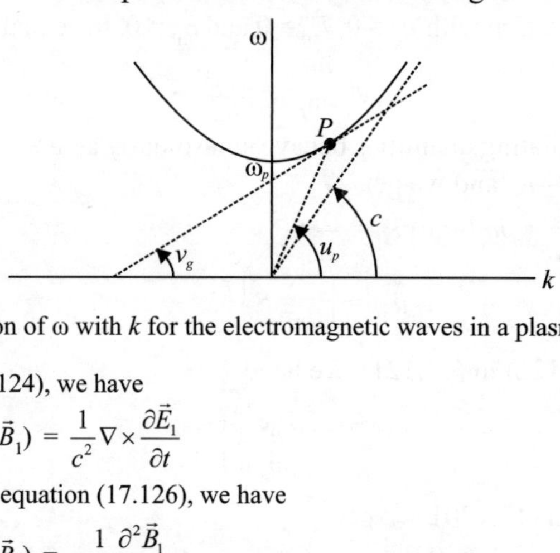

Fig. 4: Variation of ω with k for the electromagnetic waves in a plasma with $\vec{B}_0 = 0$

Taking curl of equation (17.124), we have

$$\nabla \times (\nabla \times \vec{B}_1) = \frac{1}{c^2}\nabla \times \frac{\partial \vec{E}_1}{\partial t} \tag{17.126}$$

Using equation (17.125) in equation (17.126), we have

$$\nabla \times (\nabla \times \vec{B}_1) = -\frac{1}{c^2}\frac{\partial^2 \vec{B}_1}{\partial t^2}$$

$$\nabla \times (\nabla \times \vec{B}_1) - \nabla^2 \vec{B}_1 = -\frac{1}{c^2}\frac{\partial^2 \vec{B}_1}{\partial t^2}$$

$$\nabla^2 \vec{B}_1 = \frac{1}{c^2}\frac{\partial^2 \vec{B}_1}{\partial t^2}$$

where we have used $\nabla \cdot \vec{B}_1 = 0$. We assume that oscillating quantities behave sinusoidally as $e^{i(\vec{k}\cdot\vec{r}-\omega t)}$. Then time derivative $(\partial/\partial t)$ can be replaced by $-i\omega$, and gradient ∇ by $i\vec{k}$, and we have

$$(i\vec{k})^2 \vec{B}_1 = \frac{(-i\omega)^2}{c^2}\vec{B}_1$$

$$c^2k^2 \vec{B}_1 = \omega^2 \vec{B}_1$$

Since $\vec{B}_1 \neq 0$, we have

$$c^2k^2 = \omega^2$$

$$\frac{\omega}{k} = \mathrm{c}$$

$$\frac{\mathrm{d}\omega}{\mathrm{d}k} = \mathrm{c}$$

Thus, the phase velocity and group velocity of the light waves in the plasma are c.

This figure is similar to that in case of electron plasma waves, but behaviour is quite different as the asymptotic velocity in present case is v and in that case was $\sqrt{3/2}\, v_{th}$.

14. Electromagnetic Waves Perpendicular to $\vec{B}_0$

Earlier, we have considered propagation of electromagnetic waves when there is no magnetic field. Now, we consider propagation of electromagnetic waves when magnetic field $\vec{B}_0$ is present. In the present section, we consider the case of perpendicular propagation, $\vec{k} \perp \vec{B}_0$. If we take transverse wave with $\vec{k} \perp \vec{E}_1$, we have two options for the direction of $\vec{E}_1$: (i) $\vec{E}_1$ is parallel to $\vec{B}_0$ – case of ordinary wave and (ii) $\vec{E}_1$ is perpendicular to $\vec{B}_0$ – case of extraordinary wave. Let us consider one case at a time.

14.1 Ordinary Wave

In this case, $\vec{E}_1$ is parallel to $\vec{B}_0$ and $\vec{k}$ is perpendicular to $\vec{B}_0$ as shown in Fig. 5.

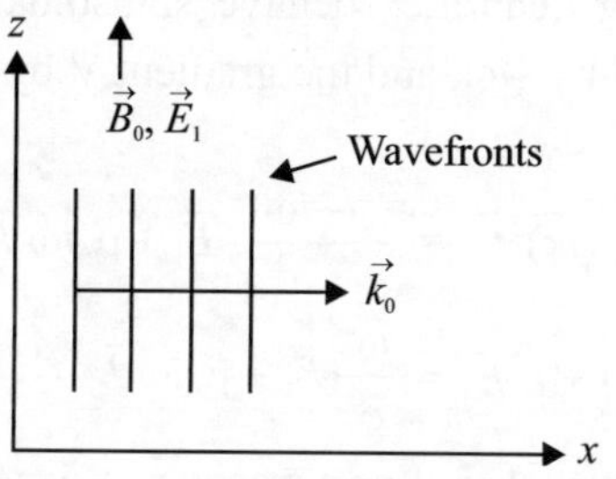

Fig. 5: Propagation of an ordinary electromagnetic wave perpendicular to $\vec{B}_0$

We have $\vec{B}_0 = \vec{B}_0\,\hat{z}$, $\vec{k} = k\,\hat{x}$, and $\vec{E}_1 = E_1\,\hat{}$. Let us consider Maxwell's equations

$$\nabla \times \vec{E} = -\frac{\partial \vec{B}}{\partial t}$$

and

$$\nabla \times \vec{B} = \frac{1}{c^2}\frac{\partial \vec{E}}{\partial t} + \mu_0 \vec{J}$$

For linearization of these equations, let us first separate variable parameters into two parts: (i) an equilibrium part, which represents the state of plasma in absence of oscillations, and is indicated by subscript 0 and (ii) perturbation part due to oscillations, indicated by subscript 1:

$$\vec{B} = \vec{B}_0 + \vec{B}_1$$

$$\vec{E} = \vec{E}_0 + \vec{E}_1$$

$$\vec{J} = \vec{J}_0 + \vec{J}_1$$

In plasma, we have $\vec{E}_0 = 0$ and $\vec{J}_0 = 0$. Equilibrium parts do not depend on time and space coordinates and we have Maxwell's equations

$$\nabla \times \vec{E}_1 = -\frac{\partial \vec{B}_1}{\partial t} \qquad (17.127)$$

and $$\nabla \times \vec{B}_1 = \frac{1}{c^2}\frac{\partial \vec{E}_1}{\partial t} + \mu_0 \vec{J}_1 \tag{17.128}$$

Differentiating equation (17.128) with respect to time t, we have

$$\nabla \times \frac{\partial \vec{B}_1}{\partial t} = \frac{1}{c^2}\frac{\partial^2 \vec{E}_1}{\partial t^2} + \mu_0 \frac{\partial \vec{J}_1}{\partial t} \tag{17.129}$$

Taking curl of equation (17.127), we have

$$\nabla \times (\nabla \times \vec{E}_1) = -\nabla \times \frac{\partial \vec{B}_1}{\partial t}$$

$$\nabla(\nabla \cdot \vec{E}_1) - \nabla^2 \vec{E}_1 = -\nabla \times \frac{\partial \vec{B}_1}{\partial t} \tag{17.130}$$

Using equation (17.129) in (17.130), we have

$$\nabla(\nabla \cdot \vec{E}_1) - \nabla^2 \vec{E}_1 = -\frac{1}{c^2}\frac{\partial^2 \vec{E}_1}{\partial t^2} - \mu_0 \frac{\partial \vec{J}_1}{\partial t} \tag{17.131}$$

We assume that oscillating quantities behave sinusoidally as $e^{i(\vec{k}\cdot\vec{r}-\omega t)}$. Then, the time derivative $(\partial/\partial t)$ can be replaced by $-i\omega$, and the gradient ∇ by $i\vec{k}$. From equation (17.131), we get

$$i\vec{k}(i\vec{k} \cdot \vec{E}_1) - (i\vec{k})^2 \vec{E}_1 = -\frac{(-i\omega)^2}{c^2}\vec{E}_1 + \mu_0 i\omega \vec{J}_1$$

$$-\vec{k}(\vec{k} \cdot \vec{E}_1) + k^2 \vec{E}_1 = \frac{\omega^2}{c^2}\vec{E}_1 + \mu_0 i\omega \vec{J}_1$$

Let us take $\vec{J}_1 = J_x\,\hat{x} + J_y\,\hat{y} + J_z\,\hat{z}$. For a transverse wave, we have $\vec{k} \cdot \vec{E}_1 = 0$ and thus, along the z-axis, we have

$$c^2 k^2 E_1 = \omega^2 E_1 + \mu_0 i\omega c^2 J_z$$

$$(\omega^2 - c^2 k^2) E_1 = -\mu_0 i\omega c^2 J_z \tag{17.132}$$

For the high frequency waves, the ions may be considered as fixed. Then the current density is due to motion of electrons along z-axis as

$$J_z = -n_{e0}\, e u_z \tag{17.133}$$

The equation of motion for electrons is

$$m_e n_e \left[\frac{\partial \vec{u}_e}{\partial t} + (\vec{u}_e \cdot \nabla)\vec{u}_e\right] = -e n_e (\vec{E} + \vec{u}_e \times \vec{B})$$

We express $\vec{u}_e = \vec{u}_{e0} + \vec{u}_{e1}$. Here, we are assuming that electron temperature T_e is zero (cold plasma). After linearisation with $\vec{u}_{e0} = 0$, and $\vec{E}_0 = 0$, we have

$$m_e \frac{\partial \vec{u}_{e1}}{\partial t} = -e\vec{E}_1 - e\vec{u}_{e1} \times \vec{B}_0$$

Since $\vec{E}_1$ is parallel to z-axis, we are interested in z-component only and therefore, the term $\vec{u}_{e1} \times \vec{B}_0$ does not contribute.

We assume that oscillating quantities behave sinusoidally as $e^{i(\vec{k}\cdot\vec{r}-\omega t)}$. Then, time derivative $(\partial/\partial t)$ can be replaced by $-i\omega$ and on linearisation along the z-axis, we get

$$m_e(-i\omega)u_z = -eE_1$$

$$u_z = \left(\frac{e}{im_e\omega}\right)E_1 \qquad (17.134)$$

From equations (17.133) and (17.134), we have

$$J_z = -\left(\frac{n_{e0}e^2}{im_e\omega}\right)E_1$$

Using this value of J_z in equation (17.132), we have

$$(\omega^2 - c^2k^2)E_1 = \omega_p^2 E_1$$

Since $E_1 \neq 0$, we have

$$\omega^2 - c^2k^2 = \omega_p^2$$

$$\omega^2 = \omega_p^2 + c^2k^2$$

where $\omega_p = (n_{e0}e^2/m_e\epsilon_0)^{1/2}$ is the plasma frequency. This is the dispersion relation for ordinary, transverse, electromagnetic waves propagating perpendicular to $\vec{B}_0$. The dispersion curve is shown in Fig. 6.

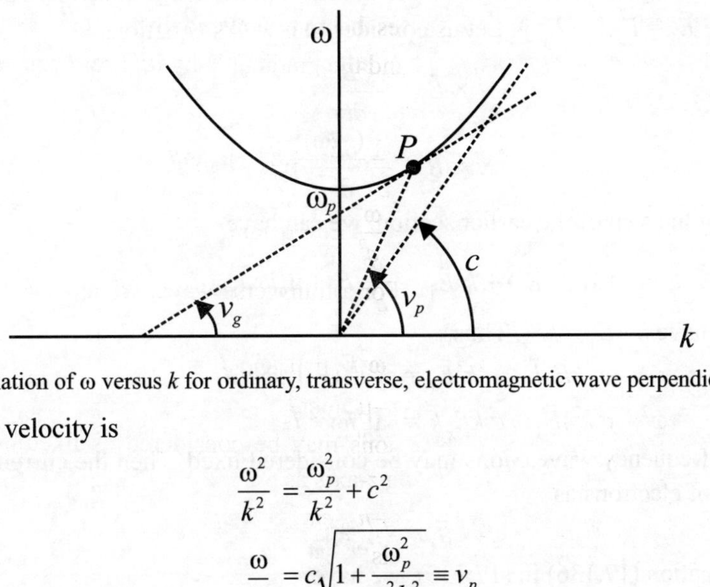

Fig. 6: Variation of ω versus k for ordinary, transverse, electromagnetic wave perpendicular to $\vec{B}_0$

The phase velocity is

$$\frac{\omega^2}{k^2} = \frac{\omega_p^2}{k^2} + c^2$$

or

$$\frac{\omega}{k} = c\sqrt{1 + \frac{\omega_p^2}{c^2k^2}} \equiv v_p$$

It shows that the phase velocity is always greater than c. The group velocity is

$$2\omega\, d\omega = 2c^2k\, dk$$

$$\frac{d\omega}{dk} = \frac{c^2}{v_p} = c\left(1 + \frac{\omega_p^2}{c^2k^2}\right)^{-1/2} \equiv v_g$$

It shows that the group velocity is always smaller than c. When $k \to \infty$, the phase velocity $v_p \to c$, and thus, the group velocity $v_g \to c$.

14.2 Extraordinary Wave

In this case, $\vec{E}_1$ is perpendicular to $\vec{B}_0$, and $\vec{k}$ is perpendicular to $\vec{B}_0$ as shown in Fig. 7.

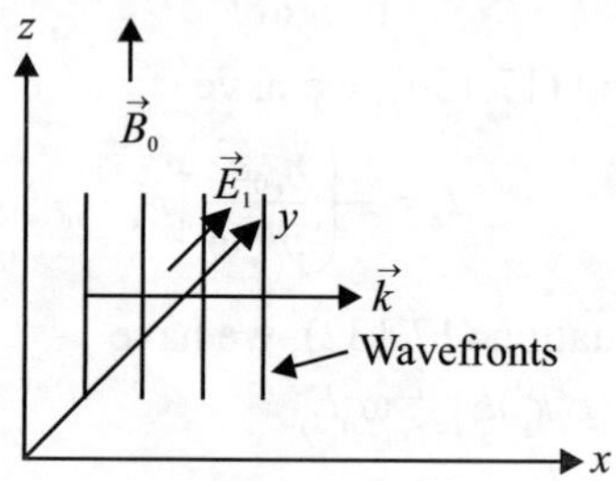

Fig. 7: Propagation of extraordinary electromagnetic wave perpendicular to $\vec{B}_0$

Here, $\vec{B}_0$ being perpendicular to $\vec{E}_1$ affects the motion of electrons. We have $\vec{B}_0 = B_0\,\hat{z}$, $\vec{k} = k\,\hat{x}$, and $\vec{E}_1 = E_1\,\hat{y}$. However, it turns out that waves with $\vec{E}_1 \perp \vec{B}_0$ are elliptically polarized instead of plane polarized. That is, when such wave propagates into plasma, it develops a component E_x along $\vec{k}$. Therefore, the wave becomes partly longitudinal and partly transverse. Thus, we have $\vec{E}_1 = E_x\,\hat{x} + E_y\,\hat{y}$. Let us consider Maxwell's equations

$$\nabla \times \vec{E} = -\frac{\partial \vec{B}}{\partial t}$$

and
$$\nabla \times \vec{B} = \frac{1}{c^2}\frac{\partial \vec{E}}{\partial t} + \mu_0 \vec{J}$$

Following the step of the earlier section, we can have

$$-\vec{k}(\vec{k}\cdot\vec{E}_1) + k^2\vec{E}_1 = \frac{\omega^2}{c^2}\vec{E}_1 + \mu_0 i\omega \vec{J}_1$$

Now, we have $\vec{k}\cdot\vec{E}_1 = kE_x$. Thus,

$$-c^2\,kE_x\vec{k} + c^2k^2\vec{E}_1 = \omega^2\vec{E}_1 + \mu_0 i\omega c^2\vec{J}_1$$

$$(\omega^2 - c^2k^2)\vec{E}_1 + c^2kE_x\vec{k} = -\mu_0 i\omega c^2\vec{J}_1 \qquad (17.135)$$

For high frequency waves, ions may be considered fixed. Then the current density is due to the motion of electrons as

$$\vec{J}_1 = -n_{e0}e\vec{u}_{e1} \qquad (17.136)$$

Using equation (17.136) in (17.135), we have

$$(\omega^2 - c^2k^2)\vec{E}_1 + c^2kE_x\vec{k} = \mu_0 i\omega c^2 n_{e0}e\vec{u}_{e1}$$

The x and y components of this equation are

$$(\omega^2 - c^2k^2)E_x + c^2k^2E_x = \mu_0 i\omega c^2 n_{e0}e\, u_x$$

$$\omega^2E_x = \mu_0 i\omega c^2 n_{e0}e\, u_x \qquad (17.137)$$

and
$$(\omega^2 - c^2k^2)E_y = \mu_0 i\omega c^2 n_{e0}e\, u_y \qquad (17.138)$$

The equation of motion for electrons is

$$m_e n_e\left[\frac{\partial \vec{u}_e}{\partial t} + (\vec{u}_e\cdot\nabla)\vec{u}_e\right] = -en_e(\vec{E} + \vec{u}_e \times \vec{B})$$

We express $\vec{u}_e = \vec{u}_{e0} + \vec{u}_{e1}$. Here, we are assuming that electron temperature T_e is zero (cold plasma). After linearisation with $\vec{u}_{e0} = 0$ and $\vec{E}_0 = 0$, we have

$$m_e \frac{\partial \vec{u}_{e1}}{\partial t} = -e(\vec{E}_1 + \vec{u}_{e1} \times \vec{B}_0)$$

We assume that oscillating quantities behave sinusoidally as $e^{i(\vec{k}\cdot\vec{r}-\omega t)}$. Then time derivative $(\partial/\partial t)$ can be replaced by $-i\omega$ and we get

$$m_e(-i\omega)\vec{u}_{e1} = -e(\vec{E}_1 + \vec{u}_{e1} \times \vec{B}_0)$$

Writing $\vec{u}_{e1} = u_x\hat{x} + u_y\hat{y} + u_z\hat{z}$, and considering for non-trivial x and y components, we have

$$im_e\omega u_x = e(E_x + u_yB_0)$$

and

$$im_e\omega u_y = e(E_y - u_xB_0)$$

We have

$$u_x = \frac{e}{im_e\omega}(E_x + u_yB_0) = \frac{eB_0}{im_e\omega}\left(\frac{E_x}{B_0} + u_y\right) \tag{17.139}$$

and

$$u_y = \frac{e}{im_e\omega}(E_y + u_xB_0) = \frac{eB_0}{im_e\omega}\left(\frac{E_y}{B_0} + u_x\right) \tag{17.140}$$

Using equation (17.140) in (17.139), we have

$$u_x = \frac{eB_0}{im_e\omega}\left[\frac{E_x}{B_0} + \frac{eB_0}{im_e\omega}\left(\frac{E_y}{B_0} + u_x\right)\right]$$

$$= \frac{eE_x}{im_e\omega} - \frac{e^2E_yB_0}{m_e^2\omega^2} + \frac{e^2B_0^2}{m_e^2\omega^2}u_x$$

Therefore,

$$u_x\left(1 - \frac{\omega_c^2}{\omega^2}\right) = \frac{e}{m_e\omega}\left[-iE_x - \frac{\omega_c}{\omega}E_y\right]$$

$$u_x = \frac{e}{m_e\omega}\left[-iE_x - \frac{\omega_c}{\omega}E_y\right]\left(1 - \frac{\omega_c^2}{\omega^2}\right)^{-1} \tag{17.141}$$

where $\omega_c = eB_0/m_e$ is cyclotron frequency. Using equation (17.139) in (17.140), we have

$$u_y = \frac{eB_0}{im_e\omega}\left[\frac{E_y}{B_0} - \frac{eB_0}{im_e\omega}\left(\frac{E_x}{B_0} + u_y\right)\right]$$

$$= \frac{eE_y}{im_e\omega} + \frac{e^2E_xB_0}{m_e^2\omega^2} + \frac{e^2B_0^2}{m_e^2\omega^2}u_y$$

Therefore,

$$u_y\left(1 - \frac{\omega_c^2}{\omega^2}\right) = \frac{e}{m_e\omega}\left[-iE_y + \frac{\omega_c}{\omega}E_x\right]$$

$$u_y = \frac{e}{m_e\omega}\left[-iE_y + \frac{\omega_c}{\omega}E_x\right]\left(1 - \frac{\omega_c^2}{\omega^2}\right)^{-1} \tag{17.142}$$

Using equation (17.141) in (17.137), we have

$$\omega^2 E_x = \frac{\mu_0 ic^2 ne^2}{m_e}\left[-iE_x - \frac{\omega_c}{\omega}E_y\right]\left(1-\frac{\omega_c^2}{\omega^2}\right)^{-1}$$

$$\omega^2\left(1-\frac{\omega_c^2}{\omega^2}\right)E_x = -i\omega_p^2\left[iE_x - \frac{\omega_c}{\omega}E_y\right]$$

$$= \omega_p^2 E_x - \frac{i\omega_p^2\omega_c}{\omega}E_y$$

$$(\omega^2 - \omega_h^2)E_x + \frac{i\omega_p^2\omega_c}{\omega}E_y = 0 \tag{17.143}$$

where $\omega_p = \sqrt{n_{e0}e^2/\epsilon_0 m_e}$ is plasma frequency, $\omega_h = \sqrt{\omega_c^2 + \omega_p^2}$ is the upper hybrid frequency. Using equation (17.142) in (17.138), we have

$$(\omega^2 - c^2k^2)E_y = \frac{\mu_0 ic^2 n_{e0}e^2}{m_e}\left[-iE_y + \frac{\omega_c}{\omega}Ex\right]\left(1-\frac{\omega_c^2}{\omega^2}\right)^{-1}$$

$$(\omega^2 - c^2k^2)\left(1-\frac{\omega_c^2}{\omega^2}\right)E_y = \omega_p^2 E_y + \frac{i\omega_p^2\omega_c}{\omega}E_x$$

$$-\frac{i\omega_p^2\omega_c}{\omega}E_x\left[(\omega^2 - c^2k^2)\left(1-\frac{\omega_c^2}{\omega^2}\right) - \omega_p^2\right]E_y = 0 \tag{17.144}$$

For convenience, let us express equations (17.143) and (17.144) as

$$AE_x + BE_y = 0$$

and

$$CE_x + DE_y = 0$$

These simultaneous equations are satisfied when

$$\begin{vmatrix} A & B \\ C & D \end{vmatrix} = 0$$

$$AD - CB = 0$$

$$AD = CB$$

Using A, B, C and D, we get

$$(\omega^2 - \omega_h^2)\left[(\omega^2 - c^2k^2)\left(1-\frac{\omega_c^2}{\omega^2}\right) - \omega_p^2\right] = \frac{iw_p^2\omega_c}{\omega}\left(-\frac{i\omega_p^2\omega_c}{\omega}\right)$$

$$(\omega^2 - \omega_h^2)\left[\omega^2 - \omega_h^2 - c^2k^2\left(1-\frac{\omega_c^2}{\omega^2}\right)\right] = \frac{\omega_p^4\omega_c^2}{\omega^2}$$

$$c^2k^2\left(1-\frac{\omega_c^2}{\omega^2}\right) = (\omega^2 - \omega_h^2) - \frac{\omega_p^4\omega_c^2/\omega^2}{\omega^2 - \omega_h^2}$$

Thus,

$$\frac{c^2k^2}{\omega^2} = \frac{(\omega^2 - \omega_h^2) - \left[(\omega_p^2\omega_c/\omega)^2/(\omega^2 - \omega_h^2)\right]}{\omega^2 - \omega_c^2}$$

$$= \frac{(\omega^2-\omega_h^2)^2-(\omega_p^4\omega_c^2/\omega^2)}{(\omega^2-\omega_c^2)(\omega^2-\omega_h^2)}$$

$$= \frac{(\omega^2-\omega_h^2)(\omega^2-\omega_c^2-\omega_p^2)-(\omega_p^4\omega_c^2/\omega^2)}{(\omega^2-\omega_c^2)(\omega^2-\omega_h^2)}$$

$$= \frac{(\omega^2-\omega_h^2)(\omega^2-\omega_c^2)}{(\omega^2-\omega_c^2)(\omega^2-\omega_h^2)} - \frac{\omega_p^2(\omega^2-\omega_h^2)+(\omega_p^4\omega_c^2/\omega^2)}{(\omega^2-\omega_c^2)(\omega^2-\omega_h^2)}$$

$$= 1-\frac{\omega_p^2(\omega^2-\omega_h^2)+(\omega_p^4\omega_c^2/\omega^2)}{(\omega^2-\omega_c^2)(\omega^2-\omega_h^2)}$$

$$= 1-\frac{\omega_p^2}{\omega^2}\left[\frac{\omega^2(\omega^2-\omega_c^2)-\omega_p^2(\omega^2-\omega_c^2)}{(\omega^2-\omega_c^2)(\omega^2-\omega_h^2)}\right]$$

$$= 1-\frac{\omega_p^2}{\omega^2}\left[\frac{\omega^2-\omega_p^2}{(\omega^2-\omega_h^2)}\right]$$

Therefore, we have

$$\frac{c^2k^2}{\omega^2} = \frac{c^2}{v_p^2} = 1-\frac{\omega_p^2}{\omega^2}\left[\frac{\omega^2-\omega_p^2}{(\omega^2-\omega_h^2)}\right] \tag{17.145}$$

This is dispersion relation for extraordinary waves. It is considerably complicated.

Cut-offs and resonances

In the dispersion relation for extraordinary waves, one can find cut-offs and resonances. When refractive index $\tilde{n} = c/v_p$ tends to zero, we have a cut-off. At the cut-off, wavelength tends to have value infinite and the wave is reflected back. When refractive index $\tilde{n}$ tends to infinite, we have a resonance. At the resonance, wavelength tends to have value zero, and the wave is absorbed. When a wave propagates through plasma in which ω_p and ω_c are varying, it may encounter cut-offs as well as resonances. Equation (17.145) shows that at $\omega = \omega_p$, we have $v_p = c$.

Resonances for extraordinary waves can be obtained from equation (17.145) by putting $k = \infty$. It occurs when $\omega \to \omega_h$. Hence, at $\omega = \omega_h = \sqrt{\omega_p^2+\omega_c^2}$ wave encounters a resonance. When the wave of frequency ω approaches the resonance point, its phase velocity and group velocity both approach to zero, and the wave energy is converted into upper hybrid oscillations. Cut-offs for extraordinary waves can be obtained from equation (17.145) by putting $k = 0$. Thus, we have

$$1-\frac{\omega_p^2(\omega^2-\omega_p^2)}{\omega^2(\omega^2-\omega_p^2)} = 0; \qquad \frac{\omega_p^2}{\omega^2} = \frac{\omega^2-\omega_p^2-\omega_c^2}{\omega^2-\omega_p^2}; \qquad 1-\frac{\omega_p^2}{\omega^2} = \frac{\omega_c^2/\omega^2}{1-\omega_p^2/\omega^2}$$

$$\left(1-\frac{\omega_p^2}{\omega^2}\right)^2 = \frac{\omega_c^2}{\omega^2}; \qquad 1-\frac{\omega_p^2}{\omega^2} = \pm\frac{\omega_c}{\omega} \qquad \omega^2 \mp \omega\omega_c - \omega_p^2 = 0$$

Thus, the cut-off frequencies are

$$\omega = \frac{\pm\omega_c \pm \sqrt{\omega_c^2 + 4\omega_p^4}}{2}$$

For convention that ω is always positive, the roots are

$$\omega_R = \frac{\omega_c + \sqrt{\omega_c^2 + 4\omega_p^4}}{2}$$

and

$$\omega_L = \frac{-\omega_c + \sqrt{\omega_c^2 + 4\omega_p^4}}{2}$$

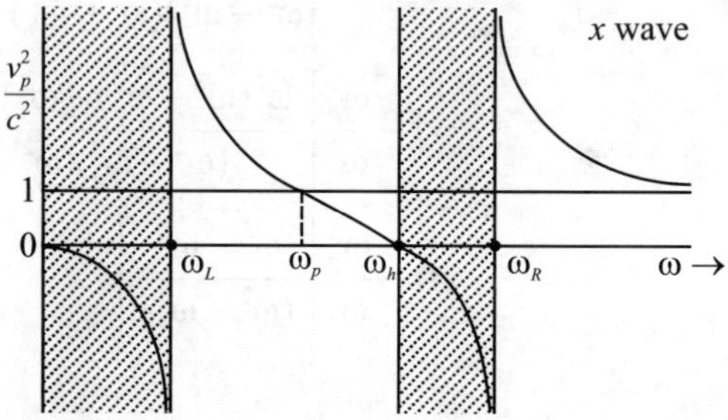

Fig. 8: Variation of phase velocity with frequency for extraordinary wave. The wave does not propagate in shaded regions.

The cut-off frequencies ω_R and ω_L are known as right-hand and left-hand cutoffs, respectively. Cut-off and resonance frequencies divide the dispersion diagram into regions of propagation and of non-propagation. Instead of ω versus k diagram, a plot of (ω^2/c^2k^2) versus ω is shown in Fig. 8, where the wave does not propagate in the shaded region.

15. Electromagnetic Waves Parallel to $\vec{B}_0$

Earlier, we have considered the case of perpendicular propagation, $\vec{k} \perp \vec{B}_0$. Now, we take the case of $\vec{k} \parallel \vec{B}_0$, as shown in Fig. 9.

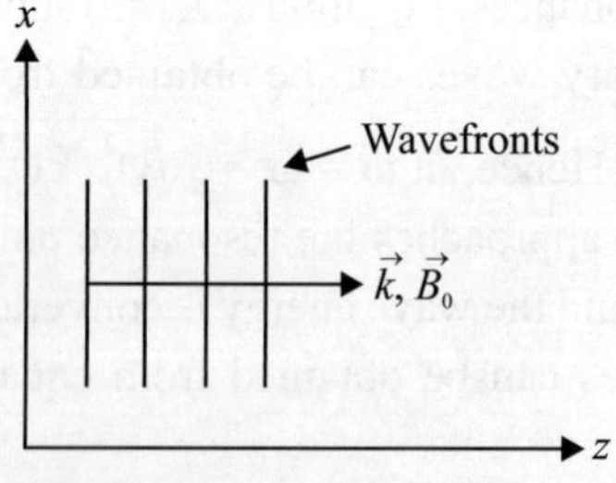

Fig. 9: Propagation of electromagnetic wave parallel to $\vec{B}_0$

We have $\vec{B}_0 = B_0\,\hat{z}$, $\vec{k} = k\,\hat{z}$. For transverse propagation, we take $\vec{E}_1 = E_x\,\hat{x} + E_y\,\hat{y}$. Let us consider Maxwell's equations

$$\nabla \times \vec{E} = -\frac{\partial \vec{B}}{\partial t}$$

and
$$\nabla \times \vec{B} = \frac{\partial \vec{E}}{\partial t} + \mu_0 \vec{J}$$

Following the step of the earlier section, we can have

$$-\vec{k}(\vec{k} \cdot \vec{E}_1) + k^2 \vec{E}_1 = \frac{\omega^2}{c^2} \vec{E}_1 + \mu_0 i\omega \vec{J}_1$$

Now, we have $\vec{k} \cdot \vec{E}_1 = 0$. Thus, we have

$$c^2 k^2 \vec{E}_1 = \omega^2 \vec{E}_1 + c^2 \mu_0 i\omega\, \vec{J}_1$$

$$(\omega^2 - c^2 k^2)\vec{E}_1 = -c^2 \mu_0 i\omega\, \vec{J}_1 \tag{17.146}$$

For high frequency waves, ions may be considered fixed. Then current density is due to the motion of electrons as

$$\vec{J}_1 = -n_{e0} e\, \vec{u}_{e1} \tag{17.147}$$

Using equation (17.147) in (17.146), we have

$$(\omega^2 - c^2 k^2)\vec{E}_1 = c^2\, \mu_0\, i\, \omega\, n_{e0} e\, \vec{u}_{e1} \tag{17.148}$$

Writing $\vec{u}_{e1} = u_x\, \hat{x} + u_y\, \hat{y} + u_z\, \hat{z}$ and $\vec{E}_1 = E_x\, \hat{x} + E_y\, \hat{y}$, the x and y components of this equation are

$$(\omega^2 - c^2 k^2)E_x = c^2\, \mu_0\, i\, \omega\, n_{e0}\, e\, u_x \tag{17.149}$$

and
$$(\omega^2 - c^2 k^2)E_y = c^2\, \mu_0\, i\, \omega\, n_{e0}\, e\, u_y \tag{17.150}$$

The equation of motion for electrons is

$$m_e n_e \left[\frac{\partial \vec{u}_e}{\partial t} + (\vec{u}_e \cdot \nabla)\vec{u}_e \right] = -e n_e (\vec{E} + \vec{u}_e \times \vec{B})$$

Here, we are assuming that electron temperature T_e is zero. We express $\vec{u}_e = \vec{u}_{e0} + \vec{u}_{e1}$. After linearisation with $\vec{u}_{e0} = 0$, and $\vec{E}_0 = 0$, we have

$$m_e \frac{\partial \vec{u}_{e1}}{\partial t} = -e(\vec{E}_1 + \vec{u}_{e1} \times \vec{B}_0)$$

Following the steps discussed in the earlier section, we get

$$u_x = \frac{e}{m_e \omega}\left[-iE_x - \frac{\omega_c}{\omega} E_y \right]\left(1 - \frac{\omega_c^2}{\omega^2} \right)^{-1} \tag{17.151}$$

and
$$u_y = \frac{e}{m_e \omega}\left[-iE_y - \frac{\omega_c}{\omega} E_x \right]\left(1 - \frac{\omega_c^2}{\omega^2} \right)^{-1} \tag{17.152}$$

where $\omega_c = eB_0/m_e$ is cyclotron frequency. Using equation (17.151) in (17.149), we have

$$(\omega^2 - c^2 k^2)E_x = \frac{c^2 \mu_0 i\, n_{e0}\, e^2}{m_e}\left[-iE_x - \frac{\omega_c}{\omega} E_y \right]\left(1 - \frac{\omega_c^2}{\omega^2} \right)^{-1}$$

$$(\omega^2 - c^2 k^2)\left(1 - \frac{\omega_c^2}{\omega^2} \right) E_x = \omega_p^2 E_x - \frac{i\omega_p^2 \omega_c}{\omega} E_y$$

$$\left[(\omega^2 - c^2 k^2)\left(1 - \frac{\omega_c^2}{\omega^2} \right) - \omega_p^2 \right] E_x + \frac{i\omega_p^2 \omega_c}{\omega} E_y = 0 \tag{17.153}$$

where $\omega_p = \sqrt{n_{e0}e^2/\epsilon_0 m_e}$ is plasma frequency. Using equation (17.152) in (17.150), we have

$$(\omega^2 - c^2k^2)E_y = \frac{c^2\mu_0 i n_{e0} e^2}{m_e}\left[-iE_y - \frac{\omega_c}{\omega}E_x\right]\left(1-\frac{\omega_c^2}{\omega^2}\right)^{-1}$$

$$(\omega^2 - c^2k^2)\left(1-\frac{\omega_c^2}{\omega^2}\right)E_y = \omega_p^2 E_y + \frac{i\omega_p^2\omega_c}{\omega}E_x$$

$$-\frac{i\omega_p^2\omega_c}{\omega}E_x + \left[(\omega^2 - c^2k^2)\left(1-\frac{\omega_c^2}{\omega^2}\right) - \omega_p^2\right]E_y = 0 \tag{17.154}$$

Defining a parameter $\alpha = \omega_p^2/(1 - \omega_c^2/\omega^2)$, equations (17.153) and (17.154) can be written as

$$(\omega^2 - c^2k^2 - \alpha)E_x + \frac{i\alpha\omega_c}{\omega}E_y = 0 \tag{17.155}$$

and
$$-\frac{i\alpha\omega_c}{\omega}E_x + (\omega^2 - c^2k^2 - \alpha)E_y = 0 \tag{17.156}$$

For convenience, let us express equations (17.155) and (17.156) as

$$AE_x + BE_y = 0$$

and
$$CE_x + DE_y = 0$$

These simultaneous equations are satisfied when

$$\begin{vmatrix} A & B \\ C & D \end{vmatrix} = 0$$

$$AD - CB = 0 \qquad AD = CB$$

Using A, B, C and D, we get

$$(\omega - c^2k^2 - \alpha)^2 = -\frac{i\alpha\omega_c}{\omega}\frac{i\alpha\omega_c}{\omega} = \left(\frac{\alpha\omega_c}{\omega}\right)^2$$

$$\omega^2 - c^2k^2 - \alpha = \pm\alpha\omega_c/\omega$$

Thus,

$$\omega^2 - c^2k^2 = \alpha\left(1 \pm \frac{\omega_c}{\omega}\right)$$

$$= \frac{\omega_p^2}{1-(\omega_c^2/\omega^2)}\left(1 \pm \frac{\omega_c}{\omega}\right)$$

$$= \omega_p^2\frac{1 \pm (\omega_c/\omega)}{[1+(\omega_c/\omega)][1-(\omega_c/\omega)]}$$

$$= \frac{\omega_p^2}{1 \mp (\omega_c/\omega)}$$

The sign $\mp$ shows that there are two possible solutions corresponding to two different waves which propagate along $\vec{B}_0$. The dispersion relations are

$$\frac{c^2k^2}{\omega^2} = 1 - \frac{\omega_p^2/\omega^2}{1-(\omega_c/\omega)} \qquad \text{R wave}$$

and

$$\frac{c^2k^2}{\omega^2} = 1 - \frac{\omega_p^2/\omega^2}{1+(\omega_c/\omega)} \qquad \text{L wave}$$

The *R* and *L* waves are circularly polarized. *R* stands for right-hand circular polarization whereas *L* stands for left-hand circular polarization.

16. Hydromagnetic Waves

In the preceding sections, we have accounted for high frequency electron waves. Now, we discuss for low frequency ion waves. Out of several possible modes, we consider here two of them: (i) hydromagnetic waves along $\vec{B}_0$, called Alfvén waves and (ii) magnetosonic waves across $\vec{B}_0$. Here, we discuss about Alfvén waves.

In a plane geometry, Alfvén wave has $\vec{k}$ along $\vec{B}_0$; $\vec{E}_1$ and $\vec{J}_1$ both perpendicular to $\vec{B}_0$; and both $\vec{B}_1$ and $\vec{u}_1$ perpendicular to both $\vec{B}_0$ and $\vec{E}_1$ as shown in Fig. 10.

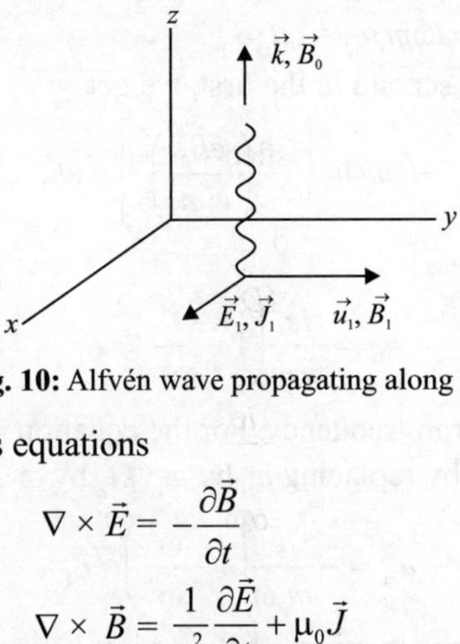

Fig. 10: Alfvén wave propagating along $\vec{B}_0$.

Let us consider Maxwell's equations

$$\nabla \times \vec{E} = -\frac{\partial \vec{B}}{\partial t}$$

and

$$\nabla \times \vec{B} = \frac{1}{c^2}\frac{\partial \vec{E}}{\partial t} + \mu_0 \vec{J}$$

Following the step of the earlier section, we can have

$$-\vec{k}(\vec{k}\cdot\vec{E}_1) + k^2\vec{E}_1 = \frac{\omega^2}{c^2}\vec{E}_1 + \mu_0 i\omega \vec{J}_1$$

Since $\vec{E}_1$ and $\vec{k}$ are perpendicular to each other, we have $\vec{k}\cdot\vec{E}_1 = 0$, and therefore,

$$(\omega^2 - c^2k^2)\vec{E}_1 = -c^2\mu_0 i\omega\, \vec{J}_1 \qquad (17.157)$$

Further, we have taken $\vec{E}_1$ along *x*-axis, and therefore only *x*-component of this equation is non-trivial. For low frequency waves, current density $\vec{J}$ has contributions from both electrons and ions as

$$\vec{J} = ne(\vec{u}_i - \vec{u}_e)$$

Considering $\vec{J} = \vec{J}_0 + \vec{J}_1$, $\vec{u}_i = \vec{u}_{i0} + \vec{u}_{i1}$, $\vec{u}_e = \vec{u}_{e0} + \vec{u}_{e1}$, the linearization process provides

$$\vec{J}_1 = n_e(\vec{u}_{i1} - \vec{u}_{e1}) \tag{17.158}$$

Considering $\vec{u}_{i1} = \vec{u}_{ix}\,\hat{x} + \vec{u}_{iy}\,\hat{y} + \vec{u}_{iz}\,\hat{z}$, $\vec{u}_{e1} = \vec{u}_{ex}\,\hat{x} + \vec{u}_{ey}\,\hat{y} + \vec{u}_{ez}\,\hat{z}$, and using equation (17.158) in (17.157) and taking x-component of resultant equation, we have

$$(\omega^2 - c^2k^2)E_1 = -c^2\mu_0 i\omega ne(u_{ix} - u_{ex}) \tag{17.159}$$

We assume also that there are no thermal motions ($T_i = 0$). The equations of motion is

$$m_i n_i\left[\frac{\partial \vec{u}_i}{\partial t} + (\vec{u}_i \cdot \nabla)\vec{u}_i\right] = en_i(\vec{E} + \vec{u}_i \times \vec{B}_0)$$

On linearisation, this equation reduces to

$$m_i\frac{\partial \vec{u}_{i1}}{\partial t} = e\vec{E}_1 + e\vec{u}_{i1} \times \vec{B}_0 \tag{17.160}$$

We assume that oscillating quantities behave sinusoidally as $e^{i(\vec{k}\cdot\vec{r}-\omega t)}$. Then time derivative $(\partial/\partial t)$ can be replaced by $-i\omega$, and gradient ∇ by $i\vec{k}$. The x and y components of equation (17.160) are

$$-i\omega m_i u_{ix} = eE_1 + eu_{iy}B_0$$

and

$$i\omega m_i u_{iy} = eu_{ix}B_0$$

Using the value of u_{iy} from second in the first, we get

$$-i\omega m_i u_{ix}\left[1 - \left(\frac{eB_0}{\omega m_i}\right)^2\right] = eE_1$$

Thus, we have

$$u_{ix} = \frac{ie}{m_i\omega}\left(1 - \frac{\Omega_c^2}{\omega^2}\right)^{-1} E_1 \tag{17.161}$$

where $\Omega_c = eB_0/m_i$ is ion cyclotron frequency. For the equation of motion for electrons, u_{ex} can be written from this expression by replacing m_i by m_e, Ω_c by ω_c, e by $-e$, and thus, we have

$$u_{ex} = -\frac{ie}{m_e\omega}\left(1 - \frac{\omega_c^2}{\omega^2}\right)^{-1} E_1$$

Obviously, $\omega_c >> \Omega_c$. For the situation, $\omega^2 << \omega_c^2$, we have

$$u_{ex} = \frac{ie}{m_e\omega}\frac{\omega^2}{\omega_c^2}E_1 \to 0 \tag{17.162}$$

For $\omega^2 << \omega_c^2$, Larmor gyrations of electrons are neglected and electrons have $\vec{E} \times \vec{B}$ drift in y-direction. Using equations (17.161) and (17.162) in (17.159), we get

$$(\omega^2 - c^2k^2)E_1 = \frac{c^2\mu_0 ne^2}{m_i}\left(1 - \frac{\Omega_c^2}{\omega^2}\right)^{-1} E_1 \tag{17.163}$$

Since $E_1 \neq 0$, and defining the ion plasma frequency $\Omega_p^2 = n_e^2/m_i\epsilon_0$, we get

$$\omega^2 - c^2k^2 = \Omega_p^2\left(1 - \frac{\Omega_c^2}{\omega^2}\right)^{-1}$$

Further, we assume that hydromagnetic waves have frequencies well below the ion cyclotron frequency ($\omega^2 << \Omega_c^2$), so that

$$\omega^2 - c^2k^2 = \Omega_p^2\left(1-\frac{\omega^2}{\Omega_c^2}\right)^{-1} = -\omega^2\frac{\Omega_p^2}{\Omega_c^2}$$

$$= -\omega^2\frac{ne^2}{m_i\epsilon_0}\frac{m_i^2}{e^2B_0^2} = -\omega^2\frac{\rho}{\epsilon_0 B_0^2}$$

where $\rho = nm_i$ is mass density. Hence,

$$\omega^2\left[1+\frac{\rho}{B_0^2\epsilon_0}\right] = c^2k^2$$

$$\frac{\omega^2}{k^2} = \frac{c^2}{1+(\rho/\epsilon_0 B_0^2)} \qquad (17.164)$$

The denominator on right side of equation (17.164) may be interpreted as relative dielectric constant for low frequency perpendicular motions. Equation (17.164) may be approximated as

$$\frac{\omega^2}{k^2} = \frac{c^2}{(\rho/\epsilon_0 B_0^2)} = \frac{B_0^2}{\mu_0\rho}$$

$$\frac{\omega}{k} = v_p = \frac{B_0}{(\mu_0\rho)^{1/2}}$$

These hydromagnetic waves travel along $\vec{B}_0$ at a constant velocity v_A, called Alfvén velocity

$$v_A = \frac{B_0}{(\mu_0\rho)^{1/2}}$$

17. Magnetosonic Waves

Here, we consider low frequency electromagnetic waves propagating across the magnetic field $\vec{B}_0$ (Fig. 11).

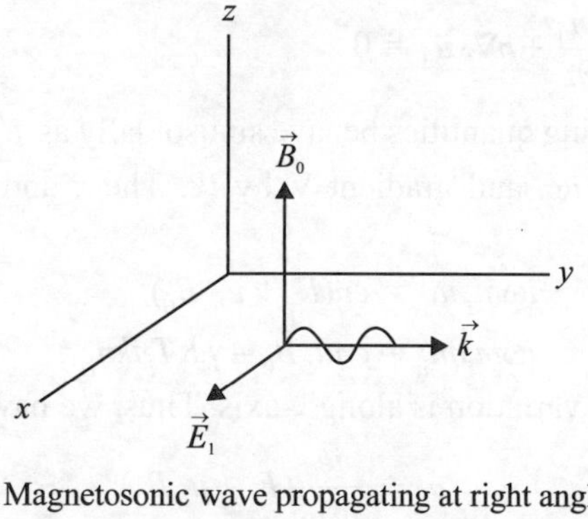

Fig. 11: Magnetosonic wave propagating at right angle to $\vec{B}_0$.

We consider $\vec{B}_0 = B_0\,\hat{z}$ and $\vec{E}_1 = E_1\,\hat{x}$, but now let $\vec{k} = k\,\hat{y}$, so that the $\vec{E}_1 \times \vec{B}_0$ drifts lie along $\vec{k}$. Hence, the plasma is compressed and released along y-axis during the course of oscillations.

Therefore, we have to include pressure term ∇p along y-axis in the equation of motion.

$$\nabla \times \vec{E} = -\frac{\partial \vec{B}}{\partial t}$$

and
$$\nabla \times \vec{B} = \frac{1}{c^2}\frac{\partial \vec{E}}{\partial t} + \mu_0 \vec{J}$$

Following the step of the earlier section, we can have

$$-\vec{k}(\vec{k}\cdot\vec{E}_1) + k^2\vec{E}_1 = \frac{\omega^2}{c^2}\vec{E}_1 + \mu_0 i\omega\,\vec{J}_1$$

Since $\vec{E}_1$ and $\vec{k}$ are perpendicular to each other, we have $\vec{k}\cdot\vec{E}_1 = 0$, and therefore,

$$(\omega^2 - c^2k^2)E_1 = -c^2\mu_0 i\omega\,\vec{J}_1 \tag{17.165}$$

Further, we have taken $\vec{E}_1$ along x-axis, and therefore only x-component of this equation is non-trivial. For low frequency waves, current density $\vec{J}_1$ has contributions from both electrons and ions as

$$\vec{J}_1 = ne(\vec{v}_{i1} - \vec{v}_{e1}) \tag{17.166}$$

Considering $\vec{u}_{i1} = \vec{u}_{ix}\,\hat{x} + \vec{u}_{iy}\,\hat{y} + \vec{u}_{iz}\,\hat{z}$, $\vec{u}_{e1} = \vec{u}_{ex}\,\hat{x} + \vec{u}_{ey}\,\hat{y} + \vec{u}_{ez}\,\hat{z}$, and using equation (17.166) in (17.165) and taking x-component of resultant equation, we have

$$(\omega^2 - c^2k^2)E_1 = -c^2\mu_0 i\omega ne(u_{ix} - u_{ex}) \tag{17.167}$$

The equations of motion and continuity are

$$m_i n_i\left[\frac{\partial \vec{u}_i}{\partial t} + (\vec{\mu}_i\cdot\nabla)\vec{u}_i\right] = en_i(\vec{E} + \vec{u}_i \times \vec{B}_0) - \gamma_i KT_i\nabla n_i$$

and
$$\frac{\partial n_i}{\partial t} + \nabla\cdot(n_i\vec{u}_i) = 0$$

On linearisation, these equations reduce to

$$m_i n\frac{\partial \vec{u}_{i1}}{\partial t} = en(\vec{E}_1 + \vec{u}_{i1} \times \vec{B}_0) - \gamma_i KT_i\nabla n_{i1} \tag{17.168}$$

and
$$\frac{\partial n_{i1}}{\partial t} + n\nabla\cdot\vec{u}_{i1} = 0 \tag{17.169}$$

We assume that oscillating quantities behave sinusoidally as $e^{i(\vec{k}\cdot\vec{r}-\omega t)}$. Then time derivative $(\partial/\partial t)$ can be replaced by $-i\omega$, and gradient ∇ by $i\vec{k}$. The x and y components of equation (17.168) are

$$-i\omega m_i nu_{ix} = en(E_1 + u_{iy}B_0)$$

and
$$i\omega m_i nu_{iy} = enu_{ix}B_0 - \gamma_i KT_i ikn_{i1}$$

Remember that density variation is along y-axis. Thus, we have

$$u_{ix} = \frac{ie}{m_i\omega}(E_1 + u_{iy}B_0) \tag{17.170}$$

and
$$u_{iy} = \frac{ie}{m_i\omega}(-u_{ix}B_0) + \frac{k}{\omega}\frac{\gamma_i KT_i}{m_i}\frac{n_{i1}}{n} \tag{17.171}$$

Linearisation of equation (17.169) gives

$$-i\omega n_{i1} = -niku_{iy}$$

$$\frac{n_{i1}}{n} = \frac{k}{\omega}u_{iy} \tag{17.172}$$

Using equation (17.172) in (17.171), we get

$$u_{iy} = -\frac{ieB_0}{m_i\omega}u_{ix} + \frac{k^2}{\omega^2}\frac{\gamma_i KT_i}{m_i}u_{iy}$$

$$= -\frac{i\Omega_c}{\omega}u_{ix} + \alpha_i u_{iy}$$

where $$\Omega_c = \frac{eB_0}{m_i}$$

and $$\alpha_i = \frac{k^2}{\omega^2}\frac{\gamma_i KT_i}{m_i}$$

$$u_{iy}(1-\alpha_i) = \frac{i\Omega_c}{\omega}u_{ix}$$

$$u_{iy} = -\frac{i\Omega_c}{\omega}u_{ix}(1-\alpha_i)^{-1} \tag{17.173}$$

Using equation (17.173) in (17.170), we get

$$u_{ix} = \frac{ie}{m_i\omega}E_1 + \frac{i\Omega_c}{\omega}\left(-\frac{i\Omega_c}{\omega}\right)(1-\alpha_i)^{-1}u_{ix}$$

$$= \frac{ie}{m_i\omega}E_1 + \frac{\Omega_c^2}{\omega^2}(1-\alpha_i)^{-1}u_{ix}$$

$$u_{ix}\left(1 - \frac{\Omega_c^2/\omega^2}{1-\alpha_i}\right) = \frac{ie}{m_i\omega}E_1$$

$$u_{ix}\left(\frac{1-\alpha_i - \Omega_c^2/\omega^2}{1-\alpha_i}\right) = \frac{ie}{m_i\omega}E_1 \tag{17.174}$$

Equation for electron, corresponding to equation (17.174) can be obtained by replacing e by $-e$, m_i by m_e, Ω_c by ω_c and α_i by α_e as

$$u_{ex}\left(\frac{1-\alpha_e - \omega_c^2/\omega^2}{1-\alpha_e}\right) = \frac{ie}{m_e\omega}E_1 \tag{17.175}$$

where $$\omega_c = \frac{eB_0}{m_e}$$

and $$\alpha_e = \frac{k^2}{\omega^2}\frac{\gamma_e KT_e}{m_e}$$

for $\omega^2 << \omega_c^2$ and $\omega^2 << k^2 u_{th}^2$, ($u_{th} = \sqrt{2KT_e/m_e}$), we have

$$u_{ex}\left(\frac{-\omega_c^2/\omega^2}{1-\alpha_e}\right) = -\frac{ie}{m_e\omega}E_1$$

$$u_{ex} = \frac{ie}{m_e \omega} \frac{\omega^?}{\omega_c^2}(1-\alpha_e)E_1 -$$

$$u_{ex} = \frac{ie}{m_e \omega} \frac{\omega^2}{\omega_c^2}\left(1 - \frac{k^2}{\omega^2}\frac{\gamma_e KT_e}{m_e}\right)E_1 \rightarrow -\frac{ik^2}{\omega B_0^2}\frac{\gamma_e KT_e}{e}E_1 \qquad (17.176)$$

Using equations (17.174) and (17.176) in (17.167), we have

$$(\omega^2 - c^2k^2)E_1 = -c^2\mu_0 i\omega ne\left[\frac{ie}{m_i\omega}\left(\frac{1-\alpha_i}{1-\alpha_i - \Omega_c^2/\omega^2}\right) + \frac{ik^2}{\omega B_0^2}\frac{\gamma_e KT_e}{e}\right]E_1$$

For $\omega^2 << \Omega_c^2$, we can neglect $1 - \alpha_i$ in comparison to Ω_c^2/ω^2 and for $E_1 \neq 0$, we have

$$\omega^2 - c^2k^2 = -c^2\mu_0 i\omega ne\left[\frac{ie}{m_i\omega}\left(\frac{1-\alpha_i}{-\Omega_c^2/\omega^2}\right) + \frac{ik^2\gamma_e KT_e}{\omega B_0^2 e}\right]$$

$$= -\frac{ne^2}{m_i\epsilon_0}\frac{\omega^2}{\Omega_c^2}(1-\alpha_i) + \frac{nk^2 m_i\gamma_e KT_e}{B_0^2 m_i\epsilon_0}$$

$$= -\frac{\Omega_p^2\omega^2}{\Omega_c^2}(1-\alpha_i) + \frac{\rho k^2\gamma_e KT_e}{B_0^2 m_i\epsilon_0}$$

$$= -\frac{\Omega_p^2\omega^2}{\Omega_c^2}(1-\alpha_i) + \frac{c^2k^2\gamma_e KT_e}{v_A^2 m_i}$$

Thus, we have

$$\omega^2 - c^2k^2 - \frac{c^2k^2}{v_A^2}\frac{\gamma_e KT_e}{m_i} + \frac{\Omega_p^2}{\Omega_c^2}(\omega^2 - \omega^2\alpha_i) = 0$$

$$\omega^2 - c^2k^2\left(1 + \frac{\gamma_e KT_e}{m_i v_A^2}\right) + \frac{\Omega_p^2}{\Omega_c^2}\left(\omega^2 - k^2\frac{\gamma_i KT_i}{m_i}\right) = 0$$

Now,

$$\frac{\Omega_p^2}{\Omega_c^2} = \frac{ne^2}{m_i\epsilon_0}\frac{m_i^2}{e^2 B_0^2} = \frac{c^2}{v_A^2}$$

Thus, we have

$$\omega^2\left(1 + \frac{c^2}{v_A^2}\right) = c^2k^2\left(1 + \frac{\gamma_e KT_e}{m_i v_A^2} + \frac{\gamma_i KT_i}{m_i v_A^2}\right)$$

$$= c^2k^2\left(1 + \frac{\gamma_e KT_e + \gamma_i KT_i}{m_i v_A^2}\right)$$

$$= c^2k^2\left(1 + \frac{v_s^2}{v_A^2}\right)$$

where $v_s^2 = (\gamma_e KT_e + \gamma_i KT_i)/m_i$. Therefore,

$$\omega^2\left(\frac{v_A^2 + c^2}{v_A^2}\right) = c^2k^2\left(\frac{v_A^2 + v_s^2}{v_A^2}\right)$$

$$\frac{\omega^2}{k^2} = c^2\frac{v_s^2 + v_A^2}{c^2 + v_A^2}$$

This is dispersion relation for magnetosonic waves propagating perpendicular to $\vec{B}_0$. In this acoustic wave, compressions and rarefactions are produced not by the motions along $\vec{E}$, but perpendicular to $\vec{E}$ by $\vec{E}_i \times \vec{B}_0$ drift.

When $\vec{B}_0 \to 0$, $v_A \to 0$, and we have $(\omega/k) = v_s$. Hence, magnetosonic wave turns into ordinary ion acoustic wave. When $KT_i \to 0$ and $KT_e \to 0$, $v_s \to 0$, and we have

$$\frac{\omega^2}{k^2} = \frac{c^2v_A^2}{c^2 + v_A^2}$$

or

$$\frac{\omega}{k} = \frac{v_A}{\sqrt{1 + v_A^2/c^2}}$$

Thus, magnetosonic wave becomes a modified Alfvén wave, whose phase velocity is smaller than that of Alfvén velocity. When v_s is finite, phase velocity of wave is

$$\frac{\omega}{k} = \sqrt{\frac{v_s^2 + v_A^2}{1 + v_A^2/c^2}}$$

It is generally larger than Alfvén velocity, and therefore, the wave is often called fast hydromagnetic wave.

Review Questions

1. Describe phase velocity of sinusoidal wave. Show that phase velocity is always larger than the speed of light.
2. Describe phase velocity and group velocity. Derive relation between them.
3. Show that velocity of a particle is the same as group velocity of corresponding wave packet.
4. For phase velocity u_p and group velocity u_g, derive following relations:

$$u_g = u_p - \lambda\frac{du_p}{d\lambda}$$

and

$$u_p u_g = c^2$$

5. Describe procedure of linearisation of an equation for plasma. Assuming that perturbation part of parameters behave sinusoidally as $e^{i(kx-\omega t)}$ linearize following equation.

$$mn\left[\frac{\partial\vec{u}}{\partial t} + (\vec{u}\cdot\nabla)\vec{u}\right] = qn(\vec{E} + \vec{u}\times\vec{B}) - \gamma KT\nabla n$$

6. Describe procedure of linearisation of an equation for plasma. Assuming that perturbation part of parameters behave sinusoidally as $e^{i(kx-\omega t)}$ linearize following equations.

$$\frac{\partial n}{\partial t} + \nabla \cdot (n\vec{u}) = 0$$

and
$$\nabla \cdot \vec{E} = \frac{e(n_i - n_e)}{\epsilon_0}$$

7. Describe procedure of linearisation of an equation for plasma. Assuming that perturbation part of parameters behave sinusoidally as $e^{i(kx-\omega t)}$ linearize following equations.

$$\nabla \cdot \vec{E} = \frac{e(n_i - n_e)}{\epsilon_0}$$

and
$$\nabla \times \vec{E} = -\frac{\partial \vec{B}}{\partial t}$$

8. Describe procedure of linearisation of an equation for plasma. Assuming that perturbation part of parameters behave sinusoidally as $e^{i(kx-\omega t)}$ linearize following equations.

$$\nabla \times \vec{E} = -\frac{\partial \vec{B}}{\partial t}$$

and
$$\nabla \times \vec{B} = \frac{1}{c^2}\frac{\partial \vec{E}}{\partial t} + \mu_0 \vec{J}$$

9. What is plasma frequency? Derive an expression for the plasma frequency. What happens when thermal motions in plasma are considered.
10. For homogeneous, isotropic, collision-less, cold plasma, derive an expression for plasma frequency.
11. For homogeneous, isotropic, collision-less, hot plasma, derive an expression for dispersion relation.
12. Show that a longitudinal wave is electrostatic and a transverse wave is electromagnetic.
13. Derive an expression for dispersion relation for electron plasma waves. Show that group velocity is never greater than the speed of light, and at a large value of wavenumber k, information travels faster than that at a small value of k.
14. Derive an expression for velocity of sound waves in ordinary gas.
15. Show that in electron plasma waves, the phase velocity is always greater than or equal to $\sqrt{3/2}\,u_{th}$ whereas the group velocity is always less than or equal to $\sqrt{3/2}\,u_{th}$, where u_{th} is thermal velocity of electrons.
16. Derive an expression for velocity of ion waves in fully ionized plasma. Estimate error introduced by plasma approximation.
17. Discuss validity of plasma approximation.
18. Describe propagation of electrostatic electron waves perpendicular to the direction of applied magnetic field.
19. Describe propagation of electrostatic ion waves perpendicular to the direction of applied magnetic field.
20. Describe about lower hybrid frequency.
21. Derive dispersion relation for transverse electromagnetic waves propagating in plasma with $\vec{B}_0 = 0$.
22. Derive dispersion relation for ordinary electromagnetic wave propagating perpendicular to $\vec{B}_0$.

23. Derive dispersion relation for extraordinary electromagnetic wave propagating perpendicular to $\vec{B}_0$.
24. Discuss about cut-offs and resonances for electromagnetic waves propagating perpendicular to $\vec{B}_0$.
25. Derive dispersion relation for electromagnetic waves propagating parallel to $\vec{B}_0$.
26. Derive an expression for velocity of Alfvén waves.
27. Derive an expression for velocity of magnetosonic waves.
28. For electromagnetic, perpendicular, transverse, extraordinary wave, derive the dispersion relation, and show that the phase velocity is always larger than the speed of light c.
29. Write short notes on the following
    (i) Phase velocity
    (ii) Group velocity
    (iii) Plasma frequency
    (iv) Plasma approximation

Multiple-Choice Questions

1. For dispersion relation $\omega^2 = c^2k^2 + c$, group velocity v_g and phase velocity v_p satisfy the relation
   A. $v_g v_p = c$ B. $v_g v_p = c^2$ C. $v_g = v_p$ D. None of these
   Ans. B
2. For dispersion relation $\omega^2 = c^2k^2 + c^3$, group velocity v_g and phase velocity v_p satisfy the relation
   A. $v_g v_p = c$ B. $v_g v_p = c^2$ C. $v_g < v_p$ D. $v_g v_p = c^3$ **Ans. B**
3. For dispersion relation $\omega^2 = ck^2 + c^3$, group velocity v_g and phase velocity v_p satisfy the relation
   A. $v_g v_p = c$ B. $v_g v_p = c^2$ C. $v_g < v_p$ D. $v_g v_p = c^3$ **Ans. A**
4. Plasma frequency is ω_p. The speed v of wave propagating through plasma and having frequency ω, such that $\omega > \omega_p$, is
   A. $v = c$ B. $v > c$ C. $v < c$ D. $v = \sqrt{c}$ **Ans. B**
5. If v is velocity of particle, which of the followings is not expression for group velocity.
   A. $\frac{d\omega}{dk}$ B. $\frac{dE}{dp}$ C. $\frac{\omega}{k}$ D. v **Ans. C**
6. Angular frequency ω and wavenumber k are related through the relation $\omega = ck$, where c is constant. Which of the following relations between phase velocity v_p and group velocity v_g is correct.
   A. $v_p < v_g$ B. $v_p = v_g$ C. $v_p > v_g$ D. None of them
   Ans. B
7. Angular frequency ω and wavenumber k are related through the relation $\omega^2 = c^2k^2$, where c is constant. Which of the following relations between phase velocity v_p and group velocity v_g is correct.
   A. $v_p < v_g$ B. $v_p = v_g$ C. $v_p > v_g$ D. None of them
   Ans. B

8. The relation between the phase velocity v_p, group velocity v_g and speed of light c is
   A. $v_p v_g = c$ B. $v_p v_g = c^2$ C. $v_p/v_g = c$ D. $v_p/v_g = c^2$ **Ans. B**
9. For a wave expressed as $y = 5\cos(6\pi t - 12x)$, the phase velocity is
   A. 1.2π B. $\pi/2$ C. 2.4 D. $2/\pi$ **Ans. B**
10. For a wave expressed as $y = 4\sin(6\pi t - 12x)$, the phase velocity is
    A. 1.5π B. $\pi/2$ C. 3 D. $2/\pi$ **Ans. B**
11. For a wave expressed as $y = 10\sin(6\pi t - 12x)$, the phase velocity is
    A. 0.6π B. $\pi/2$ C. 1.2 D. $2/\pi$ **Ans. B**
12. Suppose, ions and electrons in plasma have charges q_i and q_e, particle densities n_i and n_e, and are moving with velocities $\vec{v}_i$ and $\vec{v}_e$, respectively. The charge density ρ is
    A. $n_i q_i + n_e q_e$ B. $niqi\vec{u}_i + n_e q_e \vec{u}_e$ C. $n_i q_i - n_e q_e$ D. $n_i q_i \vec{u}_i - n_e q_e \vec{u}_e$ **Ans. A**
13. Suppose, ions and electrons in a plasma have charges q_i and q_e, particle densities n_i and n_e, and are moving with velocities $\vec{v}_i$ and $\vec{v}_e$, respectively. The current density $\vec{J}$ is
    A. $n_i q_i + n_e q_e$ B. $n_i q_i \vec{u}_i + n_e q_e \vec{u}_e$ C. $n_i q_i - n_e q_e$ D. $n_i q_i \vec{u}_i - n_e q_e \vec{u}_e$ **Ans. B**
14. For fluid plasma, the following equation is known as
    $$mn\left[\frac{\partial \vec{u}}{\partial t} + (\vec{u}\cdot\nabla)\vec{u}\right] = qn(\vec{E} + \vec{u}\times\vec{B}) - \nabla p$$
    A. equation of motion B. continuity equation
    C. equation of state D. Maxwell's equation **Ans. A**
15. Suppose, n, e, m denote the density, charge and mass of electrons in plasma. The expression for plasma frequency is
    A. $\left(\frac{ne^2}{\epsilon_0 m}\right)^{1/2}$ B. $\left(\frac{ne^2}{\epsilon_0 m}\right)^{-1/2}$ C. $\left(\frac{n^2e^2}{\epsilon_0 m}\right)^{1/2}$ D. $\left(\frac{n^2e^2}{\epsilon_0 m}\right)^{-1/2}$ **Ans. A**
16. For fluid plasma, the following equation is known as
    $$mn\left[\frac{\partial \vec{u}}{\partial t} + (\vec{u}\cdot\nabla)\vec{u}\right] = qn(\vec{E} + \vec{u}\times\vec{B}) - \gamma KT\nabla n$$
    A. equation of motion B. continuity equation
    C. equation of state D. Maxwell's equation **Ans. A**
17. For fluid plasma, the following equation is known as
    $$\frac{\partial n}{\partial t} + \nabla\cdot(n\vec{u}) = 0$$
    A. equation of motion B. continuity equation
    C. equation of state D. Maxwell's equation **Ans. C**
18. For fluid plasma, equation $p = Cn^{\gamma}$ is the
    A. equation of motion B. continuity equation
    C. equation of state D. Maxwell's equation **Ans. C**
19. In discussion of plasma, we assume that oscillating quantities behave as $e^{i(\vec{k}\cdot\vec{r}-\omega t)}$. Then time derivative ($\partial/\partial t$) is replaced by
    A. $-i\omega$ B. $i\omega$ C. $i\vec{k}$ D. $-i\vec{k}$ **Ans. A**

20. In discussion of plasma, we assume that oscillating quantities behave as $e^{i(\vec{k}\cdot\vec{r}-\omega t)}$. Then gradient ∇ is replaced by

A. $-i\omega$ B. $i\omega$ C. $i\vec{k}$ D. $-i\vec{k}$ **Ans. C**

21. In discussion of plasma, we assume that oscillating quantities behave as $e^{i(\vec{k}\cdot\vec{r}+\omega t)}$. Then time derivative $(\partial/\partial t)$ is replaced by

A. $-i\omega$ B. $i\omega$ C. $i\vec{k}$ D. $-i\vec{k}$ **Ans. B**

22. In discussion of plasma, we assume that oscillating quantities behave as $e^{i(\vec{k}\cdot\vec{r}+\omega t)}$. Then gradient ∇ is replaced by

A. $-i\omega$ B. $i\omega$ C. $i\vec{k}$ D. $-i\vec{k}$ **Ans. C**

23. In discussion of plasma, we assume that oscillating quantities behave as $e^{i(\omega t-\vec{k}\cdot\vec{r})}$. Then time derivative $(\partial/\partial t)$ is replaced by

A. $-i\omega$ B. $i\omega$ C. $i\vec{k}$ D. $-i\vec{k}$ **Ans. B**

24. In discussion of plasma, we assume that oscillating quantities behave as $e^{i(\omega t-\vec{k}\cdot\vec{r})}$. Then gradient ∇ is replaced by

A. $-i\omega$ B. $i\omega$ C. $i\vec{k}$ D. $-i\vec{k}$ **Ans. D**

25. For adiabatic plasma, pressure p and particle density n are related as (γ is ratio of two specific heats of plasma gas)

A. $\frac{\nabla p}{p}=\gamma\frac{\nabla n}{n}$ B. $\frac{\nabla p}{n}=\gamma\frac{\nabla n}{p}$ C. $\frac{\nabla p}{p}=\frac{\nabla n}{n}$ D. $\frac{\nabla p}{n}=\frac{\nabla n}{p}$ **Ans. A**

26. For adiabatic plasma, pressure p and particle density n are related as (γ is ratio of two specific heats of plasma gas)

A. $\frac{\nabla p}{p}=\gamma\frac{\nabla n}{n}$ B. $\frac{\nabla p}{n}=\gamma\frac{\nabla n}{p}$ C. $\frac{\nabla p}{p}=\frac{\nabla n}{n}$ D. $\frac{\nabla p}{n}=\gamma^2\frac{\nabla n}{p}$

Ans. A

27. For isothermal plasma, pressure p and particle density n are related as (γ is ratio of two specific heats of plasma gas)

A. $\frac{\nabla p}{p}=\gamma\frac{\nabla n}{n}$ B. $\frac{\nabla p}{n}=\gamma\frac{\nabla n}{p}$ C. $\frac{\nabla p}{p}=\frac{\nabla n}{n}$ D. $\frac{\nabla p}{n}=\frac{\nabla n}{p}$ **Ans. C**

28. For isothermal plasma, pressure p and particle density n are related as (γ is ratio of two specific heats of plasma gas)

A. $\frac{\nabla p}{p}=\gamma^2\frac{\nabla n}{n}$ B. $\frac{\nabla p}{n}=\gamma\frac{\nabla n}{p}$ C. $\frac{\nabla p}{p}=\frac{\nabla n}{n}$ D. $\frac{\nabla p}{n}=\frac{\nabla n}{p}$ **Ans. C**

29. When thermal motions are not considered, the equation of motion of electron in homogeneous, isotropic and collision-less plasma is

A. $m_e n_e\left[\frac{\partial \vec{u}_e}{\partial t}+(\vec{u}_e\cdot\nabla)\vec{u}_e\right]=en_e\vec{E}$ B. $m_e n_e\left[\frac{\partial \vec{u}_e}{\partial t}+(\vec{u}_e\cdot\nabla)\vec{u}_e\right]=-en_e\vec{E}-\nabla p_e$

C. $m_e n_e\left[\frac{\partial \vec{u}_e}{\partial t}+(\vec{u}_e\cdot\nabla)\vec{u}_e\right]=-\nabla p_e$ D. $m_e n_e\left[\frac{\partial \vec{u}_e}{\partial t}+(\vec{u}_e\cdot\nabla)\vec{u}_e\right]=-en_e\vec{E}-\gamma KT\nabla n$

Ans. A

30. Linearized form of following equation is

$$m_e\left[\frac{\partial \vec{u}_e}{\partial t}+(\vec{u}_e\cdot\nabla)\vec{u}_e\right]=-e\vec{E}$$

A. $m_e\left[\frac{\partial \vec{u}_1}{\partial t}+(\vec{u}_{e1}\cdot\nabla)\vec{u}_{e0}\right]=-e\vec{E}$ B. $m_e\frac{\partial \vec{u}_1}{\partial t}=e\vec{E}_1$

C. $m_e\left[\frac{\partial \vec{u}_1}{\partial t}+(\vec{u}_{e1}\cdot\nabla)\vec{u}_{e0}\right]=-e\vec{E}_1$ D. $m_e\frac{\partial \vec{u}_1}{\partial t}=-e\vec{E}_1$ **Ans. D**

31. Linearized form of following equation is

$$m_e\left[\frac{\partial \vec{u}_e}{\partial t}+(\vec{u}_e\cdot\nabla)\vec{u}_e\right]=-e\vec{E}$$

A. $m_e\left[\frac{\partial \vec{u}_1}{\partial t}+(\vec{u}_{e0}\cdot\nabla)\vec{u}_{e1}\right]=-e\vec{E}$ B. $m_e\frac{\partial \vec{u}_1}{\partial t}=e\vec{E}_1$

C. $m_e\left[\frac{\partial \vec{u}_1}{\partial t}+(\vec{u}_{e0}\cdot\nabla)\vec{u}_{e1}\right]=-e\vec{E}_1$ D. $m_e\frac{\partial \vec{u}_1}{\partial t}=-e\vec{E}_1$ **Ans. D**

32. Linearized form of following equation is

$$m_e\left[\frac{\partial \vec{u}_e}{\partial t}+(\vec{u}_e\cdot\nabla)\vec{u}_e\right]=e\vec{E}$$

A. $m_e\left[\frac{\partial \vec{u}_1}{\partial t}+(\vec{u}_{e1}\cdot\nabla)\vec{u}_{e0}\right]=-e\vec{E}$ B. $m_e\frac{\partial \vec{u}_1}{\partial t}=e\vec{E}_1$

C. $m_e\left[\frac{\partial \vec{u}_1}{\partial t}+(\vec{u}_{e1}\cdot\nabla)\vec{u}_{e0}\right]=-e\vec{E}_1$ D. $m_e\frac{\partial \vec{u}_1}{\partial t}=-eE_1$ **Ans. B**

33. Linearized form of following equation is

$$\frac{\partial n}{\partial t}+\nabla\cdot(n\vec{u})=0$$

A. $\frac{\partial n_1}{\partial t}+n_0\nabla\cdot\vec{u}_1=0$ B. $\frac{\partial n_1}{\partial t}+n_0\nabla\cdot\vec{u}_1$ C. $\frac{\partial n_1}{\partial t}+\vec{u}_0\nabla n_1=0$ D. $\frac{\partial n_1}{\partial t}+\vec{u}_0\nabla n_1$

Ans. A

34. Linearized form of following equation is

$$\frac{\partial n}{\partial t}-\nabla\cdot(n\vec{u})=0$$

A. $\frac{\partial n_1}{\partial t}+n_0\nabla\cdot\vec{u}_1=0$ (b) $\frac{\partial n_1}{\partial t}+n_0\nabla\cdot\vec{u}_1$ C. $\frac{\partial n_1}{\partial t}+\vec{u}_0\nabla n_1=0$ (d) $\frac{\partial n_1}{\partial t}=\vec{u}_0\nabla n_1$

Ans. B

35. Linearized form of following equation is

$$\frac{\partial E}{\partial x}=\frac{e(n_i-n_e)}{\epsilon_0}$$

A. $\frac{\partial E_1}{\partial x}=-\frac{en_{e1}}{\epsilon_0}$ B. $\frac{\partial E_1}{\partial x}=\frac{en_{e1}}{\epsilon_0}$ C. $\frac{\partial E_1}{\partial x}=-\frac{en_{i1}}{\epsilon_0}$ D. $\frac{\partial E_1}{\partial x}=\frac{en_{i1}}{\epsilon_0}$

Ans. A

36. Linearized form of following equation is

$$\frac{\partial E}{\partial x} = -\frac{en_e}{\epsilon_0}$$

A. $\frac{\partial E_1}{\partial x} = -\frac{en_{e1}}{\epsilon_0}$ B. $\frac{\partial E_1}{\partial x} = \frac{en_{e1}}{\epsilon_0}$ C. $\frac{\partial E_1}{\partial x} = -\frac{en_{e0}}{\epsilon_0}$ D. $\frac{\partial E_1}{\partial x} = \frac{en_{e0}}{\epsilon_0}$

Ans. A

37. Linearized form of following equation is

$$\nabla \times \vec{E} = -\frac{\partial \vec{B}}{\partial t}$$

A. $\nabla \times \vec{E}_1 = \frac{\partial \vec{B}_1}{\partial t}$ B. $\nabla \times \vec{E}_1 = -\frac{\partial \vec{B}_0}{\partial t} + \frac{1}{c}\frac{\partial \vec{B}_1}{\partial t}$

C. $\nabla \times \vec{E}_1 = -\frac{\partial \vec{B}_1}{\partial t}$ D. $\nabla \times \vec{E}_1 = -\frac{\partial \vec{B}_0}{\partial t} - \frac{1}{c}\frac{\partial \vec{B}_1}{\partial t}$ **Ans. C**

38. Linearized form of following equation is

$$\nabla \times \vec{E} = \frac{\partial \vec{B}_1}{\partial t} + \vec{B}$$

A. $\nabla \times \vec{E}_1 = -\frac{\partial \vec{B}_1}{\partial t} + \vec{B}_1$ B. $\nabla \times \vec{E}_1 = \frac{\partial \vec{B}_0}{\partial t} - \frac{\partial \vec{B}_1}{\partial t}$

C. $\nabla \times \vec{E}_1 = -\frac{\partial \vec{B}_1}{\partial t}$ D. $\nabla \times \vec{E}_1 = -\frac{\partial \vec{B}_1}{\partial t} + \vec{B}$ **Ans. D**

39. Linearized form of following equation is

$$\nabla \times \vec{B} = \frac{\partial \vec{E}}{\partial t} + \vec{E}$$

A. $\nabla \times \vec{B}_1 = -\frac{\partial \vec{E}_1}{\partial t} + \vec{E}_1$ B. $\nabla \times \vec{B}_1 = \frac{\partial \vec{E}_0}{\partial t} - \frac{\partial \vec{E}_1}{\partial t}$

C. $\nabla \times \vec{B}_1 = -\frac{\partial \vec{E}_1}{\partial t}$ D. $\nabla \times \vec{B}_1 = -\frac{\partial \vec{E}_1}{\partial t} + \vec{E}$

Ans. A

40. Linearized form of following equation is

$$\nabla \times \vec{E} = -\frac{\partial \vec{B}}{\partial t}$$

A. $\nabla \times \vec{E}_1 = \frac{\partial \vec{B}_1}{\partial t}$ B. $\nabla \times \vec{E}_1 = -\frac{\partial \vec{B}_0}{\partial t} - \frac{1}{c}\frac{\partial \vec{B}_1}{\partial t}$

C. $\nabla \times \vec{E}_1 = -\frac{\partial \vec{B}_1}{\partial t}$ D. $\nabla \times \vec{E}_1 = \frac{\partial \vec{B}_0}{\partial t} - \frac{\partial \vec{B}_1}{\partial t}$ **Ans. C**

41. Linearized form of following equation is

$$\nabla \times \vec{B} = \frac{\partial \vec{E}}{\partial t}$$

A. $\nabla \times \vec{B}_1 = \frac{\partial \vec{E}_1}{\partial t}$ B. $\nabla \times \vec{B}_1 = -\frac{\partial \vec{E}_0}{\partial t} - \frac{\partial \vec{E}_1}{\partial t}$

C. $\nabla \times \vec{B}_1 = -\dfrac{\partial \vec{E}_1}{\partial t}$ D. $\nabla \times \vec{B}_1 = -\dfrac{\partial \vec{E}_0}{\partial t}\dfrac{\partial \vec{E}_1}{\partial t}$ **Ans. A**

42. Linearized form of following equation for electron is

$$m\left[\frac{\partial \vec{u}}{\partial t} + (\vec{u} \cdot \nabla)\vec{u}\right] = -e(\vec{E} + \vec{u} \times \vec{B})$$

A. $m\dfrac{\partial \vec{u}_1}{\partial t} = e(\vec{E}_1 + \vec{u}_1 \times \vec{B}_0)$ B. $m\dfrac{\partial \vec{u}_1}{\partial t} = -e(\vec{E}_1 + \vec{u}_0 \times \vec{B}_1)$

C. $m\dfrac{\partial \vec{u}_1}{\partial t} = -e(\vec{E}_1 + \vec{u}_1 \times \vec{B}_0)$ D. $m\dfrac{\partial \vec{u}_1}{\partial t} = e(\vec{E}_1 - \vec{u}_1 \times \vec{B}_0)$ **Ans. C**

43. In plasma, for a perpendicular wave
A. $\vec{k} \perp \vec{B}_0$ B. $\vec{k} \parallel \vec{B}_0$ C. $\vec{k} \perp \vec{E}_1$ D. $\vec{k} \parallel \vec{E}_1$ **Ans. A**

44. In plasma, for a parallel wave
A. $\vec{k} \perp \vec{B}_0$ B. $\vec{k} \parallel \vec{B}_0$ C. $\vec{k} \perp \vec{E}_1$ D. $\vec{k} \parallel \vec{E}_1$ **Ans. B**

45. In plasma, for a transverse wave
A. $\vec{k} \perp \vec{B}_0$ B. $\vec{k} \parallel \vec{B}_0$ C. $\vec{k} \perp \vec{E}_1$ D. $\vec{k} \parallel \vec{E}_1$ **Ans. C**

46. In plasma, for a longitudinal wave
A. $\vec{k} \perp \vec{B}_0$ B. $\vec{k} \parallel \vec{B}_0$ C. $\vec{k} \perp \vec{E}_1$ D. $\vec{k} \parallel \vec{E}_1$ **Ans. D**

47. In plasma, for an electrostatic wave, the requirement is
A. $\vec{B}_1 = 0$ B. $\vec{B}_1 \neq 0$ C. $\vec{E}_1 = 0$ D. $\vec{E}_1 \neq 0$ **Ans. A**

48. In plasma, for an electromagnetic wave, the requirement is
A. $\vec{B}_1 = 0$ B. $\vec{B}_1 \neq 0$ C. $\vec{E}_1 = 0$ D. $\vec{E}_1 \neq 0$ **Ans. B**

Appendices

A. Radial and Transverse Components for a Motion Along a Curved Path

Let us consider a particle moving along a curved path (Fig. A1). Suppose at time t, the particle is at a position P (with polar coordinates r, θ). Let it reach the position Q (with polar coordinates $r + \mathrm{d}r$, $\theta + \mathrm{d}\theta$) after a time interval $\mathrm{d}t$. The velocity $\vec{v}$ of the particle can be resolved into two components: (i) radial component, v_r and (ii) transverse component, v_θ as

$$\vec{v} = v_r\hat{r} + v_\theta\hat{\theta}$$

where the radial and transverse components, respectively, of velocity are

$$v_r = \frac{\mathrm{d}r}{\mathrm{d}t}$$

and

$$v_\theta = r\frac{\mathrm{d}\theta}{\mathrm{d}t} \tag{A1}$$

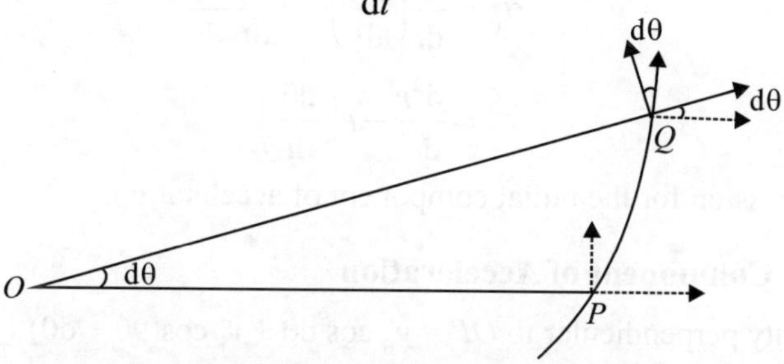

Fig. A1: A particle moving along a curved path is at the position P at a time t. After a time-interval dt, the particle reaches at the position Q.

While the particle is moving along a curved path, its direction of motion is changing continuously. Hence, the particle undergoes an acceleration, which can be resolved into two components: (i) radial component and (ii) transverse component.

(i) Radial Component of Acceleration

Let the particle move from a position P to another position Q in the time interval $\mathrm{d}t$ and the line joining the particle and the focus of the ellipse describes an angle $\mathrm{d}\theta$. Let at P, the radial and transverse components of velocity be v_r and v_θ, and the radial and transverse components of velocity at Q are v'_r and v'_θ, respectively.

At the Q, the velocity parallel to $OP = v'_r \cos \mathrm{d}\theta + v'_\theta \cos(90 + \mathrm{d}\theta)$

$$= v'_r \cos \mathrm{d}\theta - v'_\theta \sin \mathrm{d}\theta$$

The change in velocity parallel to $OP = v'_r \cos d\theta - v'_\theta \sin d\theta - v_r$

Acceleration parallel to OP,

$$\alpha_r = \frac{v'_r \cos d\theta - v'_\theta \sin d\theta - v_r}{dt} \tag{A2}$$

Let dv_r and dv_θ, respectively, denote the change in radial and transverse components of velocity during the motion from P to Q. Hence, we have

$$v'_r = v_r + dv_r$$

and

$$v'_\theta = v_\theta + dv_\theta \tag{A3}$$

Using equation (A3) in (A2), we get

$$\alpha_r = \frac{(v_r + dv_r)\cos d\theta - (v_\theta + dv_\theta)\sin d\theta - v_r}{dt}$$

Since $d\theta$ is small, we have $\cos d\theta = 1$ and $\sin d\theta = d\theta$ and thus we get

$$\alpha_r = \frac{(v_r + dv_r) - (v_\theta + dv_\theta)d\theta - v_r}{dt}$$

$$= \frac{dv_r}{dt} - v_\theta \frac{d\theta}{dt} - \frac{dv_\theta}{dt} d\theta \tag{A4}$$

The last term on the right side is very small and can be neglected. Using equation (A1) in (A4), we get

$$\alpha_r = \frac{d}{dt}\left(\frac{dr}{dt}\right) - r\frac{d\theta}{dt}\frac{d\theta}{dt}$$

$$= \frac{d^2 r}{dt^2} - r\left(\frac{d\theta}{dt}\right)^2 \tag{A5}$$

This is an expression for the radial component of acceleration.

(ii) Transverse Component of Acceleration

At the Q, the velocity perpendicular to $OP = v'_\theta \cos d\theta + v'_r \cos(90 - d\theta)$

$$= v'_\theta \cos d\theta + v'_r \sin d\theta$$

The change in velocity perpendicular to $OP = v'_\theta \cos d\theta + v' \sin d\theta - v_\theta$

Acceleration perpendicular to OP,

$$\alpha_\theta = \frac{v'_\theta \cos d\theta + v'_r \sin d\theta - v_\theta}{dt} \tag{A6}$$

Using equation (A3) in (A6), we get

$$\alpha_\theta = \frac{(v_\theta + dv_\theta)\cos d\theta + (v_r + dv_r)\sin d\theta - v_\theta}{dt}$$

Since $d\theta$ is small, we have $\cos d\theta = 1$ and $\sin d\theta = d\theta$ and thus we get

$$\alpha_\theta = \frac{(v_\theta + dv_\theta) + (v_r + dv_r)d\theta - v_\theta}{dt}$$

$$= \frac{dv_\theta}{dt} + v_r \frac{d\theta}{dt} + \frac{dv_r}{dt} d\theta \tag{A7}$$

The last term on the right side is very small and can be neglected. Using equation (A1) in (A7), we get

$$\alpha_\theta = \frac{d}{dt}\left(r\frac{d\theta}{dt}\right) + \frac{dr}{dt}\frac{d\theta}{dt}$$
$$= r\frac{d^2\theta}{dt^2} + 2\frac{dr}{dt}\frac{d\theta}{dt}$$
$$= \frac{1}{r}\frac{d}{dt}\left(r^2\frac{d\theta}{dt}\right)$$

This is expression for the transverse component of acceleration.

B. Ellipse

An ellipse (Fig. B1) is a locus of the points satisfying the equation

$$r + r' = 2a \tag{B1}$$

where a is a constant, called the semi-major axis of the ellipse, and r' and r are the distances of a point P on the ellipse from the focal points O' and O, respectively. When O' and O coincide, we have $r' = r$ and equation (B1) gives $r = a$, which is the equation of a circle. It shows that a circle is a special case of an ellipse. For $r = r'$, we have the point Q and the distance of Q from the centre C is b, known as the semi-minor axis of the ellipse. The distance of either focal point from the centre of the ellipse is ea, where e is known as the eccentricity of the ellipse ($0 \le e \le 1$). Obviously, for a circle, we have $e = 0$.

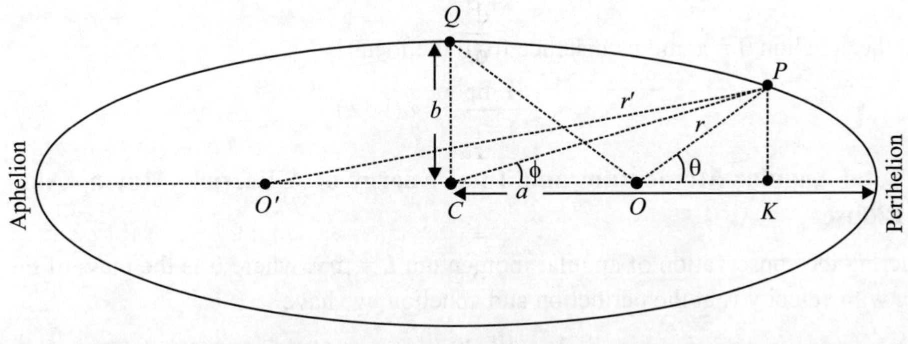

Fig. B1: The distances a and b measured from the centre C are the semi-major axis and semi-minor axis, respectively, of the ellipse. When the polar coordinates (r, θ) are measured from O, the closest point on the ellipse is known as perihelion whereas the farthest point is known as aphelion.

For getting relation between a, b and e, let us consider the point Q at the end of the minor axis. At Q, we have $r' = r$ and thus $r = a$. In the right-angle triangle OCQ, we have

$$OQ^2 = CO^2 + CQ^2$$
$$a^2 = a^2e^2 + b^2$$
$$b^2 = a^2(1 - e^2) \tag{B2}$$

Thus, the eccentricity of the ellipse is

$$e = \sqrt{1 - (b^2/a^2)}$$

and the equation of ellipse, in the Cartesian coordinates with the origin at the centre C of the ellipse, is

$$\frac{x^2}{a^2}+\frac{y^2}{b^2}=1$$

When ϕ is the angle PCO, Cartesian coordinates of a point on the ellipse are (a cos ϕ, b sin ϕ).

It is often convenient to express orbit of an object, moving on an elliptical path, in terms of the polar coordinates (r, θ), where r is the distance measured from the focus O of the ellipse and the angle θ is measured anticlockwise from the major axis of the ellipse. Using the Pythagorean theorem ($O'P^2 = O'K^2 + PK^2$) in the triangle $O'PK$, we have

$$r'^2 = (2ae + r\cos\theta)^2 + r^2\sin^2\theta$$

$$= r^2 + 4a^2e^2 + 4aer\cos\theta \qquad \text{(B3)}$$

Using equation (B1) in (B3), we get

$$(r - 2a)^2 = r^2 + 4a^2e^2 + 4aer\cos\theta$$

and therefore

$$r = \frac{a(1-e^2)}{1+e\cos\theta}$$

This is polar form of the equation of ellipse. At the perihelion $\theta = 0$ and its distance from the focus is

$$r_p = \frac{a(1-e^2)}{1+e} = a(1-e)$$

At the aphelion $\theta = \pi$ and its distance from the focus is

$$r_p = \frac{a(1-e^2)}{1-e} = a(1+e)$$

(i) Total Angular Momentum and Total Energy of A Particle Moving Along the Ellipse

Considering the conservation of angular momentum $L = \mu rv$, where μ is the mass of the body moving with velocity v, at the perihelion and aphelion, we have

$$\mu r_p v_p = \mu r_a v_a$$

and therefore

$$\frac{v_p}{v_a} = \frac{r_a}{r_p}$$

Using the expressions for r_a and r_p, we get

$$\frac{v_p}{v_a} = \frac{1+e}{1-e} \qquad \text{(B4)}$$

Considering the conservation of total energy (kinetic + potential) at the perihelion and aphelion, assuming that the body of mass μ is moving around the mass M, we have

$$\frac{1}{2}\mu v_p^2 - \frac{GM\mu}{r_p} = \frac{1}{2}\mu v_a^2 - \frac{GM\mu}{r_a}$$

or
$$v_p^2 = v_a^2 + 2GM\left[\frac{1}{r_p} - \frac{1}{r_a}\right]$$

Using here the expressions for r_a and r_p, we get

$$v_p^2 = v_a^2 + \frac{2GM}{a}\left[\frac{1}{1-e} - \frac{1}{1+e}\right]$$

$$= v_a^2 + \frac{4GMe}{a(1-e^2)}$$

Using equation (B4) here, we get

$$v_p^2 = v_p^2\frac{(1+e)^2}{(1-e)^2} + \frac{4GMe}{a(1-e^2)}$$

and thus
$$v_p^2 = \frac{GM}{a}\frac{1+e}{1-e}$$

and from equation (B4), we have

$$v_a^2 = \frac{GM}{a}\frac{1+e}{1-e}$$

Thus, the angular momentum is

$$L = \mu r_p v_p$$

$$= \mu a(1-e)\sqrt{\frac{GM}{a}\frac{1+e}{1-e}}$$

$$= \mu\sqrt{GMa(1-e^2)}$$

It shows that for a circular orbit ($e = 0$), the angular momentum would be maximum. Total energy is

$$E = \frac{1}{2}\mu v_p^2 - \frac{GM\mu}{r_p}$$

$$= \frac{\mu}{2}\frac{GM}{a}\frac{1+e}{1-e} - \frac{GM\mu}{a(1-e)}$$

$$= -\frac{GM\mu}{2a}$$

(ii) Area of the Ellipse

Area of the ellipse can be obtained by using the Cartesian equation

$$\frac{x^2}{a^2} + \frac{y^2}{b^2} = 1$$

The area A of the ellipse is

$$A = 4\int_0^a y\,\mathrm{d}x$$

$$= 4b\int_0^a \sqrt{1 - x^2/a^2}\,\mathrm{d}x$$

Putting $x = a \sin\theta$, we get

$$A = 4b\int_0^{\pi/2} \sqrt{1-\sin^2\theta}\ a\cos\theta\, d\theta$$

$$= 4ab\int_0^{\pi/2} \cos^2 d\theta$$

$$= \pi ab$$

(iii) Auxiliary Circle

A circle of radius a concentric to the ellipse is known as the auxiliary circle (Fig. B2). Cartesian coordinates of a point *P* on the ellipse are $(a\cos\phi, b\sin\phi)$, where ϕ is the angle subtended by the point *P* at the centre of the ellipse. The coordinates of a point *Q* on the auxiliary circle are $(a\cos\theta, a\sin\theta)$ where θ is the angle subtended by the point *Q* at the centre of the circle. Draw a vertical line *QPK* meeting the semi-major axis at the point *K*.

$$OK = OP\cos\phi$$

$$PK = OK\sin\phi$$

$$OK = OQ\cos\theta$$

$$QK = OQ\sin\theta$$

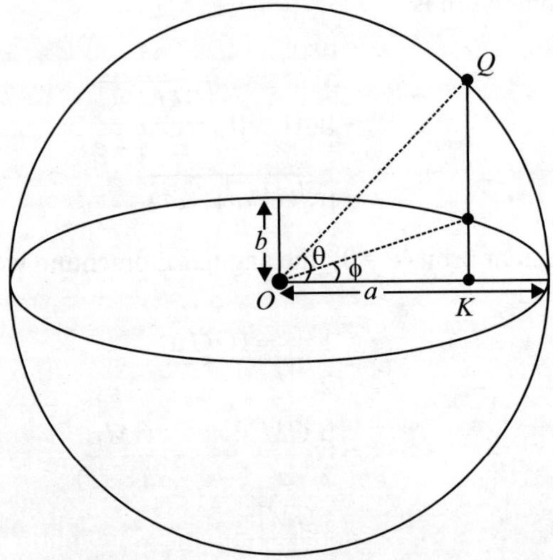

Fig. B2: Ellipse and auxiliary circle.

C. Solution of an Integral

The following integral, with *e* as a constant, can be solved by parts:

$$\int \frac{e^2\sin^2\theta}{(1+e\cos\theta)^2}d\theta = \int (e\sin\theta)\frac{e\sin\theta}{(1+e\cos\theta)^2}d\theta$$

$$= (e\sin\theta)\frac{1}{(1+e\cos\theta)} - \int (e\cos\theta)\frac{1}{(1+e\cos\theta)^2}d\theta$$

$$= \frac{e\sin\theta}{1+e\cos\theta} - \int\left\{1 - \frac{1}{(1+e\cos\theta)}\right\}d\theta$$

$$= \frac{e\sin\theta}{1+e\cos\theta} - \theta + \int\frac{1}{(1+e\cos\theta)}d\theta$$

Now, we have

$$\int\frac{1}{1+e\cos\theta}d\theta = \int\frac{1}{(\cos^2\theta/2+\sin^2\theta/2)+e(\cos^2\theta/2-\sin^2\theta/2)}d\theta$$

$$= \int\frac{\sec^2\theta/2}{(1+e)+(1-e)\tan^2\theta/2}d\theta$$

Let

$$u = \sqrt{\frac{1-e}{1+e}}\tan\theta/2$$

so that

$$du = \sqrt{\frac{1-e}{1+e}}\sec^2\theta/2\frac{d\theta}{2}$$

Hence, we have

$$\int\frac{1}{1+e\cos\theta}d\theta = \frac{2}{\sqrt{1-e^2}}\int\frac{du}{1+u^2}$$

$$= \frac{2}{\sqrt{1-e^2}}\tan^{-1}u$$

$$= \frac{2}{\sqrt{1-e^2}}\tan^{-1}\left(\sqrt{\frac{1-e}{1+e}}\tan\theta/2\right)$$

For small eccentricities ($e \ll 1$), we have

$$\sqrt{\frac{1-e}{1+e}} \approx 1$$

Thus, we have

$$\int\frac{1}{1+e\cos\theta}d\theta = \frac{\theta}{\sqrt{1-e^2}}$$

and

$$\int\frac{e^2\sin^2\theta}{(1+e\cos\theta)^2}d\theta = \frac{e\sin\theta}{1+e\cos\theta} - \theta + \frac{\theta}{\sqrt{1-e^2}}$$

Therefore,

$$\int_0^{2\pi}\frac{e^2\sin^2\theta}{(1+e\cos\theta)^2}d\theta = 2\pi\left[\frac{1}{\sqrt{1-e^2}} - 1\right]$$

D. Constants

Physical constants (c.g.s. units)

Speed of light	$c = 3.00 \times 10^{10}$ cm/s
Gravitational constant	$G = 6.67 \times 10^{-8}$ gm^{-1} cm^3 s^{-2}
Planck's constant	$h = 6.63 \times 10^{-27}$ erg s
Boltzmann constant	$k = 1.38 \times 10^{-16}$ erg/K
Stefan-Boltzmann constant	$\sigma = 5.67 \times 10^{-5}$ erg cm^{-2} s^{-1} K^{-4}
Charge of electron	$e = 4.80 \times 10^{-10}$ e.s.u.
Mass of electron	$m_e = 9.11 \times 10^{-28}$ gm
Mass of proton	$m_p = 1.67 \times 10^{-24}$ gm

Physical constants (SI units)

Speed of light	$c = 3.00 \times 10^{8}$ m/s
Gravitational constant	$G = 6.67 \times 10^{-11}$ kg^{-1} m^3 s^{-2}
Planck's constant	$h = 6.63 \times 10^{-34}$ J s
Boltzmann constant	$k = 1.38 \times 10^{-23}$ J/K
Stefan-Boltzmann constant	$\sigma = 5.67 \times 10^{-8}$ J m^{-2} s^{-1} K^{-4}
Charge of electron	$e = 1.6 \times 10^{-19}$ C
Mass of electron	$m_e = 9.11 \times 10^{-31}$ kg
Mass of proton	$m_p = 1.67 \times 10^{-27}$ kg

Bibliography

R.H. Baker and L.W. Fredrick, Astronomy IX Edition (1964) van Nostrand Reinhold Company, New York.

B. Basu, An Introduction to Astrophysics (1997) Prentice-Hall of India Pvt. Ltd., New Delhi.

J.C. Brandt and S.P. Maran, New Horizon in Astronomy (1972) W.H. Freeman & Company, San Francisco.

B.W. Carroll and D.A. Ostlie, Modern Astrophysics (1996) Addison-Wesley Publishing Company, Inc., New York.

E. Chaisson, Universe: An Evolutionary Approach to Astronomy (1988) Printice Hall, Englewood Cliffs, New Jersey.

S. Chandra, M.K. Sharma and A.K. Sharma, Beauty of our Solar System (2022) Ayan Prakashan, New Delhi.

B.T. Draine, Physics of Interstellar and Intergalactic Medium (2011) Princeton University Press, Princeton.

M. Harwit, Astrophysical Concepts, II Edition (1988) Springer-Verlag, New York.

K. Holliday, Introductory Astronomy (1999) John Wiley & Sons Ltd., New York.

A.W. Joshi and S. Chandra, Solar System for Everyone (2010) National Council for Educational Research & Training (NCERT), New Delhi.

W.J. Kaufmann, Universe (1985) W.H. Freeman & Company, New York.

K.F. Kuhn, In Quest of the Universe (1991) West Publishing Company, St. Paul.

C. Payne-Gaposchkin and K. Haramundanis, Introduction to Astronomy, 2nd Edn. (1961) Prentice Hall, Inc., New York.

W.M. Protheroe, E.R. Capriotti and G.H. Newsom, Exploring the Universe, III Ed. (1984) Bell and Howell Company, Columbus.

M.K. Sharma and S. Chandra, A Textbook of Electrodynamics (2022) I.K. International Pvt. Ltd., New Delhi.

F.H. Shu, The Physical Universe: An Introduction to Astronomy (1982) University Science Books, Sausalito, California.

T.L. Swihart, Astrophysics and Stellar Astronomy (1968) John Wiley & Sons, Inc., New York.

E.W. Woolard and G.M. Clemence, Spherical Astronomy (1966) Academic Press, New York.

M. Zeilik, Astronomy: The evolving universe, III Ed. (1982) Harper & Row, Publishers, New York.

M. Zeilik and J. Gaustand, Astronomy: The Cosmic Perspective (1983) Harper & Row, Publishers, New York.

M. Zeilik and S.A. Gregory, Introductory Astronomy and Astrophysics, IV Edition (1998) Saunders College Publishing.

References of Articles Used

A simple representation of the Gregorian calendar, Suresh Chandra and Rahul: (Bull. I.A.P.T. 5, 83 (1988).

Tide, Suresh Chandra: Physics Education (UGC, New Delhi sponsored) 17, 227 (2000).

Pluto is a planet? Suresh Chandra: Bull. I.A.P.T. 23, 191 (2006).

Pluto is no more a planet, Suresh Chandra: Bull. I.A.P.T. 23, 328 (2006).

Earth similarity index with two free parameters, Suresh Chandra, Subas Nepal & Mohit K. Sharma: Phys. Astron. Int. J. 2, 156 (2018).

Index